ALGORITHMS FOR VLSI PHYSICAL DESIGN AUTOMATION

THIRD EDITION

ALGORITHMS FOR VLSI PHYSICAL DESIGN AUTOMATION

THIRD EDITION

Naveed A. Sherwani
Intel Corporation.

KLUWER ACADEMIC PUBLISHERS
Boston / Dordrecht / London

Distributors for North, Central and South America:
Kluwer Academic Publishers
101 Philip Drive
Assinippi Park
Norwell, Massachusetts 02061 USA
Telephone (781) 871-6600
Fax (781) 871-6528
E-Mail <kluwer@wkap.com>

Distributors for all other countries:
Kluwer Academic Publishers Group
Post Office Box 322
3300 AH Dordrecht, THE NETHERLANDS
Telephone 31 78 6576 000
Fax 31 78 6576 474
E-Mail <orderdept@wkap.nl>

 Electronic Services <http://www.wkap.nl>

Library of Congress Cataloging-in-Publication

Sherwani, N.A. (Naveed A.)
 Algorithms for VLSI physical design automation / Naveed A. Sherwani.—3rd ed.
 p. cm.
 Includes bibliographical references and index.
 ISBN 0-7923-8393-1
 1.   Integrated circuits--Very large scale integration—Design and construction—
 Data processing. 2. Computer-aided design. 3. Algorithms. I. Title
 TK7874.S455   1999
 621.39'5--dc21   98-44779
                                                   CIP

Copyright © 1999 by Kluwer Academic Publishers. Sixth Printing 2004.

This printing is a digital duplication of the original edition.

To my parents
Akhter and Akram Sherwani

Contents

9 Detailed Routing   291

Foreword

Since the invention of integrated circuits thirty years ago, manufacturing of electronic systems has taken rapid strides in improvement in speed, size, and cost. For today's integrated circuit chips, switching time is on the order of nanoseconds, minimum feature size is on the order of sub-microns, transistor count is on the order of millions, and cost is on the order of a few dollars. In fact, it was estimated that the performance/cost ratio of integrated circuit chips has been increasing at the rate of one thousand-fold every ten years, yielding a total of 10^9 for the last three decades. A combination of high product performance and low per-unit cost leads to the very pervasive introduction of integrated circuit chips to many aspects of modern engineering and scientific endeavors including computations, telecommunications, aeronautics, genetics, bioengineering, manufacturing, factory automation, and so on. It is clear that the integrated circuit chip will play the role of a key building block in the *information society* of the twenty-first century.

The manufacture of integrated circuit chips is similar to the manufacture of other highly sophisticated engineering products in many ways. The three major steps are *designing* the product, *fabricating* the product, and *testing* the fabricated product. In the design step, a large number of components are to be designed or selected, specifications on how these components should be assembled are to be made, and verification steps are to be carried out to assure the correctness of the design. In the manufacturing step, a great deal of manpower, and a large collection of expensive equipment, together with painstaking care are needed to assemble the product according to the design specification. Finally, the fabricated product must be tested to check its physical functionality. As in all engineering problems, there are conflicting requirements in all these steps. In the design step, we want to obtain an optimal product design, and yet we also want the design cycle to be short. In the fabrication step, we want the product yield to be high, and yet we also need to be able to produce a large volume of the product and get them to market in time. In the testing step, we want the product to be tested thoroughly and yet we also want to be able to do so quickly.

The title of this book reveals how the issue of enormous design complexity is to be handled so that high quality designs can be obtained in a reasonable amount of design time: We use muscles (*automation*) and we use brain (*algorithms*). Professor Sherwani has written an excellent book to introduce students in computer science and electrical engineering as well as CAD engineers to the subject of physical design of VLSI circuits. Physical design is a key step in the design process. Research and development efforts in the last twenty years have led us to some very good understanding on many of the important problems in physical design. Professor Sherwani's book provides a timely, up-to-date integration of the results in the field and will be most useful both as a graduate level textbook and as a reference for professionals in the field. All aspects of the physical design process are covered in a meticulous and comprehensive manner. The treatment is enlightening and enticing. Furthermore, topics related to some of the latest technology developments such as Field Programmable Gate Arrays (FPGA) and Multi-Chip Modules (MCM) are also included. A strong emphasis is placed on the algorithmic aspect of the design process. Algorithms are presented in an intuitive manner without the obscurity of unnecessary formalism. Both theoretical and practical aspects of algorithmic design are stressed. Neither the elegance of optimal algorithms nor the usefulness of heuristic algorithms are overlooked. ¿From a pedagogical point of view, the chapters on electronic devices and on data structures and basic algorithms provide useful background material for students from computer science, computer engineering, and electrical engineering. The many exercises included in the book are also most helpful teaching aids.

This is a book on physical design algorithms. Yet, this is a book that goes beyond physical design algorithms. There are other important design steps of which our understanding is still quite limited. Furthermore, development of new materials, devices, and technologies will unquestionably create new problems and new avenues of research and development in the design process. An algorithmic outlook on design problem and the algorithmic techniques for solving complex design problems, which a reader learns through the examples drawn from physical design in this book, will transcend the confine of physical design and will undoubtedly prepare the reader for many of the activities in the field of computer-aided design of VLSI circuits. I expect to hear from many students and CAD professionals in the years to come that they have learned a great deal about physical design, computer-aided design, and scientific research from Professor Sherwani's book. I also expect to hear from many of them that Professor Sherwani's book is a source of information as well as a source of inspiration.

Urbana-Champaign, September 1992 C. L. Liu

Preface

From its humble beginning in the early 1950's to the manufacture of circuits with millions of components today, VLSI design has brought the power of the mainframe computer to the laptop. Of course, this tremendous growth in the area of VLSI design is made possible by the development of sophisticated design tools and software. To deal with the complexity of millions of components and to achieve a turn around time of a couple of months, VLSI design tools must not only be computationally fast but also perform close to optimal.

The future growth of VLSI systems depends critically on the research and development of Physical Design (PD) Automation tools. In the last two decades, the research in physical design automation has been very intense, and literally thousands of research articles covering all phases of physical design automation have been published. The development of VLSI physical design automation also depends on availability of trained manpower. We have two types of students studying VLSI physical design: students preparing for a research career and students preparing for a career in industry. Both types of students need to build a solid background. However, currently we lack courses and text books which give students a comprehensive background. It is common to find students doing research in placement, but are unaware of the latest developments in compaction. Those students seeking careers in industry will find that the VLSI physical design industry is very fast paced. They are expected to be conversant with existing tools and algorithms for all the stages of the design cycle of a VLSI chip. In industry, it is usual to find CAD engineers who work on one aspect of physical design and lack knowledge of other aspects. For example, a CAD engineer working in the development of detailed routers may not be knowledgeable about partitioning algorithms. This is again due to the lack of comprehensive textbooks which cover background material in all aspects of VLSI physical design.

Providing a comprehensive background in one textbook in VLSI physical design is indeed difficult. This is due to the fact that physical design automation requires a mix of backgrounds. Some electrical engineering and a solid undergraduate computer science background is necessary to grasp the fundamentals. In addition, some background in graph theory and combinatorics is also needed, since many of the algorithms are graph theoretic or use other combinatorial optimization techniques. This mix of backgrounds has perhaps restricted the development of courses and textbooks in this very area.

This book is an attempt to provide a comprehensive background in the principles and algorithms of VLSI physical design. The goal of this book is to serve as a basis for the development of introductory level graduate courses in VLSI physical design automation. It is hoped that the book provides self contained material for teaching and learning algorithms of physical design. All algorithms which are considered basic have been included. The algorithms are presented in an intuitive manner, so that the reader can concentrate on the basic idea of the algorithms. Yet, at the same time, enough detail is provided so that readers can actually implement the algorithms given in the text and use them.

This book grew out of a graduate level class in VLSI physical design automation at Western Michigan University. Initially written as a set of class notes, the book took form as it was refined over a period of three years.

Overview of the Book

This book covers all aspects of physical design. The first three chapters provide the background material, while the focus of each chapter of the rest of the book is on each phase of the physical design cycle. In addition, newer topics like physical design automation of FPGAs and MCMs have also been included.

In Chapter 1, we give an overview of the VLSI physical design automation field. Topics include the VLSI design cycle, physical design cycle, design styles and packaging styles. The chapter concludes with a brief historical review of the field.

Chapter 2 discusses the fabrication process for VLSI devices. It is important to understand the fabrication technology in order to correctly formulate the problems. In addition, it is important for one to understand, what is doable and what is not! Chapter 2 presents fundamentals of MOS and TTL transistors. It then describes simple NAND and NOR gates in nMOS and CMOS.

Chapter 3 presents the status of fabrication process, as well as, process innovations on the horizons and studies its impact on physical design. We also discuss several other factors such as design rules, yield, delay, and fabrication costs involved in the VLSI process.

Basic material on data structures and algorithms involved in the physical design is presented in Chapter 4. Several different data structures for layout have been discussed. Graphs which are used to model several different problems in VLSI design are defined and basic algorithms for these graphs are presented.

Chapter 5 deals with partitioning algorithms. An attempt has been made to explain all the possible factors that must be considered in partitioning the VLSI circuits. Group migration, simulated annealing and simulated evolution algorithms have been presented in detail. The issue of performance driven partitioning is also discussed.

In Chapter 6, we discuss basic algorithms for floorplanning and pin assignment. Several different techniques for placement such as, simulated annealing, simulated evolution, and force-directed are discussed in Chapter 7.

Chapter 8 deals with global routing. It covers simple routing algorithms, such as maze routing, and more advanced integer programming based methods. It also discusses Steiner tree algorithms for routing of multiterminal nets.

Chapter 9 is the longest chapter in the book and represents the depth of knowledge that has been gained in the detailed routing area in the last decade. Algorithms are classified according to the number of layers allowed for routing. In single layer routing, we discuss general river routing and the single row routing problem. All major two-layer channel and switch box routers are also presented. The chapter also discusses three-layer and multilayer routing algorithms.

Chapter 10 discusses two ways of improving layouts after detailed routing, namely, via minimization and over-the-cell routing. Basic algorithms for via minimization are presented. Over-the-cell routing is a relatively new technique for reducing routing areas. We present the two latest algorithms for over-the-cell routing.

The problems of routing clock and power/ground nets are discussed in Chapter 11. These topics play a key role in determining the layout of high performance systems. Circuit compaction is discussed in Chapter 12. One dimensional compaction, as well as two dimensional compaction algorithms are presented.

Field Programmable Gate Arrays (FPGAs) are rapidly gaining ground in many applications, such as system prototyping. In Chapter 13, we discuss physical design automation problems and algorithms for FPGAs. In particular, we discuss the partitioning and routing problems in FPGAs. Both of these problems are significantly different from problems in VLSI. Many aspects of physical design of FPGAs remain a topic of current research.

Multi-Chip Modules (MCMs) are replacing conventional printed circuit boards in many applications. MCMs promise high performance systems at a lower cost. In Chapter 14, we explore the physical design issues in MCMs. In particular, the routing problem of MCMs is a true three dimensional problem. MCMs are currently a topic of intense research.

At the end of each chapter, a list of exercises is provided, which range in complexity from simple to research level. Unmarked problems and algorithms are the simplest. The exercises marked with (†) are harder and algorithms in these exercises may take a significant effort to implement. The exercises and algorithms marked with (‡) are the hardest. In fact, some of these problems are research problems.

Bibliographic notes can be found at the end of each chapter. In these notes, we give pointers to the readers for advanced topics. An extensive bibliography is presented at the end of the text. This bibliography is complete, to the best of our knowledge, up to the September of 1998. An attempt has been made to include all papers which are appropriate for the targeted readers of this text. The readers may also find the author and the subject index at the back of the text.

Overview of the Second Edition

In 1992, when this book was originally published, the largest microprocessor had one million transistors and fabrication process had three metal layers. We have now moved into a six metal layer process and 15 million transistor microprocessors are already in advanced stages of design. The designs are moving towards a 500 to 700 Mhz frequency goal. This challenging frequency goal, as well as, the additional metal layers have significantly altered the VLSI field. Many issues such as three dimensional routing, Over-the-Cell routing, early floorplanning have now taken a central place in the microprocessor physical design flow. This changes in the VLSI design prompted us to reflect these in the book. That gave birth to the idea of the second edition.

The basic purpose of the second edition is to introduce a more *realistic* picture to the reader exposing the concerns facing the VLSI industry while maintaining the theoretical flavor of the book. New material has been added to all the chapters. Several new sections have been added to many chapters. Few chapters have been completely rewritten. New figures have been added to supplement the new material and clarify the existing material.

In summary, I have made an attempt to capture the physical design flow used in the industry and present it in the second addition. I hope that readers will find that information both useful and interesting.

Overview of the Third Edition

In 1995, when we prepared the 2nd edition of this book, a six metal layer process and 15 million transistor microprocessors were in advanced stages of design. In 1998, six metal process and 20 million transistor designs are in production. Several manufacturers have moved to 0.18 micron process and copper interconnect. One company has announced plans for 0.10 micron process and plans to integrate 200 to 400 million transistors on a chip. Operating frequency has moved from 266 Mhz (in 1995) to 650 Mhz and several Ghz experimental chips have been demonstrated. Interconnect delay has far exceeded device delay and has become a dominant theme in physical design. Process innovations such as copper, low k dielectrics, multiple threshold devices, local interconnect are once again poised to change physical design once again.

The basic purpose of the third edition is to investigate the new challenges presented by interconnect and process innovations. In particular, we wanted to identify key problems and research areas that physical design community needs to invest in order to meet the challenges. We took a task of presenting those ideas while maintaining the flavor of the book. As a result, we have added two new chapters and new material has been added to most of the chapters. A new chapter on process innovation and its impact on physical design has been added. Another focus of the book has been to promote use of Internet as a resource, so wherever possible URLs has been provided for further investigation.

Chapters 1 and 2 have been updated. Chapter 3 is a new chapter on the fabrication process and its impact. Chapter 4 (algorithms) and Chapter 5 (partitioning) have been edited for clarity. Chapter on Floorplanning, Placement and Pin Assignment has been split into Chapter 6 (Floorplanning) and Chapter 7 (Placement) to bring sharper focus to floorplanning. New sequence pair algorithms have been added to Chapter 7 (Placement) Chapter 8 and 9 have been edited for clarity and references have been updated as appropriate. New sections have been added to Chapter 10, Chapter 11 and Chapter 12. In Chapter 10, we have added material related to performance driven routing. In Chapter 11, DME algorithm has been added. In Chapter 12, we have added new compaction algorithms. Chapters 13 (FPGAs) and 14 (MCMs) have been updated. We have made an attempt to update the bibliography quite extensively and many new items have been added.

In summary, I have made an attempt to capture the impact of interconnect and process innovations on physical design flow. I have attempted to balance material on new innovations with the classical content of the 2nd edition. I hope that readers will find that information both useful and interesting.

To the Teacher

This book has been written for introductory level graduate students. It presents concepts and algorithms in an intuitive manner. Each chapter contains 3 to 4 algorithms that have been discussed in detail. This has been done so as to assist students in implementing the algorithms. Other algorithms have been presented in a somewhat shorter format. References to advanced algorithms have been presented at the end of each chapter. Effort has been made to make the book self contained.

This book has been developed for a one-semester or a two-semester course in VLSI physical design automation. In a one-semester course, it is recommended that chapters 8, 9, 11, and 12 be omitted. A half-semester algorithm development project is highly recommended. Implementation of algorithms is an important tool in making students understand algorithms. In physical design, the majority of the algorithms are heuristic in nature and testing of these algorithms on benchmarks should be stressed. In addition, the development of practical algorithms must be stressed, that is, students must be very aware of the complexity of the algorithms. An optimal $O(n^3)$ algorithm may be impractical for an input of size 10 million. Several (†) marked problems at the end of each chapter may serve as mini-projects.

In a two-semester class, it is recommended that all the chapters be included. Reading state-of-art papers must be an integral part of this class. In particular, students may be assigned papers from proceedings of DAC and ICCAD or from IEEE Transactions on CAD. Papers from Transactions typically require a little more mathematical maturity than the papers in DAC and ICCAD. An important part of this class should be a two-semester project, which may be the development of a new algorithm for some problem in physical design. A typical first part of the project may involve modifying an existing algorithm for a special application. Some (‡) problems may serve as projects.

In both the courses, a good background in hand layout is critical. It is expected that students will have access to a layout editor, such as MAGIC or LEDIT. It is very important that students actually layout a few small circuits. For examples see exercises at the end of Chapter 2.

For faculty members, a teaching aid package, consisting of a set of 400 overheads (foils) is available from the author. These are quite helpful in teaching the class, as all the important points have been summarized on section by section basis. In order to obtain these foils, please send an email (or a mail) to the author, at the address below.

To the Student

First and foremost, I hope that you will enjoy reading this book. Every effort has been made to make this book easy to read. The algorithms have been explained in an intuitive manner. The idea is to get you to develop new algorithms at the end of the semester. The book has been balanced to give a practical as well as a theoretical background. In that sense, you will find it useful, if you are thinking about a career in industry or if you are thinking about physical design as a possible graduate research topic.

What do you need to start reading this book? Some maturity in general algorithm techniques and data structures is assumed. Some electrical engineering background and mathematics background will be helpful, although not necessary. The book is self-contained to a great extent and does not need any supporting text or reference text.

If you are considering a career in this field, I have one important piece of advise for you. Research in this field moves very fast. As a result, no textbook can replace state-of-the-art papers. It is recommended that you read papers to keep you abreast of latest developments. A list of conference proceedings and journals appears in the bibliographic notes of Chapter 1. I also recommend attending DAC and ICCAD conferences every year and a membership in ACM/SIGDA, IEEE/DATC and IEEE/TC-VLSI.

To the CAD Professional

This book provides a detailed description of all aspects of physical design and I hope you have picked up this book to review your basics of physical design. While it concentrates on basic algorithms, pointers are given to advanced algorithms as well. The text has been written with a balance of theory and practice in mind. You will also find the extensive bibliography useful for finding advanced material on a topic.

Errors and Omissions

No book is free of errors and omissions. Despite our best attempt, this text may contain some errors. If you find any errors or have any constructive suggestions, I would appreciate receiving your comments and suggestions. In particular, new exercises would certainly be very helpful. You can mail your comments to:

Naveed Sherwani
Intel Corporation, Mail Stop: JFT-104
2111 N. E. 25th Avenue
Hillsboro, OR 97124-5961

or email them to `sherwani@ichips.intel.com`.

A concentrated effort has been made to include all pertinent references to papers and books that we could find. If you find omissions in the book, please feel free to remind me.

This book was typeset in Latex. Figures were made using 'xfig' and inserted directly into the text as .ps files using 'transfig'. The bibliography was generated using Bibtex and the index was generated with a program written by Siddharth Bhingarde.

Portland, March, 1998   Naveed A. Sherwani

Acknowledgments

No book is a product of one person. The same is true for this book. First I must thank all members of the *nitegroup* who worked tirelessly for days and nights (mostly nights) for the final six months. First I would like to thank Siddharth Bhingarde, Surendra Burman, Moazzem Hossain, Chandar Kamalanathan, Wasim Khan, Arun Shanbhag, Timothy Strunk and Qiong Yu. Thanks are also to due to nitegroup members who have graduated. In particular Roshan Gidwani, Jahangir Hashmi, Nancy Holmes, and Bo Wu, who helped in all stages of this project. Special thanks are due to the youngest member of nitegroup, Timothy Strunk, who made (almost) all the figures in the text and brought enthusiasm to the team. Thanks are also due to Anand Panyam, Konduru Nagesh and Aizaz Manzar for helping in the final stages of this project. Many students in my class CS520 (Introduction to VLSI design automation) suffered through earlier version of this book, and I would like to thank them for their constructive suggestions.

Several colleagues and friends contributed significantly by reviewing several chapters and using parts of the book in their courses. In this regard, I would like to thank Jeff Banker, Ajay Gupta, Mark Kerstetter, Sartaj Sahni, and Jason Cong. I would also like to especially thank Dinesh Mehta and Si-Qing Zheng. I wish to express my sincere thanks to Malgorzata Marek-Sadowska, who made very critical remarks and contributions to the improvement in the quality of the text.

Thanks are due to two special people, who have contributed very generously in my career and helped in many ways. I would like to thank Vishwani Agrawal and C. L. Liu for their constant encouragement and sound words of advise.

I would like to thank several different organizations who have contributed directly to this project. First, I would like to thank Ken Wan, and the rest of the CTS group at Advanced Micro Devices for helping with many technical details. I would also like to thank ACM SIGDA for supporting our research during the last four years. Thanks are also due to Western Michigan University, and in particular Donald Nelson and Douglas Ferraro, who, despite all costs, made the necessary facilities available to complete this book. The National Science Foundation deserves thanks for supporting the VLSI laboratory and our research at Western Michigan University. I would also like to thank Reza Rashidi and the staff of FRC laboratory for their help in printing the text and cover design.

I would also like to thank our system manager Patty Labelle, who cheerfully accepted our occasional abuse of the system and kept all the machines up when we needed them. I must also thank our department secretaries Phyllis Wolf and Sue Moorian for being very helpful during all stages of this project.

I must thank my copy editor Frank Strunk, who very carefully read the manuscript in the short time we gave him. Thanks are also due to Carl Harris, editor at Kluwer Academic Publishers for being understanding and going beyond his call of duty to help out with my requests.

Finally, I wish to thank my parents and my family for supporting me throughout my life and for being there when I needed them. They suffered as I neglected many social responsibilities to complete this book.

Kalamazoo, September, 1992   Naveed A. Sherwani

Acknowledgments for the Second Edition

The second edition project would not have been possible without the help of Siddharth Bhingarde, Aman Sureka, Rameshwar Donakanti and Anand Panyam. In particular, Siddharth worked with me for many many nights on this project. I am very grateful to these individuals for their help.

Several of my colleagues at Intel helped as reviewers of the chapters. In this regard, I would like to thank Marc Rose, John Hansen, Dave Ackley, Mike Farabee, and Niraj Bindal.

Several friends and family members helped by being copy editors. Sabahat Naveed, Shazia Asif and Akram Sherwani helped by editing many revisions. Internet played a key role, as many of these revisions were done in Pakistan and then emailed to me.

I would like to thank Intel Corporation for helping me with this project. In particular, I would like to thank Atiq Bajwa for making the time available for me to complete the project.

Portland, March, 1995   Naveed A. Sherwani

Acknowledgments for the Third Edition

The third edition would not have been possible without the help of Faran Rafiq, Srinivasa Danda, Siddharth Bhingarde, Niraj Bindal, Prashant Saxena, Peichen Pan and Anand Panyam. I am very grateful to these individuals for their help. In particular, I am indebted to Faran Rafiq, who worked tireless with me and this project would not have been possible without his dedication and hard work.

I would like to thank Intel Corporation for helping me with the third edition. In particular, I would like to thank Manpreet Khaira for the Research and Development environment, which has helped mature many ideas.

I must thank my copy editor Tawni Schlieski, who very carefully read the new chapters and turned them around in a very short time. I am very thankful to Carl Harris, editor at Kluwer Academic Publishers for encouraging me to write the third edition.

Finally, I am very thankful to my wife Sabahat and my daughter Aysel for their encouragement and support.

Portland, September, 1998   Naveed A. Sherwani

Chapter 1

VLSI Physical Design Automation

The information revolution has transformed our lives. It has changed our perspective of work, life at home and provided new tools for entertainment. The internet has emerged as a medium to distribute information, communication, event planning, and conducting E-commerce. The revolution is based on computing technology and communication technology, both of which are driven by a revolution in Integrated Circuit (IC) technology. ICs are used in computers for microprocessor, memory, and interface chips. ICs are also used in computer networking, switching systems, communication systems, cars, airplanes, even microwave ovens. ICs are now even used in toys, hearing aids and implants for human body. MEMs technology promises to develop mechanical devices on ICs thereby enabling integration of mechanical and electronic devices on a miniature scale. Many sensors, such as acceleration sensors for auto air bags, along with conversion circuitry are built on a chip. This revolutionary development and widespread use of ICs has been one of the greatest achievements of humankind.

IC technology has evolved in the 1960s from the integration of a few transistors (referred to as *Small Scale Integration* (SSI))o the integration of millions of transistors in *Very Large Scale Integration* (VLSI) chips currently in use. Early ICs were simple and only had a couple of gates or a flip-flop. Some ICs were simply a single transistor, along with a resistor network, performing a logic function. In a period of four decades there have been four generations of ICs with the number of transistors on a single chip growing from a few to over 20 million. It is clear that in the next decade, we will be able to build chips with billions of transistors running at several Ghz. We will also be able to build MEM chips with millions of electrical and mechanical devices. Such chips will enable a new era of devices which will make such exotic applications, such as tele-presence, augumented reality and implantable and wearable computers, possible. Cost effective world wide point-to-point communication will be common and available to all.

This rapid growth in integration technology has been (and continues to be) made possible by the automation of various steps involved in the design and fabrication of VLSI chips. Integrated circuits consist of a number of electronic components, built by layering several different materials in a well-defined fashion on a silicon base called a *wafer*. The designer of an IC transforms a circuit description into a geometric description, called the *layout*. A layout consists of a set of planar geometric shapes in several layers. The layout is checked to ensure that it meets all the design requirements. The result is a set of design files that describes the layout. An optical pattern generator is used to convert the design files into pattern generator files. These files are used to produce patterns called *masks*. During fabrication, these masks are used to pattern a silicon wafer using a sequence of photo-lithographic steps. The component formation requires very exacting details about geometric patterns and the separation between them. The process of converting the specification of an electrical circuit into a layout is called the *physical design process*. Due to the tight tolerance requirements and the extremely small size of the individual components, physical design is an extremely tedious and error prone process. Currently, the smallest geometric feature of a component can be as small as 0.25 micron (one micron, written as μm is equal to 1.0×10^{-6}m). For the sake of comparison, a human hair is 75 μm in diameter. It is expected that the feature size can be reduced below 0.1 micron within five years. This small feature size allows fabrication of as many as 200 million transistors on a 25 mm $\times$ 25 mm chip. Due to the large number of components, and the exacting details required by the fabrication process, physical design is not practical without the help of computers. As a result, almost all phases of physical design extensively use Computer Aided Design (CAD) tools, and many phases have already been partially or fully automated.

VLSI Physical Design Automation is essentially the research, development and productization of algorithms and data structures related to the physical design process. The objective is to investigate optimal arrangements of devices on a plane (or in three dimensions) and efficient interconnection schemes between these devices to obtain the desired functionality and performance. Since space on a wafer is very expensive real estate, algorithms must use the space very efficiently to lower costs and improve yield. In addition, the arrangement of devices plays a key role in determining the performance of a chip. Algorithms for physical design must also ensure that the layout generated abides by all the rules required by the fabrication process. Fabrication rules establish the tolerance limits of the fabrication process. Finally, algorithms must be efficient and should be able to handle very large designs. Efficient algorithms not only lead to fast turn-around time, but also permit designers to make iterative improvements to the layouts. The VLSI physical design process manipulates very simple geometric objects, such as polygons and lines. As a result, physical design algorithms tend to be very intuitive in nature, and have significant overlap with graph algorithms and combinatorial optimization algorithms. In view of this observation, many consider physical design automation the study of graph theoretic and combinatorial algorithms for manipulation of geometric objects in two and three dimensions. However, a pure geometric point of view ignores the electrical (both digital and analog) aspect of the physical design problem. In a VLSI circuit, polygons and lines have inter-related electrical properties, which exhibit a very complex behavior and depend on a host of variables. Therefore, it is necessary to keep the electrical aspects of the geometric objects in perspective while developing algorithms for VLSI physical design automation. With the introduction of Very Deep Sub-Micron (VDSM), which provides very small features and allows dramatic increases in the clock frequency, the effect of electrical parameters on physical design will play a more dominant role in the design and development of new algorithms.

In this chapter, we present an overview of the fundamental concepts of VLSI physical design automation. Section 1.1 discusses the design cycle of a VLSI circuit. New trends in the VLSI design cycle are discussed in Section 1.2. In Section 1.3, different steps of the physical design cycle are discussed. New trends in the physical design cycle are discussed in Section 1.4. Different design styles are discussed in Section 1.5 and Section 1.6 presents different packaging styles. Section 1.7 presents a brief history of physical design automation and Section 1.8 lists some existing design tools.

1.1 VLSI Design Cycle

The VLSI design cycle starts with a formal specification of a VLSI chip, follows a series of steps, and eventually produces a packaged chip. A typical design cycle may be represented by the flow chart shown in Figure 1.1. Our emphasis is on the physical design step of the VLSI design cycle. However, to gain a global perspective, we briefly outline all the steps of the VLSI design cycle.

1. **System Specification:** The first step of any design process is to lay down the specifications of the system. System specification is a high level representation of the system. The factors to be considered in this process include: performance, functionality, and physical dimensions (size of the die (chip)). The fabrication technology and design techniques are also considered. The specification of a system is a compromise between market requirements, technology and economical viability. The end results are specifications for the size, speed, power, and functionality of the VLSI system.

2. **Architectural Design:** The basic architecture of the system is designed in this step. This includes, such decisions as RISC (Reduced Instruction Set Computer) versus CISC (Complex Instruction Set Computer), number of ALUs, Floating Point units, number and structure of pipelines, and size of caches among others. The outcome of architectural design is a Micro-Architectural Specification (MAS). While MAS is a textual (English like) description, architects can accurately predict the performance, power and die size of the design based on such a description.

Such estimates are based on the scaling of existing design or components of existing designs. Since many designs (especially microprocessors) are based on modifications or extensions to existing designs, such a method can provide fairly accurate early estimates. These early estimates are critical to determine the viability of a product for a market segment. For example, for mobile computing (such as lap top computer), low power consumption is a critical factor, due to limited battery life. Early estimates based on architecture can be used to determine if the design is likely to meet its power spec.

3. **Behavioral or Functional Design:** In this step, main functional units of the system are identified. This also identifies the interconnect requirements between the units. The area, power, and other parameters of each unit are estimated. The behavioral aspects of the system are considered without implementation specific information. For example, it may specify that a multiplication is required, but exactly in which mode such multiplication may be executed is not specified. We may use a variety of multiplication hardware depending on the speed and word size requirements. The key idea is to specify behavior, in terms of input, output and timing of each unit, without specifying its internal structure. The outcome of functional design is usually a timing diagram or other relationships between units. This information leads to improvement of the overall design process and reduction of the complexity of subsequent phases. Functional or behavioral design provides quick emulation of the system and allows fast debugging of the full system. Behavioral design is largely a manual step with little or no automation help available.

4. **Logic Design:** In this step the control flow, word widths, register allocation, arithmetic operations, and logic operations of the design that represent the functional design are derived and tested. This description is called Register Transfer Level (RTL) description. RTL is expressed in a Hardware Description Language (HDL), such as VHDL or Verilog. This description can be used in simulation and verification. This description consists of Boolean expressions and timing information. The Boolean expressions are minimized to achieve the smallest logic design which conforms to the functional design. This logic design of the system is simulated and tested to verify its correctness. In some special cases, logic design can be automated using *high level synthesis* tools. These tools produce a RTL description from a behavioral description of the design.

5. **Circuit Design:** The purpose of circuit design is to develop a circuit representation based on the logic design. The Boolean expressions are converted into a circuit representation by taking into consideration the speed and power requirements of the original design. *Circuit Simulation* is used to verify the correctness and timing of each component. The circuit design is usually expressed in a detailed circuit diagram. This diagram shows the circuit elements (cells, macros, gates, transistors) and interconnec-

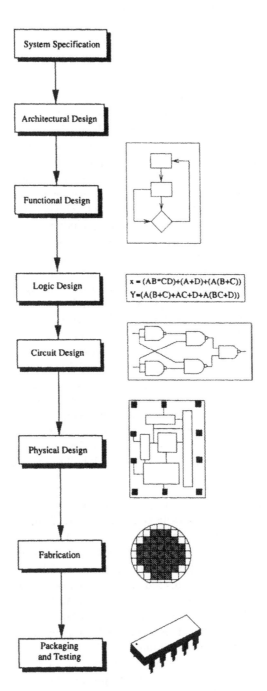

Figure 1.1: A simple VLSI design cycle.

tion between these elements. This representation is also called a *netlist*. Tools used to manually enter such description are called *schematic capture tools*. In many cases, a netlist can be created automatically from logic (RTL) description by using *logic synthesis* tools.

6. **Physical Design:** In this step the circuit representation (or netlist) is converted into a geometric representation. As stated earlier, this geometric representation of a circuit is called a *layout*. Layout is created by converting each logic component (cells, macros, gates, transistors) into a geometric representation (specific shapes in multiple layers). which perform the intended logic function of the corresponding component. Connections between different components are also expressed as geometric patterns typically lines in multiple layers. The exact details of the layout also depend on design rules, which are guidelines based on the limitations of the fabrication process and the electrical properties of the fabrication materials. Physical design is a very complex process and therefore it is usually broken down into various sub-steps. Various verification and validation checks are performed on the layout during physical design. In many cases, physical design can be completely or partially automated and layout can be generated directly from netlist by *Layout Synthesis* tools. Most of the layout of a high performance design (such as a microprocessor) may be done using manual design, while many low to medium performance design or designs which need faster time-to-market may be done automatically. Layout synthesis tools, while fast, do have an area and performance penalty, which limit their use to some designs. Manual layout, while slow and manually intensive, does have better area and performance as compared to synthesized layout. However this advantage may dissipate as larger and larger designs may undermine human capability to comprehend and obtain globally optimized solutions.

7. **Fabrication:** After layout and verification, the design is ready for fabrication. Since layout data is typically sent to fabrication on a tape, the event of release of data is called *Tape Out*. Layout data is converted (or fractured) into photo-lithographic masks, one for each layer. Masks identify spaces on the wafer, where certain materials need to be deposited, diffused or even removed. Silicon crystals are grown and sliced to produce wafers. Extremely small dimensions of VLSI devices require that the wafers be polished to near perfection. The fabrication process consists of several steps involving deposition, and diffusion of various materials on the wafer. During each step one mask is used. Several dozen masks may be used to complete the fabrication process. A large wafer is 20 cm (8 inch) in diameter and can be used to produce hundreds of chips, depending of the size of the chip. Before the chip is mass produced, a prototype is made and tested. Industry is rapidly moving towards a 30 cm (12 inch) wafer allowing even more chips per wafer leading to lower cost per chip.

8. **Packaging, Testing and Debugging:** Finally, the wafer is fabricated
   and diced into individual chips in a fabrication facility. Each chip is then
   packaged and tested to ensure that it meets all the design specifications
   and that it functions properly. Chips used in Printed Circuit Boards
   (PCBs) are packaged in Dual In-line Package (DIP), Pin Grid Array
   (PGA), Ball Grid Array (BGA), and Quad Flat Package (QFP). Chips
   used in Multi-Chip Modules (MCM) are not packaged, since MCMs use
   bare or naked chips.

It is important to note that design of a complex VLSI chip is a complex
human power management project as well. Several hundred engineers may
work on a large design project for two to three years. This includes architecture
designers, circuit designers, physical design specialists, and design automation
engineers. As a result, design is usually partitioned along functionality, and
different units are designed by different teams. At any given time, each unit
may not be at the same level of design. While one unit may be in logic design
phase, another unit may be completing its physical design phase. This imposes
a serious problem for chip level design tools, since these tools must work with
partial data at the chip level.

The VLSI design cycle involves iterations, both within a step and between
different steps. The entire design cycle may be viewed as transformations of
representations in various steps. In each step, a new representation of the
system is created and analyzed. The representation is iteratively improved to
meet system specifications. For example, a layout is iteratively improved so
that it meets the timing specifications of the system. Another example may be
detection of design rule violations during design verification. If such violations
are detected, the physical design step needs to be repeated to correct the error.
The objectives of VLSI CAD tools are to minimize the time for each iteration
and the total number of iterations, thus reducing time-to-market.

1.2 New Trends in VLSI Design Cycle

The design flow described in the previous section is conceptually simple and
illustrates the basic ideas of the VLSI design cycle. However, there are many
new trends in the industry, which seek to significantly alter this flow. The
major contributing factors are:

1. **Increasing interconnect delay:** As the fabrication process improves,
   the interconnect is not scaling at the same rate as the devices. Devices are
   becoming smaller and faster, and interconnect has not kept up with that
   pace. As a result, almost 60% of a path delay may be due to interconnect.
   One solution to interconnect delay and signal integrity issue is insertion
   of repeaters in long wires. In fact, repeaters are now necessary for most
   chip level nets. This techniques requires advanced planning since area for
   repeaters must be allocated upfront.

2. **Increasing interconnect area:** It has been estimated that a microprocessor die has only 60%-70% of its area covered with active devices. The rest of the area is needed to accommodate the interconnect. This area also leads to performance degradation. In early ICs, a few hundred transistors were interconnected using one layer of metal. As the number of transistors grew, the interconnect area increased. However, with the introduction of a second metal layer, the interconnect area decreased. This has been the trend between design complexity and the number of metal layers. In current designs, with approximately ten million transistors and four to six layers of metal, one finds about 40% of the chips real estate dedicated to its interconnect. While more metal layers help in reducing the die size, it should be noted that more metal layers (after a certain number of layers) do not necessarily mean less interconnect area. This is due to the space taken up by the vias on the lower layers.

3. **Increasing number of metal layers:** To meet the increasing needs of interconnect, the number of metal layers available for interconnect is increasing. Currently, a three layer process is commonly used for most designs, while four layer and five layer processes are used mainly for microprocessors. As a result, a three dimensional view of the interconnect is necessary.

4. **Increasing planning requirements:** The most important implication of increasing interconnect delay, area of the die dedicated to interconnect, and a large number of metal layers is that the relative location of devices is very important. Physical design considerations have to enter into design at a much earlier phase. In fact, functional design should include *chip planning*. This includes two new key steps; block planning and signal planning. Block planning assigns shapes and locations to main functional blocks. Signal planning refers to assignment of the three dimensional regions through which major busses and signals will be routed. Timing should be estimated to verify the validity of the chip plan. This plan should be used to create timing constraints for later stages of design.

5. **Synthesis:** The time required to design any block can be reduced if layout can be directly generated or *synthesized* from a higher level description. This not only reduces design time, it also eliminates human errors. The biggest disadvantage is the area used by synthesized blocks. Such blocks take larger areas than hand crafted blocks. Depending upon the level of design on which synthesis is introduced, we have two types of synthesis.

    Logic Synthesis: This process converts an HDL description of a block into schematics (circuit description) and then produces its layout. Logic synthesis is an established technology for blocks in a chip design, and for complete Application Specific Integrated Circuits (ASICs). Logic synthesis is not applicable for large regular blocks, such as RAMs, ROMs, PLAs and Datapaths, and complete microprocessor chips for two reasons;

speed and area. Logic synthesis tools are too slow and too area inefficient to deal with such blocks.

High Level Synthesis: This process converts a functional or micro-architectural description into a layout or RTL description. In high level synthesis, input is a description which captures only the behavioral aspects of the system. The synthesis tools form a spectrum. The synthesis system described above can be called general synthesis. A more restricted type synthesizes some constrained architectures. For example, Digital Signal Processing (DSP) architectures have been successfully synthesized. These synthesis systems are sometimes called *Silicon Compilers*. An even more restricted type of synthesis tools are called *Module Generators*, which work on smaller size problems. The basic idea is to simplify the synthesis task, either by restricting the architecture or restricting the size of the problem. Silicon compilers sometimes use the output of module generators. High level synthesis is an area of current research and is not used in actual chip development [GDWL92]. In summary, high level synthesis systems provide very good implementations for specialized classes of systems, and they will continue to gain acceptance as they become more generalized.

In order to accommodate the factors discussed above, the VLSI design cycle is changing. In Figure 1.2, we show a VLSI design flow which is *closer to reality*. Due to increasing interconnect delay, the physical design starts very early in the design cycle to get improved estimates of the performance of the chip. The early floor physical design activities lead to increasingly improved chip layout as each block is refined. This also allows better utilization of the chip area to distribute the interconnect in three dimensions. This distribution helps in reducing the die size, improving yield and reducing cost. Essentially, the VLSI design cycle produces increasingly better defined descriptions of the given chip. Each description is verified and, if it fails to meet the specification, the step is repeated.

1.3   Physical Design Cycle

The input to the physical design cycle is a circuit diagram and the output is the layout of the circuit. This is accomplished in several stages such as partitioning, floorplanning, placement, routing, and compaction. The different stages of physical design cycle are shown in Figure 1.3. Each of these stages will be discussed in detail in various chapters; however, to give a global perspective, we present a brief description of all the stages here.

1. **Partitioning:** A chip may contain several million transistors. Due to the limitations of memory space and computation power available it may not be possible to layout the entire chip (or generically speaking any large circuit) in the same step. Therefore, the chip (circuit) is normally partitioned into sub-chips (sub-circuits). These sub-partitions are called

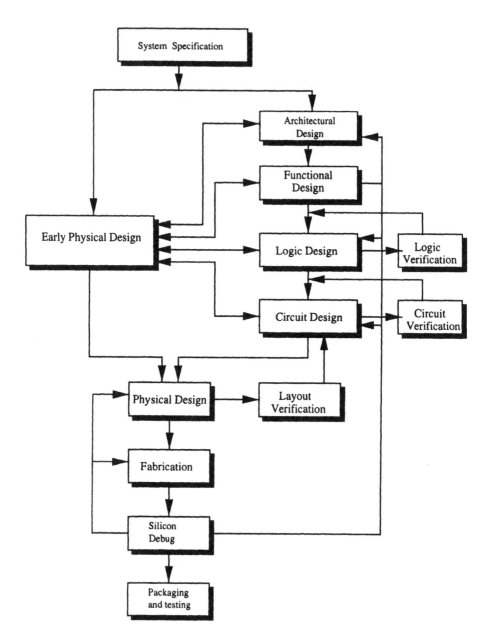

Figure 1.2: VLSI design cycle.

blocks. The actual partitioning process considers many factors such as the size of the blocks, number of blocks, and number of interconnections between the blocks. The output of partitioning is a set of blocks and the interconnections required between blocks. Figure 1.3(a) shows that the input circuit has been partitioned into three blocks. In large circuits, the partitioning process is hierarchical and at the topmost level a chip may have 5 to 25 blocks. Each block is then partitioned recursively into smaller blocks.

2. **Floorplanning and Placement:** This step is concerned with selecting good layout alternatives for each block, as well as the entire chip. The area of each block can be estimated after partitioning and is based approximately on the number and the type of components in that block. In addition, interconnect area required within the block must be considered. The actual rectangular shape of the block, which is determined by the aspect ratio may, however, be varied within a pre-specified range. Many blocks may have more general rectilinear shapes. Floorplanning is a critical step, as it sets up the ground work for a good layout. However, it is computationally quite hard. Very often the task of floorplanning is done by a design engineer, rather than a CAD tool. This is due to the fact that a human is better at 'visualizing' the entire floorplan and taking into account the information flow. Manual floorplanning is sometimes necessary as the major components of an IC need to be placed in accordance with the signal flow of the chip. In addition, certain components are often required to be located at specific positions on the chip.

During placement, the blocks are exactly positioned on the chip. The goal of placement is to find a minimum area arrangement for the blocks that allows completion of interconnections between the blocks, while meeting the performance constraints. That is, we want to avoid a placement which is routable but does not allow certain nets to meet their timing goals. Placement is typically done in two phases. In the first phase an initial placement is created. In the second phase, the initial placement is evaluated and iterative improvements are made until the layout has minimum area or best performance and conforms to design specifications. Figure 1.3(b) shows that three blocks have been placed. It should be noted that some space between the blocks is intentionally left empty to allow interconnections between blocks.

The quality of the placement will not be evident until the routing phase has been completed. Placement may lead to an unroutable design, i.e., routing may not be possible in the space provided. In that case, another iteration of placement is necessary. To limit the number of iterations of the placement algorithm, an estimate of the required routing space is used during the placement phase. Good routing and circuit performance depend heavily on a good placement algorithm. This is due to the fact that once the position of each block is fixed, very little can be done to

improve the routing and the overall circuit performance. Late placement
changes lead to increased die size and lower quality designs.

3. **Routing:** The objective of the routing phase is to complete the interconnections between blocks according to the specified netlist. First, the space
   not occupied by the blocks (called the routing space) is partitioned into
   rectangular regions called *channels* and *switchboxes*. This includes the
   space between the blocks as well the as the space on top of the blocks.
   The goal of a router is to complete all circuit connections using the shortest possible wire length and using only the channel and switch boxes.
   This is usually done in two phases, referred to as the *Global Routing* and
   Detailed Routing phases. In global routing, connections are completed
   between the proper blocks of the circuit disregarding the exact geometric
   details of each wire and pin. For each wire, the global router finds a list of
   channels and switchboxes which are to be used as a passageway for that
   wire. In other words, global routing specifies the different regions in the
   routing space through which a wire should be routed. Global routing is
   followed by detailed routing which completes point-to-point connections
   between pins on the blocks. Global routing is converted into exact routing
   by specifying geometric information such as the location and spacing of
   wires and their layer assignments. Detailed routing includes channel routing and switchbox routing, and is done for each channel and switchbox.
   Routing is a very well studied problem, and several hundred articles have
   been published about all its aspects. Since almost all problems in routing
   are computationally hard, the researchers have focused on heuristic algorithms. As a result, experimental evaluation has become an integral part
   of all algorithms and several benchmarks have been standardized. Due
   to the very nature of the routing algorithms, complete routing of all the
   connections cannot be guaranteed in many cases. As a result, a technique
   called *rip-up and re-route* is used, which basically removes troublesome
   connections and reroutes them in a different order. The routing phase of
   Figure 1.3(c) shows that all the interconnections between the three blocks
   have been routed.

4. **Compaction:** Compaction is simply the task of compressing the layout
   in all directions such that the total area is reduced. By making the
   chip smaller, wire lengths are reduced, which in turn reduces the signal
   delay between components of the circuit. At the same time, a smaller
   area may imply more chips can be produced on a wafer, which in turn
   reduces the cost of manufacturing. However, the expense of computing
   time mandates that extensive compaction is used only for large volume
   applications, such as microprocessors. Compaction must ensure that no
   rules regarding the design and fabrication process are violated during the
   process. Figure 1.3(d) shows the compacted layout.

5. **Extraction and Verification:** Design Rule Checking (DRC) is a process
   which verifies that all geometric patterns meet the design rules imposed

by the fabrication process. For example, one typical design rule is the wire separation rule. That is, the fabrication process requires a specific separation (in microns) between two adjacent wires. DRC must check such separation for millions of wires on the chip. There may be several dozen design rules, some of them are quite complicated to check. After checking the layout for design rule violations and removing the design rule violations, the functionality of the layout is verified by *Circuit Extraction*. This is a reverse engineering process, and generates the circuit representation from the layout. The extracted description is compared with the circuit description to verify its correctness. This process is called *Layout Versus Schematics* (LVS) verification. Geometric information is extracted to compute Resistance and Capacitance. This allows accurate calculation of the timing of each component, including interconnect. This process is called *Performance Verification*. The extracted information is also used to check the reliability aspects of the layout. This process is called *Reliability Verification* and it ensures that layout will not fail due to electro-migration, self-heat and other effects [Bak90].

Physical design, like VLSI design, is iterative in nature and many steps, such as global routing and channel routing, are repeated several times to obtain a better layout. In addition, the quality of results obtained in a step depends on the quality of the solution obtained in earlier steps. For example, a poor quality placement cannot be 'cured' by high quality routing. As a result, earlier steps have more influence on the overall quality of the solution. In this sense, partitioning, floorplanning, and placement problems play a more important role in determining the area and chip performance, as compared to routing and compaction. Since placement may produce an 'unroutable' layout, the chip might need to be re-placed or re-partitioned before another routing is attempted. In general, the whole design cycle may be repeated several times to accomplish the design objectives. The complexity of each step varies, depending on the design constraints as well as the design style used. Each step of the design cycle will be discussed in greater detail in a later chapter.

1.4 New Trends in Physical Design Cycle

As fabrication technology improves and process enters the deep sub-micron range, it is clear that interconnect delay is not scaling at the same rate as the gate delay. Therefore, interconnect delay is a more significant part of overall delay. As a result, in high performance chips, interconnect delay must be considered from very early design stages. In order to reduce interconnect delay several methods can be employed.

1. **Chip level signal planning:** At the chip level, routing of major signals and buses must be planned from early design stages, so that interconnect distances can be minimized. In addition, these global signals must be routed in the top metal layers, which have low delay per unit length.

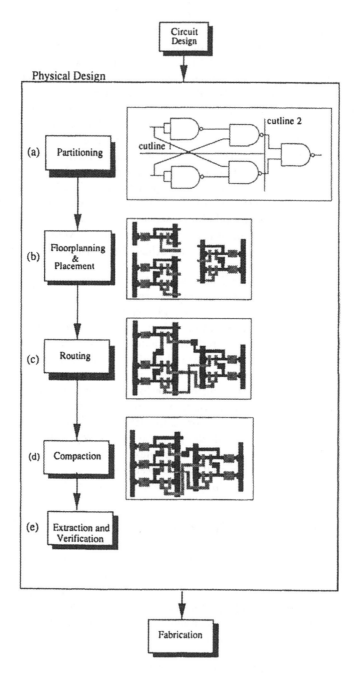

Figure 1.3: Physical design cycle.

2. **OTC routing:** Over-the-Cell (OTC) routing is a term used to describe routing over blocks and active areas. This is a departure from conventional channel and switchbox routing approach. Actually, chip level signal planning is OTC routing on the entire chip. The OTC approach can also be used within a block to reduce area and improve performance. The OTC routing approach essentially makes routing a three dimensional problem. Another effect of the OTC routing approach is that the pins are not brought to the block boundaries for connections to other blocks. Instead, pins are brought to the top of the block as a sea-of-pins. This concept, technically called the *Arbitrary Terminal Model* (ATM), will be discussed in a later chapter.

The conventional decomposition of physical design into partitioning, placement and routing phases is conceptually simple. However, it is increasingly clear that each phase is interdependent on other phases, and an integrated approach to partitioning, placement, and routing is required.

Figure 1.4 shows the physical design cycle with emphasis on timing. The figure shows that timing is estimated after floorplaning and placement, and these steps are iterated if some connections fail to meet the timing requirements. After the layout is complete, resistance and capacitance effects of one component on another can be extracted and accurate timing for each component can be calculated. If some connections or components fail to meet their timing requirements, or fail due to the effect of one component on another, then some or all phases of physical design need to be repeated. Typically, these 'repeat-or-not-to-repeat' decisions are made by experts rather than tools. This is due to the complex nature of these decisions, as they depend on a host of parameters.

1.5 Design Styles

Physical design is an extremely complex process. Even after breaking the entire process into several conceptually easier steps, it has been shown that each step is computationally very hard. However, market requirements demand quick time-to-market and high yield. As a result, restricted models and design styles are used in order to reduce the complexity of physical design. This practice began in the late 1960s and led to the development of several restricted design styles [Feu83]. The design styles can be broadly classified as either full-custom or semi-custom. In a full-custom layout, different blocks of a circuit can be placed at any location on a silicon wafer as long as all the blocks are non-overlapping. On the other hand, in semi-custom layout, some parts of a circuit are predesigned and placed on some specific place on the silicon wafer. Selection of a layout style depends on many factors including the type of chip, cost, and time-to-market. Full-custom layout is a preferred style for mass produced chips, since the time required to produce a highly optimized layout can be justified. On the other hand, to design an Application Specific Integrated Circuit (ASIC),

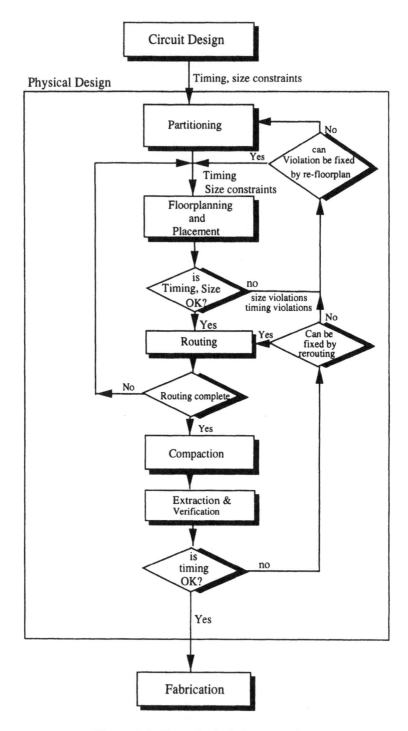

Figure 1.4: New physical design cycle.

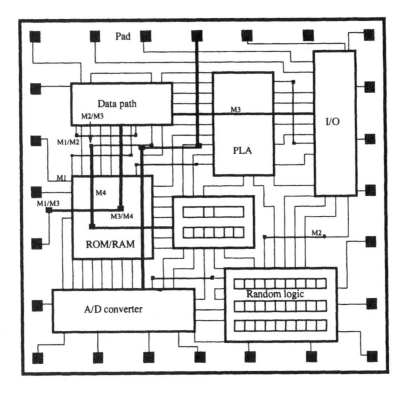

Figure 1.5: Full-custom structure.

a semi-custom layout style is usually preferred. On a large chip, each block may use a different layout design style.

1.5.1 Full-Custom

In its most general form of design style, the circuit is partitioned into a collection of sub-circuits according to some criteria such as functionality of each sub-circuit. The process is done hierarchically and thus full-custom designs have several levels of hierarchy. The chip is organized in clusters, clusters consist of units, and units are composed of *functional blocks* (in short, blocks). For sake of simplicity, we use the term blocks for units, blocks, and clusters. The full-custom design style allows functional blocks to be of any size. Figure 1.5 shows an example of a very simple circuit with few blocks. Other levels of hierarchy are not shown for this simple example. Internal routing in each block is not shown for the sake of clarity. In the full-custom design style, blocks can be placed at any location on the chip surface without any restrictions. In other words, this style is characterized by the absence of any constraints on the physical design process. This design style allows for very compact designs.

However, the process of automating a full-custom design style has a much higher complexity than other restricted models. For this reason it is used only when the final design must have minimum area and design time is less of a factor. The automation process for a full-custom layout is still a topic of intensive research. Some phases of physical design of a full-custom chip may be done manually to optimize the layout. Layout compaction is a very important aspect in full-custom design. The rectangular solid boxes around the boundary of the circuit are called *I/O pads*. Pads are used to complete interconnections between different chips or interconnections between the chip and the board. The spaces not occupied by blocks are used for routing of interconnecting wires. Initially all the blocks are placed within the chip area with the objective of minimizing the total area. However, there must be enough space left between the blocks so that routing can be completed using this space and the space on top of the blocks. Usually several metal layers are used for routing of interconnections. Currently, three metal layers are common for routing. A four metal layer process is being used for microprocessors, and a six layer process is gaining acceptance, as fabrication costs become more feasible. In Figure 1.5, note that width of the M1 wire is smaller than the width of the M2 wire. Also note that the size of the via between M1 and M2 is smaller than the size of the via between higher layers. Typically, metal widths and via sizes are larger for higher layers. The figure also shows that some routing has been completed on top of the blocks. The routing area needed between the blocks is becoming smaller and smaller as more routing layers are used. This is due to the fact that more routing is done on top of the transistors in the additional metal layers. If all the routing can be done on top of the transistors, the total chip area is determined by the area of the transistors. However, as circuits become more complex and interconnect requirements increase, the die size is determined by the interconnect area and the total transistor area serves as a lower bound on the die size of the chip.

In a hierarchical design of a circuit, each block in a full-custom design may be very complex and may consist of several sub-blocks, which in turn may be designed using the full-custom design style or other design styles. It is easy to see that since any block is allowed to be placed anywhere on the chip, the problem of optimizing area and the interconnection of wires becomes difficult. Full custom design is very time consuming; thus the method is inappropriate for very large circuits, unless performance or chip size is of utmost importance. Full custom is usually used for the layout of microprocessors and other performance and cost sensitive designs.

1.5.2 Standard Cell

The design process in the standard cell design style is somewhat simpler than full-custom design style. Standard cell architecture considers the layout to consist of rectangular cells of the same height. Initially, a circuit is partitioned into several smaller blocks, each of which is equivalent to some predefined subcircuit (cell). The functionality and the electrical characteristics of each predefined cell are tested, analyzed, and specified. A collection of these cells is called a *cell library*. Usually a cell library consists of 500-1200 cells. Terminals on cells may be located either on the boundary or distributed throughout the cell area. Cells are placed in rows and the space between two rows is called a *channel*. These channels and the space above and between cells is used to perform interconnections between cells. If two cells to be interconnected lie in the same row or in adjacent rows, then the channel between the rows is used for interconnection. However, if two cells to be connected lie in two non-adjacent rows, then their interconnection wire passes through empty space between any two cells or passes on top of the cells. This empty space between cells in a row is called a *feedthrough*. The interconnections are done in two steps. In the first step, the feedthroughs are assigned for the interconnections of non-adjacent cells. Feedthrough assignment is followed by routing. The cells typically use only one metal layer for connections inside the cells. As a result, in a two metal process, the second metal layer can be used for routing in over-the-cell regions. In a three metal layer process, almost all the channels can be removed and all routing can be completed over the cells. However, this is a function of the density of cells and distribution of pins on the cells. It is difficult to obtain a channelless layout for chips which use highly packed dense cells with poor pin distribution. Figure 1.6 shows an example of a standard cell layout. A cell library is shown, along with the complete circuit with all the interconnections, feedthroughs, and power and ground routing. In the figure, the library consists of four logic cells and one feedthrough cell. The layout shown consists of several instances of cells in the library. Note that representation of a layout in the standard cell design style is greatly simplified as it is not necessary to duplicate the cell information.

The standard cell layout is inherently non-hierarchical. The hierarchical circuits, therefore, have to undergo some transformation before this design style can be used. This design style is well-suited for moderate size circuits and medium production volumes. Physical design using standard cells is somewhat simpler as compared to full-custom, and is efficient using modern design tools. The standard cell design style is also widely used to implement the 'random or control logic' part of the full-custom design as shown in Figure 1.5.

Logic Synthesis usually uses the standard cell design style. The synthesized circuit is mapped to cell circuits. Then cells are placed and routed.

While standard cell designs are quicker to develop, a substantial initial investment is needed in the development of the cell library, which may consist of several hundred cells. Each cell in the cell library is 'hand crafted' and requires highly skilled physical design specialists. Each type of cell must be created with several transistor sizes. Each cell must then be tested by simulation and its performance must be characterized. Cell library development is a significant project with enormous manpower and financial resource requirements.

A standard cell design usually takes more area than a full-custom or a hand-crafted design. However, as more and more metal layers become available for routing and design tools improve, the difference in area between the two design styles will gradually reduce.

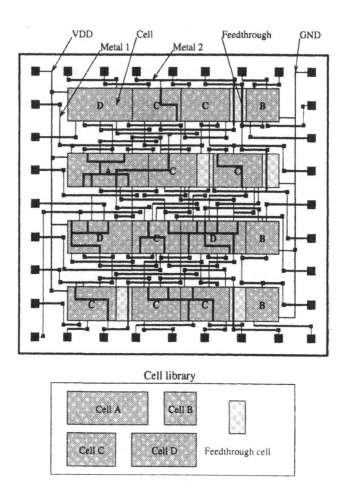

Figure 1.6: Standard cell structure.

1.5.3   Gate Arrays

This design style is a simplification of standard cell design. Unlike standard cell design, all the cells in gate array are identical. Each chip is an array of identical gates or cells. These cells are separated by both vertical and horizontal spaces called vertical and horizontal channels. The circuit design is modified such that it can be partitioned into a number of identical blocks. Each block must be logically equivalent to a cell on the gate array. The name 'gate array' signifies the fact that each cell may simply be a gate, such as a three input NAND gate. Each block in design is mapped or placed onto a prefabricated cell on the chip during the partitioning/placement phase, which is reduced to a block to cell assignment problem. The number of partitioned blocks must be less than or equal to the total number of cells on the chip. Once the circuit

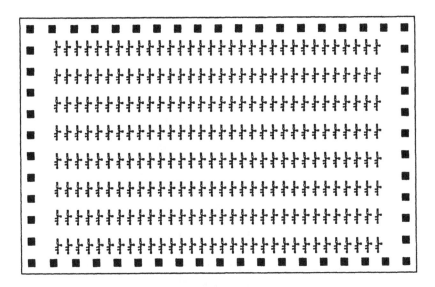

Figure 1.7: A conceptual uncommitted gate array.

is partitioned into identical blocks, the task is to make the interconnections between the prefabricated cells on the chip using horizontal and vertical channels to form the actual circuit. Figure 1.7 shows an 'uncommitted' gate array, which is simply a term used for a prefabricated chip. The gate array wafer is taken into a fabrication facility and routing layers are fabricated on top of the wafer. The completed wafer is also called a 'customized wafer'. It should be noted that the number of tracks allowed for routing in each channel is fixed. As a result, the purpose of the routing phase is simply to complete the connections rather than minimize the area. Two layers of interconnections are most common; though one and three layers are also used. Figure 1.8 illustrates a committed gate array design. Like standard cell designs, synthesis can also use the gate array style. In gate array design the entire wafer, consisting of several dozen chips, is prefabricated.

This simplicity of gate array design is gained at the cost of rigidity imposed upon the circuit both by the technology and the prefabricated wafers. The advantage of gate arrays is that the steps involved for creating any prefabricated wafer are the same and only the last few steps in the fabrication process actually depend on the application for which the design will be used. Hence gate arrays are cheaper and easier to produce than full-custom or standard cell. Similar to standard cell design, gate array is also a non-hierarchical structure.

The gate array architecture is the most restricted form of layout. This also means that it is the simplest for algorithms to work with. For example, the task of routing in gate array is to determine if a given placement is routable. The routability problem is conceptually simpler as compared to the routing

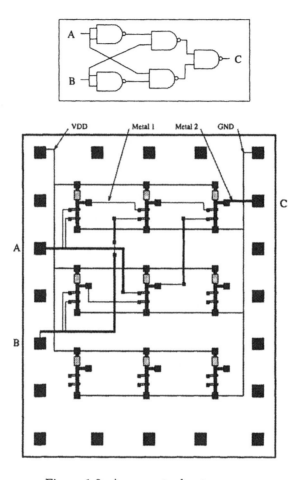

Figure 1.8: A conceptual gate array.

problem in standard cell and full-custom design styles.

1.5.4   Field Programmable Gate Arrays

The Field Programmable Gate Array (FPGA) is a new approach to ASIC design that can dramatically reduce manufacturing turn-around time and cost for low volume manufacturing [Gam89, Hse88, Won89]. In FPGAs, cells and interconnect are prefabricated. The user simply 'programs' the interconnect. FPGA designs provide large scale integration and user programmability. A FPGA consists of horizontal rows of programmable logic blocks which can be interconnected by a programmable routing network. FPGA cells are more complex than standard cells. However, almost all the cells have the same layout. In its simplistic form, a logic block is simply a memory block which can be pro-

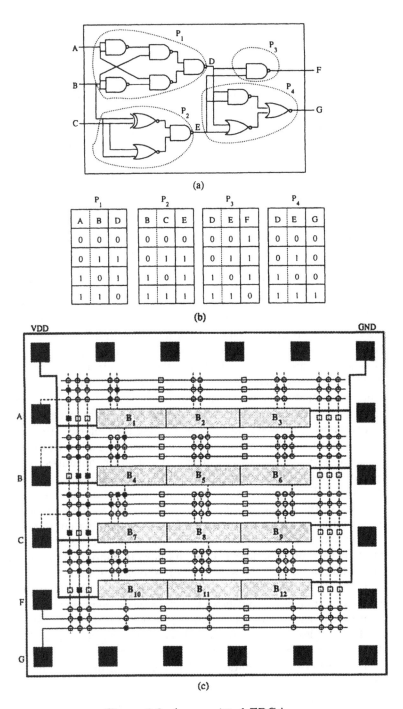

P₁			P₂			P₃			P₄		
A	B	D	B	C	E	D	E	F	D	E	G
0	0	0	0	0	0	0	0	1	0	0	0
0	1	1	0	1	1	0	1	1	0	1	0
1	0	1	1	0	1	1	0	1	1	0	0
1	1	0	1	1	1	1	1	0	1	1	1

(b)

Figure 1.9: A committed FPGA.

grammed to remember the logic table of a function. Given a certain input, the logic block 'looks up' the corresponding output from the logic table and sets its output line accordingly. Thus by loading different look-up tables, a logic block can be programmed to perform different functions. It is clear that 2^K bits are required in a logic block to represent a K-bit input, 1-bit output combinational logic function. Obviously, logic blocks are only feasible for small values of K. Typically, the value of K is 5 or 6. For multiple outputs and sequential circuits the value of K is even less. The rows of logic blocks are separated by horizontal routing channels. The channels are not simply empty areas in which metal lines can be arranged for a specific design. Rather, they contain prede-fined wiring 'segments' of fixed lengths. Each input and output of a logic block is connected to a dedicated vertical segment. Other vertical segments merely pass through the blocks, serving as feedthroughs between channels. Connec-tion between horizontal segments is provided through *antifuses*, whereas the connection between a horizontal segment and a vertical segment is provided through a *cross fuse*. Figure 1.9(c) shows the general architecture of a FPGA, which consists of four rows of logic blocks. The cross fuses are shown as circles, while antifuses are shown as rectangles. One disadvantage of fuse based FPGAs is that they are not reprogrammable. There are other types of FPGAs which allow re-programming, and use pass gates rather than programmable fuses.

Since there are no user specific fabrication steps in a FPGA, the fabrica-tion process can be set up in a cost effective manner to produce large quan-tities of generic (unprogrammed) FPGAs. The customization (programming) of a FPGA is rather simple. Given a circuit, it is decomposed into smaller subcircuits, such that each subcircuit can be mapped to a logic block. The interconnections between any two subcircuits is achieved by programming the FPGA interconnects between their corresponding logic blocks. Programming (blowing) one of the fuses (antifuse or cross fuse) provides a low resistance bidi-rectional connection between two segments. When blown, antifuses connect the two segments to form a longer one. In order to program a fuse, a high voltage is applied across it. FPGAs have special circuitry to program the fuses. The cir-cuitry consists of the wiring segments and control logic at the periphery of the chip. Fuse addresses are shifted into the fuse programming circuitry serially. Figure 1.9(a) shows a circuit partitioned into four subcircuits, P_1, P_2, P_3, and P_4. Note that each of these four subcircuits have two inputs and one output. The truth table for each of the subcircuits is shown in Figure 1.9(b). In Fig-ure 1.9(c), P_1, P_2, P_3, and P_4 are mapped to logic blocks B_1, B_4, B_7, and B_{10} respectively and appropriate antifuses and cross fuses are programmed (burnt) to implement the entire circuit. The programmed fuses are shown as filled circles and rectangles. We have described the 'once-program' type of FPGAs. Many FPGAs allow the user to re-program the interconnect, as many times as needed. These FPGAs use non-destructive methods of programming, such as pass-transistors.

The programmable nature of these FPGAs requires new CAD algorithms to make effective use of logic and routing resources. The problems involved in customization of a FPGA are somewhat different from those of other design styles; however, many steps are common. For example, the partition problem of FPGAs is different than partitioning the problem in all design style while the placement and the routing is similar to gate array approach. These problems will be discussed in detail in Chapter 11.

1.5.5 Sea of Gates

The sea of gates is an improved gate array in which the master is filled completely with transistors. The master of the sea-of-gates has a much higher density of logic implemented on the chip, and allows a designer to fabricate complex circuits, such as RAMs, to be built. In the absence of routing channels, interconnects have to be completed either by routing through gates, or by adding more metal or polysilicon interconnection layers. There are problems associated with either solution. The former reduces the gate utilization; the latter increases the mask count and increases fabrication time and cost.

1.5.6 Comparison of Different Design Styles

The choice of design style depends on the intended functionality of the chip, time-to-market and total number of chips to be manufactured. It is common to use full-custom design style for microprocessors and other complex high volume applications, while FPGAs may be used for simple and low volume applications. However, there are several chips which have been manufactured by using a mix of design styles. For large circuits, it is common to partition the circuit into several small circuits which are then designed by different teams. Each team may use a different design style or a number of design styles. Another factor complicating the issue of design style is re-usability of existing designs. It is a common practice to re-use complete or partial layout from existing chips for new chips to reduce the cost of a new design. It is quite typical to use standard cell and gate array design styles for smaller and less complex Application Specific ICs (ASICs), while microprocessors are typically full-custom with several standard cell blocks. Standard cell blocks can be laid out using logic synthesis tools.

Design styles can be seen as a continuum from very flexible (full-custom) to a rather rigid design style (FPGA) to cater to differing needs. Table 1.1 summarizes the differences in cell size, cell type, cell placement and interconnections in full-custom, standard cell, gate array and FPGA design styles. Another comparison may be on the basis of area, performance, and the number of fabrication layers needed. (See Table 1.2). As can be seen from the table, full-custom provides compact layouts for high performance designs but requires a considerable fabrication effort. On the other hand, a FPGA is completely pre-fabricated and does not require any user specific fabrication steps. However, FPGAs can only be used for small, general purpose designs.

	style			
	full-custom	standard cell	gate array	FPGA
cell size	variable	fixed height*	fixed	fixed
cell type	variable	variable	fixed	programmable
cell placement	variable	in row	fixed	fixed
interconnections	variable	variable	variable	programmable
design cost	high	medium	medium	low

Table 1.1: Comparison of different design styles.
* uneven height cells are also used.

	style			
	full-custom	standard cell	gate array	FPGA
Area	compact	compact to moderate	moderate	large
Performance	high	high to moderate	moderate	low
Fabricate	All layers	All layers	Routing layers only	No layers

Table 1.2: Area, Performance and Fabrication layers for different design styles.

1.6 System Packaging Styles

The increasing complexity and density of semiconductor devices are the key driving forces behind the development of more advanced VLSI packaging and interconnection approaches. Two key packaging technologies being used currently are Printed Circuit Boards (PCB) and Multi-Chip Modules (MCMs). Let us first start with die packaging techniques.

1.6.1 Die Packaging and Attachment Styles

Dies can be packaged in a variety of styles depending on cost, performance and area requirements. Other considerations include heat removal, testing and repair.

1.6.1.1 Die Package Styles

ICs are packaged into ceramic or plastic carriers called Dual In-Line Packages (DIPs), then mounted on a PCB. These packages have leads on 2.54 mm centers on two sides of a rectangular package. PGA (Pin Grid Array) is a package in which pins are organized in several concentric rectangular rows. DIPs and PGAs require large thru-holes to mount them on boards. As a result, thru-hole assemblies were replaced by Surface Mount Assemblies (SMAs). In SMA, pins of the device do not go through the board, they are soldered to the surface of the board. As a result, devices can be placed on both sides of the board. There are two types of SMAs; leaded and leadless. Both are available in quad packages with leads on 1.27, 1.00, or 0.635 mm centers. Yet another variation of SMA is the Ball Grid Array (BGA), which is an array of solder balls. The balls are pressed on to the PCB. When a BGA device is placed and pressed the balls melt forming a connection to the PCB. All the packages discussed above suffer from performance degradation due to delays in the package. In some applications, a naked die is used directly to avoid package delays.

1.6.1.2 Package and Die Attachment Styles

The chips need to be attached to the next level of packaging, called system level packaging. The leads of pin based packages are bent down and are soldered into plated holes which go inside the printed circuit board. (see Figure 1.10). SMAs such as BGA do not need thru holes but still require a relatively large footprint.

In the case of naked dies, die to board connections are made by attaching wires from the I/O pads on the edge of the die to the board. This is called the wire bond method, and uses a robotic wire bonding machine. The active side of the die faces away from the board. Although package delays are avoided in wire bonded dies, the delay in the wires is still significant as compared to the interconnect delay on the chip.

Controlled Collapsed Chip Connection (C4) is another method of attaching a naked die. This method aims to eliminate the delays associated with the wires in the wire bond method. The I/O pins are distributed over the die (ATM style) and a solder ball is placed over the I/O pad. The die is then turned over, such that the active side is facing the board, then pressure is applied to fuse the balls to the board.

The exact layout of chips on PCBs and MCMs is somewhat equivalent to the layout of various components in a VLSI chip. As a result, many layout problems such as partitioning, placement, and routing are similar in VLSI and packaging. In this section, we briefly outline the two commonly used packaging styles and the layout problems with these styles.

1.6.2 Printed Circuit Boards

A Printed Circuit Board (PCB) is a multi-layer sandwich of routing layers. Current PCB technology offers as many as 30 or more routing layers. Via specifications are also very flexible and vary, such that a wide variety of combinations is possible. For example, a set of layers can be connected by a single via called the *stacked via*. The traditional approach of single chip packages on a PCB have intrinsic limitations in terms of silicon density, system size, and contribution to propagation delay. For example, the typical inner lead bond pitch on VLSI chips is 0.0152 cm. The finest pitch for a leaded chip carrier is 0.0635 cm. The ratio of the area of the silicon inside the package to the package area

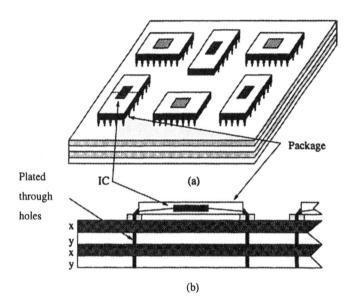

Figure 1.10: Chip placement on a printed circuit board.

is about 6%. If a PCB were completely covered with chip carriers, the board would only have at most a 6% efficiency of holding silicon. In other words, 94% or more of the board area would be wasted space, unavailable to active silicon and contributing to increased propagation delays. Thru-hole assemblies gave way to Surface Mount Assemblies (SMAs). SMAs eliminated the need for large diameter plated-thru-holes, allowing finer pitch packages and increasing routing density. SMAs reduce the package footprint and improve performance.

The SMA structure reduces package footprints, decreases chip-to-chip distances and permits higher pin count ICs. A 64 pin leadless chip carrier requires only a 12.7 mm × 12.7 mm footprint with a 0.635 mm pitch. This space conservation represents a twelve fold density improvement, or a four fold reduction in interconnection distances, over DIP assemblies.

The basic package selection parameter is the pin count. DIPs are used for chips with no more than 48 pins. PGAs are used for higher pin count chips. BGAs are used for even higher pin count chips. Other parameters include power consumption, heat dissipation and size of the system desired.

The layout problems for printed circuit boards are similar to layout problems in VLSI design, although printed circuit boards offer more flexibility and a wider variety of technologies. The routing problem is much easier for PCBs due to the availability of many routing layers. The planarity of wires in each layer is a requirement in a PCB as it is in a chip. There is little distinction between global routing and detailed routing in the case of circuit boards. In fact, due to the availability of many layers, the routing algorithm has to be modified to adapt to this three dimensional problem. Compaction has no place in PCB layout due to the constraints caused by the fixed location of the pins on packages.

For more complex VLSI devices, with 120 to 196 I/Os, even the surface mounted approach becomes inefficient and begins to limit system performance. A 132 pin device in a 635 μm pitch carrier requires a 25.4 to 38.1 mm^2 footprint. This represents a four to six fold density loss, and a two fold increase in interconnect distances as opposed to a 64 pin device. It has been shown that the interconnect density for current packaging technology is at least one order of magnitude lower than the interconnect density at the chip level. This translates into long interconnection lengths between devices and a corresponding increase in propagation delay. For high performance systems, the propagation delay is unacceptable. It can be reduced to a great extent by using SMAs such as BGAs. However, a higher performance packaging and interconnection approach is necessary to achieve the performance improvements promised by VLSI technologies. This has led to the development of multi-chip modules.

1.6.3 Multichip Modules

Current packaging and interconnection technology is not complementing the advances taking place in the IC. The key to semiconductor device improvements is the shrinking feature size, i.e., the minimum gate or line width on a device. The shrinking feature size provides increased gate density, increased gates per chip and increased clock rates. These benefits are offset by an increase in the number of I/Os and an increase in chip power dissipation. The increased clock rate is directly related to device feature size. With reduced feature sizes each on-chip device is smaller, thereby having reduced parasitics, allowing for faster switching. Furthermore, the scaling has reduced on-chip gate distances and, consequently, interconnect delays. However, much of the improvement in system performance promised by the ever increasing semiconductor device performance has not been realized. This is due to the performance barriers imposed by todays packaging and interconnection technologies.

Increasingly more complex and dense semiconductor devices are driving the development of advanced VLSI packaging and interconnection technology to meet increasingly more demanding system performance requirements. The alternative approach to the interconnect and packaging limits of conventional chip carrier/PCB assemblies is to eliminate packaging levels between the chip and PCB. One such approach uses MCMs. The MCM approach eliminates the single chip package and, instead, mounts and interconnects the chips directly onto a higher density, fine pitch interconnection substrate. Dies are wire bonded to the substrate or use a C4 bonding. In some MCM technologies, the substrate is simply a silicon wafer, on which layers of metal lines have been patterned. This substrate provides all of the chip-to-chip interconnections within the MCM. Since the chips are only one tenth of the area of the packages, they can be placed closer together on an MCM. This provides for both higher density assemblies, as well as shorter and faster interconnects. Figure 1.11 shows

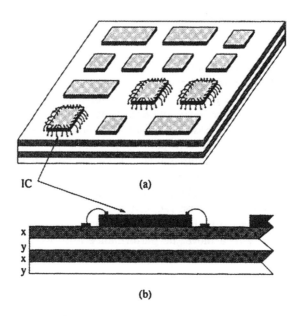

Figure 1.11: A multi-chip module with wire bonded dies.

diagram of an MCM package with wire bonded dies. One significant problem with MCMs is heat dissipation. Due to close placement of potentially several hundred chips, a large amount of heat needs to be dissipated. This may require special, and potentially expensive heat removal methods.

At first glance, it appears that it is easy to place bare chips closer and closer together. There are, however, limits to how close the chips can be placed together on the substrate. There is, for example, a certain peripheral area around the chip which is normally required for bonding, engineering change pads, and chip removal and replacement.

It is predicted that multichip modules will have a major impact on all aspects of electronic system design. Multichip module technology offers advantages for all types of electronic assemblies. Mainframes will need to interconnect the high numbers of custom chips needed for the new systems. Cost-performance systems will use the high density interconnect to assemble new chips with a collection of currently available chips, to achieve high performance without time-consuming custom design, allowing quick time-to-market.

In the long term, the significant benefits of multichip modules are: reduction in size, reduction in number of packaging levels, reduced complexity of the interconnection interfaces and the fact that the assemblies will clearly be cheaper and more efficient. However, MCMs are currently expensive to manufacture due to immature technology. As a result, MCMs are only used in high performance applications. The multichip revolution in the 1990s will have an impact on electronics as great or greater than the impact of surface mount technology in the 1980s.

The layout problems in MCMs are essentially performance driven. The partitioning problem minimizes the delay in the longest wire. Although placement in MCM is simple as compared to VLSI. global routing and detailed routing are more complex in MCM because of the large number of layers present in MCM. The critical issues in routing include the effect of cross-talk, and delay modeling of long interconnect wires. These problems will be discussed in more detail in Chapter 12.

1.6.4 Wafer Scale Integration

MCM packaging technology does not completely remove all the barriers of the IC packaging technology. Wafer Scale Integration (WSI) is considered as the next major step, bringing with it the removal of a large number of barriers. In WSI, the entire wafer is fabricated with several types of circuits, the circuits are tested, and the defect-free circuits are interconnected to realize the entire system on the wafer.

The attractiveness of WSI lies in its promise of greatly reduced cost, high performance, high level of integration, greatly increased reliability, and significant application potential. However, there are still major problems with WSI technology, such as redundancy and yield, that are unlikely to be solved in the near future. Another significant disadvantage of the WSI approach is its inability to mix and match dies from different fabrication processes. The fabrication process for microprocessors is significantly different than the one for memories. WSI would force a microprocessor and the system memory to be fabricated on the same process. This is significant sacrifice in microprocessor performance or memory density, depending on the process chosen.

1.6.5 Comparison of Different Packaging Styles

In this section, we compare different packaging styles which are either being used today or might be used in future. In [Sag89] a figure of merit has been derived for various technologies, using the product of the propagation speed (inches/10^{-9}sec) and the interconnection density (inches/sq. in). The typical figures are reproduced here in Table 1.3. The figure of merit for VLSI will need to be partially adjusted (downward) to account for line resistance and capacitance. This effect is not significant in MCMs due to higher line conductivity, lower drive currents, and lower output capacitance from the drivers.

MCM technology provides a density, performance, and cost comparable to or better than, WSI. State-of-the-art chips can be multiple-sourced and technologies can be mixed on the same substrate in MCM technology. Another advantage of MCM technology is that all chips are pretestable and replaceable. Furthermore, the substrate interconnection matrix itself can be pretested and repaired before chip assembly; and test, repair, and engineering changes are possible even after final assembly. However, MCM technology is not free of all problems. The large number of required metallurgical bonds and heat removal

Technology	Figure of Merit (inches/psec . density inches/sq in)
WSI	28.0
MCM	14.6
PCB	2.2

Table 1.3: Figure of merit speed-density.

are two of the existing problems. While WSI has higher density than MCM, its yield problem makes it currently unfeasible. The principal conclusion that can be drawn from this comparison is that WSI cannot easily compete with technology already more or less well established in terms of performance, density, and cost.

1.7 Historical Perspectives

During the 1950s the photolithographic process was commonly used in the design of circuits. With this technology, an IC was created by fabricating transistors on crystalline silicon. The design process was completely manual. An engineer would create a circuit on paper and assemble it on a breadboard to check the validity of the design. The design was then given to a layout designer, who would draw the silicon-level implementation. This drawing was cut out on rubylith plastic, and carefully inspected for compliance with the original design. Photolithographic masks were produced by optically reducing the rubylith design and these masks were used to fabricate the circuit [Feu83].

In the 1970s there was a tremendous growth in circuit design needs. The commonly used rubylith patterns became too large for the laboratories. This technology was no longer useful. Numerically controlled pattern generation machinery was implemented to replace the rubylith patterns. This was the first major step towards design automation. The layouts were transferred to data tapes and for the first time, design rule checking could be successfully automated [Feu83].

By the 1970s a few large companies developed interactive layout software which portrayed the designs graphically. Soon thereafter commercial layout systems became available. This interactive graphics capability provided rapid layout of IC designs because components could quickly be replicated and edited, rather than redrawn as in the past [Feu83]. For example, L-Edit is one such circuit layout editor commercially available. In the next phase, the role of computers was explored to help perform the manually tedious layout process. As the layout was already in the computer, routing tools were developed initially to help perform the connections on this layout, subject to the design rules specified for that particular design.

As the technology and tools are improving, the VLSI physical design is moving towards high performance circuit design. The high-performance circuit design is of highest priority in physical design. Current technology allows us to interconnect over the cells/blocks to reduce the total chip area, thereby reducing the signal delay for high performance circuits. Research on parallel algorithms for physical design has also drawn great interest since the mid 80s. The emergence of parallel computers promises the feasibility of automating many time consuming steps of physical design.

In the early decades, most aspects of VLSI design were done manually. This elongated the design process, since any changes to improve any design step would require a revamping of the previously performed steps, thus resulting in a very inefficient design. The introduction of computers in this area accelerated some aspects of design, and increased efficiency and accuracy. However, many other parts could not be done using computers, due to the lack of high speed computers or faster algorithms. The emergence of workstations led to the development of CAD tools which made designers more productive by providing the designers with 'what if' scenarios. As a result, the designers could analyze various options for a specific design and choose the optimal one. But there are some features of the design process which are not only expensive, but also too difficult to automate. In these cases the use of certain knowledge based systems is being considered. VLSI design became interactive with the availability of faster workstations with larger storage and high-resolution graphics, thus breaking away from the traditional batch processing environment. The workstations also have helped in the advancement of integrated circuit technology by providing the capabilities to create complex designs. Table 1.4 lists the development of design tools over the years.

1.8 Existing Design Tools

Design tools are essential for the *correct-by-construction* approach, that is get the design right the very first time. Any design tool should have the following capabilities.

- layout the physical design for which the tool should provide some means of schematic capture of the information. For this either a textual or interactive graphic mode should be provided.

- physical verification which means that the tool should have design rule checking capability.

- some form of simulation to verify the behavior of the design.

There are tools available with some of the above mentioned capabilities. For example, BELLE (Basic Embedded Layout Language) is a language embedded in PASCAL in which the layout can be designed by textual entry. ABCD (A Better Circuit Description) is also a language for CMOS and nMOS designs. The graphical entry tools, on the other hand, are very convenient for the designers, since such tools operate mostly through menus. KIC, developed at

Year	Design Tools
1950-1965	Manual design
1965-1975	Layout editors Automatic routers (for PCB) Efficient partitioning algorithm
1975-1985	Automatic placement tools Well defined phases of design of circuits Significant theoretical development in all phases
1985-1990	Performance driven placement and routing tools Parallel algorithms for physical design Significant development in underlying graph theory Combinatorial optimization problems for layout
1990-1995	Over-the-Cell Routing tools Three dimensional interconnect based physical design Synthesis tools mature and gain widespread acceptance
1995 - Present	Interconnect design and Modelling dominates physical design Process related tools (reliability, Electro-migration)

Table 1.4: Design Tools Development.

the University of California, Berkeley and PLAN, developed at the University of Adelaide, are examples of such tools. Along with the workstations came peripherals, such as plotters and printers with high-resolution graphics output facilities which gave the designer the ability to translate the designs generated on the workstation into hardcopies.

The rapid development of design automation has led to the proliferation of CAD tools for this purpose. Some tools are oriented towards the teaching of design automation to the educational community, while the majority are designed for actual design work. Some of the commercially available software is also available in educational versions, to encourage research and development in the academic community. Some of the design automation CAD software available for educational purposes are L-Edit, MAGIC, SPICE etc. We shall briefly discuss some of the features of L-Edit and MAGIC.

L-Edit is a graphical layout editor that allows the creation and modification of IC mask geometry. It runs on most PC-family computers with a Graphics adapter. It supports files, cells, instances, and mask primitives. A file in L-Edit is made up of cells. An independent cell may contain any number of combinations of mask primitives and instances of other cells. An instance is a copy of a cell. If a change is made in an instanced cell, the change is reflected in all instances of that cell. There may be any number of levels in the hierarchy.

In L-Edit files are self-contained, which means that all references made in a file relate only to that file. Designs made by L-Edit are only limited by the memory of the machine used. Portability of designs is facilitated by giving a facility to convert designs to CIF (Caltech Intermediate Format) and vice versa. L-Edit itself uses a SLY (Stack Layout Format) which can be used if working within the L-Edit domain. The SLY is like the CIF with more information about the last cell edited, last view and so on. L-edit exists at two levels, as a low-level full-custom mask editor and a high-level floor planning tool.

MAGIC is an interactive VLSI layout design software developed at the University of California, Berkeley. It is now available on a number of systems, including personal computers. It is based on the Mead and Conway design style. MAGIC is a fairly advanced editor. MAGIC allows automatic routing, stretching and compacting cells, and circuit extraction to name a few. All these functions are executed, as well as concurrent design rule checking which identifies violations of design rules when any change is made to the circuit layout. This reduces design time as design rule checking is done as an event based checking rather than doing it as a lengthy post-layout operation as in other editors. This carries along with it an overhead of time to check after every operation, but this is certainly very useful when a small change is introduced in a large layout and it can be known immediately if this change introduces errors in the layout rather than performing a design rule check for the whole layout.

MAGIC is based on the corner stitched data structure proposed by Ousterhout [SO84]. This data structure greatly reduces the complexity of many editing functions, including design rule checking. Because of the ease of design using MAGIC, the resulting circuits are 5-10% denser than those using conventional layout editors. This density tradeoff is a result of the improved layout editing which results in a lesser design time. MAGIC permits only Manhattan designs and only rectilinear paths in designing circuits. It has a built-in hierarchical circuit extractor which can be used to verify the design, and has an on-line help feature.

1.9 Summary

The sheer size of the VLSI circuit, the complexity of the overall design process, the desired performance of the circuit and the cost of designing a chip dictate that CAD tools should be developed for all the phases. Also, the design process must be divided into different stages because of the complexity of entire process. Physical design is one of the steps in the VLSI design cycle. In this step, each component of a circuit is converted into a set of geometric patterns which achieves the functionality of the component. The physical design step can further be divided into several substeps. All the substeps of physical design step are interrelated. Efficient and effective algorithms are required to solve different problems in each of the substeps. Good solutions at each step are required, since a poor solution at an earlier stage prevents a good solution at a later stage. Despite significant research efforts in this field, CAD tools still lag behind the technological advances in fabrication. This calls for the development of efficient algorithms for physical design automation.

Bibliographic Notes

Physical design automation is an active area of research where over 200 papers are published each year. There are several conferences and journals which deal with all aspects physical design automation in several different technologies. Just like in other fields, the Internet is playing a key role in Physical design research and development. We will indicate the URL of all key conferences, journals and bodies in the following to faciliate the search for information.

The key conference for physical design is *International Symposium on Physical Design* (ISPD), held annually in April. ISPD covers all aspects of physical design. The most prominent conference is EDA is the *ACM/IEEE Design Automation Conference* (DAC), (www.dac.com) which has been held annually for the last thirtyfive years. In addition to a very extensive technical program, this conference features an exhibit program consisting of the latest design tools from leading companies in VLSI design automation. The *International Conference on Computer Aided Design* (ICCAD) (www.iccad.com) is held yearly in Santa Clara and is more theoretical in nature than DAC. Several other conferences, such as the *IEEE International Symposium on Circuits and Systems* (ISCAS) (www.iscas.nps.navy.mil) and the *International Conference on Computer Design* (ICCD), include significant developments in physical design automation in their technical programs. Several regional conferences have been introduced to further this field in different regions of the world. These include the *IEEE Midwest Symposium on Circuits and Systems* (MSCAS), the *IEEE Great Lakes Symposium on VLSI* (GLSVLSI) (www.eecs.umich.edu/glsvlsi/) the *European Design Automation Conference* (EDAC), and the *International Conference on VLSI Design* (vcapp.csee.usf.edu/vlsi99/) in India. There are several journals which are dedicated to the field of VLSI Design Automation which include broad coverage of all topics in physical design. The premier journal is the *IEEE Transactions on CAD of Circuits and Systems* (akebono.stanford.edu/users/nanni/tcad). Other journals such as, *Integration*, the *IEEE Transactions on Circuits and Systems*, and the *Journal of Circuits, Systems and Computers* also publish significant papers in physical design automation. Many other journals occasionally publish articles of interest to physical design. These journals include *Algorithmica*, *Networks*, the *SIAM journal of Discrete and Applied Mathematics*, and the *IEEE Transactions on Computers*.

The access to literature in Design automation has been recently enhanced by the availability of the Design Automation Library (DAL), which is developed by the ACM Special interest Group on Design Automation (SIGDA). This library is available on CDs and contains all papers published in DAC, ICCAD, ICCD, and IEEE Transactions on CAD of Circuits and Systems.

An important role of the Internet is through the forum of newsgroups. comp.lsi.cad is a newsgroup dedicated to CAD issues, while specialized groups such as comp.lsi.testing and comp.cad.synthesis discuss testing and synthesis topics. Since there are very large number of newsgroups and they keep evolving, the reader is encouraged to search the Internet for the latest topics.

Several online newslines and magazines have been started in last few years. EE Times (www.eet.com) provides news about EDA industry in general. Integrated system design (www.isdmag.com) provides articles on EDA tools in general, but covers physical design as well.

ACM SIGDA (www.acm.org/sigda/) and Design Automation Technical Committee (DATC) (www.computer.org/tab/DATC) of IEEE Computer Society are two representative societies dealing with professional development of the people involved, and technical aspects of the design automation field. These committees hold conferences, publish journals, develop standards, and support research in VLSI design automation.

Chapter 2

Design and Fabrication of VLSI Devices

VLSI chips are manufactured in a fabrication facility usually referred to as a "fab". A fab is a collection of manufacturing facilities and "clean rooms", where wafers are processed through a variety of cutting, sizing, polishing, deposition, etching and cleaning operations. Clean room is a term used to describe a closed environment where air quality must be strictly regulated. The number and size of dust particles allowed per unit volume is specified by the classification standard of the clean room. Usually space-suit like overalls and other dress gear is required for humans, so they do not contaminate the clean room. The cleanliness of air in a fab is a critical factor, since dust particles cause major damage to chips, and thereby affect the overall yield of the fabrication process. The key factor which describes the fab in terms of technology is the minimum feature size it is capable of manufacturing. For example, a fab which runs a 0.25 micron fabrication process is simply referred to as a 0.25 micron fab.

A chip consists of several layers of different materials on a silicon wafer. The shape, size and location of material in each layer must be accurately specified for proper fabrication. A *mask* is a specification of geometric shapes that need to be created on a certain layer. Several masks must be created, one for each layer. The actual fabrication process starts with the creation of a silicon wafer by crystal growth. The wafer is then processed for size and shape with proper tolerance. The wafer's size is typically large enough to fabricate several dozen identical/different chips. Masks are used to create specific patterns of each material in a sequential manner, and create a complex pattern of several layers. The order in which each layer is defined, or 'patterned' is very important. Devices are formed by overlapping a material of certain shape in one layer by another material and shape in another layer. After patterning all the layers, the wafer is cut into individual chips and packaged. Thus, the VLSI physical design is a process of creating all the necessary masks that define the sizes and location of the various devices and the interconnections between them.

The complex process of creating the masks requires a good understanding of the functionality of the devices to be formed, and the rigid rules imposed by the fabrication process. The manufacturing tolerances in the VLSI fabrication process are so tight that misalignment of a shape in a layer by a few microns can render the entire chip useless. Therefore, shapes and sizes of all the materials on all the layers of a wafer must conform to strict design rules to ensure proper fabrication. These rules play a key role in defining the physical design problems, and they depend rather heavily on the materials, equipment used and maturity of the fabrication process. The understanding of limitations imposed by the fabrication process is very important in the development of efficient algorithms for VLSI physical design.

In this chapter we will study the basic properties of the materials used in the fabrication of VLSI chips, and details of the actual fabrication process. We will also discuss the layout of several elementary VLSI devices, and how such elementary layouts can be used to construct the layout of larger circuits.

2.1 Fabrication Materials

The electrical characteristics of a material depend on the number of 'available' electrons in its atoms. Within each atom electrons are organized in concentric shells, each capable of holding a certain number of electrons. In order to balance the nuclear charge, the inner shells are first filled by electrons and these electrons may become inaccessible. However, the outermost shell may or may not be complete, depending on the number of electrons available. Atoms organize themselves into molecules, crystals, or form other solids to completely fill their outermost shells by sharing electrons. When two or more atoms having incomplete outer shells approach close enough, their accessible outermost or valence electrons can be shared to complete all shells. This process leads to the formation of covalent bonds between atoms. Full removal of electrons from an atom leaves the atom with a net positive charge, of course, while the addition of electrons leaves it with a net negative charge. Such electrically unbalanced atoms are called *ions*.

The current carrying capacity of a material depends on the distribution of electrons within the material. In order to carry electrical current, some 'free' electrons must be available. The resistance to the flow of electricity is measured in terms of the amount of resistance in ohms (Ω) per unit length or resistivity. On the basis of resistivity, there are three types of materials, as described below:

1. **Insulators:** Materials which have high electrical resistance are called insulators. The high electric resistance is due to strong covalent bonds which do not permit free movement of electrons. The electrons can be set free only by large forces and generally only from the surface of the solid. Electrons within the solids cannot move and the surface of the stripped insulator remains charged until new electrons are reintroduced. Insulators have electrical resistivity greater than millions of GΩ-cm. The principle insulator used in VLSI fabrication is silicon dioxide. It is used to

electrically isolate different devices, and different parts of a single device to satisfy design requirements.

2. **Conductors:** Materials with low electrical resistance are referred to as conductors. Low resistance in conductors is due to the existence of valence electrons. These electrons can be easily separated from their atoms. If electrons are separated from their atoms, they move freely at high speeds in all directions in the conductor, and frequently collide with each other. If some extra electrons are introduced into this conductor, they quickly disperse themselves throughout the material. If an escape path is provided by an electrical circuit, then electrons will move in the direction of the flow of electricity. The movement of electrons, in terms of the number of electrons pushed along per second, depends on how hard they are being pushed, the cross-sectional area of the conductive corridor, and finally the electron mobility factor of the conductor. Conductors can have resistivity as low as $1 \ \mu\Omega$-cm, and are used to make connections between different devices on a chip. Examples of conductors used in VLSI fabrication include aluminum and gold. A material that has almost no resistance, i.e., close to zero resistance, is called a *superconductor*. Several materials have been shown to act as superconductors and promise faster VLSI chips. Unfortunately, all existing superconductors work at very low temperatures, and therefore cannot be used for VLSI chips without specialized refrigeration equipment.

3. **Semiconductors:** Materials with electrical resistivity at room temperature ranging from 10 mΩ-cm to 1 GΩ-cm are called semiconductors. The most important property of a semiconductor is its mode of carrying electric current. Current conduction in semiconductors occurs due to two types of carriers, namely, *holes* and *free electrons*. Let us explain these concepts by using the example of semiconductor silicon, which is widely used in VLSI fabrication. A silicon atom has four valence electrons which can be readily bonded with four neighboring atoms. At room temperatures the bonds in silicon atoms break randomly and release electrons, which are called free electrons. These electrons make bonds with bond deficient ionized sites. These bond deficiencies are known as holes. Since the breaking of any bond releases exactly one hole and one free electron, while the opposite process involves the capture of one free electron by one hole, the number of holes is always equal to number of free electrons in pure silicon crystals (see Figure 2.1). Holes move about and repel one another, just as electrons do, and each moving hole momentarily defines a positive ion which inhibits the intrusion of other holes into its vicinity. In silicon crystals, the mobility of such holes is about one third that of free electrons, but charge can be 'carried' by either or both. Since these charge carriers are very few in number, 'pure' silicon crystal will conduct weakly. Although there is no such thing as completely pure crystalline silicon, it appears that, as pure crystals, semiconductors seem to have no electrical properties of great utility.

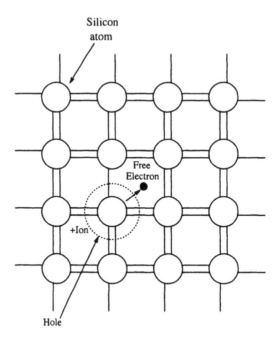

Figure 2.1: Electrons and Holes.

Semiconductor crystals can be enriched either in holes or electrons by embedding some atoms of certain other elements. This fact makes it possible to build useful devices. An atom of phosphorus has five valence electrons, whereas an atom of boron has three valence electrons. Atoms of either kind can be locked into the silicon lattice, and even a few atoms can make a dominant contribution. Once placed into a silicon lattice, the fifth valence electron of each phosphorus atom is promptly freed. On the other hand, if a boron atom is placed into a silicon lattice, the covalence deficit of a boron atom is no less promptly covered by a neighborly silicon atom, which takes a hole in exchange and passes it on. With enough phosphorus (or boron) atoms per crystal, the number of free electrons or holes in general circulation can be increased a million fold.

This process of substituting other atoms for some of the semiconductor atoms is called *doping*. Semiconductors doped with electron donors such as phosphorus are said to be of the *n-type*, while boron doping, which results in extra holes, produces *p-type* semiconductors. Though the doping elements give semiconductors desirable characteristics, they are referred to as impurities. Doping of silicon is easily accomplished by adding just the right amount of the doping element to molten silicon and allowing the result to cool and crystallize. Silicon is also doped by diffusing the dopant as a vapor through the surface of the crystalline solid at high temperature. At such temperatures all atoms are vibrating significantly in all directions. As a result, the dopant atoms can find accommodations in minor lattice defects without greatly upsetting the overall structure. The conductivity is directly related to the level of doping. The heavily doped material is referred to as n^+ or p^+. Heavier doping leads to higher conductivity of the semiconductor.

In VLSI fabrication, both silicon and germanium are used as semiconductors. However, silicon is the dominant semiconductor due to its ease of handling and large availability. A significant processing advantage of silicon lies in its capability of forming a thermally grown silicon dioxide layer which is used to isolate devices from metal layers.

2.2 Transistor Fundamentals

In digital circuits, a 'transistor' primarily means a 'switch'- a device that either conducts charge to the best of its ability or does not conduct any charge at all, depending on whether it is 'on' or 'off'. Transistors can be built in a variety of ways, exploiting different phenomenon. Each transistor type gives rise to a circuit family. There are many different circuit families. A partial list would include TTL (Transistor-Transistor Logic), MOS (Metal-Oxide-Semiconductor), and CMOS (Complimentary MOS) families, as well as the CCD (Charge-Coupled Device), ECL (Emitter-Coupled Logic), and I²L (Integrated Injection Logic) families. Some of these families come in either p or n flavor (in CMOS both at once), and some in both high-power and low-power versions. In addition, some families are also available in both high and low speed versions. We restrict our discussion to TTL and MOS (and CMOS), and start with basic device structures for these types of transistors.

2.2.1 Basic Semiconductor Junction

If two blocks, one of n-type and another of p-type semiconductor are joined together to form a semiconductor junction, electrons and holes immediately start moving across the interface. Electrons from the n-region leave behind a region which is rich in positively charged immobile phosphorus ions. On the other hand, holes entering the interface from the p-region leave behind a region with a high concentration of uncompensated negative boron ions. Thus we have three different regions as shown in Figure 2.2. These regions establish a device with a remarkable one-way-flow property. Electrons cannot be introduced in the p-region, due to its strong repulsion by the negatively charged ions. Similarly, holes cannot be introduced in the n-region. Thus, no flow of electrons is possible in the p-to-n direction. On the other hand, if electrons are introduced in the n-region, they are passed along towards the middle region. Similarly, holes introduced from the other side flow towards the middle, thus establishing a p-to-p flow of holes.

The one-way-flow property of a semiconductor junction is the principle of

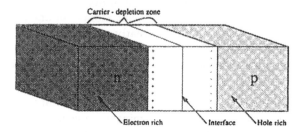

Figure 2.2: The three regions in a n-p junction.

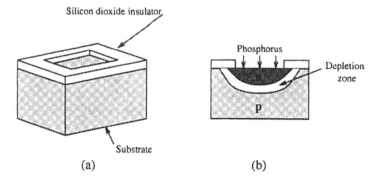

Figure 2.3: Formation of a diffused junction.

the diode, and it can be used to develop two types of devices: unipolar and bipolar. Unipolar devices are created by using the semiconductor junction, under suitable external conditions, to modulate the flow of charge between two regions of opposite polarity. On the other hand, bipolar devices are created, under suitable external conditions, by isolating one semiconductor region from another of opposite majority-carrier polarity, thus permitting a charge to flow within the one without escaping into the other.

Great numbers of both types of devices can be made rather easily by doping a silicon wafer with diffusion of either phosphorus or boron. The silicon wafer is pre-doped with boron and covered with silicon dioxide. Diffused regions are created near the surface by cutting windows into a covering layer of silicon dioxide to permit entry of the vapor, as shown in Figure 2.3. Phosphorus vapor, for example, will then form a bounded n-region in a boron doped substrate wafer if introduced in sufficient quantity to overwhelm the contribution of the boron ions in silicon. All types of regions, with differing polarities, can be formed by changing the diffusion times and diffusing vapor compositions, therefore creating more complex layered structures. The exact location of regions is determined by the mask that is used to cut windows in the oxide layer.

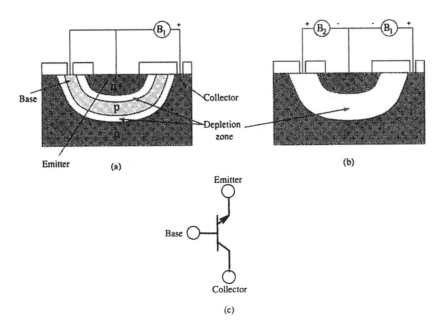

Figure 2.4: TTL transistor.

In the following, we will discuss how the semi-conductor junction is used in the formation of both TTL and MOS transistors. TTL is discussed rather briefly to enable more detailed discussion of the simpler, unipolar MOS technology.

2.2.2 TTL Transistors

A TTL transistor is an n-p-n device embedded in the surface of p-type semiconductor substrate (see Figure 2.4(a), the p-substrate is not shown). There are three regions in a TTL transistor, namely the emitter, the base, and the collector. The main idea is to control the flow of current between collector (n-region) and emitter (n^+-region) by using the base (p-region). The basic construction of a TTL transistor, both in its 'on' state and 'off' state, along with its symbol is shown in Figure 2.4. In order to understand the operation of a TTL transistor, consider what happens if the regions of the transistor are connected to a battery B_1 as shown in Figure 2.4(a). A few charge carriers are removed from both base and collector; however, the depletion zones at the emitter-base and base-collector interfaces prevent the flow of currents of significant size along any pathway. Now if another battery with a small voltage B_2 is connected as shown in Figure 2.4(b), then two different currents begin to flow. Holes are introduced into the base by B_2, while electrons are sent into the emitter by both B_1 and B_2. The electrons in the emitter cross over into the base region. Some of these electrons are neutralized by some holes, and since the base region is rather thin, most of the electrons pass through the base and move into the collector. Thus a flow of current is established from emitter to collector. If B_2 is disconnected from the circuit, holes in the base cause the flow to stop. Thus the flow of a very small current in the B_2-loop modulates the flow of a current many times its size in the B_1-loop.

2.2.3   MOS Transistors

MOS transistors were invented before bipolar transistors. The basic principle of a MOS transistor was discovered by J. Lilienfeld in 1925, and O. Heil proposed a structure closely resembling the modern MOS transistor. However, material problems failed these early attempts. These attempts actually led to the development of the bipolar transistor. Since the bipolar transistor was quite successful, interest in MOS transistors declined. It was not until 1967 that the fabrication and material problems were solved, and MOS gained some commercial success. In 1971, nMOS technology was developed and MOS started getting wider attention.

MOS transistors are unipolar and simple. The field-induced junction provides the basic unipolar control mechanism of all MOS integrated circuits. Let us consider the n-channel MOS transistor shown in Figure 2.5(a). A p-type semiconductor substrate is covered with an insulating layer of silicon dioxide or simply oxide. Windows are cut into oxide to allow diffusion. Two separate n-regions, the source and the drain, are diffused into the surface of a p-substrate through windows in the oxide. Notice that source and drain are insulated from each other by a p-type region of the substrate. A conductive material (polysilicon or simply poly) is laid on top of the gate.

If a battery is connected to this transistor as shown in Figure 2.5(b), the poly acquires a net positive charge, as some of its free electrons are conducted away to the battery. Due to this positive charge, the holes in the substrate beneath the oxide are forced to move away from the oxide. As a result, electrons begin to accumulate beneath the oxide and form an n-type channel if the battery pressure, or more precisely the gate voltage V_g, is increased beyond a threshold value V_t. As shown in Figure 2.5(b), this channel provides a pathway for the flow of electrons from source to drain. The actual direction of flow depends on the source voltage (V_s) and the drain voltage (V_d). If the battery is now disconnected, the charge on the poly disappears. As a result, the channel disappears and the flow stops. Thus a small voltage on the gate can be used to control the flow of current from source to drain. The symbols of an n-channel MOS gate are shown in Figure 2.5(c). A p-channel MOS transistor is a device complementary to the n-channel transistor, and can be formed by using an n-type substrate and forming two p-type regions.

Integrated systems in metal-oxide semiconductor (MOS) actually contain three or more layers of conducting materials, separated by intervening layers of insulating material. As many as four (or more) additional layers of metal are used for interconnection and are called *metal1*, *metal2*, *metal3* and so on.

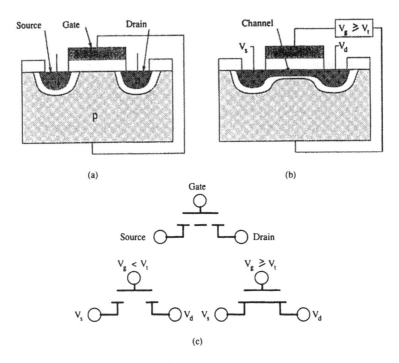

Figure 2.5: A nMOS transistor.

Different patterns for paths on different levels, and the locations for contact cuts through the insulating material to connect certain points between levels, are transferred into the levels during the fabrication process from masks. Paths on the metal level can cross poly or diffusion levels in the absence of contact cuts with no functional effects other than a parasitic capacitance. However, when a path on the poly level crosses a path on the diffusion level, a transistor is formed.

The nMOS transistor is currently the preferred form of unipolar integration technology. The name MOS survives the earlier period in which gates were made of metal (instead of poly). Aluminum is the metal of choice for all conductivity pathways, although unlike the *aluminum/oxide/semiconductor* sandwich that provides only *two* topological levels on which to make interconnections, the basic *silicon-gate* structures provide three levels, and are therefore more compact and correspondingly faster. Recent advances in fabrication have allowed the use of up to four (or more) layers of metal. However, that process is expensive and is only used for special chips, such as microprocessors. Two or three metal technology is more commonly used for general purpose chips.

The transistors that are non-conducting with zero gate bias (gate to source voltage) are called *enhancement mode transistors*. Most MOS integrated circuits use transistors of the enhancement type. The transistors that conduct with zero gate bias are called *depletion mode transistors*. For a depletion mode transistor to turn off, its gate voltage V_g must be more negative than its threshold voltage (see Figure 2.6). The channel is enriched in electrons by an implant step; and thus an n-channel is created between the source and the drain. This channel allows the flow of electrons, hence the transistor is normally in its 'on' state. This type of transistor is used in nMOS as a resistor due to poor conductivity of the channel as shown in Figure 2.6(d).

The MOS circuits dissipate DC power i.e., they dissipate power even when the output is low. The heat generated is hard to remove and impedes the performance of these circuits. For nMOS transistors, as the voltage at the gate increases, the conductivity of the transistor increases. For pMOS transistors, the p-channel works in the reverse, i.e., as the voltage on the gate increases, the conductivity of the transistor decreases. The combination of pMOS and nMOS transistors can be used in building structures which dissipate power only while switching. This type of structure is called CMOS (Complementary Metal-Oxide Semiconductor). The actual design of CMOS devices is discussed in Section 2.5.

CMOS technology was invented in the mid 1960's. In 1962, P. K. Weimer discovered the basic elements of CMOS flip-flops and independently in 1963, F. Wanlass discovered the CMOS concept and presented three basic gate structures. CMOS technology is widely used in current VLSI systems. CMOS is an inherently low power circuit technology, with the capability of providing a lower power-delay product comparable in design rules to nMOS and pMOS technologies. For all inputs, there is always a path from '1' or '0' to the output and the full supply voltage appears at the output. This 'fully restored' condition simplifies circuit design considerably. Hence the transistors in the CMOS gate do not have to be 'ratioed', unlike the MOS gate where the lengths of load and driver transistors have to be adjusted to provide proper voltage at the output. Another advantage of CMOS is that there is no direct path between VDD and GND for any combination of inputs. This is the basis for the low static power dissipation in CMOS. Table 2.1 illustrates the main differences between nMOS and CMOS technology. As shown in the table, the major drawback of CMOS circuits is that they require more transistors than nMOS circuits. In addition, the CMOS process is more complicated and expensive. On the other hand, power consumption is critical in nMOS and bipolar circuits, while it is less of a concern in CMOS circuits. Driver sizes can be increased in order to reduce net delay in CMOS circuits without any major concern of power. This difference in power consumption makes CMOS technology superior to nMOS and bipolar technologies in VLSI design.

2.3   Fabrication of VLSI Circuits

Design and Layout of VLSI circuits is greatly influenced by the fabrication process; hence a good understanding of the fabrication cycle helps in designing efficient layouts. In this section, we review the details of fabrication.

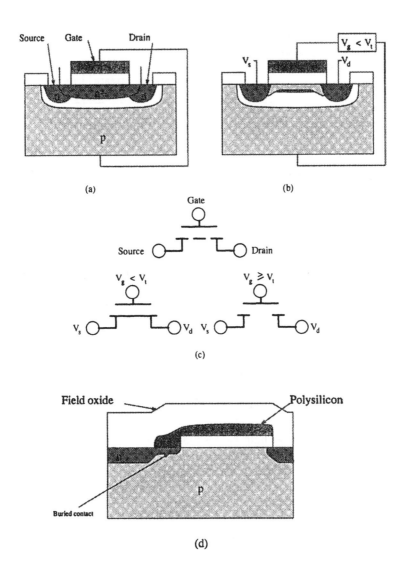

Figure 2.6: A depletion mode transistor.

CMOS	MOS
Zero static power dissipation	Power is dissipated in the circuit with output of gate at '0'
Power dissipated during logic transition	Power dissipated during logic transition
Requires 2N devices for N inputs for complementary static gates	Requires (N+1) devices for N inputs
CMOS encourages regular layout styles	Depletion, load and different driver transistors create irregularity in layout

Table 2.1: Comparison of CMOS and MOS characteristics.

Fabrication of a VLSI chip starts by growing a large silicon crystal ingot about 20 centimeters in diameter. The ingot is sliced into several wafers, each about a third of a millimeter thick. Under various atmospheric conditions, phosphorus is diffused, oxide is grown, and polysilicon and aluminum are each deposited in different steps of the process. A complex VLSI circuit is defined by 6 to 12 separate layer patterns. Each layer pattern is defined by a mask. The complete fabrication process, which is a repetition of the basic three-step process(shown in Figure 2.7), may involve up to 200 steps.

1. **Create:** This step creates material on or in the surface of the silicon wafer using a variety of methods. Deposition and thermal growth are used to create materials on the wafer, while ion implantation and diffusion are used to create material (actually they alter the characteristics of existing material) in the wafer.

2. **Define:** In this step, the entire surface is coated with a thin layer of light sensitive material called *photoresist.* Photoresist has a very useful property. The ultraviolet light causes molecular breakdown of the photoresist in the area where the photoresist is exposed. A chemical agent is used to remove the dis-integrated photoresist. This process leaves some regions of the wafer covered with photoresist. Since exposure of the photoresist occurred while using the mask, the pattern of exposed parts on the wafer is exactly the same as in the mask. This process of transferring a pattern from a mask onto a wafer is called *photolithography* and it is illustrated in Figure 2.8.

3. **Etch:** Wafers are immersed in acid or some other strong chemical agent to etch away either the exposed or the unexposed part of the pattern, depending on whether positive or negative photoresist has been used. The photoresist is then removed to complete the pattern transfer process.

This three step process is repeated for all the masks. The number of masks and actual details of each step depend on the manufacturer as well as the

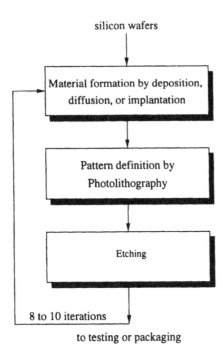

Figure 2.7: Basic steps in MOS fabrication process.

technology. In the following analysis, we will briefly review the basic steps in nMOS and CMOS fabrication processes.

2.3.1 nMOS Fabrication Process

The first step in the n-channel process is to grow an oxide layer on lightly doped p-type substrate (wafer). The oxide is etched away (using the diffusion mask) to expose the active regions, such as the sources and drains of all transistors. The entire surface is covered with poly. The etching process using the poly mask removes all the poly except where it is necessary to define gates. Phosphorus is then diffused into all uncovered silicon, which leads to the formation of source and drain regions. Poly (and the oxide underneath it) stops diffusion into any other part of the substrate except in source and drain areas. The wafer is then heated, to cover the entire surface with a thin layer of oxide. This layer insulates the bare semiconductor areas from the pathways to be formed on top. Oxide is patterned to provide access to the gate, source, and drain regions as required. It should be noted that the task of aligning the poly and diffusion masks is rather easy, because it is only their intersections that define transistor boundaries. This self-alignment feature is largely responsible for the success of silicon-gate technology. The formation of a depletion mode

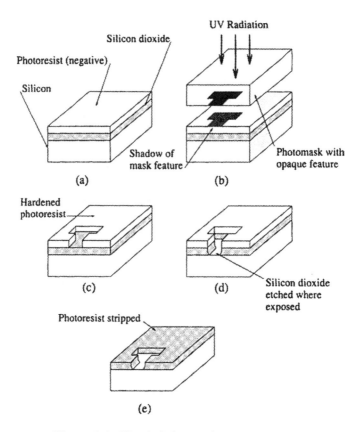

Figure 2.8: Photolithographic process.

transistor requires an additional step of ion implantation.

A thin covering of aluminum is deposited over the surface of the oxide, which now has 'hills' and 'valleys' in its structure. Etching then removes all but the requisite wires and contacts. Additional metal layers may be laid on top if necessary. It is quite common to use two layers of metal. At places where connections are to be made, areas are enlarged somewhat to assure good interlevel contact even when masks are not in perfect alignment. All pathways are otherwise made as small as possible, in order to conserve area. In addition to normal contacts, an additional contact is needed in nMOS devices. The gate of a depletion mode transistor needs to be connected to its source. This is accomplished by using a *buried contact*, which is a contact between diffusion and poly.

The final steps involve covering the surface with oxide to provide mechanical and chemical protection for the circuit. This oxide is patterned to form windows which allow access to the aluminum bonding pads, to which gold wires will be attached for connection to the chip carrier. The windows and pads are very

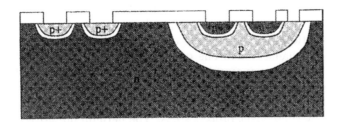

Figure 2.9: A p-well CMOS transistor.

large as compared to the devices.

2.3.2 CMOS Fabrication Process

CMOS transistors require both a p-channel and an n-channel device. However, these two types of devices require two different substrates. nMOS transistors require a p-type substrate, while pMOS transistors require a n-type substrate. CMOS transistors are created by diffusing or implanting an *n-type well* in the original p-substrate. The well is also called a *tub* or an *island*. The p-channel devices are placed in the n-well. This is called the *n-well CMOS process*. A complementary process of *p-well CMOS* starts with an n-type substrate for pMOS devices, and creates a p-well for nMOS devices. The structure of a CMOS transistor is shown in Figure 2.9 (p-substrate is not shown). A *twin-tub CMOS* process starts with a lightly doped substrate and creates both a n-well and a p-well.

As compared to the nMOS process, the CMOS process requires a additional mask for ion implanting to create the deep p-wells, n-wells, or both.

2.3.3 Details of Fabrication Processes

The complete fabrication cycle consists of the following steps: crystal growth and wafer preparation, epitaxy, dielectric and polysilicon film deposition, oxidation, diffusion, ion implantation, lithography and dry etching. These steps are used in a generic silicon process, however they are similar to those of other technologies. Below, we discuss details of specific fabrication processes.

1. **Crystal Growth and Wafer Preparation:**   Growing crystals essentially involves a phase change from solid, liquid, or gas phases to a crystalline solid phase. The predominant method of crystal growth is Czochralski (CZ) growth, which consists of crystalline solidification of atoms from the liquid phase. In this method, single-crystal ingots are pulled from molten silicon contained in a fused silica crucible.

   Electronic-grade silicon, which is a polycrystalline material of high purity, is used as the raw material for the preparation of a single crystal. The

conditions and the parameters during the crystal-pulling operation define many properties of the wafer, such as dopant uniformity and oxygen
concentration. The ingots are ground to a cylindrical shape of precisely
controlled diameter and one or more flats are ground along its length.
The silicon slices are sawed from the ingot with an exact crystallographic
orientation. Crystal defects adversely affect the performance of the device. These defects may be present originally in the substrate or may be
caused by subsequent process steps. The harmful impurities and defects
are removed by careful management of the thermal processes.

2. **Epitaxy:**   This is the process of depositing a thin single-crystal layer
on the surface of a single-crystal substrate. The word epitaxy is derived
from two Greek words: *epi*, meaning 'upon', and *taxis*, meaning 'ordered'.
Epitaxy is a Chemical Vapor Deposition (CVD) process in which a batch
of wafers is placed in a heated chamber. At high temperatures (900^0 to
1250^0C), deposition takes place when process gases react at the wafer
surface. A typical film growth rate is about 1 μm/min. The thickness
and doping concentration of the epitaxial layer is accurately controlled
and, unlike the underlying substrate, the layer can be made oxygen- and
carbon-free. A limitation of epitaxy is that the degree of crystal perfection
of the deposited layer cannot be any better than that of the substrate.
Other process-related defects, such as slip or impurity precipitates from
contamination can be minimized.

In bipolar device technology, an *epi-layer* is commonly used to provide a
high-resistivity region above a low-resistivity buried layer, which has been
formed by a previous diffusion or 'implant and drive-in' process. The
heavily doped buried layer serves as a low-resistance collector contact,
but an additional complication arises when epitaxial layers are grown
over patterned buried layer regions. To align the subsequent layers in
relation to the pattern of the buried layer, a step is produced in the
pre-epitaxial processing.

3. **Dielectric and Polysilicon film deposition:**   The choice of a particular reaction is often determined by the deposition temperature(which
must be compatible with the device materials), the properties, and certain
engineering aspects of deposition (wafer throughput, safety, and reactor
maintenance).

The most common reactions for depositing silicon dioxide for VLSI circuits are:

- Oxidizing silane (silicon hydrate) with oxygen at 400^0-450^0C.
- Decomposing tetra-ethoxysilane at 650^0 to 750^0C, and reacting
  dichlorosilane with nitrous oxide at 850^0 to 900^0C.

Doped oxides are prepared by adding a dopant to the deposition reaction.
The hydrides arsine, phosphine, or diborane are often used because they

are readily available gases. However, halides and organic compounds can also be used. Polysilicon is prepared by pyrolyzing silane at 600^0 to 650^0C.

4. **Oxidation:** The process of oxidizing silicon is carried out during the entire process of fabricating integrated circuits. The production of high-quality IC's requires not only an understanding of the basic oxidization mechanism, but also the electrical properties of the oxide. Silicon dioxide has several properties:

   - Serves as a mask against implant or diffusion of dopant into silicon.

   - Provides surface passivation.

   - Isolates one device from another.

   - Acts as a component in MOS structures.

   - Provides electrical isolation of multilevel metalization systems.

   Several techniques such as thermal oxidation, wet anodization, CVD etc. are used for forming the oxide layers.

   When a low charge density level is required between the oxide and the silicon, *Thermal oxidation* is preferred over other techniques. In the thermal oxidation process, the surface of the wafer is exposed to an oxidizing ambient of O_2 or H_2O at elevated temperatures, usually at an ambient pressure of one atmosphere.

5. **Diffusion:** The process in which impurity atoms move into the crystal lattice in the presence of a chemical gradient is called diffusion. Various techniques to introduce dopants into silicon by diffusion have been studied with the goals of controlling the dopant concentration, uniformity, and reproducibility, and of processing a large number of device wafers in a batch to reduce the manufacturing costs. Diffusion is used to form bases, emitters, and resistors in bipolar device technology, source and drain regions, and to dope polysilicon in MOS device technology. Dopant atoms which span a wide range of concentrations can be introduced into silicon wafers in the following ways:

   - Diffusion from a chemical source in vapor form at high temperatures.

   - Diffusion from doped oxide source.

   - Diffusion and annealing from an ion implanted layer.

6. **Ion Implantation:** Ion implantation is the introduction of ionized projectile atoms into targets with enough energy to penetrate beyond surface regions. The most common application is the doping of silicon during device fabrication. The use of 3-keV to 500-keV energy for doping of boron, phosphorus, or arsenic dopant ions is sufficient to implant the ions from about 100 to $10,000A^o$ below the silicon surface. These depths

place the atoms beyond any surface layers of $30A^o$ native SiO_2, and therefore any barrier effect of the surface oxides during impurity introduction is avoided. The depth of implantation, which is nearly proportional to the ion energy, can be selected to meet a particular application.

With ion implantation technology it is possible to precisely control the number of implanted dopants. This method is a low-temperature process and is compatible with other processes, such as photoresist masking.

7. **Lithography:**   As explained earlier, lithography is the process delineating the patterns on the wafers to fabricate the circuit elements and provide for component interconnections. Because the polymeric materials resist the etching process they are called *resists* and, since light is used to expose the IC pattern, they are called *photoresists*.

   The wafer is first spin-coated with a photoresist. The material properties of the resist include (1) mechanical and chemical properties such as flow characteristics, thickness, adhesion and thermal stability, (2) optical characteristics such as photosensitivity, contrast and resolution and (3) processing properties such as metal content and safety considerations. Different applications require more emphasis on some properties than on others. The mask is then placed very close to the wafer surface so that it faces the wafer. With the proper geometrical patterns, the silicon wafer is then exposed to ultraviolet (UV) light or radiation, through a photomask. The radiation breaks down the molecular structure of areas of exposed photoresist into smaller molecules. The photoresist from these areas is then removed using a solvent in which the molecules of the photoresist dissolve so that the pattern on the mask now exists on the wafer in the form of the photoresist. After exposure, the wafer is soaked in a solution that develops the images in the photosensitive material. Depending on the type of polymer used, either exposed or nonexposed areas of film are removed in the developing process. The wafer is then placed in an ambient that etches surface areas not protected by polymer patterns. Resists are made of materials that are sensitive to UV light, electron beams, X-rays, or ion beams. The type of resist used in VLSI lithography depends on the type of exposure tool used to expose the silicon wafer.

8. **Metallization:**   Metal is deposited on the wafer with a mechanism similar to spray painting. Motel metal is sprayed via a nozzle. Like spray painting, the process aims for an even application of metal. Unlike spray painting, process aims to control the thickness within few nanometers. An uneven metal application may require more CMP. Higher metal layers which are thick may require several application of the process get the desired height. Copper, which has better interconnect properties is increasing becoming popular as the choice material for interconnect. Copper does require special handling since a liner material must be provide between copper and other layers, since copper atoms may migrate into other layers due to electr-migration and cause faults.

9. **Etching:** Etching is the process of transferring patterns by selectively removing unmasked portions of a layer. Dry etching techniques have become the method of choice because of their superior ability to control critical dimensions reproducibly. Wet etching techniques are generally not suitable since the etching is isotropic, i.e., the etching proceeds at the same rate in all directions, and the pattern in the photoresist is undercut. Dry etching is synonymous with plasma-assisted etching, which denotes several techniques that use plasmas in the form of low pressure gaseous discharges. The dominant systems used for plasma-assisted etching are constructed in either of two configurations: parallel electrode (planar) reactors or cylindrical batch (hexode) reactors. Components common to both of these include electrodes arranged inside a chamber maintained at low pressures, pumping systems for maintaining proper vacuum levels, power supplies to maintain a plasma, and systems for controlling and monitoring process parameters, such as operating pressure and gas flow.

10. **Planarization:** The Chemical Mechanical Planarization (CMP) of silicon wafers is an important development in IC manufacturing. Before the advent of CMP, each layer on the wafer was more un-even then the lower layer, as a result, it was not possible to icrease the number of metal layers. CMP provides a smooth surface after each metalization step. CMP has allowed essentially unlimited number of layers of interconnect. The CMP process is like "Wet Sanding" down the surface until it is even. Contact and via layers are filled with tungsten plugs and planarized by CMP. ILD layers are also planarized by CMP.

11. **Packaging:** VLSI fabrication is a very complicated and error prone process. As a result, finished wafers are never perfect and contain many 'bad' chips. Flawed chips are visually identified and marked on the wafers. Wafers are then diced or cut into chips and the marked chips are discarded. 'Good' chips are packaged by mounting each chip in a small plastic or ceramic case. Pads on the chip are connected to the legs on the case by tiny gold wires with the help of a microscope, and the case is sealed to protect it from the environment. The finished package is tested and the error prone packages are discarded. Chips which are to be used in an MCM are not packaged, since MCM uses unpackaged chips.

The VLSI fabrication process is an enormous scientific and engineering achievement. The manufacturing tolerances maintained throughout the process are phenomenal. Mask alignment is routinely held to 1 micron in 10 centimeters, an accuracy of one part in 10^5, which is without precedent in industrial practice. For comparison, note that a human hair is 75 microns in diameter.

2.4   Design Rules

The constraints imposed on the geometry of an integrated circuit layout, in order to guarantee that the circuit can be fabricated with an acceptable yield, are called *design rules*. The purpose of design rules is to prevent unreliable, or hard-to-fabricate (or unworkable) layouts. More specifically, layout rules are introduced to preserve the integrity of topological features on the chip and to prevent separate, isolated features from accidentally short circuiting with each other. Design rules must also ensure thin features from breaking, and contact cuts from slipping outside the area to be contacted. Usually, design rules need to be re-established when a new process is being created, or when a process is upgraded from one generation to the next. The establishment of new design rules is normally a compromise between circuit design engineers and process engineers. Circuit designers want smaller and tighter design rules to improve performance and decrease chip area, while process engineers want design rules that lead to controllable and reproducible fabrication. The result is a set of design rules that yields a competitive circuit designed and fabricated in a cost effective manner.

Design rules must be simple, constant in time, applicable in many processes and standardized among many fabrication facilities. Design rules are formulated by observing the interactions between features in different layers and limitations in the design process. For example, consider a contact window between a metal wire and a polysilicon wire. If the window misses the polysilicon wire, it might etch some lower level or the circuit substrate, creating a fatal fabrication defect. One should, undoubtedly, take care of basic width, spacing, enclosure, and extension rules. These basic rules are necessary parts of every set of design rules. Some conditional rules depend on electrical connectivity information. If, for instance, two metal wires are part of the same electrical node, then a short between them would not affect the operation of circuit. Therefore, the spacing requirement between electrically connected wires can be smaller than that between disconnected wires.

The design rules are specified in terms of microns. However, there is a key disadvantage of expressing design rules in microns. A chip is likely to remain in production for several years; however newer processes may be developed. It is important to produce the chip on the newer processes to improve yield and profits. This requires modifying or shrinking the layout to obey the newer design rules. This leads to smaller die sizes and the operation is called *process shifting*. If the layout is specified in microns, it may be necessary to rework the entire layout for process shifting. To overcome this scaling problem, Mead and Conway [MC79] suggested the use of a single parameter to design the entire layout. The basic idea is to characterize the process with a single scalable parameter called *lambda* (λ), defined as the maximum distance by which a geometrical feature on any one layer can stray from another feature, due to over-etching, misalignment, distortion, over or underexposure, etc, with a suitable safety factor included. λ is thus equal to the maximum misalignment of a feature from its intended position in the wafer. One can think of λ as either

Diffusion Region Width	2λ
Polysilicon Region Width	2λ
Diffusion-Diffusion Spacing	3λ
Poly-Poly Spacing	2λ
Polysilicon Gate Extension	2λ
Contact Extension	λ
Metal Width	3λ

Table 2.2: Basic nMOS design rules.

some multiple of the standard deviation of the process or as the resolution of the process. Currently, λ is approximately 0.25×10^{-6} m (0.25 μm). In order to simplify our presentation, we will use lambda.

Design rules used by two different fabrication facilities may be different due to the availability of different equipment. Some facilities may not allow use of a fourth or a fifth metal layer, or they may not allow a 'stacked via'. Usually, design rules are very conservative (devices take larger areas) when a fabrication process is new and tend to become tighter when the process matures. Design rules are also conservative for topmost layers (metal4 and metal5 layers) since they run over the roughest terrain.

The actual list of design rules for any particular process may be very long. Our purpose is to present basic ideas behind design rules, therefore, we will analyze simplified nMOS design rules. Table 2.2 lists basic nMOS design rules. We have omitted several design rules dealing with buried contact, implant, and others to simply our discussion.

As stated earlier, design rules are specified in fractions of microns. For example, separation for five metal layers may be 0.35 μm, 0.65 μm, 0.65 μm, 1.25 μm, and 1.85 μm respectively. Similar numbers are specified for each rule. Such rules do make presentation rather difficult, explaining our motivation to use the simpler lambda system. Although the lambda system is simple, sometimes it can become over-simplifying or even misleading. At such places we will indicate the problems caused by our simplified design rules.

In order to analyze design rules it is helpful to envision the design rules as a set of constraints imposed on the geometry of the circuit layout. We classify the rules in three types.

1. **Size Rules:** The minimum feature size of a device or an interconnect line is determined by the line patterning capability of lithographic equipment used in the IC fabrication. In 1998, the minimum feature size is 0.25 μm. Interconnect lines usually run over a rough surface, unlike the smooth surface over which active devices are patterned. Consequently, the minimum feature size used for interconnects is somewhat larger than the one used for active devices, based on patternability considerations. However, due to advances in planarization techniques, roughness problem

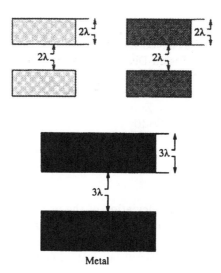

Figure 2.10: Size and separation rules.

of higher layers is essentially a solved problem.

The design rule must specify the minimum feature sizes on different layers to ensure a valid design of a circuit. Figure 2.10 shows different size rules for feature sizes in different layers.

2. **Separation Rules:** Different features on the same layer or in different layers must have some separation from each other. In ICs, the interconnect line separation is similar to the size rule. The primary motivation is to maintain good interconnect density. Most IC processes have a spacing rule for each layer and overlap rules for vias and contacts. Figure 2.10 also shows the different separation rules in terms of λ.

3. **Overlap Rules:** Design rules must protect against fatal errors such as a short-circuited channel caused by the mismigration of poly and diffusion, or the formation of an enhancement-mode FET in parallel with a depletion-mode device, due to the misregistration of the ion-implant area and the source/drain diffusion as shown in Figure 2.11. The overlap rules are very important for the formation of transistors and contact cuts or vias.

Figure 2.12 shows the overlap design rules involved in the formation of a contact cut.

In addition to the rules discussed above, there are other rules which do not scale. Therefore they cannot be reported in terms of lambda and are reported in terms of microns. Such rules include:

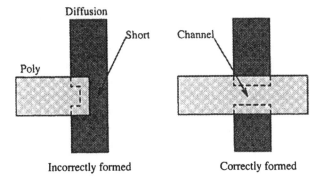

Figure 2.11: (a) Incorrectly formed channel; (b) Correctly formed channel.

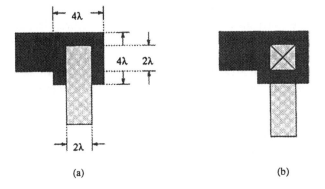

Figure 2.12: overlap rules for contact cuts.

1. The size of bonding pads, determined by the diameter of bonding wire and accuracy of the bonding machine.

2. The size of cut in overglass (final oxide covering) for contacts with pads.

3. The scribe line width (The line between two chips, which is cut by a diamond knife).

4. The feature distance from the scribe line to avoid damage during scribing.

5. The feature distance from the bonding pad, to avoid damage to the devices during bonding.

6. The bonding pitch, determined by the accuracy of bonding machine.

We have presented a simple overview of design rules. One must study actual design rules provided by the fabrication facility rather carefully before one starts the layout. CMOS designs rules are more complicated than nMOS design rules, since additional rules are needed for tubs and pMOS devices.

The entire layout of a chip is rarely created by minimum design rules as discussed above. For performance and/or reliability reasons devices are designed with wider poly, diffusion or metal lines. For example, long metal lines are sometimes drawn using using two or even three times the minimum design rule widths. Some metal lines are even tapered for performance reasons. The purpose of these examples is to illustrate the fact that layout in reality is much more complex. Although we will maintain the simple rules for clarity of presentation, we will indicate the implications of complexity of layout as and when appropriate.

2.5   Layout of Basic Devices

Layout is a process of translating schematic symbols into their physical representations. The first step is to create a plan for the chip by understanding the relationships between the large blocks of the architecture. The layout designer partitions the chip into relatively smaller subcircuits (blocks) based on some criteria. The relative sizes of blocks and wiring between the blocks are both estimated and blocks are arranged to minimize area and maximize performance. The layout designer estimates the size of the blocks by computing the number of transistors times the area per transistor. After the top level 'floorplan' has been decided, each block is individually designed. In very simple terms, layout of a circuit is a matter of picking the layout of each subcircuit and arranging it on a plane. In order to design a large circuit, it is necessary to understand the layout of simple gates, which are the basic building blocks of any circuit. In this section, we will discuss the structure of various VLSI devices such as the Inverter, NAND and NOR gates in both MOS and CMOS technologies.

2.5.1   Inverters

The basic function of an *inverter* is to produce an output that is complement of its input. The logic table and logic symbol of a basic inverter are shown in Figure 2.13(a) and (d) respectively. If the inverter input voltage A is less than the transistor threshold voltage V_t then the transistor is switched off and the output is pulled up to the positive supply voltage VDD. In this case the output is the complement of the input. If A is greater than V_t, the transistor is switched on and current flows from the supply voltage through the resistor R to GND. If R is large, V_{out} could be pulled down well below V_t, thus again complementing the input.

The main problem in the design of an inverter layout is the creation of the resistor. Using a sufficiently large resistor R would require a very large area compared to the area occupied by the transistor. This problem of large resistor can be solved by using a *depletion mode* transistor. The depletion mode transistor has a threshold voltage which is less than zero. Negative voltage is required to turn off a depletion mode transistor. Otherwise the gate is always

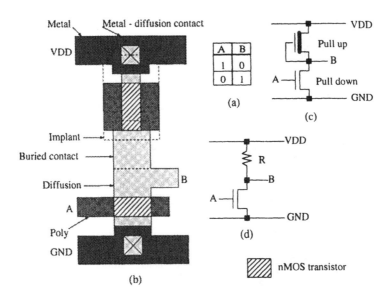

Figure 2.13: An nMOS inverter.

turned on. The circuit diagram of an inverter is shown in Figure 2.13(c). The basic inverter layout on the silicon surface in MOS is given in Figure 2.13(b). It consists of two polysilicon (*poly*) regions overhanging a path in the diffusion level that runs between VDD and GND. This forms the two MOS transistors of the inverter. The transistors are shown by hatched regions in Figure 2.13(b). The upper transistor is called pull-up transistor, as it pulls up the output to 1. Similarly, the lower transistor is called the pull-down transistor as it is used to pull-down the output to zero. The inverter input A is connected to the poly that forms the gate of the lower of the two transistors. The pull-up is formed by connecting the gate of the upper transistor to its drain using a buried contact. The output of the inverter is on the diffusion level, between the drain of the pull-down and the source of the pull-up. The pull-up is the depletion mode transistor, and it is usually several times longer than the pull-down in order to achieve the proper inverter logic threshold. VDD and GND are laid out in metal1 and contact cuts or vias are used to connect metal1 and diffusion.

The CMOS inverter is conceptually very simple. It can be created by connecting a p-channel and a n-channel transistor. The n-channel transistor acts as a pull-down transistor and the p-channel acts as a pull-up transistor. Figure 2.14 shows the layout of a CMOS inverter. Depending on the input, only one of two transistors conduct. When the input is low, the p-channel transistor between VDD and output is in its "on" state and output is pulled-up to VDD, thus inverting the input. During this state, the n-channel transistor does not conduct. When input is high, the output goes low. This happens due to the "on" state of the n-channel transistor between GND and output. This pulls the

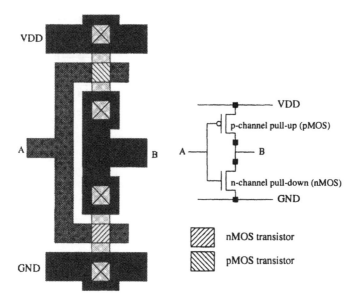

Figure 2.14: A CMOS inverter.

output down to GND, thus inverting the input. During this state, p-channel transistor does not conduct. The design rules for CMOS are essentially same as far as poly, diffusion, and metal layers are concerned. Additional CMOS rules deal with tub formation.

2.5.2   NAND and NOR Gates

NAND and NOR logic circuits may be constructed in MOS systems as a simple extension of the basic inverter circuit. The circuit layout in nMOS, truth tables, and logic symbols of a two-input NAND gate are shown in Figure 2.15 and NOR gate is shown in Figure 2.16.

In the NAND circuit, the output will be high only when both of the inputs A and B are high. The NAND gate simply consists of a basic inverter with an additional enhancement mode transistor in series with the pull-down transistor (see Figure 2.15). NAND gates with more inputs may be constructed by adding more transistors in series with the pull-down path. In the NOR circuit, the output is low if either of the inputs, A and B is high or both are high. The layout (Figure 2.16) of a *two-input* NOR gate shows a basic inverter with an additional enhancement mode transistor in parallel with the *pull-down* transistor. To construct additional inputs, more transistors can be placed in parallel on the pull-down path. The logic threshold voltage of an n-input NOR circuit decreases as a function of the number of active inputs (inputs moving together from *logic-0* to *logic-1*). The delay time of the NOR gate with one input active

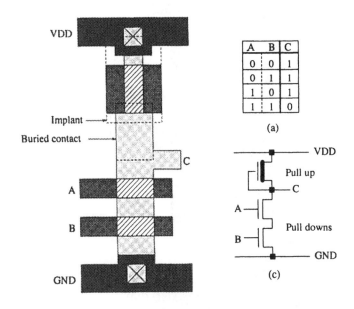

A	B	C
0	0	1
0	1	1
1	0	1
1	1	0

(a)

(c)

Figure 2.15: A nMOS NAND gate.

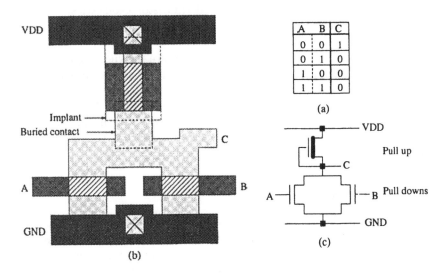

A	B	C
0	0	1
0	1	0
1	0	0
1	1	0

(a)

(b)

(c)

Figure 2.16: A nMOS NOR gate.

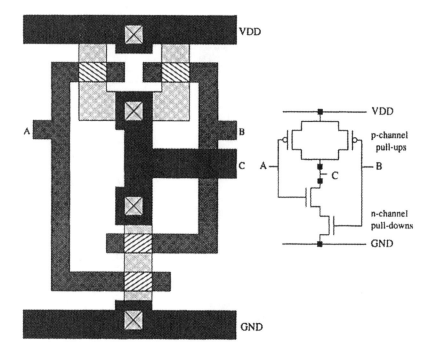

Figure 2.17: A CMOS NAND gate.

is the same as that of an inverter of equal transistor geometries, except for added stray capacitance. In designing such simple combined circuits, a single pull-up resistor must be fixed above the point of output.

The layouts of CMOS NAND and NOR gates are shown in Figure 2.17 and Figure 2.18 respectively. It is clear from Figure 2.17, that both inputs must be high in order for the output to be pulled down. In all other cases, the output will be high and therefore the gate will function as a NAND. The CMOS NOR gate can be pulled up only if both of the inputs are low. In all other cases, the output is pulled down by the two n-channel transistors, thus enabling this device to work as a NOR gate.

2.5.3   Memory Cells

Electronic memory in digital systems ranges from fewer than 100 bits from a simple four-function pocket calculator, to $10^5 - 10^7$ bits for a personal computer. Circuit designers usually speak of memory capacities in terms of bits since a unit circuit (for example a filp-flop) is used to store each bit. System designers in the other hand state memory capacities in the terms of *bytes* (typically 8-9 bits) or *words* representing alpha-numeric characters, or scientific numbers. A key characteristic of memory systems is that only a single byte or word is stored

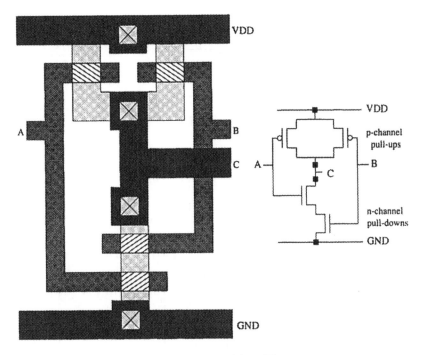

Figure 2.18: A CMOS NOR gate.

or retrieved during each cycle of memory operation. Memories which can be accessed to store or retrieve data at a fixed rate, independent of the physical location of the memory address are called Random Access Memories or RAMs. A typical RAM cell is shown in Figure 2.19.

2.5.3.1 Static Random Access Memory (SRAM)

SRAMs use static CMOS circuits to store data. A common CMOS SRAM is built using cross-coupled inverters to form a bi-stable latch as shown in Figure 2.20. The memory cell can be accessed using the pass transistors P1 and P2 which are connected to the BIT and the BIT' lines respectively. Consider the read operation. The n-MOS transistors are poor at passing a *one* and the p-transistors are generally quite small (large load resistors). To overcome this problem, the BIT and the BIT' lines are precharged to a n-threshold below VDD before the pass transistors are switched on. When the SELECT lines (word line) are asserted, the RAM cell will try to pull down either the BIT or the BIT' depending on the data stored. In the write operation, data and it's complement are fed to the BIT and the BIT' lines respectively. The word line is then asserted and the RAM cell is charged to store the data. A low on the SELECT lines, decouples the cell from the data lines and corresponds

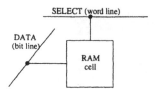

Figure 2.19: Block diagram of a generic RAM cell.

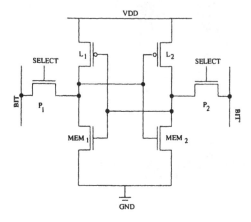

Figure 2.20: A CMOS Static RAM cell.

to a hold state. The key aspect of the precharged RAM read cycle is the timing relationship between the RAM addresses, the precharge pulse and the row decoder (SELECT line). A word line assertion preceding the precharge cycle may cause the RAM cell to flip state. On the other hand, if the address changes after the precharge cycle has finished, more than one RAM cell will be accessed at the same time, leading to erroneous *read* data.

The electrical considerations in such a RAM are simple to understand as they directly follow the CMOS Inverter characteristics. The switching time of the cell is determined by the output capacitance and the feedback network. The time constants which control the charging and discharging are

$$\tau_{ch} = \frac{C_L}{\beta_p(VDD - |V_{Tp}|)}$$

$$\tau_{dis} = \frac{C_L}{\beta_n(VDD - V_{Tn})}$$

where C_L is the total load capacitance on the output nodes and β_n and β_p are the transconductance parameters for the n and p transistors respectively.

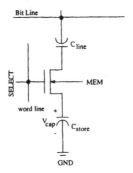

Figure 2.21: A CMOS Dynamic RAM cell.

Minimizing the time constants within the constraints of the static noise margin requirements gives a reasonable criterion for the initial design of such cells.

2.5.3.2 Dynamic Random Access Memory (DRAM)

A DRAM cell uses a dynamic charge storage node to hold data. Figure 2.21 shows a basic 1-Transistor cell consisting of an access nMOS MEM, a storage capacitor C_{store} and the input capacitance at the *bit line* C_{line}.

When the *Select* is set to high, C_{store} gets charged up to the bit line voltage according to the formula,

$$V_{cap}(t) = V_{max}[\frac{t/\tau_{ch}}{1 + t/\tau_{ch}}]$$

where $\tau_{ch} = 2C_{store}/\beta_n V_{max}$ is the charging time constant. The 90% voltage point ($0.9V_{max}$) is reached in a low-high time of $t_{LH} = 9\tau_{ch}$ and is the minimum logic 1 loading interval. Thus a logic one is stored into the capacitor. When a logic zero needs to be stored, the *Select* is again set to high and the charge on C_{store} decays according to the formula,

$$V_{cap}(t) = V_{max}[\frac{2e^{-t/\tau_{dis}}}{1 + e^{-t/\tau_{dis}}}]$$

where $\tau_{dis} = C_{store}/\beta_n V_{max}$ is discharge time constant. The 10% voltage point ($0.1V_{max}$) requires a high-low time of $t_{HL} = 2.94\tau_{dis}$. Thus it takes longer to load a logic 1 ($t_{LH} = 6.11t_{HL}$) than to load a logic 0. This is due to the fact that the gate-source potential difference decreases during a logic 1 transfer.

The read operation for a dynamic RAM cell corresponds to a charge sharing event. The charge on C_{store} is partly transferred onto C_{line}. Suppose C_{store} has an initial voltage of V_C. The bit line capacitance C_{line} is initially charged to a voltage V_{pre} (typically 3V). The total system charge is thus given by $Q_T = V_C C_{store} + V_{pre} C_{line}$. When the *SELECT* is set to a high voltage, M

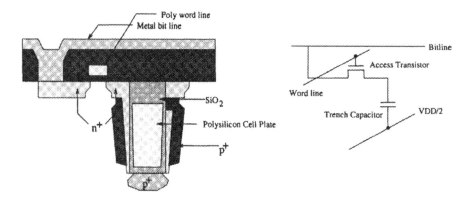

Figure 2.22: A deep trench CMOS Dynamic RAM cell.

becomes active and conducts current. After the transients have decayed, the capacitors are in parallel and equilibrate to the same final voltage V_f such that

$$V_f = \frac{V_C C_{store} + V_{pre} C_{line}}{C_{store} + C_{line}}.$$

Defining the capacitance ratio $r = C_{line}/C_{store}$ yields the final voltage V_f as

$$V_f = \frac{V_C + r V_{pre}}{1 + r}.$$

If a logic 1, is initially stored in the cell, then $V_C = V_{max}$ and

$$V_1 = \frac{V_{max} + r V_{pre}}{1 + r}.$$

Similarly for $V_C = 0$ volts,

$$V_0 = \frac{r V_{pre}}{1 + r}.$$

Thus the difference between a logic 1 and a logic 0 is

$$\Delta V = \frac{V_{max}}{1 + r}.$$

The above equation clearly shows that a small r is desirable. In typical 16 Mb designs, $C_{store} \approx 30 fF$ and $C_{line} \approx 250 fF$ giving $r \approx 8$.

Dynamic RAM cells are subject to charge leakage, and to ensure data integrity, the capacitor must be *refreshed* periodically. Typically a dynamic refresh operation takes place at the interval of a few milliseconds where the peripheral logic circuit reads the cell and re-writes the bit to ensure integrity of the stored data.

High-value/area capacitors are required for dynamic memory cells. Recent processes use three dimensions to increase the capacitance/area. One such

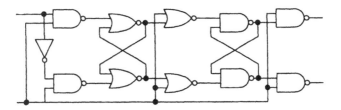

Figure 2.23: Circuit 1.

structure is the trench capacitor shown in Figure 2.22. The sides of the trench are doped n^+ and coated with a thin 10 nm oxide. Sometimes a thin oxynitride is used because its high dielectric constant increases the capacitance. The cell is filled with a polysilicon plug which forms the bottom plate of the cell storage capacitor. This is held at $VDD/2$ via a metal connection at the edge of the array. The sidewall n^+ forms the other side of the capacitor and one side of the pass transistor that is used to enable data onto the bit lines. The bottom of the trench has a p^+ plug that forms a channel-stop region to isolate adjacent capacitors.

2.6 Summary

The fabrication cycle of VLSI chips consists of a sequential set of basic steps which are crystal growth and wafer preparation, epitaxy, dielectric and polysilicon film deposition, oxidation, lithography, and dry etching. During the fabrication process, the devices are created on the chip. When some fixed size material crosses another material, devices are formed. While designing the devices, a set of design rules has to be followed to ensure proper function of the circuit.

2.7 Exercises

1. Draw the layout using nMOS gates with minimum area for the circuit shown in Figure 2.23. For two input nMOS NAND gates, assume the length of pull-up transistor to be eight times the length of either pull-down. (Unless stated otherwise, use two metal layers for routing and ignore delays in vias).

2. Compute the change in area if CMOS gates are used in a minimum area layout of the circuit in Figure 2.23.

3. For the circuit shown in Figure 2.24, generate a layout in which the longest wire does not have a delay more than 0.5 psec. Assume that the

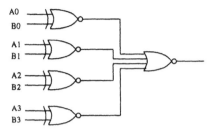

Figure 2.24: A 4-bit comparator.

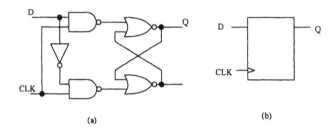

Figure 2.25: A D flip-flop.

width of the wire is 2 μm, the height is 0.5 μm, the thickness of the oxide below the wire is 1.0 μm. $\epsilon_o = 3.9$ and $\epsilon_s = 8.845 \times 10^{-14}$ F/cm.

4. Layout the circuit given in Figure 2.23 so that the delay in the longest path from input to output is minimized. Assume 2 μm CMOS process and assume each gate delay to be 2 nsec.

†5. In order to implement a memory, one needs a circuit element, which can store a bit value. This can be done using flip-flops. A D flip-flop is shown in Figure 2.25. Memories can be implemented using D flip-flops as shown in Figure 2.26. A 2 × 2 bit memory is shown in the figure. The top two filp-flops store word 0, while the bottom two flip flops store the word 1. A indicates the address line, if $A = 0$ the top two bits consisting of top word 0 are selected, otherwise the bottom word is selected. CS, RD and OE indicate chip select, read and output enable signals, respectively. I_0 and I_1 are two input lines, while O_0 and O_1 indicate output lines.

   (a) Layout the 4 bit memory shown in Figure 2.26.

   (b) Calculate the read and write times for this memory in 2 μm CMOS process. Assume gate delay to be 2 nsec.

   (c) Estimate the total area (in microns) for a 256 KByte memory using 2 μm CMOS process.

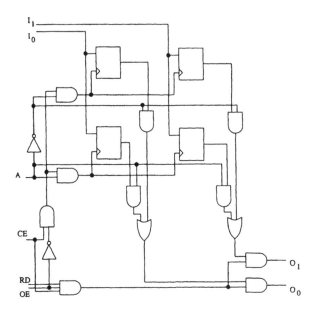

Figure 2.26: A 2×2 bit memory.

6. (a) Draw the circuit diagram of a half-adder.

   (b) Draw the layout of the of the half-adder with minimum area.

   (c) How many such chips may be fabricated on a wafer of diameter 10 cm ? (Assume scrap edge distance, $\alpha = 5$ mm.)

7. For the function $F = \bar{A}BC + A\bar{B}C + AB\bar{C} + ABC$

   (a) Draw the logic circuit diagram.

   (b) Generate a minimum area layout for the circuit.

8. Estimate the number of transistors needed to layout a k-bit full adder. Compute the area required to layout such a chip in nMOS and CMOS.

9. *Skew* is defined as the difference in the signal arrival times at two difference devices. Skew arises in the interconnection of devices and in routing of a clock signal. Skew must be minimized if the system performance is to be maximized. Figure 2.27 shows a partial layout of a chip. Complete the layout of chip by connecting signal source to all the terminals as shown in the Figure 2.27. All paths from the source to the terminals must be of equal length so as to have zero skew.

10. Suppose a new metal layer is added using present design rules. How many design rules would be needed ?

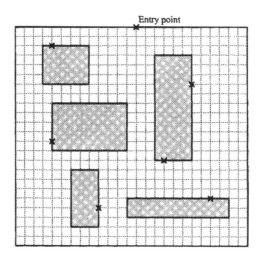

Figure 2.27: Routing with Minimum skew.

11. Compute the number of masks needed to produce a full custom chip using k-metal layer CMOS technology.

Bibliographic Notes

Weste and Eshraghian[WE92] cover CMOS design from a digital system level to the circuit level. In addition to the traditional VLSI design technologies it covers the emerging BiCMOS technology. Mead and Conway [MC79] discuss the physical design of VLSI devices. The details about the design rules can also be found in [MC79]. Advanced discussion on VLSI circuit and devices can be found in [Gia89]. The book by Bakoglu [Bak90] covers interconnects and the effects of scaling on device dimensions in detail. The basic functions of nMOS and CMOS VLSI devices are discussed in [Muk86].

Chapter 3

Fabrication Process and its Impact on Physical Design

The biggest driving force behind growth in the semiconductor, computer, networking, consumer electronics, and software industries in the last half century has been the continuous scaling, or miniaturization, of the transistor. Computers and other electronic devices have become smaller, more portable, cheaper, easier to use, and more accessible to everyone. As long as we can make the transistor faster and smaller, make the wires that interconnect them less resistive to electrical current, and make each chip denser, the digital revolution will continue.

The manufacture of ICs, like any other high volume manufacturing business, is very cost sensitive. The yield of the fab must be very high to be profitable. So in any given process generation, semiconductor manufacturers use process equipment and critical dimensions that allow them acceptable yields. As more and more chips are manufactured and tested in a process, the process matures and the yield of the process increases. When the yield increases, more aggressive (that is, smaller) critical dimensions can be used to shrink the layout. This process of shrinking the layout, in which every dimension is reduced by a factor is called *scaling*. In general, scaling refers to making the transistors, and the technology that connects them, smaller. As a transistor becomes smaller, it becomes faster, conducts more electricity, and uses less power, the cost of producing each transistor goes down, and more of them can be packed on a chip.

If a chip is scaled, it leads to a smaller die size, increased yield, and increased performance. Suppose a chip in 0.25 micron process generation is x microns wide and x microns high. Assume a shrink factor of 0.7 from 0.25 to 0.18 micron process. Therefore, on 0.18 process, we can essentially produce a $0.7x$ micron wide and $0.7x$ micron high chip. That is, the scaled chip is half the size of the original chip, in terms of area. It will have better yield, due to smaller die size and it will have better performance due to faster transistors and shorter interconnect.

As transistors are scaled, their characteristics (such a delay, leakage current, threshold voltage, etc) also scale but not uniformly. For example, power may not scale with device size. In particular, by middle of next decade, it is expected that the leakage current of a transistor will be of the same value whether the transistor is on or off.

The biggest concern in scaling, is the mismatch in the scaling of devices and interconnect. Interconnect delay is not scaling at the same rate as the device delay. As a result, it has become a more dominant factor in overall delay. This has fundamentally changed the perspective of physical design. Earlier it was possible to lay down the devices and interconnect them and be sure that the design would work. It was possible to ignore the delay in the interconnect, as it was $5 - 10\%$ of the overall delay. As interconnect has now become $50 - 70\%$ of the overall delay, it was necessary to find optimal locations for devices so that interconnect delay is as small as possible. This has led to the consideration of physical design (in particular interconnect planning) at very early stages of design (even at architectural level), as well as throughout the VLSI design cycle.

As a result of the potential side effects, it is important to be aware of process technology and innovations, so as to understand the impact on physical design. The purpose of this chapter is to explain the process scaling, process innovations, process side-effects and their impact on physical design. Our focus is to identify potential future physical design problems due to process. Section 3.1 discusses the scaling methods, Section 3.2 presents the status of fabrication process (circa 1998). Section 3.3 is dedicated to issues with the process such as the parasitics effects, interconnect delay and noise, power dissipation, and yield, among others. In Section 3.4, we discuss the future of the process and innovations that might solve the current and future process related problems. Section 3.5 discusses the solutions and options for interconnect problems. Finally, Section 3.6 discusses CAD tools needed to help in process development.

3.1   Scaling Methods

There are two basic types of scaling to consider. One is *full scaling*, in which all device dimensions, both surface and vertical, and all voltages are reduced by the same factor. Other type is called the *constant-voltage scaling*, wherein only device dimensions are scaled, while maintaining voltage levels. These two methods are compared in Table 3.1, where the scaling factor is specified as S.

In full scaling, devices improve in speed, dissipate less power and can be packed closer together. As a result, high speed chips that integrate large numbers of transistors are made possible. However, devices cannot be scaled down beyond a certain limit. This limit is imposed by a number of second order effects which arise due to simple scaling technique.

In constant voltage scaling, ICs can be used in existing systems without multiple power supplies. In addition, the gate delay is reduced by an additional factor of S. On the other hand, constant voltage scaling leads to higher power

Parameter	Full scaling	CV scaling
Dimensions: width, length, oxide thickness	$1/S$	$1/S$
Voltages: Power, threshold	$1/S$	1
Gate capacitance	$1/S$	$1/S$
Current	$1/S$	S
Propagation delay	$1/S$	$1/S^2$

Table 3.1: Scaling effect on device parameters.

dissipation density, and increments in electric fields, which may lead to oxide breakdown problems.

Except for features related to bonding pads and scribe lines, all other features can be scaled. If the design is not limited by the pad-size, the layout can be scaled and then pads of original size can be placed around the shrunken layout.

3.2 Status of Fabrication Process

In 1998, the standard production fabrication process is the 0.25 micron CMOS process. It has a 1.8V VDD and 4.2 nm oxide. Transistors use complementary doped poly and Shallow Trench Isolation (STI). In terms of interconnect, process supports 5-6 layers of aluminum interconnect using (typically) a Ti/Al-Cu/Ti/TiN stack. Some manufacturers provide a local interconnect layer as well. The key feature of interconnect is the high aspect ratio metal lines for improved resistance and electro-migration. Contact and via layers are filled with tungsten plugs and planarized by CMP. ILD layers are also planarized by CMP. New processes support copper layers for interconnect.

3.2.1 Comparison of Fabrication Processes

In this section, we compare 0.25 micron processes of leading manufacturers. Table 3.2 shows the key features of production fabrication processes of five leading semiconductor manufacturers. The five processes are listed from IBM (International Business Machines), AMD (Advanced Micro Devices), DEC (Digital Equipment Corporation, which is now part of Intel and Compaq), TI (Texas Instruments) and Intel Corporation.

Several interesting observations can be made about the processes.

1. Synergy between Metal Layers: Some manufacturers have syngerized the metal layers. For example, DEC's CMOS-7 process has (1:2) ratio between two lower metal layers and three higher metal layers. Similarly, TI's CO7 has (1:3) ratio between first four layers and the last layer. While Intel's process favors syngergy between M2 and M3 and there is

Company	IBM	AMD	DEC	TI	Intel
Process	CMOS-6x	CS-44	CMOS-7	C07	P856
No. of metal layers	6	5	6	5	5
M0	Yes	Yes	No	No	No
Stacked Vias	Yes	Yes	Yes	Yes	Yes
Voltage	1.8V	2.5V	1.8V	1.8V	1.8V
M1	0.7	0.88	0.84	0.85	0.64
M2	0.9	0.88	0.84	0.85	0.93
M3	0.9	0.88	1.7	0.85	0.93
M4	0.9	1.13	1.7	0.85	1.6
M5	0.9	3.0	1.7	2.5	2.56

Table 3.2: Comparison of Fabrication Processes.

no obvious relationship between these layers and higher layers. The other other extreme is IBM's essentially gridded approach for four higher layers. Only M1 is os smaller pitch, possibly to help cell density. Metal synergy may help in routing, as routes are on certain tracks and vias can be only on the routing grids. The importance of routing grid is related to methodology for wire sizing. In particular, the method, by which routers provide wide wire and tapering capability.

2. Local interconnect: While some manufacturers have provided local interconnect for better cell density (50%), other have opted to forego it since it may cost almost as much as a full layer but provides limited routing capability.

3. Use of higher metal layers: Table 3.2 clearly shows a divergence in the width of the higher metal layers. Note that AMD, TI, and Intel use very wide wires on higher metal layers. These layers provide high performance for long global interconnect. While IBM and DEC provide smaller width lines, possibly leaving the option for the designer to use a wider than minimum width wire, if necessary. However, this does allow process to provide optimal performance from wires in higher layers.

Table 3.3 shows details about the spacing between interconnect and aspect ratio of the interconnect for Intel's P856 process. First note that the thickness of metal lines varies widely. M1 is quite narrow at 0.48 um, while M5 is quite thick at 1.9 um. Aspect ratio of a wire is the ratio of its thickness to its width. Note that the interconnect aspect ratio is almost as high as 2.0 for some layers. That is, M2 (and M3) is twice as thick as it is wide. Higher aspect ratio provides better interconnect performance on smaller widths but also introduces the wall-to-wall capacitance.

Layer	Pitch(um)	Thickness(um)	Aspect Ratio
M1	0.64	0.48	1.5
M2	0.93	0.89	1.9
M3	0.93	0.89	1.9
M4	1.60	1.33	1.7
M5	2.56	1.90	1.5

Table 3.3: Intel's P856 Interconnect dimensions

3.3 Issues related to the Fabrication Process

Process scaling has allowed a high level of integration, better yields (for a constant die size), lower costs and allowed larger die sizes (for a constant yield). As a result, process has also introduced several problems and issues that need to be addressed. In this section, we will discuss these process related issues.

The first set of issues is related to parasitics effects, such as stray capacitances. The second set of issues is related to interconnect, which poses two type of problems, delay/noise and signal integrity which may limit maximum frequency. Other interconnect problems are associated with size and complexity of interconnect. The amount of interconnect needed to connect millions of transistors runs into hundreds of thousands of global signals and millions of local and mid-distance signals. The sheer size of interconnect needs to be addressed, otherwise the die size grows to accommodate the interconnect. This larger die size may make the design project more costly or even infeasible. This is due to that fact that larger die may have longer interconnect and may not allow the chip to reach its targeted frequency. Other issues include power dissipation and yield.

3.3.1 Parasitic Effects

The proximity of circuit elements in VLSI allows the inter-component capacitances to play a major role in the performance of these circuits. The stray capacitance and the capacitance between the signal paths and ground are two major parasitic capacitances. Another parasitic capacitance is the inherent capacitance of the MOS transistor. This capacitance has to be accounted for, as it has more effect than the parasitic capacitance to the ground. All MOS transistors have a parasitic capacitance between the drain edge of the gate and drain node. In an inverter, this capacitance will be charged in one direction for one polarity input and in the opposite direction for the opposite polarity input. Thus, on a gross scale its effect on the system is twice that of an equivalent parasitic capacitance to ground. Therefore, gate-to-drain capacitances should be doubled, and added to the gate capacitance and the stray capacitances, to account for the total capacitance of the node and thus for the effective delay of the inverter.

Interconnect capacitance is of two types; between wires across layers and between wires within layers. The former is more significant than the later. The interconnect capacitance within layers can be reduced by increasing the wire spacing and by using power lines shielding. Whereas, the interconnect capacitance across layers can be reduced by connecting wires in adjacent layers perpendicular to each other.

3.3.2 Interconnect Delay

The calculation of delay in a layout depends on a variety of parameters. The delays in a circuit can be classified as gate delay and interconnect delay. Both the gate and the interconnect delay depend on parameters such as the width and length of poly, thickness of oxide, width and length of metal lines, etc. Details on this topic can be found in [Bak90]. The process of extracting these parameters is called *extraction*. The tool that computes the delay using these parameters and a suitable delay model is often referred as an *RC-extractor*. We will restrict this discussion to the calculation of interconnect delays.

Historically, interconnect delay was considered to be electrically negligible. Currently interconnections are becoming a major concern in high performance ICs and the RC delay due to interconnect is the key factor in determining the performance of a chip. The resistance of wires increases rapidly as chip size grows larger and minimum feature size reduces. Resistance plays a vital role in determining RC delay of the interconnection.

The relative resistance values of metal, diffusion, poly, and drain-to-source paths of transistors are quite different. Diffused layers and polysilicon layers have more than one hundred times the resistance per square area of the metal layer. The resistance of a uniform slab of conducting material is given by:

$$R = \frac{\rho l_c}{h_c w_c}$$

where ρ is the resistivity, and w_c, h_c, and l_c are the width, thickness, and length of the conductor. The empirical formula for the interconnection capacitance is given by:

$$C = \left[1.15 \left(\frac{w_c}{t_o} \right) + 2.80 \left(\frac{h_c}{t_o} \right)^{0.222} \right.$$
$$\left. + \left[0.06 \left(\frac{w_c}{t_o} \right) + 1.66 \left(\frac{h_c}{t_o} \right) - 0.14 \left(\frac{h_c}{t_o} \right)^{0.222} \right] \left(\frac{t_o}{w_{ic}} \right)^{1.34} \right] \epsilon_s \, \epsilon_o \, l_c$$

where, C is the capacitance of the conductor, w_{ic} is the spacing of chip interconnections, t_o is the thickness of the oxide, $\epsilon_o = 3.9$ is the dielectric constant of the insulator, and $\epsilon_s = 8.854 \times 10^{-14}$ F/cm is the permittivity of free space. Various analytical models for two-dimensional interconnection capacitance calculations can be found in [Cha76, ST83]. Path capacitance could be computed by adding via capacitances to the net capacitances.

The expressions given above show that the interconnect delay is the more dominant delay in current technology. Consider a 2 cm long, 0.5 μm thick wire, having a 1.0 μm thick oxide beneath it in an chip fabricated using 2 μm technology. The resistance of such a wire is 600 Ω and its capacitance is approximately 4.0 pF. As a result, it has a distributed RC constant of 2.4 nsec. This delay is equivalent to a typical gate delay of 1 to 2 nsec in 2 μm technology. In this technology, the maximum die size was limited to 1 cm × 1 cm and therefore the gate delays dominated the interconnect delays. On the other hand, a similar calculation for 0.5-0.7 μm technology shows that if only sub-nanosecond delays are allowed, the maximum wire length can be at most 0.5 cm. Since, the gate delays are sub-nanosecond and the die size is 2.5 cm × 2.5 cm, it is clear that interconnect delays started dominating the gate delay in 0.5 μm process generation.

The delay problem is more significant for signals which are global in nature, such as clock signals. A detailed analysis of delay for clock lines is presented in Chapter 11.

3.3.3 Noise and Crosstalk

When feature sizes and signal magnitudes are reduced, circuits become more susceptible to outside disturbances, resulting in *noise*. Noise principally stems from resistive and capacitive coupling. Smaller feature sizes result in reduced node capacitances. This helps to improve circuit delays; however, these nodes also become more vulnerable to external noise, especially if they are dynamically charged. The coupling between neighboring circuits and interconnections and the inductive noise generated by simultaneous switching of circuits are most prevalent forms of internal noise. As chip dimensions and clock frequency increase, the wavelengths of the signals become comparable to interconnection lengths, and this makes interconnections better 'antennas.'

Noise generated by off-chip drivers and on-chip circuitry is a major problem in package and IC design for high-speed systems. The noise performance of a VLSI chip has three variables: noise margin, impedance levels, and characteristics of noise sources. Noise margin is a parameter closely related to the input-output voltage characteristics. This is a measure of the allowable noise voltage in the input of a gate such that the output will not be affected. Noise margin is defined in terms of two parameters: Low Noise Margin(LNM) and High Noise Margin(HNM). The LNM and HNM are given by:

$$LNM = \max(V_{IL}) - \max(V_{OL})$$

$$HNM = \min(V_{OH}) - \min(V_{IH})$$

Where V_{IL} and V_{IH} are low and high input voltages and V_{OL} and V_{OH} are low and high output voltages respectively.

One of the forms of noise is *crosstalk*, which is a result of mutual capacitance and inductance between neighboring lines. The amount of coupling between two lines depends on these factors. The closeness of lines, how far they are

from the ground plane, and the distance they run close to each other. As a
result of crosstalk, propagation delay increases and logic faults occur. The
delay increases because the overall capacitance of the line increases which in
turn augments the RC delay of the line.

3.3.4   Interconnect Size and Complexity

The number of nets on a chip increases as the number of transistors are increased. Rent's rule is typically used to estimated the number of pins in a block
(unit, cluster, or chip) and number of transistors in that block (unit, cluster
or chip). Rent's rule state that the number of I/Os needed are proportional to
the number of transistors N and a constant K, which depends on the ability
to share signals. Rent's rule is expressed as:

$$C = KN^n$$

where C is the average number of signal and control I/Os. K is typically 2.5
for high performance systems, and n is a constant in the range of 1.5 to 3.0.
Originally, Rent's rule was observed by plotting I/Os versus transistors count
of real systems. Since that time, several stochastic and geometric arguments
have also been proposed that support Rent's rule.

It is quite clear from Rent's rule that the signal complexity at all levels of
the chip stays ahead of the integration level (number of transistors, sub-circuits,
etc).

3.3.5   Other Issues in Interconnect

Several other issues in interconnect may also cause some problems. Higher
aspect ratio wires that are used to provide better performance for higher layers
also cause more cross (wall to wall) capacitance. As a result, signals that may
conflict need to be routed spaced from each other. Another issue is inductance
modeling and design. As chip frequency reaches GHz and beyond, wires start
acting like transmission lines and circuits behave like RLC circuits. In addition,
use of different dielectrics, that are used on different layers complicates the
delay, noise and inductance modeling.

3.3.6   Power Dissipation

Heat generation and its removal from a VLSI chip is a very serious concern.
Heat generated is removed from the chip by heat transfer. The heat sources
in a chip are the individual transistors. The heat removal system must be
efficient and must keep the junction temperature below a certain value, which
is determined by reliability constraints. With higher levels of integration, more
and more transistors are being packed into smaller and smaller areas. As a
result, for high levels of integration heat removal may become the dominant
design factor. If all the generated heat is not removed, the chip temperature
will rise and the chip may have a thermal breakdown. In addition, chips must

be designed to avoid *hotspots*, that is, the temperature must be as uniform as possible over the entire chip surface.

CMOS technology is known for using low power at low frequency with high integration density. There are two main components that determine the power dissipation of a CMOS gate. The first component is the static power dissipation due to leakage current and the second component is dynamic power dissipation due to switching transient current and charging/discharging of load capacitances. A CMOS gate uses 0.003 mW/MHz/gate in 'off' state and 0.8 mW/MHz/gate during its operation. It is easy to see that with one million gates, the chip will produce less than a watt of heat. In order to accurately determine the heat produced in a chip, one must determine the power dissipated by the number of gates and the number of off chip drivers and receivers. For CMOS circuits, one must also determine the average percent of gates active at any given time, since heat generation in the 'off' state is different than that of 'on' state. In ECL systems, power consumption is typically 25 mW/gate irrespective of state and operating frequency. Current heat removal system can easily handle 25 to 100W in a high performance package. Recently, MCM systems have developed, which can dissipate as much as 600W per module.

Power dissipation has become a topic of intense research and development. A major reason is the development of lap-top computers. In lap-top computers, the battery life is limited and low power systems are required. Another reason is the development of massively parallel computers, where hundreds (or even thousands) of microprocessors are used. In such systems, power dissipation and corresponding heat removal can become a major concern if each chip dissipates a large amount of power.

In recent years, significant progress has been in made in development of low power circuits and several research projects have now demonstrated practical lower power chips operating at 0.5 V. In some microprocessors, 25-35% power is dissipated in the clock circuitry, so low power dissipation can be achieved by literally 'switching-off' blocks which are not needed for computation in any particular step.

3.3.7 Yield and Fabrication Costs

The cost of fabrication depends on the yield. Fabrication is a very complicated combination of chemical, mechanical and electrical processes. Fabrication process requires very strict tolerances and as a result, it is very rare that all the chips on a wafer are correctly fabricated. In fact, sometimes an entire wafer may turn out to be non-functional. If a fabrication process is new, its yield is typically low and design rules are relaxed to improve yield. As the fabrication process matures, design rules are also improved, which leads to higher density. In order to ensure that a certain number of chips per wafer will work, an upper limit is imposed on the chip dimension X. Technically speaking, an entire wafer can be a chip (wafer scale integration). The yield of such a process would however be very low, in fact it might be very close to zero.

Wafer yield accounts for wafers that are completely bad and need not be tested. The prediction of the number of good chips per wafer can be made on the basis of how many dies (chips) fit into a wafer (N_d) and the probability of a die being functional after processing (Y). The cost of an untested die C_{ud} is given by

$$C_{ud} = \frac{C_w}{N_d * Y}$$

where, C_w is the cost of wafer fabrication. The number of dies per wafer depends on wafer diameter and the maximum dimension of the chip. It should be noted that product $N_d * Y$ is equal to total number of "good" dies per wafer N_y. The number of dies of a wafer N_d is given by

$$N_d = \pi \frac{(D - \alpha)^2}{4X^2}$$

where D is the diameter of the wafer (usually 10 cm), and α is the useless scrap edge width of a wafer (mm). The yield is given by:

$$Y = (1 - A\delta/c)^c$$

where, A is the area of a single chip, δ is the defect density, that is, the *defects per square millimeter*, and c is a parameter that indicates defect clustering.

The cost of packaging depends on the material used, the number of pins, and the die area. The ability to dissipate the power generated by the die is the main factor which determines the cost of material used.

Die size depends on technology and gates required by the function and maximum number of dies on the chip, but it is also limited by the number of pins that can be placed on the border of a square die.

The number of gates N_g in a single IC is given by:

$$N_g = \frac{(X^2 - P * A_{io})}{A_g}$$

where, P is the total number of pads on the chip surface, A_{io} is the area of an I/O cell and A_g is the area of a logic gate. It should be noted that a gate is a group of transistors and depending on the architecture and technology, the number of transistors required to form a gate will vary. However, on the average there are 3 to 4 transistors per gate.

The number of pads required to connect the chip to the next level of interconnect, assuming that pads are only located at the periphery of the chip is

$$P = 4(X/S - 1)$$

where, S is the minimum pad to pad pitch.

An optimal design should have the necessary number of gates well distributed about the chip's surface. In addition, it must have minimum size, so as to improve yield.

The total fabrication cost for a VLSI chip includes costs for wafer preparation, lithography, die packaging and part testing. Again, two key factors determine the cost: the maturity of process and the size of the die. If the process is new, it is likely to have a low yield. As a result, price per chip will be higher. Similarly, if a chip has a large size, the yield is likely to be low due to uniform defect density, with corresponding increase in cost.

Fabrication facilities can be classified into three categories, prototyping facilities (fewer than 100 dies per lot), moderate size facilities (1,000 to 20,000 dies per lot) and large scale facilities, geared towards manufacturing 100,000+ parts per lot.

Prototyping facilities such as MOSIS, cost about $150 for tiny chips (2.22 mm × 2.26 mm) in a lot of 4, using 2.0 μm CMOS process. For bigger die sizes (7.9 mm × 9.2 mm) the cost is around $380 per die, including packaging. The process is significantly more expensive for 0.8 μm CMOS. The total cost for this process is around $600 per square mm. MOSIS accepts designs in CIF and GDS-II among other formats, and layouts may be submitted via electronic mail. For moderate lot sizes (1,000 to 20,000) the cost for tiny chips (2 mm × 2 mm) is about $27 in a lot size of 1,000. For large chips (7 mm × 7 mm), the cost averages $65 per chip in a lot of 1000. In addition, there is a lithography charge of $30,000 and a charge of $5,000 for second poly (if needed). These estimates are based on a 1.2 μm, two metal, single poly CMOS process and include the cost of packaging using a 132 Pin Grid Array (PGA) and the cost of testing. The large scale facilities are usually inexpensive and cost between $5 to $20 per part depending on yield and die size. The costs are included here to give the reader a real world perspective of VLSI fabrication. These cost estimates should not be used for actual budget analysis.

3.4 Future of Fabrication Process

In this section, we discuss the projected future for fabrication process. We also discuss several innovations in lithography, interconnect and devices.

3.4.1 SIA Roadmap

Fabrication process is very costly to develop and deploy. A production fab costs upwards of two billion dollars (circa 1998). In the early 1990's, it became clear that process innovations would require joint collaboration and innovations from all the semiconductor manufacturers. This was partly due to the cost of the process equipment and partly due to the long time it takes to innovate, complete research and development and use the developed equipment or methodologies in a real fab. Semiconductor Industry Association (SIA) and SRC started several projects to further research and development in fabrication process.

In 1994, SIA started publishing the National Technology Roadmap for Semiconductors. The roadmap provides a vision for process future. In 1997, it was revised and key features are listed in table 3.4. (note that 1000 nanometers = 1 micron).

Feature Size (nm)	250	180	130	100
Time Frame	1997	1999	2003	2006
Logic Transistors per area (Millions/sq. cm.)	3.7	6.2	18	39
Chip Frequency (MHz): Cost/Perf designs	> 300	> 400	> 500	> 650
Chip Frequency (MHz): High Perf designs	> 500	> 750	> 1100	> 1500
Chip Size (sq. mm)	300	360	430	520
Wiring Levels	6	6-7	7	7-8
Package Pins per Chip	512	512	768	768
Power Supply Voltage (desktop)	2.5	1.8	1.5	1.2
Power Supply Voltage (portable)	1.8-2.5	0.9-1.8	0.9	0.9
Interconnect	planar	planar	planar	planar
Min. Interconnect CD (nm)	250	180	130	100
Min. Contacted Interconnect Pitch (logic) nm	640	460	340	260
Contact/Via Critical Dimension(nm)	280/400	200/280	140/200	110/140
Via Aspect Ratio : Logic	2.0	2.0	2.3	2.7
Metal Aspect Ratio	1.8	1.8	2.1	2.4
Max. Interconnect Length (meters/chip)	820	1480	2840	5140

Table 3.4: Feature Size and Integrated Circuit Capability, 1997-2006. National Technology Roadmap for Semiconductors, 1997

3.4.2   Advances in Lithography

Currently, it is possible to integrate about 20 million transistors on a chip. Advances in X-ray lithography and electron-beam technology indicate that these technologies may soon replace photolithography. For features larger than 0.020 μm, X-ray beam lithography can be used. For smaller devices, a scanning tunneling electron microscope may be used. It has been shown that this technique has the possibility of constructing devices by moving individual atoms. However, such fabrication methods are very slow. Just based on X-ray lithography, there seems to be no fundamental obstacle to the fabrication of one billion transistor integrated circuits. For higher levels of integration there are some limits. These limits are set by practical and fundamental barriers. Consider a MOS transistor with a square gate 0.1 μm on a side and 0.005 μm oxide layer thickness. At 1V voltage, there are only 300 electrons under the gate, therefore, a fluctuation of only 30 electrons changes the voltage by 0.1 V. In addition, the wave nature of electrons poses problems. When device features reach 0.005 μm, electrons start behaving more like waves than particles. For such small devices, the wave nature as well as the particle nature of electrons has to be taken into account.

3.4.3 Innovations in Interconnect

As discussed earlier, interconnect poses a serious problem as we attempt to achieve higher levels of integration. As a result, several process innovations are targeted towards solution of the interconnect problems, such as delay, noise, and size/complexity.

3.4.3.1 More Layers of Metal

Due to planarization achieved by CMP, from a pure technology point of view, any number of metal layers is possible. Hence, it is purely a cost/benefit trade-off which limits addition of layers. An increasing number of metal layers has a significant impact on the physical design tools. In particular, large numbers of layers stresses the need for signal planning tools.

3.4.3.2 Local Interconnect

The polysilicon layer used for the gates of the transistor is commonly used as an interconnect layer. However the resistance of doped polysilicon is quite high. If used as a long distance conductor, a polysilicon wire can represent significant delay. One method to improve this, that requires no extra mask levels, is to reduce the polysilicon resistance by combining it with a refractory metal. In this approach a Sillicide (silicon and tantalum) is used as the gate material. The sillicide itself can be used as a local interconnect layer for connections within the cells. Local interconnect allows a direct connection between polysilicon and diffusion, thus alleviating the need for area intensive contacts and metal. Also known as Metal 0, local interconnect is not a true metal layer. Cell layout can be improved by 25 to 50% by using local interconnect. However, CAD tools need to comprehend restrictions to use M0 effectively.

3.4.3.3 Copper Interconnect

Aluminum has long been the conductor of choice for interconnect, but as the chip size shrinks it is contributing to interconnect delay problem, due to its high resistance. Although Copper is a superior conductor of electricity, it was not being used earlier for interconnect because not only does copper rapidly diffuse into silicon, it also changes the electrical properties of silicon in such a way as to prevent the transistors from functioning. However, one by one, the hurdles standing in the way of this technology have been overcome. These ranged from a means of depositing copper on silicon, to the development of an ultra thin barrier to isolate copper wires from silicon. Several manufacturers have introduced a technology that allows chip makers to use copper wires, rather than the traditional aluminum interconnects, to link transistors in chips.

This technique has several advantages. Copper wires conduct electricity with about 40 percent less resistance than aluminum which translates into a speedup of as much as 15 percent in microprocessors that contain copper wires. Copper wires are also far less vulnerable than those made of aluminum to electro-migration, the movement of individual atoms through a wire, caused by high electrical currents, which creates voids and ultimately breaks the wires. Another advantage of copper becomes apparent when interconnect is scaled down. At small dimensions, the conventional aluminum alloys can't conduct electricity well enough, or withstand the higher current densities needed to make these circuits switch faster. Copper also has a significant problem of electro-migration. Without suitable barriers, copper atoms migrate into silicon and corrupt the whole chip.

Some manufacturers claim that the chips using Copper interconnects are less expensive than aluminum versions, partly because copper is slightly cheaper, but mainly because the process is simpler and the machinery needed to make the semiconductors is less expensive. However other chip manufacturing companies are continuing with aluminum for the time being. They believe that the newer dual Damascene process with copper requires newer, and more expensive equipment. The equipment is required to put down the barrier layer - typically tantalum or tantalum nitride - then a copper-seed layer, and the electroplating of the copper fill.

There a several possible impacts of copper on physical design. One impact could be to reduce the impact of interconnect on design. However, most designers feel that copper only postpones the interconnect problem by two years. Copper interconnect also may lead to reduction in total number of repeaters thereby reducing the impact on floorplan and overall convergence flow of the chip.

3.4.3.4   Unlanded Vias

The concept of unlanded via is quite simple. The via enclosures were required since wires were quite narrow and alignment methods were not accurate. For higher layers, where wire widths are wider, it is now possible to have via enclosure within the wire and, as a result, there is no need for an extended landing pad for vias. This simplifies the routing and related algorithms.

3.4.4   Innovations/Issues in Devices

In addition to performance gains in devices due to scaling, several new innovations are in store for devices as well. One notable innovation is Multi-Threshold devices. It is well known that the leakage current is inversely proportionate to the threshold voltage. As operating voltage drops, leakage current (as a ratio to operating voltage) increases. At the same time, lower V_t devices do have better performance, as a result these can be used to improve speed. In

Process Technology	0.1 u
Transistors	200 M
Logic Transistors	40 M
Size	520 mm^2
Clock frequency	2 - 3.5 GHz
Chip I/O's	4,000
Wiring levels (metals)	7 - 8
Voltage	0.9 - 1.2
Power	160 W
Supply current	160 Amps

Table 3.5: Aggressive projections for a chip in 2006.

critical paths, which might be limiting to design frequency, devices of lower V_t can be used. Process allows dual V_t devices. This is accomplished by multiple implant passes to create different implant densities. Most manufacturers plan to provide dual V_t rather than full adjustable multiple V_t, although techniques for such devices are now known.

3.4.5 Aggressive Projections for the Process

Several manufacturers have released roadmaps, which are far more aggressive than the SIA roadmap. In this section, we discuss two such projections.

Several manufacturers have indicated that the frequency projections of SIA are too conservative. Considering that several chips are already available at 600-650 MHz range and some experimental chips have been demonstrated in the one Gigahertz frequency range. It is quite possible that frequencies in the range of 2-3.5 Ghz may be possible by year the 2006. A more aggressive projection based on similar considerations is presented in Table 3.5.

Texas Instruments has recently announced a 0.10 micron process well ahead of the SIA roadmap. Drawn using 0.10-micron rules, the transistors feature an effective channel length of just 0.07-micron and will be able to pack more than 400 million transistors on a single chip, interconnected with up to seven l ayers of wiring. Operating frequencies exceeding 1 gigahertz (GHz), internal voltages as low as 1 V and below. The process uses copper and low K dielectrics. TI has also developed a series of ball grid array (BGA) packages that use fine pitch wire bond and flip chip interconnects and have pin counts ranging from 352 to 1300 pins. Packages are capable of high frequency operations in the range of 200 megahertz through more than one gigahertz. Power dissipation in these packages ranges from four watts to 150 watts. TI plans to initiate designs in the new 0.07-micron CMOS process starting in the year 2000, with volume production beginning in 2001.

3.4.6   Other Process Innovations

The slowing rate of progress in CMOS technology has made process technologists investigate its variants. The variants discussed below have little or no impact on physical design. These are being discussed here to provide a perspective on new process developments.

3.4.6.1   Silicon On Insulator

One important variant is Silicon On Insulator (SOI) technology. The key difference in SOI as compared to bulk CMOS process is the wafer preparation. In SOI process, oxygen is implanted on the wafer in very heavy doses, and then the wafer is annealed at a high temperature until a thin layer of SOI film is formed. It has been shown that there is no difference in yield between bulk and SOI wafers. Once the SOI film is made on the wafer, putting the transistor on the SOI film is straightforward. It basically goes through the same process as a similar bulk CMOS wafer. There are minor differences in the details of the process, but it uses the exact same lithography, tool set, and metalization.

There are two key benefits of SOI chips: Performance and power.

1. Performance: SOI-based chips have 20 to 25% cycle time and 25 to 35% improvement over equivalent bulk CMOS technology. This is equivalent to about two years of progress in bulk CMOS technology. The sources of increased SOI performance are elimination of area junction capacitance and elimination of "body effect" in bulk CMOS technology.

2. Low power: The ability of SOI as a low-power source originates from the fact that SOI circuits can operate at low voltage with the same performance as a bulk technology at high voltage.

Recently, fully-functional microprocessors and large static random access memory chips utilizing SOI have been developed. SOI allows designers to achieve a two-year performance gain over conventional bulk silicon technology.

3.4.6.2   Silicon Germaniun

Wireless consumer products have revolutionized the communications marketplace. In order to service this new high-volume market, faster, more powerful integrated circuit chips have been required. For many of these applications, silicon semiconductors have been pushed to the 1 to 2 GHz frequency domain. However, many new RF applications require circuit operation at frequencies up to 30 GHz, a regime well out of the realm of ordinary silicon materials. Compound III-V semiconductors (those made from elements from groups III and V in the periodic table) such as GaAs have been used successfully for such chips. These materials, however, are very expensive. Moreover, silicon technology has long proven to be very high-yielding by comparison to devices made with GaAs materials. So the chip manufacturers have chosen to enhance the performance of silicon to capitalize on the cost advantages and compete with

III-V compound semiconductor products. One such enhancement technique is to add small amounts of germanium to silicon in order to create a new material (SiGe) with very interesting semiconducting properties. Unfortunately, the germanium atom is 4% larger than the silicon atom; as a result, growing this material has been very tricky. A technique of using ultra-high vacuum chemical vapor deposition (UHV-CVD) to grow these films has proven to be key in overcoming these difficulties. The result has been the ability to produce high-performance SiGe heterojunction bipolar transistors. Conventional silicon devices have a fixed band gap of 1.12 eV (electron-Volt) which limits switching speed, compared to III-V compound materials such as GaAs. However, with the addition of germanium to the base region of a bipolar junction transistor (BJT) and grading the Ge concentration across the transistor base region, it is possible to modify the band gap to enhance performance of the silicon transistor. The potential benefits of SiGe technology are significant. One key benefit is that Silicon Germanium chips are made with the same tools as silicon chips. This means millions of dollars won't have to be invested in new semiconductor tools, as is typically the case when a shift is made from one generation of chip technology to the next. A number of circuit designs have been fabricated with SiGe technology in order to demonstrate its capability for RF chips. Among the circuits that have been measured are: voltage-controlled oscillators (VCOs), low-noise amplifiers (LNAs), power amplifiers, mixers, digital delay lines.

3.5 Solutions for Interconnect Issues

In this section, we briefly review possible options for the solution of interconnect problems.

1. **Solutions for Delay and Noise:** Several solutions exist for delay and noise problems. These include use of better interconnect material, such as copper, better planning to reduce long interconnect and use of wire size and repeaters.

   (a) Better interconnect materials: Table 3.6 gives the values of l_c at different w_c for two different materials when the value of R is 1000 Ω. The resistivity values for aluminum and polysilicon are 3 $\mu\Omega$-cm and 1000 $\mu\Omega$-cm respectively. From this table, it is clear that as compared to poly, aluminum is a far superior material for routing. Interconnect delays could also be reduced by using wider wires and inserting buffers. There is a wide range of possible values of polysilicon resistance (as shown above) for different commercial purposes. If a chip is to be run on a variety of fabrication lines, it is desirable for the circuit to be designed so that no appreciable current is drawn through a long thin line of polysilicon.

   As stated earlier, copper is the best option in terms of delay and incorporation into the leading processes.

w_c	Aluminum	Polysilicon
2.00 μm	44	0.13
1.00 μm	11	3.33 $\times 10^{-2}$
0.75 μm	6.2	1.87 $\times 10^{-2}$
0.50 μm	2.8	8.33 $\times 10^{-3}$
0.25 μm	0.69	2.08 $\times 10^{-3}$

Table 3.6: Comparison of wire length l_c (mm).

(b) Early Planning/Estimation: If interconnects needs are comprehended at an early stage, it is possible to avoid long interconnect and channels. Interconnect planning should plan all major busses and identify areas of congestion.

(c) Interconnect Sizing: Typically minimum size wire for a given layer (especially higher layer) are not capable of meeting the performance needs and they needs to be sized. Lower metal layers need to be sized typically for reliability (current carrying capacity) reasons. Several algorithms have now been proposed to optimally size (or taper) wires. Some router can size and taper on the fly.

(d) Repeater Insertion and design: If interconnect is too long, repeaters have to be inserted to meet delay or signal shape constraints. Typically, repeaters need to be planned, since they need to be placed.

(e) Shielding and Cross Talk avoidance: Crosstalk, and hence noise, can be reduced by making sure that no two conflicting lines are laid out parallel for longer than a fixed length. The other method to reduce crosstalk is to place grounded lines between signals. The effect of crosstalk and noise in MCMs is discussed in Chapter 14.

2. **Solutions for size and complexity of Interconnect** There are fundamentally three ways to deal with the sheer amount of interconnect.

(a) More metal layers: Industry has steadily increased number of routing layers. Due to CMP, it is virtually possible to add as many layers as needed. Addition of a new layer is purely a cost/benefit trade-off question and not a technology question. Addition more layers have a negative effect on the lower layers, as vias cause obstacles in lower layers. In general, benefit of each additional layer decreases, as more and more layers are added.

(b) Local interconnect: As pointed out earlier, cell density can be improved by providing local interconnect layer. Local interconnect still remain questionable due to the routing limitations. Many manufacturers prefer to add lower layers for routing.

(c) Metal Synergy: To improve utilization of metal system, metal grids may be aligned and strictly enforced. This will allow maximum number of wires routable in a given die area. However, this scheme is applicable to lower frequency designs, where performance or reliability needs do not dictate wide ranges of wire sizes.

In addition to these ideas, one may consider more exotic ideas:

(a) Pre-Routed Grids: If all the wiring is pre-assigned on a six or eight layer substrate and routing problems is reduced to selection of appropriate tracks (and routes), then we can optimize placement to maximize usage of routing resources. This is akin to programmable MCM discussed in Chapter 14. However, unlike programmable MCM, such a routing grid will be used for planning and detailing the layout and only the used part of the grid will be actually fabricated.

(b) Encoding of signals: We can encode a signal in less number of bits for long busses. This will reduce the number of bits in long busses at the cost of time to transmit the message and overhead for encoding and decoding.

(c) Optical Interconnect: Although not feasible in the near future, optical interconnect may one day provide reliable means for distributing clock, thereby freeing up routing resources.

3.6 Tools for Process Development

In earlier sections of this chapter, we have discussed the impact of process innovations on CAD tools (in particular physical design tools). However, an emerging area of research and development is related to tools that help in design of the process. Given the complexity of choices and range of possibilities that technology offers, process decisions are very complex.

Currently, process decisions are made by a set of experts based on extrapolations from previous process generations. This extra-polation methods has worked up to this point mainly because the technology did not offer too many choices. It is significantly easier to decide if we want to migrate to a four layer process from a three layer process, if the bonding method, type of vias and dielectrics are not changing.

In general, two types of tools are needed.

1. Tools for Interconnect Design: The key decisions that need to be made for any process interconnect include: number of metal layers, line widths, spacing and thickness for each layer, type and thickness of ILD, type of vias, and bonding method. CAD tools are needed that can re-layout (or estimate re-layout of) a design, based on the options listed above. For example, should we have a seven layer process or a better designed six metal layer process can only be answered if a CAD tool can re-layout a target design with six and seven metal layers and compare the results.

2. Tools for Transistor design: Complex parameters are involved in transistor design. Lack of tools to help design the transistor may lead to non-optimal design. CAD tools are needed which can simulate the operation of a transistor by changing the settings of these parameters. Such tools may allow use of less efficient transistors for better area utilization.

3.7 Summary

Physical design is deeply impacted by the fabrication process. For example, chip size is limited by a consideration of yield: the larger the chip, the greater the probability of surface flaws and therefore the slimmer the chances of finding enough good chips per wafer to make batch production profitable. A good deal of physical design effort thus goes into the development of efficient structures that lead to high packing densities. As a result, physical designer must be very educated about the fabrication process. A designer must take into account many factors, including the total number of gates, the maximum size of die, number of I/O pads and power dissipation, in order to develop chips with good yield.

The fabrication process has made tremendous strides in the past few decades and it is continuing to reduce feature size and increase both the chip size and performance. While the SIA roadmap calls for 0.01 micron process in the year 2006, many feel that it may happen much sooner. Such aggressive approaches have kept the semi-conductor industry innovating for the last three decades and promise to continue to motivate them in the next decade.

In this chapter, we have reviewed scaling methods, innovations in fabrications process, parasitic effects, and the future of fabrication process. Interconnect is the most significant problem that needs to be solved. We need faster interconnect and more interconnect to complete all the wiring required by millions of transistors. In this direction, we have noted that innovations like copper interconnect, unlanded vias, and local interconnect have a significant effect on physical design.

3.8 Exercises

1. Given a die of size 25 mm × 25 mm and $\lambda = 0.7$ μm, estimate the total number of transistors that can be fabricated on the die to form a circuit.

2. Estimate the maximum number of transistors that can be fabricated on a die of size 25 mm × 25 mm when $\lambda = 0.1$ μm.

3. Estimate the total power required (and therefore heat that needs to be removed) by a maximally packed 19 mm × 23 mm chip in 0.75 μm CMOS technology. (Allow 10% area for routing and assume 500 MHz clock frequency).

4. Assuming that the heat removal systems can only remove 80 watts from a 19 mm × 23 mm chip, compute the total number of CMOS transistors possible on such a chip, and compute the value of λ for such a level of integration. (Assume 500 MHz clock frequency.)

5. Assuming a 15 mm × 15 mm chip in 0.25 micron process where interconnect delay is 50% of the total delay, consider a net that traverses the full length of the chip diagonally. What is maximum frequency this chip can operate on ?

6. What will happen to the frequency of the chip in problem 3 if we migrate (process shift) it to 0.18 micron process? (assume 0.7 shrink factor and discuss assumptions made about the bonding pads and scribe lines which do not scale).

7. What will happen to the frequency of the chip in problem 3 if we migrate (process shift) it to 0.18 process?

Bibliographic Notes
The *International Solid-State Circuits Conference* (ISSCC) is the premier conference which deals with new developments in VLSI devices, microprocessors and memories. The *IEEE International Symposium on Circuits and Systems* also includes many papers on VLSI devices in its technical program. The *IEEE Journal of Solid-State Circuits* publishes papers of considerable interest on VLSI devices and fabrication. Microprocessor reports is a valuable source of information about process, comparisons between microprocessors and other related news of the semi-conductor industry. Several companies (such as IBM, DEC, Intel, AMD, TI) have internet sites that provide significant information about their process technology and processors.

Chapter 4

Data Structures and Basic Algorithms

VLSI chip design process can be viewed as transformation of data from HDL code in logic design, to schematics in circuit design, to layout data in physical design. In fact, VLSI design is a significant database management problem. The layout information is captured in a symbolic database or a polygon database. In order to fabricate a VLSI chip, it needs to be represented as a collection of several layers of planar geometric elements or polygons. These elements are usually limited to Manhattan features (vertical and horizontal edges) and are not allowed to overlap within the same layer. Each element must be specified with great precision. This precision is necessary since this information has to be communicated to output devices such as plotters, video displays, and pattern-generating machines. Most importantly, the layout information must be specific enough so that it can be sent to the fab for fabrication. Symbolic database captures net and transistor attributes. It allows a designer to rapidly navigate throughout the database and make quick edits while working at a higher level. The symbolic database is converted into a polygon database prior to tapeout. In the polygon database, the higher level relationship between the objects is somewhat lost. This process is analogous to conversion of a higher level programming language (say FORTRAN) code to a lower level programming language (say Assembly) code. While it is easier to work at symbolic level, it cannot be used by the fab directly. In some cases, at late stages of the chip design process, some edits have to be made in the polygon database. The major motivation for use of the symbolic database is technology independence. Since physical dimensions in the symbolic database are only relative, the design can be implemented using any process. However, in practice, complete technology independence has never be reached.

The layouts have historically been drawn by human layout designers to conform to the design rules and to perform the specified functions. A physical design specialist typically converted a small circuit into layout consisting of a set of polygons. These manipulations were time consuming and error prone, even for small layouts. Rapid advances in fabrication technology in recent years have dramatically increased the size and complexity of VLSI circuits. As a result, a single chip may include several million transistors. These technological advances have made it impossible for layout designers to manipulate the layout databases without sophisticated CAD tools. Several physical design CAD tools have been developed for this purpose, and this field is referred to as Physical Design Automation. Physical design CAD tools require highly specialized algorithms and data structures to effectively manage and manipulate layout information. These tools fall in three categories. The first type of tools help a human designer to manipulate a layout. For example, a layout editor allows designers to add transistors or nets to a layout. The second type of tools are designed to perform some task on the layout automatically. Example of such tools include channel routers and placement tools. It is also possible to invoke a tool of second type from the layout editor. The third type of tools are used for checking and verification. Example of such tools include; DRC (design rule checker) and LVS verifier (layout versus schematics verifier). The bulk of the research of physical design automation has focused on tools of the last two types. However, due to broad range and significant impact the tools of second type have received the most attention. The major accomplishment in that area has been decomposition of the physical design problem into several smaller (and conceptually easier) problems. Unfortunately, even these problems are still computationally very hard. As a result, the major focus has been on development on design and analysis of heuristic algorithms for partitioning, placement, routing and compaction. Many of these algorithms are based on graph theory and computational geometry. As a result, it is important to have a basic understanding of these two fields. In addition, several special classes of graphs are used in physical design. It is important to understand properties and algorithms about these classes of graphs to develop effective algorithms in physical design.

This chapter consists of three parts. First we discuss the basic algorithms and mathematical methods used in VLSI physical design. These algorithms form the basis for many of the algorithms presented later in this book. In the second part of this chapter, we shall study the data structures used in layout editors and the algorithms used to manipulate these data structures. We also discuss the formats used to represent VLSI layouts. In the third part of this chapter, we will focus on special classes of graphs, which play a fundamental role in development of these algorithms. Since most of the algorithms in VLSI physical design are graph theoretic in nature, we devote a large portion of this chapter to graph algorithms. In the following, we will review basic graph theoretic and computation geometry algorithms which play a significant role in many different VLSI design algorithms. Before we discuss the algorithms, we shall review the basic terminology.

4.1   Basic Terminology

A *graph* is a pair of sets $G = (V, E)$, where V is a set of vertices, and E is a set of pairs of distinct vertices called *edges*. We will use $V(G)$ and $E(G)$ to refer to the vertex and edge set of a graph G if it is not clear from the context. A vertex u is adjacent to a vertex v if (u, v) is an edge, i.e., $(u, v) \in E$. The set of vertices adjacent to v is $Adj(v)$. An edge $e = (u, v)$ is *incident* on the vertices u and v, which are the ends of e. The degree of a vertex u is the number of edges incident with the vertex u.

A complete graph on n vertices is a graph in which each vertex is adjacent to every other vertex. We use K_n to denote such a graph. A graph H is called the *complement* of graph $G = (V, E)$ if $H = (V, F)$, where, $F = E(K_{|V|}) - E$.

A graph $G' = (V', E')$ is a *subgraph* of a graph G if and only if $V' \subseteq V$ and $E' \subseteq E$. If $E' = \{(u, v) \mid (u, v) \in E \text{ and } u, v \subseteq V'\}$ then G' is a *vertex induced subgraph* of G. Unless otherwise stated, by subgraph we mean vertex induced subgraph.

A *walk* P of a graph G is defined as a finite alternating sequence $P = v_0, e_1, \ldots, e_k, v_k$ of vertices and edges, beginning and ending with vertices, such that each edge is incident with the vertices preceding and following it.

A *tour* is a walk in which all edges are distinct. A walk is called an *open walk* if the terminal vertices are distinct. A *path* is a open walk in which no vertex appears more than once.

The *length* of a path is the number of edges in it. A path is a (u, v) path if $v_0 = u$ and $v_k = v$. A *cycle* is a path of length $k, k > 2$ where $v_0 = v_k$. A cycle is called *odd* if it's length k is odd, otherwise it is an *even* cycle. Two vertices u and v in G are *connected* if G has a (u, v) path. A graph is connected if all pairs of vertices are connected. A *connected component* of G is a maximal connected subgraph of G. An edge $e \in E$ is called an *cut edge* in G if its removal from G increases the number of connected components of G by at least one. A *tree* is a connected graph with no cycles. A complete subgraph of a graph is called a *clique*.

A *directed graph* is a pair of sets $(V, \vec{E})$, where V is a set of vertices and $\vec{E}$ is a set of ordered pairs of distinct vertices, called *directed edges*. We use the notation $\vec{G}$ for a directed graph, unless it is clear from the context. A directed edge $\vec{e} = (u, v)$ is incident on u and v and the vertices u and v are called the *head* and *tail* of $\vec{e}$, respectively. $\vec{e}$ is an *in-edge* of v and an *out-edge* of u. The *in-degree* of u denoted by $d^-(u)$ is equal to the number of in-edges of u, similarly the *out-degree* of u denoted by $d^+(u)$ is equal to the number of out-edges of u. An *orientation* for a graph $G = (V, E)$ is an assignment of direction for each edge. An orientation is called *transitive* if, for each pair of edges (u, v) and (v, w), there exists an edge (u, w). If such a transitive orientation exists for a graph G, then G is called a *transitively orientable graph*. Definitions of subgraph, path, walk are easily extended to directed graphs. A *directed acyclic graph* is a directed graph with no directed cycles. A vertex u is an *ancestor* of v (and v is a *descendent* of u) if there is a (u, v) directed path in G. A *rooted tree* (or *directed tree*) is a directed acyclic graph in which all vertices have in-degree 1 except the *root*, which has in-degree 0. The root of a rooted tree T is denoted by *root(T)*. The *subtree* of tree T rooted at v is the subtree of T induced by the descendents of v. A *leaf* is a vertex in a directed acyclic graph with no descendents.

A *hypergraph* is a pair (V, E), where V is a set of vertices and E is a family of sets of vertices. A *hyperpath* is a sequence $P = v_0, e_1, \ldots, v_{k-1}, e_k, v_k$ of distinct vertices and distinct edges, such that vertices v_{i-1} and v_i are elements of the edge $e_i, 1 \le i \le k$. Two vertices u and v are connected in a hypergraph if the hypergraph has a (u, v) hyperpath. A hypergraph is connected if every pair of vertices are connected.

A *bipartite graph* is a graph G whose vertex set can be partitioned into two subsets X and Y, so that each edge has one end in X and one end in Y; such a partition (X, Y) is called *bipartition* of the graph. A *complete bipartite graph* is a bipartite graph with bipartition (X, Y) in which each vertex of X is adjacent to each vertex of Y; if $|X| = m$ and $|Y| = n$, such a graph is denoted by $K_{m,n}$. An important characterization of bipartite graphs is in terms of odd cycles. A graph is bipartite if and only if it does not contain an odd cycle.

A graph is called *planar* if it can be drawn in the plane without any two edges crossing. Notice that there are many different ways of 'drawing' a planar graph. A drawing may be obtained by mapping a vertex to a point in the plane and mapping edges to paths in the plane. Each such drawing is called an *embedding* of G. An embedding divides the plane into finite number of regions. The edges which bound a region define a *face*. The unbounded region is called the *external* or *outside face*. A face is called an odd face if it has odd number of edges. Similarly a face with even number of edges is called an even face. The *dual* of a planar embedding T is a graph $G_T = (V_T, E_T)$, such the $V_T = \{v | \text{ v is a face in } T\}$ and two vertices share an edge if their corresponding faces share an edge in T.

4.2   Complexity Issues and NP-hardness

Several general algorithms and mathematical techniques are frequently used to develop algorithms for physical design.   While the list of mathematical and algorithmic techniques is quite extensive, we will only mention the basic techniques. One should be very familiar with the following techniques to appreciate various algorithms in physical design.

1. Greedy Algorithms

2. Divide and Conquer Algorithms

3. Dynamic Programming Algorithms

4. Network Flow Algorithms

5. Linear/Integer Programming Techniques

Since these techniques and algorithms may be found in a good computer science or graph algorithms text, we omit the discussion of these techniques and refer the reader to an excellent text on this subject by Cormen, Leiserson and Rivest [CLR90].

The algorithmic techniques mentioned above have been applied to various problems in physical design with varying degrees of success. Due to the very large number of components that we must deal with in VLSI physical design automation, all algorithms must have low time and space complexities. For algorithms which must operate on the entire layout ($n \geq 10^6$), even quadratic algorithms may be intolerable. Another issue of great concern is the constants in the time complexity of algorithms. In physical design, the key idea is to develop practical algorithms, not just polynomial time complexity algorithms. As a result, many linear and quadratic algorithms are infeasible in physical design due to large constants in their complexity.

The major cause of concern is absence of polynomial time algorithms for majority of the problems encountered in physical design automation. In fact, there is evidence that suggests that no polynomial time algorithm may exist for many of these problems. The class of solvable problems can be partitioned into two general classes, P and NP. The class P consists of all problems that can be solved by a deterministic turing machine in polynomial time. A conventional computer may be viewed as such a machine. Minimum cost spanning tree, single source shortest path, and graph matching problems belong to class P. The other class called NP, consists of problems that can be solved in polynomial time by a nondeterministic turing machine. This type of turing machine may be viewed as a parallel computer with as many processors as we may need. Essentially, whenever a decision has several different outcomes, several new processors are started up, each pursuing the solution for one of the outcomes. Obviously, such a model is not very realistic. If every problem in class NP can be reduced to a problem **P**, then problem **P** is in class NP-complete. Several thousand problems in computer science, graph theory, combinatorics, operations research, and computational geometry have been proven to be NP-complete. We will not discuss the concept of NP-completeness in detail, instead, we refer the reader to the excellent text by Garey and Johnson on this subject [GJ79]. A problem may be stated in two different versions. For example, we may ask does there exist a subgraph H of a graph G, which has a specific property and has size k or bigger? Or we may simply ask for the largest subgraph of G with a specific property. The former type is the decision version while the latter type is called the optimization version of the problem. The optimization version of a problem **P**, is called NP-hard if the decision version of the problem **P** is NP-complete.

4.2.1 Algorithms for NP-hard Problems

Most optimization problems in physical design are NP-hard. If a problem is known to be NP-complete or NP-hard, then it is unlikely that a polynomial time algorithm exists for that problem. However, due to practical nature of the

physical design automation field, there is an urgent need to solve the problem
even if it cannot be solved optimally. In such cases, algorithm designers are
left with the following four choices.

4.2.1.1   Exponential Algorithms

If the size of the input is small, then algorithms with exponential time complexity may be feasible. In many cases, the solution of a certain problem be
critical to the performance of the chip and therefore it is practical to spend
extra resources to solve that problem optimally. One such exponential method
is integer programming, which has been very successfully used to solve many
physical design problems. Algorithms for solving integer programs do not have
polynomial time complexity, however they work very efficiently on moderate
size problems, while worst case is still exponential. For large problems, algorithms with exponential time complexity may be used to solve small sub-cases,
which are then combined using other algorithmic techniques to obtain the global
solution.

4.2.1.2   Special Case Algorithms

It may be possible to simplify a general problem by applying some restrictions
to the problem. In many cases, the restricted problem may be solvable in
polynomial time. For example, the graph coloring problem is NP-complete for
general graphs, however it is solvable in polynomial time for many classes of
graphs which are pertinent to physical design.

Layout problems are easier for simplified VLSI design styles such as standard cell, which allow usage of special case algorithms. Conceptually, it is
much easier to place cells of equal heights in rows, rather than placing arbitrary sized rectangles in a plane. The clock routing problem, which is rather
hard for full-custom designs, can be solved in $O(n)$ time for symmetric structures such as gate arrays. Another example may be the Steiner tree problem
(see Section 4.3.1.6). Although the general Steiner tree problem is NP-hard,
a special case of the Steiner tree problem, called the single trunk steiner tree
problem (see exercise 4.7), can be solved in $O(n)$ time.

4.2.1.3   Approximation Algorithms

When exponential algorithms are computationally infeasible due to the size
of the input and special case algorithms are ruled out due to absence of any
restriction that may be used, designers face a real challenge. If optimality is not
necessary and near-optimality is sufficient, then designers attempt to develop an
approximation algorithm. Often in physical design algorithms, near-optimality
is good enough. Approximation algorithms produce results with a guarantee.
That is, while they may not produce an optimal result, they guarantee that the
result would never be worse than a lower bound determined by the *performance
ratio* of the algorithm. The performance ratio γ of an algorithm is defined as
$\frac{\Phi}{\Phi^*}$, where Φ is the solution produced by the algorithm and Φ^* is the optimal

Algorithm AVC
begin
    $S = \phi$;
    $E' = E$;
    $R = \phi$;
    while $E' \neq \phi$ **do**
        (* Select an arbitrary edge $e = (u, v)$ from E' *)
        $e = \text{Select}(E')$;
        $S = S \cup \{u, v\}$;
        $R = R \cup e$;
        $E' = E' - \{(p, q) \mid (p = u \text{ or } v) \text{ or } (q = u \text{ or } v)\}$;
end.

Figure 4.1: Algorithm AVC.

solution for the problem. Recently, many algorithms have been developed with γ very close to 1. We will use *vertex cover* as an example to explain the concept of an approximation algorithm.

Vertex cover is basically a problem of covering all edges by using as few vertices as possible. In other words, given an undirected graph $G = (V, E)$, select a subset $V' \subseteq V$, such that for each edge $(u, v) \in E$, either u or v or both are in V' and V' has minimum size among all such sets. The vertex cover problem is known to be NP-complete for general graphs. However, the simple algorithm given in Figure 4.1 achieves near optimal results. The basic idea is to select an arbitrary edge (u, v) and delete it and all edges incident on u and v. Add u and v to the vertex cover set S. Repeat this process on the new graph until all edges are deleted. The selected edges are kept in a set R.

Since no edge is checked more than once, it is easy to see that the algorithm AVC runs in $O(|E|)$ time.

Theorem 1 *Algorithm AVC produces a solution with a performance ratio of 0.5.*

Proof: Note that no two edges in R have a vertex in common, and $|S| = 2 \times |R|$. However, since R is set of vertex disjoint edges, at least $|R|$ vertices are needed to cover all the edges. Thus $\gamma \geq \frac{|R|}{2|R|} = 0.5$.

In Section 4.5.6.2, we will present an approximation algorithm for finding maximum k-partite ($k \geq 2$) subgraph in circle graphs. That algorithm is used in topological routing, over-the-cell routing, via minimization and several other physical design problems.

4.2.1.4 Heuristic Algorithms

Faced with NP-complete problems, heuristic algorithms are frequently the answer. A heuristic algorithm does produce a solution but does not guarantee the optimality of the solution. Such algorithms must be tested on various benchmark examples to verify their effectiveness. The bulk of research in physical design has concentrated on heuristic algorithms. An effective heuristic algorithm must have low time and space complexity and must produce an optimal or near optimal solution in most realistic problem instances. Such algorithms must also have good average case complexities. Usually good heuristic algorithms are based on optimal algorithms for special cases and are capable of producing optimal solutions in a significant number of cases. A good example of such algorithms are the channel routing algorithms (discussed in chapter 7), which can solve most channel routing problems using one or two tracks more than the optimal solution. Although the channel routing problem in general is NP-complete, from a practical perspective, we can consider the channel routing problem as *solved*.

In many cases, an $O(n)$ time complexity heuristic algorithm has been developed, even if an optimal $O(n^3)$ or $O(n^2)$ time complexity algorithm is known for the problem. One must keep in mind that optimal solutions may be hard to obtain and may be practically insignificant if a solution close to optimal can be produced in a reasonable time. Thus the major focus in physical design has been on the development of practical heuristic algorithms which can produce close to optimal solutions on real world examples.

4.3   Basic Algorithms

Basic algorithms which are frequently used in physical design as subalgorithms can be categorized as: graph algorithms and computational geometry based algorithms. In the following, we review some of the basic algorithms in both of these categories.

4.3.1   Graph Algorithms

Many real-life problems, including VLSI physical design problems, can be modeled using graphs. One significant advantage of using graphs to formulate problems is that the graph problems are well-studied and well-understood. Problems related to graphs include graph search, shortest path, and minimum spanning tree, among others.

4.3.1.1   Graph Search Algorithms

Since many problems in physical design are modeled using graphs, it is important to understand efficient methods for searching graphs. In the following, we briefly discuss the three main search techniques.

1. **Depth-First Search:** In this graph search strategy, graph is searched 'as deeply as possible'. In Depth-First Search (DFS), an edge is selected for further exploration from the most recently visited vertex v. When all the edges of v have been explored, the algorithm back tracks to the

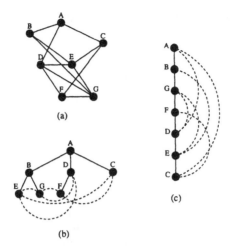

Figure 4.2: Examples of graph search algorithms.

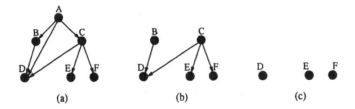

Figure 4.3: Example of the topological search.

previous vertex, which may have an unexplored edge. Figure 4.4 is an outline of a depth-first search algorithm. The algorithm uses an array MARKED(n) which is initialized to zero before calling the algorithm to keep track of all the visited vertices.

It is easy to see that the time complexity of depth-first search is $O(|V| + |E|)$. Figure 4.2(c) shows an example of the depth first search for the graph shown in Figure 4.2(a).

2. **Breadth-First Search:** The basic idea of Breadth-First Search (BFS) is to explore all vertices adjacent to a vertex before exploring any other vertex. Starting with a source vertex v, the BFS first explores all edges of v, puts the reachable vertices in a queue, and marks the vertex v as visited. If a vertex is already marked visited then it is not enqueued. This process is repeated for each vertex in the queue. This process of visiting edges produces a BFS tree. The BFS algorithm can be used to search both directed and undirected graphs. Note that the main difference be-

Algorithm DEPTH-FIRST-SEARCH(v)
begin
   MARKED(v) = 1;
   for each vertex u, such that $(u, v) \in E$ **do**
      if MARKED(u) = 0 **then**
         DEPTH-FIRST-SEARCH(u);
end.

Figure 4.4: Algorithm DEPTH-FIRST-SEARCH.

tween the DFS and the BFS is that the DFS uses a stack (recursion is implemented using stacks), while the BFS uses a queue as its data structure. The time complexity of breadth first search is also $O(|V| + |E|)$. Figure 4.2(b) shows an example of the BFS of the graph shown in Figure 4.2(a).

3. **Topological Search:** In a directed acyclic graph, it is very natural to visit the parents, before visiting the children. Thus, if we list the vertices in the topological order, if G contains a directed edge (u, v), then u appears before v in the topological order. Topological search can be done using depth first search and hence it has a time complexity of $O(|V| + |E|)$. Figure 4.3 shows an example of the topological search. Figure 4.3(a) shows an entire graph. First vertex A will be visited since it has no parent. After visiting A, it is deleted (see Figure 4.3(b)) and we get vertices B and C as two new vertices to visit. Thus, one possible topological order would be A, B, C, D, E, F.

4.3.1.2 Spanning Tree Algorithms

Many graph problems are subset selection problems, that is, given a graph $G = (V, E)$, select a subset $V' \subseteq V$, such that V' has property $\mathcal{P}$. Some problems are defined in terms of selection of edges rather than vertices. One frequently solved graph problem is that of finding a set of edges which spans all the vertices. The Minimum Spanning Tree (MST) is an edge selection problem. More precisely, given an edge-weighted graph $G = (V, E)$, select a subset of edges $E' \subseteq E$ such that E' induces a tree and the total cost of edges $\sum_{e_i \in E'} wt(e_i)$, is minimum over all such trees, where $wt(e_i)$ is the cost or weight of the edge e_i.

There are basically three algorithms for finding a MST:

1. Boruvka's Algorithm

2. Kruskal's Algorithm

3. Prim's Algorithm.

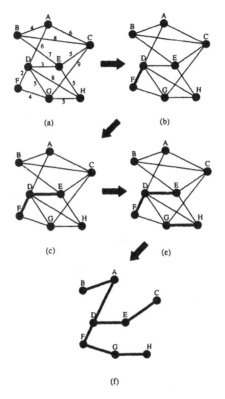

Figure 4.5: Kruskal's algorithm used to find a MST.

We will briefly explain Kruskal's algorithm [Kru56], whereas the details of other algorithms can be found in [Tar83]. Kruskal's algorithm starts by sorting the edges by nondecreasing weight. Each vertex is assigned to a set. Thus at the start, for a graph with n vertices, we have n sets. Each set represents a partial spanning tree, and all the sets together form a spanning forest. For each edge (u, v) from the sorted list, if u and v belong to the same set, the edge is discarded. On the other hand, if u and v belong to disjoint sets, a new set is created by union of these two sets. This edge is added to the spanning tree. In this way, algorithm constructs partial spanning trees and connects them whenever an edge is added to the spanning tree. The running time of Kruskal's algorithm in $O(|E| \log |E|)$. Figure 4.5 shows an example of Kruskal's algorithm. First edge (D, F) is selected since it has the lowest weight (See Figure 4.5(b)). In the next step, there are two choices since there are two edges with weight equal to 2. Since ties are broken arbitrarily, edge (D, E) is selected. The final tree is shown in Figure 4.5(f).

4.3.1.3   Shortest Path Algorithms

Many routing problems in VLSI physical design are in essence shortest path problems in special graphs. Shortest path problems, therefore, play a significant role in global and detailed routing algorithms.

1. **Single Pair Shortest Path:** This problem may be viewed as a vertex or edge selection problem. Precisely stated, given an edge-weighted graph $G = (V, E)$ and two vertices $u, v \in V$, select a set of vertices $P \subseteq V$ including u, v such that P induces a path of minimum cost in G. Let $wt(p, q)$ be the weight of edge (p, q), we assume that $wt(p, q) \geq 0$ for each $(p, q) \in E$.

   Dijkstra [Dij59] developed an $O(n^2)$ algorithm for single pair shortest path, where n is the number of vertices in the graph. In fact, Dijkstra's algorithm finds shortest paths from a given vertex to all the vertices in the graph. See Figure 4.6 for a description of the shortest path algorithm. Figure 4.7(a) shows an edge weighted graph while Figure 4.7(b) shows the shortest path between vertices B and F found by Dijkstra's algorithm.

2. **All Pairs Shortest Paths:** This problem is a variant of SPSP, in which the shortest path is required for all possible pairs in the graph. There are a few variations of all pairs shortest path algorithms for directed graphs. Here we discuss the Floyd-Warshall algorithm which runs in $O(|V|^3)$ time and is based on a dynamic programming technique.

   The algorithm is based on the following observation. Given a directed graph $G = (V, E)$, let $V = \{v_1, v_2, \ldots, v_n\}$. Consider a subset $V' = \{v_1, v_2, \ldots, v_k\} \subseteq V$ for some k. For any pair of vertices $v_i, v_j \in V$, consider all paths from v_i to v_j with intermediate vertices from V' and let p be the the one with minimum weight (an *intermediate vertex* of a path $p = (v_1, v_2, \ldots, v_l)$ is any vertex of p other than v_1 and v_l). The Floyd-Warshall algorithm exploits the relationship between path p and the shortest path from v_i to v_j with intermediate vertices from $\{v_1, v_2, \ldots, v_{k-1}\}$. Let $d_{ij}^{(k)}$ be the weight of a shortest path from v_i to v_j with all intermediate vertices from $\{v_1, v_2, \ldots, v_k\}$. For $k = 0$, a path from v_i to v_j is one with no intermediate vertices, thus having at most one edge, hence $d_{ij}^{(0)} = wt(v_i, v_j)$. A recursive formulation of all pairs shortest path problem can therefore be given as:

$$d_{ij}^{(k)} = \begin{cases} wt(v_i, v_j) & \text{if } k = 0 \\ \min\{d_{ij}^{(k-1)}, d_{ik}^{(k-1)} + d_{kj}^{(k-1)}\} & \text{if } k \geq 1 \end{cases}$$

   The all pairs shortest path problem allows many paths to share an edge. If we restrict the number of paths that can use a particular edge, then the all pairs shortest path problem becomes NP-hard. The all pairs shortest path problem plays a key role in the global routing phase of physical design.

Algorithm SHORTEST-PATH(u)
begin
for $i = 1$ **to** n **do**
 if $((u, i) \in E)$ **then** $D[i] = wt(u, i)$;
 else $D[i] = +\infty$;
 $P[i] = u$;
$V' = V - u$; $D[u] = 0$;
while $(|V'| > 0)$ **do**
 Select v such that $D[v] = \min_{w \in V'} D[w]$;
 $V' = V' - v$;
 for $w \in V'$ **do**
  if $(D[w] > D[v] + wt(v, w))$ **then**
   $D[w] = D[v] + wt(v, w)$;
   (* $D[w]$ is the length of the shortest path from u to w. *)
   $P[w] = v$;
   (* $P[w]$ is the parent of w. *)
for $w \in V$ **do**
 (* print the shortest path from w to u. *)
 $q = w$;
 print q;
 while $(q \neq u)$ **do**
  $q = P[q]$;
  print q;
 print q;
end.

Figure 4.6: Algorithm SHORTEST-PATH.

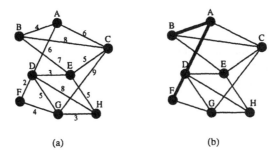

(a)   (b)

Figure 4.7: Single pair shortest path between vertices B and F.

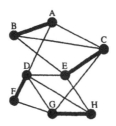

Figure 4.8: Matching.

4.3.1.4 Matching Algorithms

Given an undirected graph $G = (V, E)$, a *matching* is a subset of edges $E' \subseteq E$ such that for all vertices $v \in V$, at most one edge of E' is incident on v. A vertex is said to be *matched* by matching E' if some edge in E' is incident on v; otherwise v is unmatched. A *maximum matching* is a matching with maximum cardinality among all matchings of a graph, i.e., if E' is a maximum matching in G, then for any other matching E'' in G, $|E'| \geq |E''|$. Figure 4.8 shows an example of matching. A matching is called a *bipartite matching* if the underlying graph is a bipartite graph. Both matching and bipartite matching have many important applications in VLSI physical design. The details of different matching algorithms may be found in [PS82].

4.3.1.5 Min-Cut and Max-Cut Algorithms

Min-cut and Max-cut are two frequently used graph problems which are related to partitioning the vertex set of a graph.

The simplest Min-cut problem can be defined as follows: Given a graph $G = (V, E)$, partition V into subsets V_1 and V_2 of equal sizes, such that the number of edges $E' = \{(u, v) | u \in V_1, v \in V_2\}$ is minimized. The set E' is also referred to as a *cut*. A more general min-cut problem may specify the sizes of subsets, or it may require partitioning V into k different subsets. The min-cut problem is NP-complete [GJ79]. Min-cut and many of its variants have several applications in physical design, including partitioning and placement.

The Max-cut problem can be defined as follows: Given a graph $G = (V, E)$, find the maximum bipartite graph of G. Let $G' = (V, E')$ be the maximum bipartite of G, which is obtained by deleting K edges of G, then G has a max-cut of size $|E| - K$.

Max-cut problem is NP-complete [GJ79]. Hadlock [Had75] presented an algorithm which finds max-cut of a planar graph. The algorithm is formally presented in Figure 4.9. Procedure PLANAR-EMBED finds a planar embedding of G, and CONSTRUCT-DUAL creates a dual graph for the embedding. Procedure CONSTRUCT-WT-GRAPH constructs a complete weighted graph by using only vertices corresponding to odd faces. The weight on the edge (u, v) indicates the length of the shortest path between vertices u and v in G_F.

> **Algorithm** MAXCUT
> **begin**
> F = PLANAR-EMBED(G);
>   Let $f_1, f_2, \ldots, f_k$ be the faces of the F;
> G_F = CONSTRUCT-DUAL(F);
>   Let R be the set of vertices of odd degree in G_F and let
> Q be the set of all pairs consisting of vertices in R;
> G_W = CONSTRUCT-WT-GRAPH(R, Q);
> M = MIN-WT-MATCHING(G_W);
>   Using M find a set of paths, one for each of the matched pairs
>   in G_F;
>   Each path determines a set of edges whose deletion leaves the
>   graph G bipartite;
> **end.**

Figure 4.9: Algorithm MAXCUT.

Note that the number of odd faces in any planar graph is even. Procedure MIN-WT-MATCHING pairs up the vertices in R. Each edge in matching represents a path in G. This path actually passes through even faces and connects two odd faces. All edges on the path are deleted. Notice that this operation creates a large even face. This edge deletion procedure is repeated for each matched edge in M. In this way, all odd faces are removed. The resulting graph is bipartite.

Consider the example graph shown in Figure 4.10(a). The dual of the graph is shown in Figure 4.10(b). The minimum weight matching of cost 4 is (3, 13) and (5, 10). The edges on the paths corresponding to the matched edges, in M, have been deleted and the resultant bipartite graph is shown in Figure 4.10(d).

4.3.1.6 Steiner Tree Algorithms

Minimum cost spanning trees and single pair shortest paths are two edge selection problems which can be solved in polynomial time. Surprisingly, a simple variant of these two problems, called the Steiner minimum tree problem, is computationally hard.

The Steiner Minimum Tree (SMT) problem can be defined as follows: Given an edge weighted graph $G = (V, E)$ and a subset $D \subseteq V$, select a subset $V' \subseteq V$, such that $D \subseteq V'$ and V' induces a tree of minimum cost over all such trees.

The set D is referred to as the set of *demand points* and the set $V' - D$ is referred to as *Steiner points*. In terms of VLSI routing the demand points are the net terminals. It is easy to see that if $D = V$, then SMT is equivalent to MST, on the other hand, if $|D| = 2$ then SMT is equivalent to SPSP. Unlike MST and SPSP, SMT and many of its variants are NP-complete [GJ77]. In view of the NP-completeness of the problem, several heuristic algorithms have

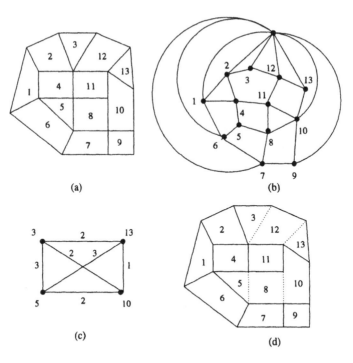

Figure 4.10: An example of finding a max-cut for a planar graph.

been developed.

Steiner trees arise in VLSI physical design in routing of multi-terminal nets. Consider the problem of interconnecting two points in a plane using the shortest path. This problem is similar to the routing problem of a two terminal net. If the net has more than two terminals then the problem is to interconnect all the terminals using minimum amount of wire, which corresponds to the minimization of the total cost of edges in the Steiner tree. The global and detailed routing of multi-terminal nets is an important problem in the layout of VLSI circuits. This problem has traditionally been viewed as a Steiner tree problem [CSW89, HVW85]. Due to their important applications, Steiner trees have been a subject of intensive research [CSW89, GJ77, Han76, HVW85, HVW89, Hwa76b, Hwa79, LSL80, SW90]. Figure 4.11(b) shows a Steiner tree connecting vertices A, I, F, E, and G of Figure 4.11(a).

The *underlying grid graph* is the graph defined by the intersections of the horizontal and vertical lines drawn through the demand points. The problem is then to connect terminals of a net using the edges of the underlying grid graph. Figure 4.12 shows the underlying grid graph for a set of four points. Therefore Steiner tree problems are defined in the Cartesian plane and edges are restricted to be rectilinear. A Steiner tree whose edges are constrained to rectilinear shapes is called a *Rectilinear Steiner Tree* (RST). A *Rectilinear*

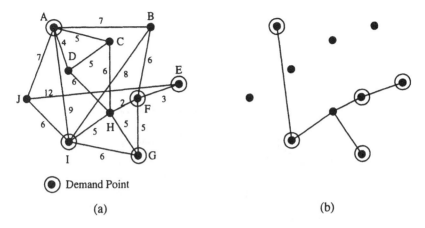

Figure 4.11: A Steiner tree.

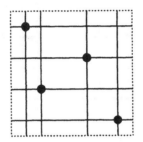

Figure 4.12: The formation of the underlying grid.

Steiner Minimum Tree (RSMT) is an RST with minimum cost among all RSTs. In fact, the MST and RSMT have an interesting relationship. In the MST, the cost of the edges are evaluated in the rectilinear metric. The following theorem was proved by Hwang [Hwa76b].

Theorem 2 *Let $COST_{MST}$ and $COST_{RSMT}$ be the costs of a minimum cost rectilinear spanning tree and rectilinear Steiner minimum tree, respectively. Then*

$$\frac{COST_{MST}}{COST_{RSMT}} \leq \frac{3}{2}$$

As a result, many heuristic algorithms use MST as a starting point and apply local modifications to obtain an RST. In this way, these algorithms can guarantee that weight of the RST is at most $\frac{3}{2}$ of the weight of the optimal tree [HVW85, Hwa76a, Hwa79, LSL80]. Consider the example shown in Figure 4.13. In Figure 4.13(a), we show a minimum spanning tree for the set of

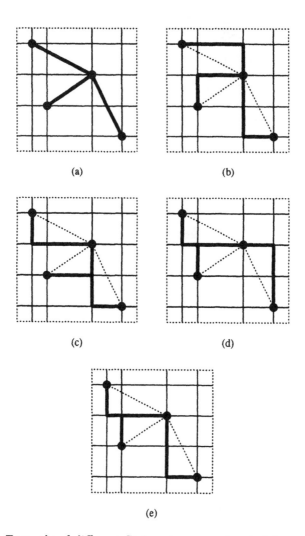

Figure 4.13: Example of different Steiner trees constructed from a minimum cost spanning.

four points. Figure 4.13(b), (c), (d), and (e) show different Steiner trees that can be obtained by using different layouts of edges of spanning tree. Layout of edges should be selected so as to maximize the overlap between layouts and hence minimize the total length of the tree. Figure 4.13(e) shows a minimum cost Steiner tree.

4.3.2 Computational Geometry Algorithms

One of the basic tasks in computational geometry is the computation of the line segment intersections. Investigations to solve this problem have continued for several decades, with the domain expanding from simple segment intersections to intersections between geometric figures. The problem of detecting these types of intersections has practical applications in several areas, including VLSI physical design and motion-planning in robotics.

4.3.2.1 Line Sweep Method

The detailed description of line sweep method and its many variations can be found in [PS85]. A brief description of the line sweep method is given in this section. The n line segments are represented by their $2n$ endpoints, which are sorted by increasing x-coordinate values. An imaginary vertical *sweep line* traverses the endpoint set from left to right, halting at each x-coordinate in the sorted list. This sweep line represents a listing of segments at a given x-coordinate, ordered according to their y-coordinate value. If the point is a left endpoint, the segment is inserted into a data structure which keeps track of the ordering of the segment with respect to the vertical line. The inserted segment is checked with its immediate top and bottom neighbors for an intersection. An intersection is detected when two segments are consecutive in order. If the point is a right endpoint, a check is made to determine if the segments immediately above and below it intersect; then this segment is deleted from the ordering. This algorithm halts when it detects one intersection, or has traversed the entire set of endpoints and no intersection exists. Consider the example shown in Figure 4.15. The intersection between segments A and C will be detected when segment B is deleted and A and C will become consecutive. A description of the line sweep algorithm is given in Figure 4.14.

The sorting of $2n$ endpoints can be done in $O(n \log n)$ time. A balanced tree structure is used for T, which keeps the y-order of the active segments; this allows the operations *INSERT, DELETE, ABOVE,* and *BELOW* to be performed in $O(\log n)$ time. Since the **for** loop executes at most $2n$ times, the time complexity of the algorithm is $O(n \log n)$.

4.3.2.2 Extended Line Sweep Method

The line sweep algorithm can be extended to report all K intersecting pairs found among n line segments. The extended line sweep method performs the line sweep with the vertical line, inserting and deleting the segments in the tree R; but when a segment intersection is detected, the point of intersection

Algorithm LINE-SWEEP
begin
    Sort the endpoints lexicographically on
        x-coordinates so that Point[1] is leftmost and
        Point[2n] is rightmost;
    for $i = 1$ to $2n$ **do**
        (* Let S be the segment of which P is an endpoint *)
        $P = $ Point[i];
        if P is the left endpoint of S **then**
            INSERT (S, T);
            $A = $ ABOVE(S, T);
            $B = $ BELOW(S, T);
            if A intersects S **then** return (A, S);
            if B intersects S **then** return (B, S);
        else   (* P is the right endpoint of S *)
            $A = $ ABOVE(S, T);
            $B = $ BELOW(S, T);
            if A intersects B **then** return (A, B);
            Delete(S, T);
end.

Figure 4.14: Algorithm LINE-SWEEP.

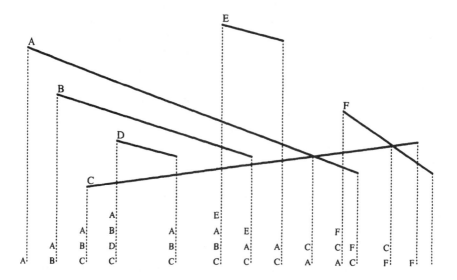

Figure 4.15: Line sweep example.

is inserted into the heap Q of sorted endpoints, in its proper x-coordinate value order. The sweep line will then also halt at this point, the intersection is reported, and the order of the intersecting segments in R is swapped. New intersections between these swapped segments and their nearest neighbors are checked and points inserted into Q if this intersection occurs. This algorithm halts when all endpoints and intersections in Q have traversed by the sweep line, and all intersecting pairs are reported.

The special case in which the line segments are either all horizontal or vertical lines was also discussed. A horizontal line is specified by its y-coordinate and by the x-coordinates of its left and right endpoints. Each vertical line is specified by its x-coordinate and by the y-coordinates of its upper and lower endpoints. These points are sorted in ascending x-coordinates and stored in Q.

The line sweep proceeds from left to right; when it encounters the left endpoint of a horizontal segment S, it inserts the left endpoint into the data structure R. When a vertical line is encountered, we check for intersections with any horizontal segments in R, which lie within the y-interval defined by the vertical line segment.

4.4 Basic Data Structures

A layout editor is a CAD tool which allows a human designer to create and edit a VLSI layout. It may have some semi-automatic features to speed up the layout process. Since layout editors are interactive, their underlying data structures should enable editing operations to be performed within an acceptable response time. The data structures should also have an acceptable space complexity, as the memory on workstations is limited.

A layout can be represented easily if partitioned into a collection of *tiles*. A tile is a rectangular section of the layout within a single layer. The tiles are not allowed to overlap within a layer. The elements of a layout are referred to as *block tiles*. A block tile can be used to represent p-diffusion, n-diffusion, poly segment, etc. For ease of presentation, we will refer to a block tile simply as a *block*. The area within a layout that does not contain a block is referred to as *vacant space*. Figure 4.16 shows a simple layout containing several blocks. Later, in the chapter we will introduce a method that partitions the vacant space into a series of *vacant tiles*.

4.4.1 Atomic Operations for Layout Editors

The basic set of operations that give a designer the freedom to fully manipulate a layout, is referred to as the *Atomic Operations*. The following is the list of atomic operations that a layout editor must support.

1. **Point Finding:** Given the coordinate of a point $p = (x, y)$, determine whether p lies within a block and, if so, identify that block.

2. **Neighbor Finding:** This operation is used to determine all blocks touching a given block B.

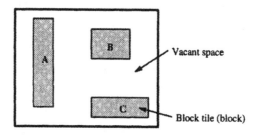

Figure 4.16: Block tile (block) representation in a layout.

3. **Block Visibility:** This operation is used to determine all blocks visible, in the x and y direction, from a given block B. Note that this operation is different from the neighbor finding operation.

4. **Area Searching:** Given a fixed area A, defined by its upper left corner (x, y), the length l, and the width w, determine whether A intersects with any blocks B. This operation is very useful in the placement of blocks. Given a block to be placed in a particular area, we need to check if other blocks are currently residing in that area.

5. **Directed Area Enumeration:** Given a fixed area A, defined by its upper left corner (x, y), length l, and width w, visit each block intersecting A exactly once in sorted order according to its distance from a given side (top, bottom, left, or right) of A.

6. **Block Insertion:** Block insertion refers to the act of inserting a new block B into the layout such that B does not intersect with any existing block.

7. **Block Deletion:** This operation is used to remove an existing block B from the layout. In an iterative approach to placement, blocks are moved from one location to another. This is done by inserting the block into a new location and then deleting the block at its previous location.

8. **Plowing:** Given an area A and a direction d, remove all blocks B_i from A by shifting each B_i in direction d while preserving ordering between the blocks.

9. **Compaction:** Compaction refers to plowing or "compressing" of the entire layout. If compaction is along the x-axis or y-axis then the compaction is called *1-dimensional compaction*. When the compaction is carried out in both x- and y-direction then it is called *2-dimensional compaction*.

10. **Channel Generation:** This operation refers to determining the vacant space in the layout and partitioning it into tiles.

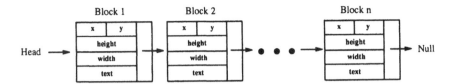

Figure 4.17: Linked list representation.

4.4.2   Linked List of Blocks

The simplest data structure used to store the components of a layout is a linked list, where each node in the list represents a block. The linked list representation is shown in Figure 4.17. Notice that the blocks are not stored in any particular order, as none was specified in the original description of the data structure; however, it is clear that a sorted or self-organizing list will improve average algorithmic complexity for some of the atomic operations. The space complexity of the linked list method is $O(n)$ where n is the number of blocks in the layout.

For illustration purpose, we now present an algorithm for neighbor finding using the linked list data structure. Given a block B, the neighbors of B are all the blocks that share a side with B. Each block is represented in the list by its location (coordinate of upper left corner), height, width, text to describe the block, and a pointer to the next node (block) in the list. The algorithm finds the neighbors on the right side of the given block, however, the algorithm can be easily modified to find all the neighbors of a given block. The input to the algorithm is a specific block B, and the linked list L of all blocks. A formal description of the algorithm for neighbor finding is shown in Figure 4.18.

Linked list data structures are suitable for a hierarchical system since each level of hierarchy contains few blocks. However, this data structure is not suitable for non-hierarchical systems and for hierarchical systems with a large number of blocks in each level. The major disadvantage of this structure is that it does not explicitly represent the vacant space. However, the data structure can be altered to form a new representation of the layout in which the vacant space is stored as a collection of *vacant tiles*. A vacant tile maintains the same geometric restrictions that are implied by the definition of a tile. Converting the vacant space into a collection vacant tiles can be done by extending the upper and lower boundaries of each block horizontally to the left and to the right until it encounters another block or the boundary of the layout as shown in figure 4.19. This partitions the entire area into a collection of tiles (block and vacant), organizing the vacant tiles into *maximal horizontal strips*, thus allowing the entire area of the layout to be represented in the linked list data structure. We call this data structure the *modified linked list*.

Algorithm NEIGHBOR-FIND1(B, L)
begin
    $neighbor\text{-}list = \phi$;
    $x1 = B.x + B.width$;
    $y11 = B.y$;
    $y12 = B.y + B.height$;
    for each block $R \in L$ such that $R \neq B$ **do**
        $y21 = R.y$;
        $y22 = R.y + R.height$;
        if ($x1 = R.x$ **and** ($y11 \leq y21 \leq y12$ **or**
                $y11 \leq y22 \leq y12$ **or**
                $y21 < y11 < y12 < y22$)) **then**
            INSERT(R, $neighbor\text{-}list$);
    return $neighbor\text{-}list$;
end.

Figure 4.18: Algorithm NEIGHBOR-FIND1.

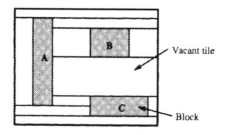

Figure 4.19: Vacant tile creation.

4.4.3   Bin-Based Method

The bin-based data structure does not keep information pertaining to the vacant space and does not create vacant tiles. In the bin-based system, a virtual grid is superimposed on the layout area as shown in Figure 4.20. The grid divides the area into a series of bins which can be represented using a two-dimensional array. Each element in the array ($\mathcal{B}$(row,col)) contains all the blocks that intersect with the corresponding bin. In Figure 4.20, $\mathcal{B}(2,3)$ contain blocks D,E,F,G, and H while $\mathcal{B}(2,4)$ contain blocks G and H. The space complexity for the bin-based data structure is $O(bn)$, where b denotes the total number of bins and n is the number of blocks.

Clearly, the bin-based data structure can be viewed as an augmented version of the linked list data structure in which a time-space compromise has been made to improve the average case performance on several of the atomic

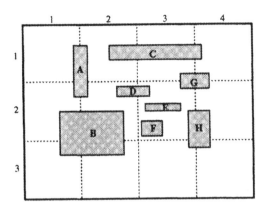

Figure 4.20: Bin-based representation.

operations. However, it is easy to construct pathological examples which cause the worst-case performance of the bin-based structure to degenerate to that of the linked list. This is possible since the bin size is fixed, while the block size may vary. If we insert n blocks into the layout such that all the blocks fall in the same bin, the performance of the bin-based data structure is equivalent to that of the linked list for most atomic operations, and worse than the linked list in the neighbor finding, area searching, and directed area enumeration operations, since bins containing no blocks must be tested. The worst case complexity for these operations is $O(b + n)$.

As in the linked list data structure, we now present an algorithm to find the neighbors of a given block. This algorithm finds the neighbors on the right side of the given block. However, the algorithm can easily be extended to find all the neighbors of a given block. The input to the algorithm is a specific block A and the set of bins $\mathcal{B}$. A formal description of the algorithm is shown in Figure 4.21.

In general, the bin-based data structure is highly sensitive to the time-space tradeoff. For instance, if the bins are small with respect to the average size of a block, the blocks are likely to intersect with more than one bin, thereby increasing storage requirements. Furthermore, many bins may remain empty, creating wasted storage space. If the bins are too large the average case performance will be reduced since the linked lists used to store blocks in each bin will be very long. Obviously, the best case is when each bin contains exactly $\frac{n}{b}$ blocks, and no block is stored in more than one bin.

Even though the bin-based method can be used to locate all blocks within an area (bin), it does not allow for any representation of locality. In order to find the block closest to another, it may be necessary to search other surrounding bins, and in the worst case all the bins. Because the bin-based data structure does not represent the vacant space, operations such as compaction are tedious and time consuming.

Algorithm NEIGHBOR-FIND2(A, $\mathcal{B}$)
begin
    $neighbor\text{-}list = \phi$;
    $x1 = A.x + A.width$;
    $y11 = A.y$;
    $y12 = A.y + A.height$;
    let $\mathcal{B}' \subseteq \mathcal{B}$ be set of bins which contain A;
    for all bins $X \in \mathcal{B}'$ **do**
        for each block $R \in X$ **do**
            $y21 = R.y$;
            $y22 = R.y + R.height$;
            if ($x1 = R.x$ **and** ($y11 \le y21 \le y12$ **or**
                       $y11 \le y22 \le y12$ **or**
                       $y21 < y11 < y12 < y22$)) **then**
                INSERT($R, neighbor\text{-}list$);
    return $neighbor\text{-}list$;
end.

Figure 4.21: Algorithm NEIGHBOR-FIND2.

4.4.4   Neighbor Pointers

Most operations in a layout system require local block information to perform efficiently. Both the linked list and bin-based data structures do not keep local information, such as neighboring blocks. To overcome this limitation, the *neighbor pointer* data structure was developed. The neighbor pointer data structure represents each block by its size (upper left hand corner, length, and width) as well as the pointers to all of its neighbors. The space complexity of the data structure is bounded by $O(n^2)$. Figure 4.22 shows how neighbor pointers are maintained for Block A.

The neighbor pointer data structure is designed to perform well on plowing and compaction operations, unlike the linked list and bin-based structures. Plowing operation can be performed easily since each block directly stores information about its neighbors. In other words, for any block B_i, all blocks affected by moving B_i can be referenced directly. Since compaction is a form of plowing, it can also be performed easily using the neighbor pointer data structure. Figure 4.23, shows how the neighbor pointers of block A are updated when block B is moved.

The primary disadvantage of neighbor pointers is that the data structure is difficult to maintain. A simple modification to the layout may require all the pointers in the data structure be updated. For instance, a plow operation may modify the neighbors of each block B_i, and, in this case, updating the pointers could take as much as $O(n^2)$ time. Furthermore, block insertion and deletion operations each take $O(n)$ time. Since vacant space is not explicitly represented,

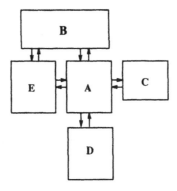

Figure 4.22: Neighbor pointers.

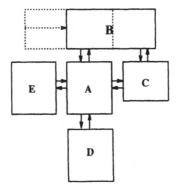

Figure 4.23: Update of neighbor pointers.

channel generation cannot be performed without extensive modification to the data structure.

4.4.5 Corner Stitching

Corner stitching is a radically different data structure used for IC layout editing [Ous84]. Corner stitching is novel in the sense that it is the first data structure to represent both vacant and block tiles in the design. As in the neighbor pointer structure, information about the relative locations of blocks is stored; however, unlike neighbor pointers, the corner stitch data structure can be updated rapidly.

The corner-stitch data structuring technique provides various powerful operations such as stretching, compaction, neighbor-finding, and channel finding. These operations are possible in the order of the number of neighbors, which

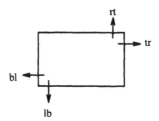

Figure 4.24: Corner stitches.

in the worst case would be the order of the size of the layout; i.e., the number
of objects in the layout. The advantage of corner stitch data structure is that
it permits easy modification to the layout.

The two main features of the corner stitch data structure pertain to the
way in which it keeps track of vacant tiles and how the tiles are linked. To
partition the vacant space into a collection of vacant tiles, the vacant space
must be divided into maximal horizontal strips (as discussed in section 4.4.2).
Hence the whole layout is represented as tiles (vacant and block).

Tiles are linked by a set of pointers called *corner stitches*. Each tile contains
four stitches, two at its top right corner and two at the bottom left corner as
shown in Figure 4.24. The pointers at these two corners are sufficient to perform
all operations. The corner stitch method stores both the vertical and horizontal
pointers. Each tile also stores the same number of pointers, irrespective of the
number of neighbors it has. In this structure, vacant tiles can assume any
size, and this helps in naturally adapting to the variations in the size of the
blocks. In other words, a new block, created over a set of a vacant tiles, will
result in a number of vacant tiles to be split, thus enabling the layout to be
updated easily. The pointers in each of the four directions provide a type of
sorting similar to that of the neighbor pointers. Figure 4.25, shows how corner
stitches link the tiles of a layout. The stitches exceeding the boundary of the
layout have been omitted in the figure, but the data structure represents them
as NULL pointers.

The record structure used in this section to represent a tile is the same as
the one shown in Figure 4.17 with the exception that the single pointer used
to link the tiles will be replaced with the corner stitch pointers (rt, tr, bl, lb).
It should also be noted that we define the upper left corner of the layout to be
point (0,0).

In the following we present the various operations that can be performed
on a layout using the corner stitch data structure.

1. **Point Finding:** a point p_2, the following sequence of steps finds a path
   through the corner stitches from the current point p_1 to p_2 traversing the
   minimum number of tiles.

   (1) The first step is to move up or down, using **rt** and **lb** pointers until

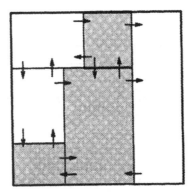

Figure 4.25: An example of the corner stitch layout.

Algorithm POINT-FIND(B, x, y)
begin
    do
        while $(y < B.y)$ **or** $(y > B.y + B.height)$ **do**
          if $y < B.y$ **then** $B = B.rt$; **else** $B = B.lb$;
        while $(x < B.x)$ **or** $(x > B.x + B.width)$ **do**
          if $x < B.x$ **then** $B = B.bl$; **else** $B = B.tr$;
      while $(y < B.y)$ **or** $(y > B.y + B.height)$;
end.

Figure 4.26: Algorithm POINT-FIND.

a tile is found whose vertical range contains the destination point.

(2) Then, a tile is found whose horizontal range contains the destination
point by moving left or right using **tr** or **bl** pointers.

(3) Whenever there is a misalignment (the search goes out of the vertical
range of the tile that contains the destination point) due to the above
operations, steps 1 and 2 has to be iterated several times to locate
the tile containing the point.

This operation is illustrated in Figure 4.27. In worst case, this algorithm
traverses all the tiles in the layout. On an average though, $\sqrt{N}$ tiles will
be visited. This algorithm handles the inherent asymmetry in designs by
readjusting the misalignments that occur during the search.

Figure 4.26 shows a formal description of the algorithm for point finding.
The input to the algorithm is a specific block B, and the coordinates of
the desired point (x,y).

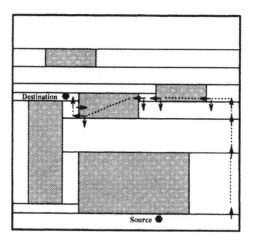

Figure 4.27: Point finding using corner stitches.

2. **Neighbor Finding:**   Following algorithm finds all the tiles that touch a given side of a given tile. This is also illustrated in Figure 4.28.

   (1) From the **tr** pointer of the given tile the algorithm starts traversing using the **lb** pointer downwards until it reaches a tile which does not completely lie within the vertical range of the given tile.

3. **Area Search:**   Given an area, the following algorithm reports if there are any blocks in the area. This is illustrated in Figure 4.29.

   (1) First the tile in which the upper left corner of the given area is located.

   (2) If the tile corresponding to this corner is a space tile, then if its right edge is within the area of interest, the adjacent tile must be a block.

   (3) If a block was found in step 2, then the search is complete. If no block was found, then the next tile touching the right edge of the area of interest is found, by traversing the **lb** stitches down and then traversing right using the **tr** stitches.

   (4) Steps 2 and 3 are repeated until either the area has been searched or a block has been found.

4. **Enumerate all Tiles:**   Given an area, the following algorithm reports all tiles intersecting that area. This is illustrated in Figure 4.30.

   (1) The algorithm first finds the tile in which the upper left corner of the given area is located. Then it steps down through all the tiles along the left edge, using the same technique as in area searching.

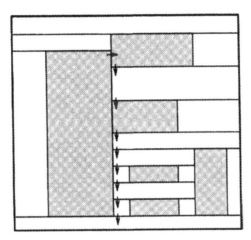

Figure 4.28: Neighbor finding using corner stitches.

(2) The algorithm enumerates all the tiles found in step 1 recursively (one tile at a time) using the procedure given in lines (R1) through (R5).

(R1) Number the current tile (this will generally involve some application specific processing).

(R2) If the right edge of the tile is outside of the search area, then the algorithm returns from the R procedure.

(R3) Otherwise, the algorithm uses the neighbor-finding algorithm to locate all the tiles that touch the right side of the current tile and also intersect the search area.

(R4) For each of these neighbors, if the bottom left corner of the neighbor touches the current tile then it calls R to enumerate the neighbor recursively (for example, this occurs in Figure 4.30 when tile 1 is the current tile and tile 2 is the neighbor).

(R5) Or, if the bottom edge of the search area cuts both the current tile and the neighbor, then it calls R to enumerate the neighbor recursively (in Figure 4.30, this occurs when tile 8 is the current tile and tile 9 is the neighbor).

5. **Block Creation:** The algorithm given below creates a block of given height and width on a certain given location in the plane. An illustration of how the vacant tiles will change when block E is added to the layout is given in Figure 4.31. Notice that the vacant tiles remain in maximal horizontal strips after the block has been added.

(1) First of all the algorithm checks if a block already exists by using the area search algorithm.

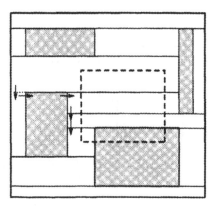

Figure 4.29: Area search using corner stitches.

(2) It then finds the vacant tile containing the top edge of the area occupied by new tile.

(3) The tile found in step (2) is split using the horizontal line along the top edge of the new tile. In addition, the corner stitches of the tiles adjoining the new tile are updated.

(4) The vacant tile containing the bottom edge of the new block is found and split in the same fashion as in step (3) and corner stitches of the tiles adjoining adjoining the new tile are updated.

(5) The algorithm traverses down along the left side and right side of the area of the new tile respectively, using the same technique in step (3) and updates corner stitches of the tiles as necessary.

6. **Block Deletion:**   The following algorithm deletes a block at a given location in the plane. An illustration of how the vacant tiles will change when block C is deleted from the layout is given in Figure 4.32. Notice that the vacant tiles remain in maximal horizontal strips after the block has been deleted.

(1) First, the block to be deleted is changed to a vacant tile.

(2) Second, using the neighbor finding algorithm for the right edge of the deleted tile find all the neighbors. For each vacant tile neighbor, the algorithm either splits the deleted tile or the neighbor tile so that the two tiles have the same vertical span and then merges them horizontally.

(3) Third, find all the neighbors for the left edge of the deleted tile. For each vacant tile neighbor the algorithm either splits the deleted tile or the neighbor tile so that the two tiles have the same vertical span and then merges them horizontally. After each horizontal merge, it

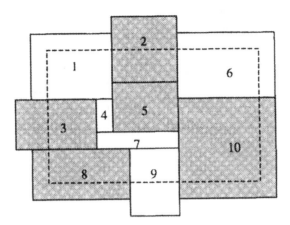

Figure 4.30: Enumerate all tiles using corner stitches.

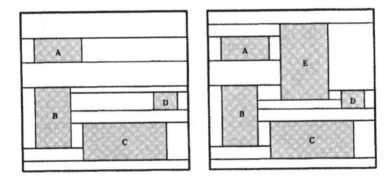

Figure 4.31: Block creation using corner stitches.

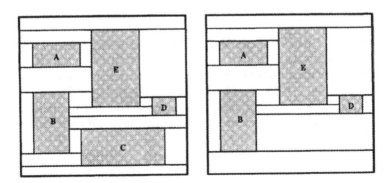

Figure 4.32: Block deletion using corner stitches.

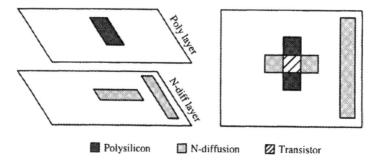

Figure 4.33: A simple two-layer layout.

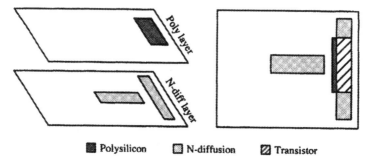

Figure 4.34: Simple two-layer layout plowed along the x-axis.

performs a vertical merge if possible with the tiles above and below it.

4.4.6   Multi-layer Operations

Thus far, the operations have been limited to those that concern only a single layer. It is important to realize that layouts contain many layers. More importantly, the functionality of the layout depends on the relationship between the position of the blocks on different layers. For instance, Figure 4.33 shows a simple transistor formed on two layers (polysilicon and n-diffusion) accompanied by a separate segment on the n-diffusion layer. If each layer is plowed along the positive x-axis, as shown in Figure 4.34, the original function of the layout has been altered as a transistor has been created on another part of the layout.

Design rule checking is also multilayer operation. In Figure 4.34, an illegal transistor has been formed. Even though the design rules were maintained while plowing on a single layer, the overall layout has an obvious design flaw.

If proper design rule checking is to take place, then the position of the blocks on each layer must be taken into consideration.

4.4.7 Limitations of Existing Data Structures

When examining the different data structures described in this chapter, it is easy to see that there is no single data structure that performs all operations with good space and time complexity. For example, the linked list and bin-based data structures are not suitable for use in a system in which a large database needs to be updated very frequently. Although the simpler data structures, like the linked list, are easy to understand, they are not suitable for most layout editors. On the other hand, more advanced data structures, such as the corner stitch, perform well where the simpler data structures become inefficient. Yet, they do not allow for the full flexibility needed in managing complicated layouts.

A limitation that all the data structures discussed share, is that they only work on rectangular objects. For example, the data structures do not support objects that are circular or L-shaped. New data structures, therefore, need to be developed that handle non-rectangular objects. Also as parallel computation is becoming more popular, new data structures need to be developed that can adapt to a parallel computation environment.

4.4.8 Layout Specification Languages

A common and simple method of producing system layouts is to draw them manually using a layout editor. This is done on one lambda grid using familiar color codes to identify various systems layers. Once the layout has been drawn, it can then be digitized or translated into machine-readable form by encoding it into a symbolic layout language. The function of a symbolic layout language, in its simplest form, is similar to that of macro-assembler. The user defines *symbols* (macros) that describe the layout of basic system cells. The function of the assembler for such a language is to scan and decode the statements and translate them into design files in intermediate form. The effectiveness of such languages could be further increased by constructing an assembler capable of handling nested symbols. Through the use of nested symbols, system layouts may be described in a hierarchical manner, leading to very compact descriptions of structured designs.

Caltech Intermediate Form (CIF) is one of the popular intermediate forms of layout description. Its purpose is to serve as a standard machine-readable representation from which other forms can be constructed for specific output devices such as plotters, video displays, and pattern-generation machines. CIF provides participating design groups easy access to output devices other than their own, enables the sharing of designs, and allows combining several designs to form a larger chip. A CIF file is composed of a sequence of commands, each being separated by a semi-colon (;). Each command is made up of a sequence of characters which are from a fixed character set. Table 4.1 lists the command

Command	Representation
B length width center direction	A Box of length and width(specified by integers) with center and direction(specified by points). Default direction is (1,0)
C symbol transformation	Call symbol(specified by integers)
DD symbol	Delete definition denoted by symbol(specified by integers)
DF	Finish definition
DS symbol index-1 index-2	Start definition denoted by Symbol with index-1 and index-2(specified by integers)
E	End of CIF marker
L shortname	Layer name
P path	Polygon with path
R diameter center	Circle of diameter(specified by integer) with center(specified by a point)
W width path	Wire with width(specified by integer) and path
(comments)	comments enclosed in parenthesis
number usercommand	user extension

Table 4.1: Command symbols and their representations.

symbols and their forms.

A more formal listing of the commands is given in Figure 4.35. The syntax for CIF is specified using a recursive language definition as proposed by [Wir77]. The notation used is similar to the one used to express rules in programming languages and is as follows: the production rules use equals (=) to relate identifiers to expressions; vertical bar (|) for *or*; double quotes (" ") around terminal characters; curly braces ({ }) indicate repetition any number of times including zero; square brackets ([ ]) indicate optional factors (i.e., zero or one repetition); parentheses ( ) are used for grouping; rules are terminated by a period (.).

The number of objects in a design and the representation of the primitive elements make the size of the CIF file very large. Symbolic definition is a way of reducing the file size by equating a commonly used command to a symbol. Layer names have to be unique, which will ensure the integrity of the design while combining several layers which represent one design.

CIF uses a right-handed coordinate system where x increases to the right and y increases upward. CIF represents the entire layout in the first quadrant. The unit of measurement of the distance is usually a micrometer(μm). Below are several examples of geometric shapes expressed in CIF. Note that the corresponding example correspond to the respective shape in Figure 4.36.

(a) **Boxes:** A box of length 30, width 50, and which is centered at (15,25), in CIF is B 30 50 15 25; (See Figure 4.36(a))

In this form the length of the box is the measurement of the side that is parallel to the x-axis and the width of the box is the measurement of the side that is parallel to the y-axis.

(b) **Polygons:** Polygon with vertices at (0,0), (20,50), (50,30), and (40,0) in

Alphabet	Rules
cifFile	= {{blank}[command]semi} endCommand{blank}
command	= primCommmand | defDeleteCommand | defStart Command
	semi{{blank}[primCommand]semi}defFinishCommand.
primCommand	= polygonCommand | boxCommand | roundFlashCommand | wireCommand | layerCommand | callCommand | userExtensionCommand | commentCommand.
polygonCommand	= "P" path
boxCommand	= "B" integer sep integer sep point [sep point]
roundFlashCommand	= "R" integer sep point
wireCommand	= "W" integer sep path
layerCommand	= "L" {blank} shortname
defStartCommand	= "D" {blank} "S" integer [sep integer sep integer]
defFinishCommand	= "D" {blank} "F"
defDeleteCommand	= "D" {blank} "D" integer
callCommand	= "C" integer transformation
userExtensionCommand	= digit userText
commentCommand	= "("comment Text")"
endCommand	= "E"
transformation	= {{blank} ("T" point | "M" {blank}"X"|"M" {blank} "Y" | "R" point)}
path	= point {sep point}
point	= sInteger sep sInteger
sInteger	= {sep}["-"] integerD
integer	= {sep} integerD
integerD	= digit {digit}
shortname	= c[c][c][c]
c	= digit | upperChar
userText	= {userChar}
commentText	= {commentChar} | commentText "("commentText")"commentText
semi	= {blank} ";"{blank}
sep	= upperChar | blank
digit	= "0"| "1"| "2"| "3"| "4"| "5"| "6"| "7"| "8"| "9"
upperChar	= "A"| "B"| "C"| "D"| "E"| "F"| "G"| "H"| "I"| "J"| "K"| "L"| "M"| "N"| "O"| "P"| "Q"| "R"| "S"| "T"| "U"| "V"| "W"| "X"| "Y"| "Z"
blank	= any ASCII character except digit, upperChar, "-", "(", ")", or";"
userChar	= any ASCII character except ";"
commentChar	= any ASCII character except "(" or ")"

Figure 4.35: Formal list of commands and their protocols.

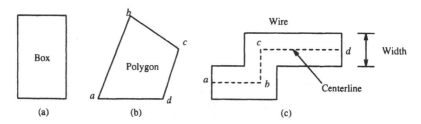

Figure 4.36: CIF terminologies for different geometric shapes.

CIF is P 0 0 20 50 50 30 40 00; (See Figure 4.36(b))

For a polygon with n sides, the coordinates of n vertices must be specified through the path of the edges.

(c) **Wires**: A wire with width 20 and which follows the path specified by the coordinates (0,10), (30,10), (30,30), (80,30) in CIF is W 20 0 10 30 10 30 30 80 30; (See Figure 4.36(c))

For a wire, the width must be given first and then the path of the wire is specified by giving the coordinates of the path along its center. For an object to qualify as a wire, it must have a uniform width.

As shown by the representation of a polygon, CIF will describe shapes that do not have Manhattan or rectilinear features. It is actually possible to represent a box that does not have Manhattan features. This is done using a direction vector. This eliminates the need for any trigonometric functions such as sin, cos, tan, etc. It is also easy to incorporate in the box description. The direction vector is made up two integer components, the first being the component of the direction vector along the x-axis, and the second being the same along the y-axis. The direction vector (1 1) will rotate the box 45^0 counterclockwise as will (2 2), (50 50), etc. The direction vector pointing to the x-axis can be represented as direction (1 0). With this new information a new descriptor can be added to box called the direction. Figure 4.37 shows a box with length 25, width 60, center 80,40 and direction -20, 20. When using direction, the length is the measure of the side parallel to the direction vector, and width is the measure of the side perpendicular to the direction vector. The direction vector is optional and if not used defaults to the positive x-axis.

B 25 60 80 40 -20 20;

To maintain the integrity of the layers for these geometric objects they must be labeled with the exact name of the fabrication mask (layer) on which it belongs. Rather than repeating the layers specified for each object, it is specified once and all objects defined after it belong to the same layer.

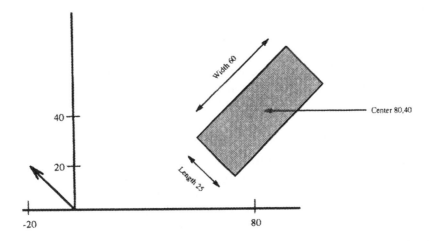

Figure 4.37: Box representation in CIF format.

Using CIF, a cell can be defined as a symbol by using the DS and DF commands. If an instance of a cell is required, the call command for that cell is used. The entire circuit is usually described as a group of nested symbols. A final call command is used to instantiate the circuit.

One of the popular layout description language used in the industry is GDSII.

4.5 Graph Algorithms for Physical design

The basic objects in VLSI design are rectangles and the basic problems in physical design deal with arrangement of these rectangles in a two or three dimensional space. The relationships between these objects, such as overlap and distances, are very critical in development of physical design algorithms. Graphs are a well developed tool used to study relationships between objects. Naturally, graphs are used to model many VLSI physical design problems and they play a very pivotal role in almost all VLSI design algorithms. In this section, we will define various graphs which are used in modeling of physical design problems.

4.5.1 Classes of Graphs in Physical Design

A layout is a collection of rectangles. Rectangles, which are used for routing, are thin and long and the width of these rectangles can be ignored for the sake of simplicity. In VLSI routing problems, such simple models are frequently used where the routing wires are represented as lines. In such cases, one needs to optimally arrange lines in two and three dimensional space. As a result, there are several different graphs which have been defined on lines and their relationships. Rectangles, which do not allow simplifying assumptions about the width, must also be modeled. For placement and compaction problems, it is common to use a graph which represents a layout as a set of rectangles and their adjacencies and relationships. As a result, a graph may be defined to represent the relationships between the rectangles. Thus we have two types of graphs dealing with lines and rectangles. Complex layouts with non-rectilinear objects require more involved modeling techniques and will not be discussed.

4.5.1.1 Graphs Related to a Set of Lines

Lines can be classified into two types depending upon the alignment with axis. We prefer to use the terminology of line interval or simply interval for lines which are aligned to axis. An interval I_i is represented by its left and right endpoints, denoted by l_i and r_i, respectively. Given a set of intervals $\mathcal{I} = \{I_1, I_2, \ldots, I_n\}$, we define three graphs on the basis of the different relationships between them.

We define an *overlap graph* $G_O = (V, E_O)$, as

$$V = \{v_i \mid v_i \text{ represents interval } I_i \}$$

$$E_O = \{(v_i, v_j) \mid l_i < l_j < r_i < r_j\}$$

In other words, each vertex in the graph corresponds to an interval and an edge is defined between v_i and v_j if and only if the interval I_i overlaps with I_j but does not completely contain or reside within I_j.

We define a *containment graph* $G_C = (V, E_C)$, where the vertex set V is the same as defined above and E_C a set of edges defined below:

$$E_C = \{(v_i, v_j) \mid l_i < l_j, r_i > r_j\}$$

In other words an edge is defined between v_i and v_j if and only if the interval I_i completely contains the interval I_j.

We also define an interval graph $G_I = (V, E_I)$ where the vertex set V is the same as above, and two vertices are joined by an edge if and only if their corresponding intervals have a non-empty intersection. It is easy to see that $E_I = E_O \cup E_C$. An example of the overlap graph for the intervals in Figure 4.38(a) is shown in Figure 4.38(b) while the containment graph and the interval graph are shown in Figure 4.38(c) and Figure 4.38(d) respectively. Interval graphs form a well known class of graphs and have been studied extensively [Gol80].

Overlap, containment and interval graphs arise in many routing problems, including channel routing, single row routing and over-the-cell routing.

If lines are non-aligned, then it is usually assumed (for example, in channel routing) that all the lines originate at a specific y-location and terminate at a specific y-location. An instance of such a set of lines is shown in Figure 4.39(a). This type of diagram is sometimes called a *matching diagram.*

Permutation graphs are frequently used in routing and can be defined by matching diagram. We define a *permutation graph* $G_P = (V, E_P)$, where the

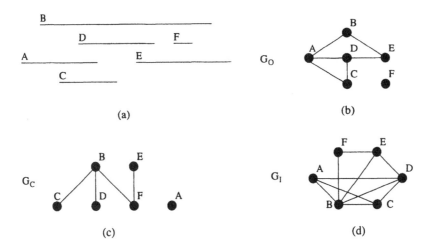

(a)   (b)

(c)   (d)

Figure 4.38: Various graphs associated with a set of intervals.

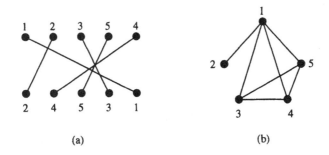

(a)   (b)

Figure 4.39: Matching diagram and permutation graph.

vertex set V is the same as defined above and E_P a set of edges defined below:

$$E_P = \{(v_i, v_j) \mid \text{if line } i \text{ intersects line } j \}$$

An example of permutation graph for the matching diagram in Figure 4.39(a) is shown in Figure 4.39(b). It is well known that the class of containment graphs is equivalent to the class of permutation graphs [Gol80].

Two sided box defined above is called a channel. The channel routing problem, which arises rather frequently in VLSI design, uses permutation graphs to model the problem. A more general type of routing problem, called the switch-box routing problem, uses a four-sided box (see Figure 4.40(a)). The graph defined by the intersection of lines in a switchbox is equivalent to a *circle graph* shown in Figure 4.40(b). Overlap graphs are equivalent to circle graphs. Circle graphs were originally defined as the intersection graph of chords of a circle,

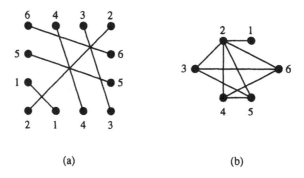

(a)   (b)

Figure 4.40: Switchbox and its circle graph.

can be recognized in polynomial time [GHS86].

4.5.1.2 Graphs Related to Set of Rectangles

As mentioned before, rectangles are used to represent circuit blocks in a layout design. Note that no two rectangles in a plane are allowed to overlap. Rectangles may share edges, i.e., two rectangles may be neighbors to each other. Given a set of rectangles $R = \{R_1, R_2, \ldots, R_m\}$ corresponding to a layout in a plane, a *neighborhood graph* is a graph $G = (V, E)$, where

$$V = \{v_i | v_i \text{ represents the rectangle } R_i \text{ and}\}$$

$$E = \{(v_i, v_j) | R_i \text{ and } R_j \text{ are neighbors}\}$$

The neighborhood graph is useful in the global routing phase of the design automation cycle where each channel is defined as a rectangle, and two channels are neighbors if they share a boundary. Figure 4.41 gives an example of a neighborhood graph, where for example, rectangles A and B are neighbors in Figure 4.41(a), and as a result there is an edge between vertices A and B in the corresponding neighborhood graph shown in Figure 4.41(b).

Similarly, given a graph $G = (V, E)$, a rectangular dual of the graph is a set of rectangles $R = \{R_1, R_2, \ldots, R_m\}$ where each vertex $v_i \in V$ corresponds to the rectangle $R_i \in R$ and two rectangles share an edge if their corresponding vertices are adjacent. Figure 4.42(b) shows an example of a rectangular dual of a graph shown in Figure 4.42(a). This graph is particularly important in floorplanning phase of physical design automation. It is important to note that not all graphs have a rectangular dual.

4.5.2 Relationship Between Graph Classes

The classes of graphs used in physical design are related to several well known classes of graphs, such as triangulated graphs, comparability graphs, and co-comparability graphs, which are defined below.

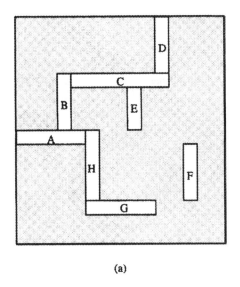

(a)

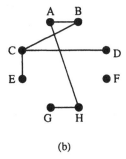

(b)

Figure 4.41: Neighborhood graphs.

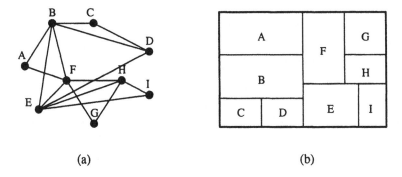

(a)   (b)

Figure 4.42: Rectangular duals.

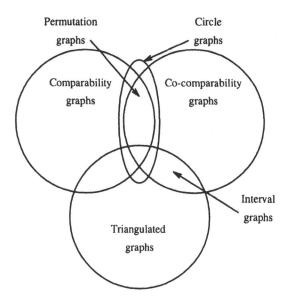

Figure 4.43: Relationship between different classes of graphs.

An interesting class of graphs based on the notion of cycle length is *triangulated graphs*. If $C = v_0, e_1, \ldots, e_{k-1}, v_k$ is a cycle in G, a *chord* of C is an edge e in $E(G)$ connecting vertices v_i and v_j such that $e \neq e_i$ for any $i = 1, \ldots, k$. A graph is *chordal* if every cycle containing at least four vertices has a chord. Chordal graphs are also known as *triangulated graphs*. A graph $G = (V, E)$ is a *comparability graph* if it is *transitively orientable*. A graph is called a *co-comparability graph* if the complement of G is transitively orientable.

Triangulated and comparability graphs can be used to characterize interval graphs. A graph G is called an *interval* graph if and only if G is triangulated and the complement of G is a comparability graph. Similarly, comparability and co-comparability graphs can be used to characterize permutation graphs. A graph G is called a *permutation graph* if and only if G is a comparability graph and the complement of G is also a comparability graph.

The classes of graphs mentioned above are not unrelated, in fact, interval graphs and permutation graphs have a non-empty intersection. Similarly the classes of permutation and bipartite graphs have a non-empty intersection. On the other hand, the class of circle graphs properly contains the class of permutation graphs. In Figure 4.43 shows the relationship between these classes.

4.5.3   Graph Problems in Physical Design

Several interesting problems related to classes of graphs discussed above arise in VLSI physical design. We will briefly state the definitions of these problems.

An extensive list of problems and related results may be found in [GJ79].

Independent Set Problem

Instance: Graph $G = (V, E)$, positive integer $K \leq |V|$.
Question: Does G contain an independent set of size K or more, i.e., a subset $V' \subset V$ such that $|V'| \geq K$ and such that no two vertices in V' are joined by an edge in E ?

Maximum Independent Set (MIS) problem is the optimization version of the Independent Set problem. The problem is NP-complete for general graphs and remains NP-complete for many special classes of graphs [GJ79, GJS76, MS77, Pol74, YG78]. The problem is solvable in polynomial time for interval graphs, permutation graphs and circle graphs. Algorithms for maximum independent set for these classes are presented later in this chapter.

An interesting variant of the MIS problem, called the k-MIS, arises in various routing problems. The objective of the k-MIS is to select a subset of vertices, which can be partitioned into k independent sets. That is, the selected subset is k-colorable.

Clique Problem

Instance: Graph $G = (V, E)$, positive integer $K \leq |V|$.
Question: Does G contain a clique of size K or more, i.e., a subset $V' \subset V$ such that $|V'| \geq K$ and such that every two vertices in V' are joined by an edge in E ?

Maximum clique problem is the optimization version of the clique problem. The problem is NP-complete for general graphs and for many special classes of graphs. However, the problem is solvable in polynomial time for chordal graphs [Gav72] and therefore also for interval graphs, comparability graphs [EPL72], and circle graphs [Gav73], and therefore for permutation graphs.

The maximum clique problem for interval graphs arises in the channel routing problem.

Graph K-Colorability

Instance: Graph $G = (V, E)$, positive integer $K \leq |V|$.
Question: Is G K-colorable, i.e., does there exist a function $f : V \rightarrow \{ 1, 2, ..., K \}$ such that $f(u) \neq f(v)$ whenever $\{u, v\} \in E$?

The minimization of the above problem is more frequently used in physical design of VLSI. The minimization version asks for the minimum number of colors needed to properly color a given graph. The minimum number of colors needed to color a graph is called the *chromatic number* of the graph. The problem is NP-complete for general graphs and remain so for all fixed $K \geq 3$. It is polynomial for $K = 2$, since that is equivalent to bipartite graph recognition. It also remains NP-complete for $K = 3$ if G is the intersection graph for straight line segments in the plane [EET89]. For arbitrary K, the problem is NP-complete for circle graphs. The general problem can be solved in polynomial time for comparability graphs [EPL72], and for chordal graphs [Gav72].

As discussed earlier, many problems in physical design can be transformed into the problems discussed above. Most commonly, these problems serve as sub-problems and as a result, it is important to understand how these problems are solved. We will review the algorithms for solving these problems for several classes of graphs in the subsequent subsections.

It should be noted that most of the problems have polynomial time complexity algorithms for comparability, co-comparability, and triangulated graphs. This is due to the fact these graphs are *perfect graphs* [Gol80]. A graph $G = (V, E)$ is called perfect, if the size of the maximum clique in G is equal to the chromatic number of G and this is true for all subgraphs H of G. Perfect graphs admit polynomial time complexity algorithms for maximum clique, maximum independent set, among other problems. Note that chromatic number and maximum clique problems are equivalent for perfect graphs.

Interval graphs and permutation graphs are defined by the intersection of different classes of perfect graphs, and are therefore themselves perfect graphs. As a result, many problems which are NP-hard for general graphs are polynomial time solvable for these graphs. On the other hand, circle graphs are not perfect and generally speaking are much harder to deal with as compared to interval and permutation graphs. To see that circle graphs are not perfect, note that an odd cycle of five or more vertices is a circle graph, but it does not satisfy the definition of a perfect graph.

4.5.4   Algorithms for Interval Graphs

Among all classes of graphs defined on a set of lines, interval graphs are perhaps the most well known. It is very structured class of graphs and many algorithms which are NP-hard for general graphs are polynomial for interval graphs [Gol77]. Linear time complexity algorithms are known for recognition, maximum clique, and maximum independent set problems among others for this class of graphs [Gol80]. The maximal cliques of an interval graph can be linearly ordered such that for every vertex $v \in V$, the cliques containing v occur consecutively [GH64]. Such an ordering of maximal cliques is called a *consecutive linear ordering*. An $O(|V| + |E|)$ algorithm for interval graph recognition that produces a consecutive linear ordering of maximal cliques is presented in [BL76]. In this section, we review algorithms for finding maximum independent set and maximum clique in an interval graph.

4.5.4.1   Maximum Independent Set

An optimal algorithm for computing maximum independent set of an interval graph was developed by Gupta, Lee, and Leung [GLL82]. The algorithm they presented is greedy in nature and is described below in an informal fashion. The algorithm first sorts the $2n$ end points in ascending order of their values. It then scans this list from left to right (i.e., in ascending order of their values) until it first encounters a right endpoint. It then outputs the interval having this right endpoint as a member of a maximum independent set and deletes

> **Algorithm** MAX-CLIQUE($\mathcal{I}$)
> **begin**
>   SORT-INTERVAL($\mathcal{I}, A$);
> $cliq = 0$;
> $max\_cliq = 0$;
>   **for** $i = 1$ to $2n$ **do**
>   **if** $A[i] = $ L **then** $cliq = cliq + 1$;
>   **if** $cliq > max\_cliq$ **then** $max\_cliq = cliq$;
>   **else** $cliq = cliq - 1$;
>   **return** max_cliq;
> **end.**

Figure 4.44: Algorithm MAX-CLIQUE.

all intervals containing this point. This process is repeated until there is no interval left in the list. It can be easily shown that the algorithm produces a maximum independent set in a interval graph and the time complexity of the algorithm is dominated by sorting the intervals that is $O(n \log n)$. The time complexity of the algorithm is thus $O(n \log n)$, where n is the total number of intervals.

Theorem 3 *Given an interval graph, the MIS can be found in $O(n \log n)$ time, where n is the total number vertices in the graph.*

In [YG87], an optimal algorithm for finding the maximum k-colorable subgraph in an interval graph has been presented. We present an outline of that algorithm.

The set of the interval is processed from left to right in increasing order of endpoints. For a vertex v_i let I_i denote its corresponding interval, having a maximum k-colorable subgraph $G(U')$ for a set of nodes already processed. The next node v is added to U' if $G(U' \cup \{v\})$ contains no clique with more than k nodes, and is discarded otherwise.

It can be easily shown that this greedy algorithm indeed finds the optimal k-colorable independent set in an interval graph. For details, refer to [YG87].

4.5.4.2 Maximum Clique and Minimum Coloring

Since interval graphs are perfect, the cardinality of a minimum coloring is the same as that of maximum clique in interval graphs. The algorithm shown in Figure 4.44 finds a maximum clique in a given interval graph. The input to the algorithm is a set of intervals $\mathcal{I} = \{I_1, I_2, \ldots, I_n\}$ representing an interval graph. Each interval I_i is represented by its left end point l_i and right end point r_i.

In the algorithm shown in Figure 4.44, SORT-INTERVAL sorts the list of end points of all the intervals and generates an array $A[i]$ to denote whether the endpoint at the position i in the sorted list is a left endpoint or right endpoint. $A[i] = $ L if the corresponding end point is a left endpoint. Note that the algorithm finds the size of the maximum clique in a given interval graph. However, the algorithm can easily be extended to find the maximum clique. It is easy to see that the worst case complexity of the algorithm is $O(n^2)$, where n is the total number of intervals. The time complexity of the algorithm can be reduced to $O(n \log n)$ by keeping track of the minimum of the right end points of all the intervals. The left edge algorithm (LEA) described in detailed routing (chapter 7) is a simple variation of the algorithm in Figure 4.44.

4.5.5   Algorithms for Permutation Graphs

The class of permutation graphs was introduced by Pnnueli, Lempel, and Even [PLE71]. They also showed that the class of permutation graphs is transitive and introduced an $O(n^2)$ algorithm to find the maximum clique [EPL72]. In [Gol80], Golumbic showed an $O(n \log n)$ time complexity algorithm for finding the chromatic number χ in a permutation graph.

Permutation graphs are also a structured class of graphs similar to interval graphs. Most problems, which are polynomial for permutation graphs, are also polynomial for interval graphs. In this section, we present an outline of several important algorithms related to permutation graphs.

4.5.5.1   Maximum Independent Set

The maximum independent set in a permutation graph can be found in $O(n \log n)$ time [Kim90]. As mentioned before, permutation graphs can be represented using matching diagrams as shown in Figure 4.39.

The binary insertion technique can be used on a matching diagram to find a maximum independent set in a permutation graph. Given a permutation $P = (P_1, P_2, \ldots, P_n)$ of n numbers $N = (1, 2, \ldots, n)$ corresponding to a permutation graph, note that an increasing subsequence of P represents an independent set in the permutation graph. Similarly, a decreasing subsequence of P represents a clique in the permutation graph. Therefore, to find a maximum independent set, we need to find a maximum increasing subsequence of P. It is necessary to know the the relations of the positions of numbers in the permutation. A stack is used to keep track of the relations. The algorithm works as follows:

The sequence N is scanned in increasing order. In the jth iteration, j is placed on the top of the stack i whenever j does not intersect with the front entries of the stack q, but intersects with the front entry of stack r, where $1 \leq i < j$ and $i \leq r \leq m$, and m is the total number of stacks during jth iteration. If j does not intersect with any of the front entries of the stacks $1, 2, \ldots, m$, then the stack $m + 1$ is created and j is placed on top stack $m + 1$. It is easy to see that the stack search and insertion can be done using binary search in $O(\log n)$ time.

Once the numbers are placed in stacks, stacks can be scanned from bottom up to get a maximum increasing subsequence. We illustrate the algorithm by

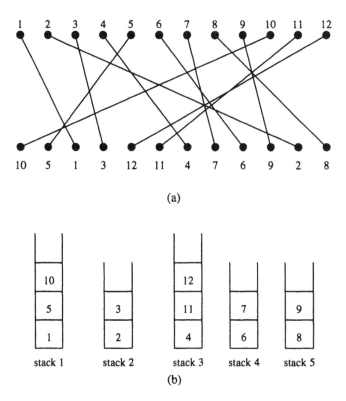

(a)

(b)

Figure 4.45: Example of Maximum independent set in permutation graph.

means of an example shown in Figure 4.45(a). Initially, the permutation P is given as $(10, 5, 1, 3, 12, 11, 4, 7, 6, 9, 2, 8)$. The top row is processed from left to right. First 1 is placed on the stack 1. Then 2 is placed on stack 2, because 2 does not intersect with 1. After that the 3 goes on top of stack 2, since it intersects with 2 but does not intersect with 1 on top of stack 1. The 4 is placed in front of a new stack 3. Then the 5 intersects with all of the front entries of all the stacks, thus 5 is placed in front of the stack 1. In this way, all the numbers are placed in 5 stacks as shown Figure 4.45(b). Now the stacks are scanned starting from stack 5 to stack 1. One number from each stack is selected so that the numbers are in decreasing order. If 9 from stack 5, 7 from stack 4, 4 from stack 3, 3 from stack 2, and 1 from stack 1 is selected then the generated set $\{9, 7, 4, 3, 1\}$ is a maximum independent set of the corresponding permutation graph. Note the total number of stacks is equal to the chromatic number of the permutation graph.

In [LSL90], Lou, Sarrafzadeh, and Lee presented a $\Theta(n \log n)$ time complexity algorithm for finding a maximum two-independent set in permutation graphs. Cong and Liu [CL91] presented an $O(n^2 \log n + nm)$ time complexity algorithm to compute a maximum weighted k-independent set in permutation graphs where m is bounded by n^2. In fact, their algorithm is very general and applicable to any comparability graph.

4.5.5.2 Maximum k-Independent Set

The complement of a permutation graph is a permutation graph. Hence, MKIS problem in graph G is equivalent to maximum k-clique problem in $\bar{G}$. In this section, we discuss an $O(kn^2)$ time algorithm for finding the maximum k-clique in a permutation graph presented by Gavril [Gav87]. In fact, this algorithm is very general and applicable to any comparability graph.

The basic idea of the algorithm is to convert the maximum k-clique problem in a comparability graph into network flow problem. (See [Tar83] for an excellent survey of network flow algorithms.) First a transitive orientation is constructed for a comparability graph $G = (V, E)$, resulting in a directed graph $\vec{G} = (V, \vec{E})$. A directed path in $\vec{G}$ is also called as a chain. Note that each chain in $\vec{G}$ corresponds to a clique in G since G is a comparability graph. Next, each vertex in V is split into two vertices. Assume that $V = \{v_1, v_2, \ldots, v_n\}$. Then each vertex v_i corresponds to two vertices x_i and y_i in a new directed graph $\vec{G_1} = (V_1, \vec{E_1})$. There is a directed edge between x_i and y_i for all $1 \leq i \leq n$. A cost of -1 and capacity of 1 are assigned to the edge (x_i, y_i) for all $1 \leq i \leq n$. In addition, there is a directed edge between y_i and x_j if there exists a directed edge (v_i, v_j) in $\vec{G}$. A cost of 0 and capacity of 1 are assigned to the edge (y_i, x_j). Four new vertices s (source), t (sink), s' and t' are introduced as well as the directed edges (s', x_i) and (y_i, t') for all $i \in V$, (s, s'), and (t', t) are added. A cost of 0 and capacity of 1 are assigned to the edge (s', v_i) and (v_i, t'). A cost of 0 and capacity of k are assigned to the edges (s, s') and (t', t). The graph $\vec{G_1} = (V_1, \vec{E_1})$ so constructed is called a *network* where $V_1 = \{x_i, y_i | 1 \leq i \leq n\} \cup \{s, s', t', t\}$ and $\vec{E_1} = \{(x_i, y_i) | 1 \leq i \leq n\} \cup \{(y_i, x_j) | (v_i, v_j) \in \vec{E}\} \cup \{(s', x_i), (y_i, t') | 1 \leq i \leq n\} \cup \{(s, s'), (t', t)\}$.

Then the maximum k-clique problem in the graph G is equivalent to the min-cost max-flow problem in the network $\vec{G_1}$. The flow in a directed graph has to satisfy the following.

1. The flow $f(e)$ associated with each edge of the graph, can be assigned a value no more than the capacity of the edge.

2. The net flow that enters a vertex is equal to the net flow that leaves the vertex.

The absolute value of flow that leaves the source, e.g. $|f(s, s')|$, is called the flow of $\vec{G_1}$. The min-cost max-flow problem in the directed graph $\vec{G_1}$ is to find the assignment of $f(e)$ for each edge $e \in \vec{E_1}$ such that the flow of $\vec{G_1}$ is maximum and the total cost on the edges that the flows pass is minimum. Notice that the capacity on the directed edge (s, s') is k. Thus, the maximum flow of $\vec{G_1}$ is k. In addition, flow that passes x_i or y_i has value 1, since the capacity on the directed edge (x_i, y_i) is one for each of $1 \leq i \leq n$. Note

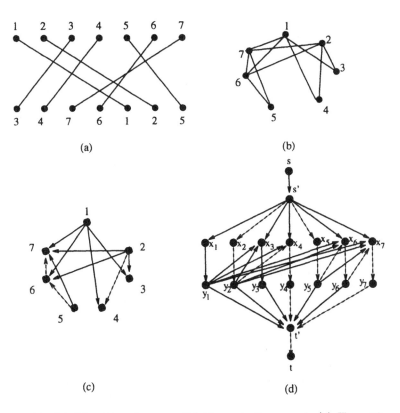

Figure 4.46: (a) Matching diagram (b) Permutation graph (c) Transitive orientation of the graph (d) Network flow graph

that a flow in $\vec{G}_1$ corresponds to a chain in $\vec{G}$. The maximum flow in $\vec{G}_1$ is k, thus the maximum number of the chains in $\vec{G}$ is k, and vice versa. The absolute value of the cost on each flow is equal to the number of vertices on the chain corresponding to the flow. Thus the minimum cost on all flows results in maximum number of vertices in the chains in $\vec{G}$, and hence maximum number of vertices in the cliques in G.

An example of a permutation graph G is given in Figure 4.46(b). The transitive orientation of G is given in Figure 4.46(c) while the network $\vec{G}_1$ is shown in Figure 4.46(d). The min-cost max-flow while $k = 2$ is highlighted in Figure 4.46(d). The chains corresponding to the min-cost max-flow are highlighted in Figure 4.46(c). The maximum 2-clique is $\{5, 6, 7\}$ and $\{2, 4\}$.

The time complexity of the algorithm is dominated by the time complexity of the algorithm to find the min-cost max-flow in a network which is $O(kn^2)$ where n is the number of vertices in the graph [Law76]. The weighted version of the MKIS problem can be solved by $O(kn^2)$ algorithm presented in [SL93].

4.5.6   Algorithms for Circle Graphs

Circle graphs are used for solving certain problems in channel routing and switchbox routing. Circle graphs are not prefect and less structured than interval and permutation graphs. Many problems, such as the maximum bipartite subgraph problem, which are polynomial for interval and permutation graphs are NP-complete for circle graphs. However, there are still many problems that can be solved in polynomial time for circle graphs which are NP-complete for general graphs. For example, polynomial time complexity algorithms are known for maximum clique and maximum independent set problems on circle graphs [Gav73], as well as for the weighted maximum clique problem [Hsu85], but the chromatic number problem for circle graphs remains NP-complete [GJMP78]. In the following, we review the circle graph algorithms used in VLSI design.

4.5.6.1   Maximum Independent Set

The problem of finding maximum independent set in a circle graph can also be solved in polynomial time. In [Sup87], Supowit presented a dynamic programming algorithm of time complexity $O(n^2)$ for finding maximum independent set in a circle graph.

Given is a set C of n chords of a circle, without loss of generality, it is assumed that no two chords share the same endpoint. Number these endpoints of the chords from 0 to $2n - 1$ clockwise around the circle. Let $G = (V, E)$ denote a circle graph where $V = \{v_{ab} | a < b, ab$ is a chord$\}$. Let G_{ij} denote the subgraph of the circle graph $G = (V, E)$, induced by the set of vertices

$$\{v_{ab} \in V : i \le a, b \le j\}.$$

Let $M(i, j)$ denote a maximum independent set of G_{ij}. If $i \ge j$, then G_{ij} is the empty graph and, hence $M(i, j) = \phi$. The algorithm is an application of dynamic programming. In particular, $M(i, j)$ is computed for each pair i, j; $M(i, j_1)$ is computed before $M(i, j_2)$ if $j_1 < j_2$. To compute $M(i, j)$, let k be the unique number such that $kj \in C$ or $jk \in C$. If k is not in the range $[i, j-1]$, then $G_{ij} = G_{i,j-1}$ and hence $M(i, j) = M(i, j-1)$. If k is in the range $[i, j-1]$, then there are two cases to consider:

1. If $v_{kj} \in M(i, j)$, then by definition of an independent set, $M(i, j)$ contains no vertices v_{ab} such that $a \in [i, k-1]$ and $b \in [k+1, j]$.

   Therefore, $M(i, j) = M(i, k-1) \cup \{v_{kj}\} \cup M(k+1, j-1)$.

2. If $v_{kj} \notin M(i, j)$, then

   $$M(i, j) = M(i, j-1).$$

Thus $M(i, j)$ is set to the larger of the two sets $M(i, j-1)$ and $M(i, k-1) \cup \{v_{kj}\} \cup M(k+1, j-1)$. The algorithm is more formally stated in Figure 4.47.

Algorithm MIS(V)
begin
    for $j = 0$ to $2N - 1$ **do**
        find k such that $kj \in C$ or $jk \in C$;
        for $i = 0$ to $j - 1$ **do**
            if $i \leq k \leq j\text{-}1$ and $|M(i, k - 1)| + 1 +$
                $|M(k + 1, j - 1)| > |M(i, j - 1)|$ **then**
              $M(i, j) = M(i, k - 1) \cup \{v_{kj}\} \cup$
                    $M(k + 1, j - 1)$;
            else $M(i, j) = M(i, j - 1)$
end.

Figure 4.47: Algorithm MIS.

Theorem 4 *The algorithm MIS finds a maximum independent set in a circle graph in time $O(n^2)$.*

4.5.6.2 Maximum k-Independent Set

In general a *k-independent set* can be defined as a set consisting of k disjoint independent sets, and a *maximum k-independent set* (k-MIS) has the maximum number of vertices among all such k-independent sets. Although, a MIS in circle graphs can be found in polynomial time, the problem of finding a k-MIS is NP-complete even for $k = 2$ [SL89a]. Since the problem has many important applications in routing and via minimization described in the later chapters, it is required to develop some provably good approximation algorithm for this problem.

In [CHS93], Cong, Hossain, and Sherwani present an approximation algorithm for a maximum k-independent set in the context of planar routing problem in an arbitrary routing region. The problem is equivalent to finding a maximum k-independent set in a circle graph. The approximation algorithm for $k = 2$, was first presented by Holmes, Sherwani and Sarrafzadeh [HSS93] and later extended to the case of $k = 4$ in [HSS91]. In this section, we present the approximation result in the context of a circle graph.

Given a graph $G_1 = (V_1, E_1)$, the algorithm finds k independent sets one after another denoted by $S_1, S_2, \ldots, S_k$, such that S_1 is a maximum independent set in G_1, and S_i is a maximum independent set in G_i, for $2 \leq i \leq k$, where $G_i = (V_i, E_i)$ is inductively defined as:

$$V_i = V_{i-1} - S_i \text{ and } E_i = E_{i-1} - \{(v_1, v_2) | v_1 \in S_i \text{ and } v_2 \notin S_i \text{ and } (v_1, v_2) \in E_{i-1}\}$$

Clearly, the algorithm reduces the problem of k-MIS to a series computations of MIS in a circle graph. Since in circle graphs, the complexity of computing 1-MIS is $O(n^2)$, the total time complexity of this approximation algorithm is $O(kn^2)$.

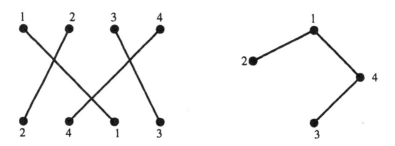

Figure 4.48: Example to show non-optimality of K-MIS algorithm.

Algorithm MKIS $(G(V, E), k)$
begin
    1. $V' = V$;
    2. **for** $i = 1$ to k **do**
        $S_i = \text{MIS}(V')$;
        $V' = V' - S_i$;
    3. return $S_1 \cup S_2 \cup \ldots \cup S_k$;
end.

Figure 4.49: Algorithm MKIS.

Consider the circle graph shown in Figure 4.48. Clearly, the 2-MIS of the graph is $\{(1,3),(2,4)\}$. The maximum independent sets in the graph are $\{(1,3),(2,4),(2,3)\}$. In the MKIS algorithm, the MIS is chosen randomly, and a *bad* selection of a MIS ($\{(2, 3)\}$ in this case) may not lead to an optimal 2-MIS for the graph. If $\{(2,3)\}$ is chosen then either we can choose 1 or 4. Thus, the total number of nets chosen is three while the optimal has four nets. A similar reasoning would show that the algorithm is non-optimal for the k-MIS problem.

The algorithm is formally stated in Figure 4.49.

For any heuristic algorithm H_k for k-MIS, the *performance ratio of H_k* is defined to be $\frac{\Psi_k}{\Psi_k^*}$, where Ψ_k is the size of the k-independent set obtained by the algorithm H_k and Ψ_k^* is the k-MIS in the same graph. The lower bound on the performance ratio is established based on the following theorem.

Theorem 5 *Let γ_k be the performance ratio of the algorithm MKIS for k-MIS. Then,*

$$\gamma_k \geq 1 - (1 - \frac{1}{k})^k$$

Corollary 1 *Given a circle graph G, MKIS can be used to approximate a maximum bipartite set of G with a performance bound of at least 0.75.*

It is easy to see that the function $f(x) = 1 - (1 - \frac{1}{x})^x$ is a decreasing function. Moreover,

$$\lim_{x \to \infty} [1 - (1 - \frac{1}{x})^x] = 1 - e^{-1}$$

where $e \approx 2.718$. Therefore, we have

Corollary 2 *For any integer k, the performance ratio of the algorithm MKIS for k-MIS is at least*

$$\gamma_k \geq 1 - e^{-1} \approx 63.2\%$$

Although the approximation result presented above is for circle graphs, this equally applicable to any class of graphs where the problem of finding MIS is polynomial time solvable.

Another variation of the MIS problem in circle graphs is called *k-density MIS*. Given a set of intervals, the objective of k-density MIS is to find an independent set of intervals with respect to overlap property such that the interval graph corresponding to that set has a clique of size at most k.

4.5.6.3 Maximum Clique

Given a circle graph $G = (V, E)$, it is easy to show that for every vertex $v \in V$, the induced subgraph $G_v = (V_v, E_v)$ is a permutation graph, where,

$$V_v = \{v\} \cup \{Adj(v)\}$$

$$E_v = \{(u, v) | u \in V_v\}$$

For each G_v, maximum clique can be found using the algorithm presented for maximum clique in a permutation graph. Let C_v be the maximum clique in G_v, then the maximum clique in G is given by $\max\{C_v\}$, for all $v \in V$. It is easy to see that the time complexity of this algorithm is $O(n^2 \log n)$.

4.6 Summary

A VLSI layout is represented as a collection of tiles on several layers. A circuit may consists of millions of such tiles. Since layout editors are interactive, their underlying data structures should enable editing operations to be performed within an acceptable response time. Several data structures have been proposed for layout systems. The most popular data structure among these is corner stitch. However, none of the data structures is equally good for all the operations. The main limitation of all the existing data structures is that they only work on rectangular objects. In other words, the data structures do not support any other shaped objects such as circular, L-shaped. Therefore, development of new data structure is needed to handle different shaped objects.

Also, as parallel computation becomes practical, new data structures need to be developed to adapt to the parallel computation environment.

Due to sheer size of VLSI circuits, low time complexity is necessary for algorithms to be practical. In addition, due to NP-hardness of many problems, heuristic and approximation algorithms play a very important role in physical design automation.

Several special graphs are used to represent VLSI layouts. The study of algorithms of these graphs is essential to development of efficient algorithms for various phases in VLSI physical design cycle.

4.7   Exercises

1. Design an algorithm to insert a block in a given area using the modified linked list data structure. Note that you need to use area searching operation to insert a block and a modified linked list to keep track of the vacant tiles.

2. Design an algorithm to delete a given block from a given set of blocks using modified linked list data structure. Note that once a block is deleted the area occupied by that block becomes vacant tile and the linked list must be updated to take care of this situation.

3. Design algorithms using a linked list data structure to perform the plowing and compaction operations.

4. Solve the problem 3 using a modified linked list.

5. The problem of connectivity extraction is very important in circuit extraction phase of physical design. It is defined as follows. Given a set of blocks $B = \{B_1, B_2, \ldots, B_n\}$ in an area, let us assume that there is a type associated to each circuit. For example, all the blocks can of one type and all the vacant tiles could be of another type. Two blocks B_i and B_j are called connected if there is a sequence of k $(k \leq n)$ distinct blocks of the same type $B_{\pi(1)}, B_{\pi(2)}, \ldots, B_{\pi(k)}$, such that $\pi(i) \leq n$, $B_{\pi(1)} = B_i$, $B_{\pi(k)} = B_j$, and $B_{\pi(1)}$ is a neighbor of $B_{\pi(2)}$, $B_{\pi(2)}$ is a neighbor of $B_{\pi(3)}$, and so on and finally $B_{\pi(k-1)}$ is a neighbor of $B_{\pi(k)}$.

   (a) Design an algorithm using a modified linked list data structure to extract the connectivity of two blocks.

   (b) Design an algorithm using a corner stitch data structure to find the connectivity of two blocks.

†6. The existing data structure can be modified to handle layouts in a multilayer environment. Consider the following data items associated to a tiles in a multilayer environment:

   record Tile =

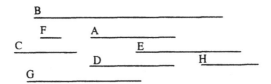

Figure 4.50: A set of intervals.

    coordinate;

    height;

    width;

    type;

    text;

    layer;

end record

(a) Design an algorithm to move the entire layout in one direction.

(b) Following problem 5, find the connectivity of any two blocks in a multilayer environment using corner stitch data structure.

7. Modify algorithm NEIGHBOR-FIND1 to find all the neighbors of a given block using a linked list data structure.

8. Modify algorithm NEIGHBOR-FIND2 to find all the neighbors of a given block using bin-based data structure.

†9. Assume a layout system that allows 45^0 segments, i.e., the blocks could be 45^0 angled parallelogram as well as rectangular. Modify the corner stitch data structure to handle this layout system. Are four pointers still sufficient in this situation ?

10. Given a family of sets of segments $S = \{S_1, S_2, \ldots, S_n\}$, where S_i is the set of segments belonging to net N_i on a layer.

Determine if there is a connectivity violation by developing an algorithm which finds all such violations.

11. For the set of intervals shown in Figure 4.50, find maximum independent set, maximum clique, and maximum bipartite subgraph in the interval graph defined by the intervals.

12. For the matching diagram shown in Figure 4.51, find its permutation graph. Find maximum independent set, minimum number of colors required to color it, and maximum bipartite subgraph in this permutation graph.

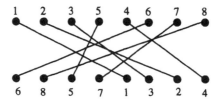

Figure 4.51: Channel problem.

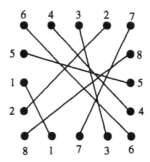

Figure 4.52: Switchbox problem.

13. For the switchbox shown in Figure 4.52, find maximum independent set, maximum clique, and maximum bipartite subgraph in the permutation graph defined by the matching diagram.

†14. Prove that the algorithm MAX-CLIQUE correctly finds the size of the maximum clique in an interval graph.

†15. Improve the time complexity of the algorithm MAX-CLIQUE to $O(n \log n)$. The algorithm should also be able to report a maximum clique.

†16. Prove that the algorithm MIS for finding a maximum independent set in circle graphs does indeed find the optimal solution.

17. Develop a heuristic algorithm for finding a maximum bipartite subgraph in circle graphs.

†18. Implement the approximation algorithm for finding a k-independent set in circle graphs. Experimentally evaluate the performance of the algorithm by implementing an exponential time complexity algorithm for finding a k-independent set.

19. Develop an efficient algorithm to find a k-density MIS in circle graphs.

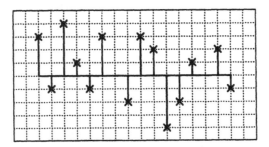

Figure 4.53: A single trunk Steiner tree.

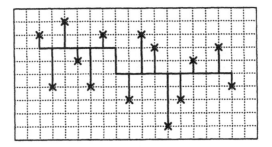

Figure 4.54: A two-trunk Steiner tree.

20. Steiner trees play a key role in global and detail routing problems. Consider the following *Single Trunk Steiner Tree* problem. A single trunk Steiner tree consists of a single horizontal line segment and all the points are joined by short vertical line segments. An example of a single trunk Steiner tree is shown in Figure 4.53.

Given a set of n points in a plane, develop an $O(n)$ algorithm for the minimum cost single trunk Steiner tree.

†21. Prove that for $n = 3$, single trunk Steiner tree is indeed an optimal rectilinear Steiner tree.

22. For $n = 4$, give an example which shows that single trunk Steiner tree is not an optimal rectilinear Steiner tree.

23. Single trunk Steiner tree can be easily generalized to k-trunk Steiner tree problem, which consists of k non-overlapping horizontal trunks. An example of a two trunk Steiner tree is shown in Figure 4.54.

Develop an efficient algorithm for 2-trunk Steiner tree problem.

†24. Does there exist an $O(n^{ck})$ algorithm for the k-trunk Steiner tree problem, for a small constant c?

†25. Implement Hadlock's Algorithm for finding max-cut in a planar graph.

†26. Prove that Hadlock's algorithm is optimal by showing it deletes minimum number of edges.

27. Given a set of rectangles in a plane, develop an efficient algorithm to detect if any two rectangles intersect or contain each other.

†28. Given a switch box, develop an efficient algorithm to find the minimum diameter of rectilinear Steiner trees. The diameter of a tree is the maximum distance between any two of its vertices.

Bibliographic Notes

The paper by John Ousterhout on the corner stitch data structure [Ous84] gives details of different algorithms used to manipulate a layout. The corner stitch data structure has been extended in various ways to account for nonrectilinear shapes and interaction of objects in different layers. In [Meh94] D. P. Mehta presented a technique for estimating the storage requirements of the Rectangular Corner Stitching data structure and the L-shaped Corner Stitching Data Structure on a given circuit by studying the circuit's geometric properties.

However, there are no efficient data structures to express the true three dimensional nature of a VLSI layout. The details of CIF can be found in Mead & Conway [MC79].

Cormen, Leiserson and Rivest [CLR90], present an in depth analysis of graph algorithms. Tarjan [Tar83] provides excellent reference for graph matchings, minimum spanning trees, and network flow algorithms. Computational geometry algorithms are discussed in detail by Preparata and Shamos [PS85]. The theory of NP-completeness is discussed in great detail in Garey and Johnson [GJ79].

General graph concepts have been described in detail in [CL86]. Algorithms and concepts for the perfect graphs, interval graphs, permutation graphs, and circle graphs can be found in [Gol80].

Chapter 5

Partitioning

Efficient designing of any complex system necessitates decomposition of the same into a set of smaller subsystems. Subsequently, each subsystem can be designed independently and simultaneously to speed up the design process. The process of decomposition is called *partitioning*. Partitioning efficiency can be enhanced within three broad parameters. First of all, the system must be decomposed carefully so that the original functionality of the system remains intact. Secondly, an interface specification is generated during the decomposition, which is used to connect all the subsystems. The system decomposition should ensure minimization of the interface interconnections between any two subsystems. Finally, the decomposition process should be simple and efficient so that the time required for the decomposition is a small fraction of the total design time.

Further partitioning may be required in the events where the size of a subsystem remains too large to be designed efficiently. Thus, partitioning can be used in a hierarchical manner until each subsystem created has a manageable size. Partitioning is a general technique and finds application in diverse areas. For example, in algorithm design, the *divide and conquer* approach is routinely used to partition complex problems into smaller and simpler problems. The increasing popularity of the parallel computation techniques brings in its fold promises in terms of provision of innovative tools for solution of complex problems, by combining partitioning and parallel processing techniques.

Partitioning plays a key role in the design of a computer system in general, and VLSI chips in particular. A computer system is comprised of tens of millions of transistors. It is partitioned into several smaller modules/blocks for facilitation of the design process. Each block has *terminals* located at the periphery that are used to connect the blocks. The connection is specified by a *netlist*, which is a collection of *nets*. A net is a set of terminals which have to be made electrically equivalent. Figure 5.1(a) shows a circuit, which has been partitioned into three subcircuits. Note that the number of interconnections between any two partitions is four (as shown in Figure 5.1(b)).

A VLSI system is partitioned at several levels due to its complexity. At the highest level, it is partitioned into a set of sub-systems whereby each sub-system can be designed and fabricated independently on a single PCB. High performance systems use MCMs instead of PCBs. At this level, the criterion for partitioning is the functionality and each PCB serves a specific task within a system. Consequently, a system consists of I/O (input /output) boards, memory boards, mother board (which hosts the microprocessor and its associated circuitry), and networking boards. Partitioning of a system into PCBs enhances the design efficiency of individual PCBs. Due to clear definition of the interface specified by the net list between the subsystems, all the PCBs can be designed simultaneously. Hence, significantly speeding up the design process.

If the circuit assigned to a PCB remains too large to be fabricated as a single unit, it is further partitioned into subcircuits such that each subcircuit can be fabricated as a VLSI chip. However, the layout process can be simplified and expedited by partitioning the circuit assigned to a chip into even smaller subcircuits. The partitioning process of a process into PCBs and an PCB into VLSI chips is physical in nature. That is, this partitioning is mandated by the limitations of fabrication process. In contrast, the partitioning of the circuit on a chip is carried out to reduce the computational complexity arising due to the sheer number of components on the chip. The hierarchical partitioning of a computer system is shown in Figure 5.2.

The partitioning of a system into a group of PCBs is called the *system level* partitioning. The partitioning of a PCB into chips is called the *board level* partitioning while the partitioning of a chip into smaller subcircuits is called the *chip level* partitioning. At each level, the constraints and objectives of the partitioning process are different as discussed below.

- **System Level Partitioning**: The circuit assigned to a PCB must satisfy certain constraints. Each PCB usually has a fixed area, and a fixed number of terminals to connect with other boards. The number of terminals available in one board (component) to connect to other boards (components) is called the *terminal count* of the board (component). For example, a typical board has dimensions 32 cm×15 cm and its terminal count is 64. Therefore, the subcircuit allocated to a board must be manufacturable within the dimensions of the board. In addition, the number of nets used to connect this board to the other boards must be within the terminal count of the board.

  The reliability of the system is inversely proportional to the number of boards in the system. Hence, one of the objectives of partitioning is to minimize the number of boards. Another important objective is the optimization of the system performance. Partitioning must minimize any degradation of the performance caused by the delay due to the connections between components on different boards. The signal carried by a net that is cut by partitioning at this level has to travel from one board to another board through the system bus. The system bus is very slow as the bus has to adhere to some strict specifications so that a variety

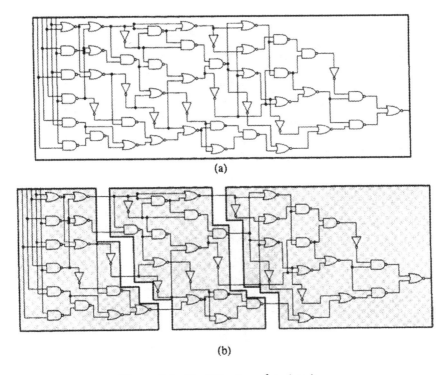

(a)

(b)

Figure 5.1: Partitioning of a circuit.

of different boards can share the same bus. The delay caused by signals traveling between PCBs (off-board delay) plays a major role in determining the system performance as this delay is much larger than the on-board or the on-chip delay.

- **Board Level Partitioning**: The board level partitioning faces a different set of constraints and fulfills a different set of objectives as opposed to system level partitioning. Unlike boards, chips can have different sizes and can accommodate different number of terminals. Typically the dimensions of a chip range from 2 mm×2 mm to 25 mm×25 mm. The terminal count of a chip depends on the package of the chip. A Dual In-line Package (DIP) allows only 64 pins while a Pin Grid Array (PGA) package may allow as many as 300 pins.

While system level partitioning is geared towards satisfying the area and the terminal constraints of each partition, board level partitioning ventures to minimize the area of each chip. The shift of emphasis is attributable to the cost of manufacturing a chip that is proportional to its area. In addition, it is expedient that the number of chips used for each board be minimized for enhanced board reliability. Minimization of the

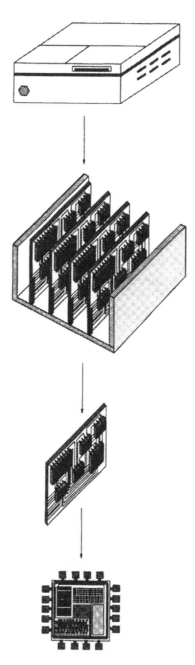

Figure 5.2: System hierarchy.

number of chips is another important determinant of performance because the off-chip delay is much larger than the on-chip delay. This differential in delay arises because the distance between two adjacent transistors on a chip is a few μm while the distance between two adjacent chips is in mm. In addition to traversing a longer distance, the signal has to travel between chips, and through the connector. The connector used to attach the chip to the board typically has a high resistance and contributes significantly to the signal delay. Figure 5.3 shows the different kinds of delay in a computer system. In Figure 5.3(b), the off-board delay is compared with the on-board delay while the off-chip delay is compared with the on-chip delay in Figure 5.3(c).

- **Chip Level Partitioning**: The circuit assigned to a chip can be fabricated as a single unit, therefore, partitioning at this level is necessary. A chip can accommodate as many as three million or more transistors. The fundamental objective of chip level partitioning is to facilitate efficient design of the chip.

After partitioning, each subcircuit, which is also called a *block*, can be designed independently using either full custom or standard cell design style. Since partitioning is not constrained by physical dimensions, there is no area constraint for any partition. However, the partitions may be restrained by user specified area constraints for optimization of the design process.

The terminal count for a partition is given by the ratio of the perimeter of the partition to the terminal pitch. The minimum spacing between two adjacent terminals is called *terminal pitch* and is determined by the design rules. The number of nets which connect a partition to other partitions cannot be greater than the terminal count of the partition. In addition, the number of nets cut by partitioning should be minimized to simplify the routing task. The minimization of the number of nets cut by partitioning is one of the most important objectives in partitioning.

A disadvantage of the partitioning process is that it may degrade the performance of the final design. Figure 5.4(a) shows two components A and B which are critical to the chip performance, and therefore, must be placed close together. However, due to partitioning, components A and B may be assigned to different partitions and may appear in the final layout as shown in Figure 5.4(b). It is easy to see that the connection between A and B is very long, leading to a very large delay and degraded performance. Thus, during partitioning, these critical components should be assigned to the same partition. If such an assignment is not possible, then appropriate timing constraints must be generated to keep the two critical components close together. Chip performance is determined by several components forming a critical path. Assignment of these components to different partitions extends the length of the critical path. Thus, a major challenge for improvement of system performance is minimization of the length of critical path.

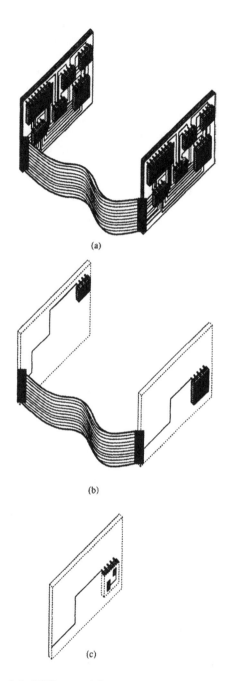

(a)

(b)

(c)

Figure 5.3: Different delays in a computer system.

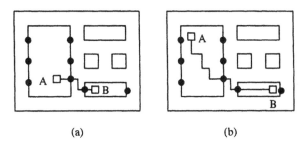

(a)   (b)

Figure 5.4: Bad partitioning increases the delay of circuit.

After a chip has been partitioned, each of the subcircuits has to be placed on a fixed plane and the nets between all the partitions have to be interconnected. The placement of the subcircuits is done by the placement algorithms and the nets are routed by using routing algorithms.

At any level of partitioning, the input to the partitioning algorithm is a set of components and a netlist. The output is a set of subcircuits which when connected, function as the original circuit and terminals required for each subcircuit to connect it to the other subcircuits. In addition to maintaining the original functionality, partitioning process optimizes certain parameters subject to certain constraints. The constraints for the partitioning problem include area constraints and terminal constraints. The objective functions for a partitioning problem include the minimization of the number of nets that cross the partition boundaries, and the minimization of the maximum number of times a path crosses the partition boundaries. The constraints and the objective functions used in the partitioning problem vary depending upon the partitioning level and the design style used. The actual objective function and constraints chosen for the partitioning problem may also depend on the specific problem.

5.1   Problem Formulation

The partitioning problem can be expressed more naturally in graph theoretic terms. A hypergraph $G = (V, E)$ representing a partitioning problem can be constructed as follows. Let $V = \{v_1, v_2, \ldots, v_n\}$ be a set of vertices and $E = \{e_1, e_2, \ldots, e_m\}$ be a set of *hyperedges*. Each vertex represents a component. There is a hyperedge joining the vertices whenever the components corresponding to these vertices are to be connected. Thus, each hyperedge is a subset of the vertex set i.e., $e_i \subseteq V$, $i = 1, 2, \ldots, m$. In other words, each net is represented by a hyperedge. The area of each component is denoted as $a(v_i), 1 \le i \le n$. The modeling of partitioning problem into hypergraphs allows us to represent the circuit partitioning problem completely as a hypergraph partitioning problem. The partitioning problem is to partition V into

$V_1, V_2, \ldots, V_k$, where

$$V_i \cap V_j = \phi, \quad i \neq j$$
$$\cup_{i=1}^{k} V_i = V$$

Partition is also referred to as a *cut*. The cost of partition is called the *cut-size*, which is the number of hyperedges crossing the cut. Let c_{ij} be the cut-size between partitions V_i and V_j. Each partition V_i has an area $Area(V_i) = \sum_{v \in V_i} a(v)$, and a terminal count $Count(V_i)$. The maximum and the minimum areas, that a partition V_i can occupy, are denoted as $A_i^{\max}$ and $A_i^{\min}$, respectively. The maximum number of terminals that a partition V_i can have is denoted as T_i. Let $P = \{p_1, p_2, \ldots, p_m\}$ be a set of hyperpaths. Let $H(p_i)$ be the number of times a hyperpath p_i is cut, and let $K_{\min}$ and $K_{\max}$ represent the minimum and the maximum number of partitions that are allowed for a given subcircuit.

The constraints and the objective functions for the partitioning algorithms vary for each level of partitioning and each of the different design styles used. This makes it very difficult to state a general partitioning problem which is applicable to all levels of partitioning or all design styles used. Hence in this section we will list all the constraints and the objective functions and the level to which they are applicable. The partitioning problem at any level or design style deals with one or more of the following parameters.

1. **Interconnections between partitions:**   The number of interconnections at any level of partitioning have to be minimized. Reducing the interconnections not only reduces the delay but also reduces the interface between the partitions making it easier for independent design and fabrication. A large number of interconnections increase the design area as well as complicate the task of the placement and routing algorithms. Minimization of the number of interconnections between partitions is called the *mincut* problem. The minimization of the cut is a very important objective function for partitioning algorithms for any level or any style of design. This function can be stated as:

$$Obj_1 : \sum_{i=1}^{k} \sum_{j=1}^{k} c_{ij}, (i \neq j) \quad \text{is minimized}$$

2. **Delay due to partitioning:**   The partitioning of a circuit might cause a critical path to go in between partitions a number of times. As the delay between partitions is significantly larger than the delay within a partition, this is an important factor which has to be considered while partitioning high performance circuits. This is an objective function for partitioning algorithms for all levels of design. This objective function can be stated mathematically as:

$$Obj_2 : \max_{p_i \in P}(H(p_i)) \quad \text{is minimized}$$

3. **Number of terminals:** Partitioning algorithms at any level must partition the circuit so that the number of nets required to connect a subcircuit to other subcircuits does not exceed the terminal count of the subcircuit. In case of system level partitioning, this limit is decided by the maximum number of terminals available on a PCB connector which connects the PCB to the system bus. In case of board level partitioning, this limit is decided by the pin count of the package used for the chips. In case of chip level partitioning, the number of terminals of a subcircuit is determined by the perimeter of the area used by the subcircuit. At any level, the number of terminals for a partition is a constraint for the partitioning algorithm and can be stated as:

$$Cons_1 : Count(V_i) \leq T_i, \quad 1 \leq i \leq k$$

4. **Area of each partition:** In case of system level partitioning, the area of each partition (board) is fixed and hence this factor appears as a constraint for the system level partitioning problem. In case of board level partitioning, although it is important to reduce the area of each partition (chip) to a minimum to reduce the cost of fabrication, there is also an upper bound on the area of a chip. Hence, in this case also, the area appears as a constraint for the partitioning problem. At chip level, the size of each partition is not so important as long as the partitions are balanced. The area constraint can be stated as:

$$Cons_2 : A_i^{\min} \leq Area(V_i) \leq A_i^{\max}, \quad i = 1, 2, \ldots, k$$

5. **Number of partitions:** The number of partitions appears as a constraint in the partitioning problem at system level and board level partitioning. This prevents a system from having too many PCBs and a PCB from having too many chips. A large number of partitions may ease the design of individual partitions but they may also increase the cost of fabrication and the number of interconnections between the partitions. At the same time, if the number of partitions is small, the design of these partitions might still be too complex to be handled efficiently. At chip level, the number of partitions is determined, in part, by the capability of the placement algorithm. The constraint on the number of partitions can be stated as,

$$Cons_3 : K_{\min} \leq k \leq K_{\max}$$

Multiway partitioning is normally reduced to a series of two-way or *bipartitioning* problem. Each component is hierarchically bipartitioned until the desired number of components is achieved. In this chapter, we will restrict ourselves to bipartitioning. When the two partitions have the same size, the partitioning process is called *bisectioning* and the partitions are called *bisections*.

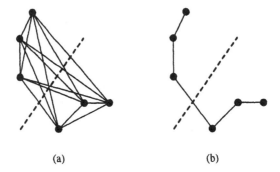

(a)   (b)

Figure 5.5: A net represented as a clique and a spanning tree.

An actual model representing the partitioning problem to be solved at system level or board level requires that the area constraint, interconnection constraint and constraint on the number of partitions be satisfied. Therefore, constraints $Cons_1, Cons_2$, and $Cons_3$ apply. If the performance of the system is also a criterion, then the objective function Obj_2 is also applicable. At chip level, the partitioning algorithms usually have Obj_1 as an objective function. In case of high performance circuits, objective function Obj_2 is also applicable.

An important factor, not discussed above, is modeling of a net. So far, we have assumed that a net is modeled as a hyperedge. However, hyperedges are hard to handle and the model is sometimes simplified. One way of simplifying the model is to represent each hyperedge by a clique of its vertices. However using this method increases the number of times the edges cross boundaries substantially as shown in Figure 5.5(a). There are other ways to represent hyperedges. For example, we can use a tree to represent a hyperedge as shown in Figure 5.5(b), but doing this destroys the symmetric property of the clique model. In general, net modeling is a hard problem and no satisfactory solution has been proposed.

5.1.1   Design Style Specific Partitioning Problems

The problems formulated above represent a general approach to partitioning. However, partitioning algorithms for different design styles have different objectives. In this section, we will discuss the partitioning problems for each design style. Partitioning problems for FPGAs and MCM will be discussed in Chapters 11 and 12, respectively.

1. **Full custom design style:**   In a full custom design style, partitions can be of different sizes and hence there are no area constraints for the partitioning algorithms. Thus, the partitioning in full custom design style has the most flexibility. During chip level partitioning, the number of terminals allowed for each partition is determined by the perimeter

of the block corresponding to a partition. Thus, the estimated terminal count for a partition i is given by

$$T_i = \frac{p_i}{d}, \quad i = 1, 2, \ldots, k$$

where, p_i is the perimeter of the block corresponding to the partition i and d is the terminal pitch. Since, the cost of manufacturing a circuit is directly proportional to the layout size, it is essential to keep the area of the layout to a minimum. The area of circuit layout is the sum of the areas occupied by components, areas used for routing the nets, and the unused areas. Since the areas occupied by the components are fixed, it is only possible to minimize the routing areas and unused areas. The routing area will be largely used by the nets that go across the boundaries of the blocks. The amount of unused areas will be determined by the placement. Therefore in addition to the terminal constraints, partitioning algorithms have to minimize the total number of nets that cross the partition boundaries. A partitioning algorithm for full custom design has objective function Obj_1 subject to the constraints $Cons_1$ and $Cons_2$. The full custom design style is typically used for the design of high-performance circuits, e.g., design of microprocessors. The delay for high-performance circuits is of critical importance. Therefore, an additional objective function Obj_2 is added to the partitioning problem for the full custom design style.

2. **Standard cell design style:** The primary objective of the partitioning algorithms in standard cell design style is to partition the circuit into a set of disjoint subcircuits such that each subcircuit corresponds to a cell in a standard cell library. In addition, the partitioning procedure is non-hierarchical. The complexity of partitioning depends on the type of the standard cells available in the standard cell library. If the library has only a few simple cell types available, there are few options for the partitioning procedure and the partitioning problem has to satisfy constraints $Cons_1$ and $Cons_2$. However, if there are many cell types available, some of which are complex, then the partitioning problem is rather complicated. The objective function to be optimized by the partitioning algorithms for standard cell design is Obj_1. For high performance circuits, Obj_1 and Obj_2 are used as combined objective functions.

3. **Gate array design style:** The circuit is bipartitioned recursively until each resulting partition corresponds to a gate on the gate array. The objective for each bipartitioning is to minimize the number of nets that cross the partition boundaries.

In future VLSI chips, the terminals may be on top of the chip and therefore terminal counts have to be computed accordingly. In addition, due to ever-reducing routing areas, the transistors will get packed closer together and thermal constraints may become dominant, as they are in MCM partitioning problems.

5.2   Classification of Partitioning Algorithms

The mincut problem is NP-complete, it follows that general partitioning problem is also NP-complete [GJ79]. As a result, variety of heuristic algorithms for partitioning have been developed. Partitioning algorithms can be classified in three ways. The first method of classification depends on availability of initial partitioning. There are two classes of partitioning algorithms under this classification scheme:

1. Constructive algorithms and

2. Iterative algorithms.

The input to a *constructive* algorithms is the circuit components and the netlist. The output is a set of partitions and the new netlist. Constructive algorithms are typically used to form some initial partitions which can be improved by using other algorithms. In that sense, constructive algorithms are used as preprocessing algorithms for partitioning. They are usually fast, but the partitions generated by these algorithms may be far from optimal.

Iterative algorithms, on the other hand, accept a set of partitions and the netlist as input and generate an improved set of partitions with the modified netlist. These algorithms iterate continuously until the partitions cannot be improved further.

The partitioning algorithms can also be classified based on the nature of the algorithms. There are two types of algorithms:

1. Deterministic algorithms and

2. Probabilistic algorithms.

Deterministic algorithms produce repeatable or deterministic solutions. For example, an algorithm which makes use of deterministic functions, will always generate the same solution for a given problem. On the other hand, the *probabilistic* algorithms are capable of producing a different solution for the same problem each time they are used, as they make use of some random functions.

The partitioning algorithms can also be classified on the basis of the process used for partitioning. Thus we have the following categories:

1. Group Migration algorithms,

2. Simulated Annealing and Evolution based algorithms and

3. Other partitioning algorithms.

The *group migration algorithms* [FM82, KL70] start with some partitions, usually generated randomly, and then move components between partitions to improve the partitioning. The group migration algorithms are quite efficient. However, the number of partitions has to be specified which is usually not known when the partitioning process starts. In addition, the partitioning of an entire system is a multi-level operation and the evaluation of the partitions obtained by the partitioning depends on the final integration of partitions at all levels, from the basic subcircuits to the whole system. An algorithm used to find a minimum cut at one level may sacrifice the quality of cuts for the following levels. The group migration method is a deterministic method which is often trapped at a local optimum and can not proceed further.

The *simulated annealing/evolution* [CH90, GS84, KGV83, RVS84] algorithms carry out the partitioning process by using a cost function, which classifies any feasible solution, and a set of moves, which allows movement from solution to solution. Unlike deterministic algorithms, these algorithms accept moves which may adversely effect the solution. The algorithm starts with a random solution and as it progresses, the proportion of adverse moves decreases. These degenerate moves act as a safeguard against entrapment in local minima. These algorithms are computationally intensive as compared to group migration and other methods.

Among all the partitioning algorithms, the group migration and simulated annealing or evolution have been the most successful heuristics for partitioning problems. The use of both these types of algorithms is ubiquitous and extensive research has been carried out on them. The following sections include a detailed discussion of these algorithms. The remaining methods will be discussed briefly later in the chapter.

5.3 Group Migration Algorithms

The *group migration* algorithms belong to a class of iterative improvement algorithms. These algorithms start with some initial partitions, formed by using a constructive algorithm. Local changes are then applied to the partitions to reduce the cutsize. This process is repeated until no further improvement is possible. Kernighan and Lin (K-L) [KL70] proposed a graph bisectioning algorithm for a graph which starts with a random initial partition and then uses pairwise swapping of vertices between partitions, until no improvement is possible. Schweikert and Kernighan [SK72] proposed the use of a net model so that the algorithm can handle hypergraphs. Fiduccia and Mattheyses [FM82] reduced time complexity of K-L algorithm to $O(t)$, where t is the number of terminals. An algorithm using vertex-replication technique to reduce the number of nets that cross the partitions was presented by Kring and Newton [KN91]. Goldberg and Burstein [GB83] suggested an algorithm which improves upon the original K-L algorithm using graph matchings. One of the problems with the K-L algorithm is the requirement of prespecified sizes of partitions. Wei and Cheng [WC89] proposed a ratio-cut model in which the sizes of the par-

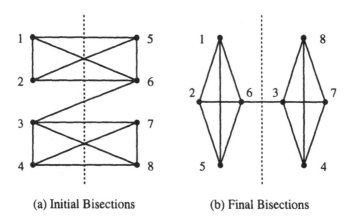

(a) Initial Bisections   (b) Final Bisections

Figure 5.6: A graph bisected by K-L algorithm.

titions do not need to be specified. The algorithms based on group migration
are used extensively in partitioning VLSI circuits. In the following sections,
these algorithms are discussed in detail.

5.3.1   Kernighan-Lin Algorithm

The K-L algorithm is a bisectioning algorithm. It starts by initially partitioning the graph $G = (V, E)$ into two subsets of equal sizes. Vertex pairs
are exchanged across the bisection if the exchange improves the cutsize. The
above procedure is carried out iteratively until no further improvement can be
achieved.

Let us illustrate the basic idea of the K-L algorithm with the help of an
example before presenting the algorithm formally. Consider the example given
in Figure 5.6(a). The initial partitions are

$$A = \{1, 2, 3, 4\}$$
$$B = \{5, 6, 7, 8\}$$

Notice that the initial cutsize is 9. The next step of K-L algorithm is
to choose a pair of vertices whose exchange results in the largest decrease of
the cutsize *or* results in the smallest increase, if no decrease is possible. The
decrease of the cutsize is computed using gain values $D(i)$ of vertices v_i. The
gain of a vertex v_i is defined as

$$D(i) = inedge(i) - outedge(i)$$

where $inedge(i)$ is the number of edges of vertex i that do not cross the bisection
boundary and $outedge(i)$ is the number of edges that cross the boundary. The
amount by which the cutsize decreases, if vertex v_i changes over to the other

partition, is represented by $D(i)$. If v_i and v_j are exchanged, the decrease of cutsize is $D(i) + D(j)$. In the example given in Figure 5.6, a suitable vertex pair is $(3, 5)$ which decreases the cutsize by 3. A tentative exchange of this pair is made. These two vertices are then locked. This lock on the vertices prohibits them from taking part in any further tentative exchanges. The above procedure is applied to the new partitions, which gives a second vertex pair of $(4, 6)$. This procedure is continued until all the vertices are locked. During this process, a log of all tentative exchanges and the resulting cutsizes is stored. Table 5.1 shows the log of vertex exchanges for the given example. Note that the partial sum of cutsize decrease $g(i)$ over the exchanges of first i vertex pairs is given in the table e.g., $g(1) = 3$ and $g(2) = 8$. The value of k for which $g(k)$ gives the maximum value of all $g(i)$ is determined from the table. In this example, $k = 2$ and $g(2) = 8$ is the maximum partial sum. The first k pairs of vertices are actually exchanged. In the example, the first two vertex pairs $(3, 5)$ and $(4, 6)$ are actually exchanged, resulting in the bisection shown in Figure 5.6(b). This completes an iteration and a new iteration starts. However, if no decrease of cutsize is possible during an iteration, the algorithm stops. Figure 5.7 presents the formal description of the K-L algorithm.

The procedure INITIALIZE finds initial bisections and initializes the parameters in the algorithm. The procedure IMPROVE tests if any improvement has been made during the last iteration, while the procedure UNLOCK checks if any vertex is unlocked. Each vertex has a status of either *locked* or *unlocked*. Only those vertices whose status is *unlocked* are candidates for the next tentative exchanges. The procedure TENT-EXCHGE tentatively exchanges a pair of vertices. The procedure LOCK locks the vertex pair, while the procedure LOG stores the log table. The procedure ACTUAL-EXCHGE determines the maximum partial sum of $g(i)$, selects the vertex pairs to be exchanged and fulfills the actual exchange of these vertex pairs.

The time complexity of Kernighan-Lin algorithm is $O(n^3)$. The Kernighan-Lin algorithm is, however, quite robust. It can accommodate additional constraints, such as a group of vertices requiring to be in a specified partition. This feature is very important in layout because some blocks of the circuit are to be kept together due to the functionality. For example, it is important to keep all components of an adder together. However, there are several disadvantages of K-L algorithm. For example, the algorithm is not applicable for hypergraphs, it cannot handle arbitrarily weighted graphs and the partition sizes have to be specified before partitioning. Finally, the complexity of the algorithm is considered too high even for moderate size problems.

5.3.2 Extensions of Kernighan-Lin Algorithm

In order to overcome the disadvantages of Kernighan-Lin Algorithm, several algorithms have been developed. In the following, we discuss several extensions of K-L algorithm.

Algorithm KL
begin
    INITIALIZE();
    while(IMPROVE(*table*) = TRUE) **do**
    (* if an improvement has been made during last iteration,
    the process is carried out again. *)
        while (UNLOCK(*A*) = TRUE) **do**
        (* if there exists any unlocked vertex in *A*,
        more tentative exchanges are carried out. *)
            for (each $a \in A$) **do**
                if ($a = unlocked$) **then**
                    for(each $b \in B$) **do**
                        if ($b = unlocked$) **then**
                            if ($D_{max} < D(a) + D(b)$) **then**
                                $D_{max} = D(a) + D(b)$;
                                $a_{max} = a$;
                                $b_{max} = b$;
            TENT-EXCHGE(a_{max}, b_{max});
            LOCK(a_{max}, b_{max});
            LOG(*table*);
            $D_{max} = -\infty$;
        ACTUAL-EXCHGE(*table*);
end.

Figure 5.7: Algorithm K-L

i	Vertex Pair	$g(i)$	$\sum_{j=1}^{i} g(i)$	Cutsize
0	-	-	-	9
1	(3,5)	3	3	6
2	(4,6)	5	8	1
3	(1,7)	-6	2	7
4	(2,8)	-2	0	9

Table 5.1: The log of the vertex exchanges.

5.3.2.1 Fiduccia-Mattheyses Algorithm

Fiduccia and Mattheyses [FM82] developed a modified version of Kernighan-Lin algorithm. The first modification is that only a single vertex is moved across the cut in a single move. This permits the handling of unbalanced partitions and nonuniform vertex weights. The other modification is the extension of the concept of cutsize to hypergraphs. Finally, the vertices to be moved across the cut are selected in such a way so that the algorithm runs much faster. As in Kernighan-Lin algorithm, a vertex is locked when it is tentatively moved. When no moves are possible, only those moves which give the best cutsize are actually carried out.

The data structure used for choosing the next vertex to be moved is shown in Figure 5.8. Each component is represented as a vertex. The vertex (component) gain is an integer and each vertex has its gain in the range $-pmax$ to $+pmax$, where $pmax$ is the maximum vertex degree in the hypergraph. Since vertex gains have restricted values, 'bucket' sorting can be used to maintain a sorted list of vertex gains. This is done using an array BUCKET[-pmax, ..., pmax], whose kth entry contains a doubly-linked list of free vertices with gains currently equal to k. Two such arrays are needed, one for each block. Each array is maintained by moving a vertex to the appropriate bucket whenever its gain changes due to the movement of one of its neighbors. Direct access to each vertex, from a separate field in the VERTEX array, allows removal of a vertex from its current list and its movement to the head of its new bucket list in constant time. As only free vertices are allowed to move, therefore, only their gains are updated. Whenever a base vertex is moved, it is locked, removed from its bucket list, and placed on a FREE VERTEX LIST, which is later used to reinitialize the BUCKET array for the next pass. The FREE VERTEX LIST saves a great deal of work when a large number of vertices (components) have permanent block assignments and are thus not free to move. For each BUCKET array, a MAXGAIN index is maintained which is used to keep track of the bucket having a vertex of highest gain. This index is updated by decrementing it whenever its bucket is found to be empty and resetting it to a higher bucket whenever a vertex moves to a bucket above MAXGAIN. Experimental results on real circuits have shown that gains tend to cluster sharply around the origin and that MAXGAIN moves very little, making the above implementation exceptionally fast and simple.

The total run time taken to update the gain values in one pass of the above algorithm is $O(n)$, where n is the number of terminals in the graph G. The F-M algorithm is much faster than Kernighan-Lin algorithm. A significant weakness of F-M algorithm is that the gain models the effect of a vertex move upon the size of the net cutsize, but not upon the gain of the neighboring vertices. Thus the gain does not differentiate between moves that may increase the probability of finding a better partition by improving the gains of other vertices and moves that reduce the gains of neighboring vertices. Krishnamurthy [Kri84] has proposed an extension to the F-M algorithm that accounts for high-order gains to get better results and a lower dependence upon the initial partition.

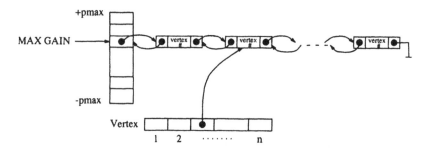

Figure 5.8: The data structure for choosing vertices.

5.3.2.2   Goldberg and Burstein Algorithm

Experimental results have shown that the quality of the final bisection obtained by iterative bisection algorithms, such as K-L algorithm, depends heavily on the ratio of the number of edges to the number of vertices [GB83]. The K-L algorithm yields good bisection if the ratio is higher than 5. However, if the ratio is less than 3, the algorithm performs poorly. The ratio in a typical problem of VLSI design is between 1.8 and 2.5. As a result, Goldberg and Burstein suggested an improvement to the original K-L algorithm or other bisection algorithms by using a technique of contracting edges to increase that ratio.

The basic idea of Goldberg-Burstein algorithm is to find a matching M in graph G, as shown in Figure 5.9(a). The thick lines indicate the edges which form matching. Each edge in the matching is contracted (and forms a vertex) to increase the density of graph. Contraction of edges in M is shown in Figure 5.9(b). Any bisection algorithm is applied to the modified graph and finally, edges are uncontracted within each partition.

5.3.2.3   Component Replication

Recall that the partitioning problem is to partition V into $V_1, V_2, \ldots, V_k$, such that

$$V_i \cap V_j = \phi, \quad i \neq j$$
$$\cup_{i=1}^{k} V_i = V$$

In component (vertex) replication technique, the condition that

$$V_i \cap V_j = \phi, \quad i \neq j$$

is dropped. That is, some vertices are allowed to be duplicated in two or more partitions. The vertex replication technique, presented by Kring and Newton [KN91], can substantially reduce the number of nets that cross boundaries of partitions. Figure 5.10(a) shows a partitioning of a circuit without vertex replication. However, when the inverters are replicated, as in Figure 5.10(b),

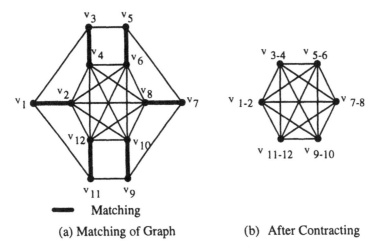

(a) Matching of Graph   (b) After Contracting

Figure 5.9: Matching and edge contraction in graph.

the cutsize is reduced. When a component is replicated, it is copied into both subcircuits and its output are generated locally and do not contribute to the cutsize. Replication does require the inputs to the component to be available on both sides of the partition. If inputs are not available on both sides, the inputs must be propagated across the partition and will contribute to the cutsize.

Once a vertex has been replicated, it tends to remain so and nets connected to the components remain in both subcircuits. Thus, while vertex replication does reduce the cutsize, it tends to reduce the ability to further improve the partition. To achieve good results with this technique, it is critical to limit component replication to where it is most useful by actively limiting the number of replicated components.

The replications of vertices must be done very carefully as in some situations, vertex replication may outweigh the benefit of a reduced cutsize. For example, the added redundancy may increase the circuit area, fault rate and testing. Also, vertex replication cannot be adopted by an arbitrary algorithm. Only those algorithms which carry out partitioning at component level can combine vertex replication techniques to reduce the cutsize. When vertex replication is used in algorithms which deal with more than one components at a time [Kri84], the vertex replication technique can actually increase the cutsize. However, there are cases, especially at the system level, where vertex replication is of great advantage. The algorithm has been tested for two types of circuits, combinational circuits and industrial circuits. The results are summarized in Table 5.2 in which the net cutsize reduction is the percentage reduction in the total number of partitioned nets when compared to partitions obtained without component replication. The component replication is the percentage of the total number of replicated components to the total number of compo-

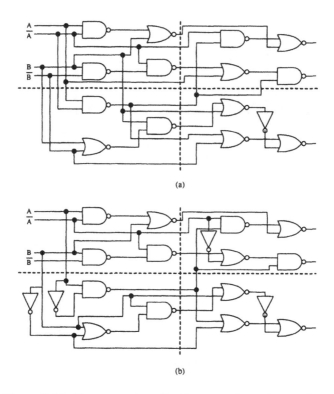

(a)

(b)

Figure 5.10: Component replication to reduce the cutsize.

nents. Table 5.2 clearly shows that vertex replication can substantially reduce
the number of partitioned nets without significantly increasing the size of the
circuit.

5.3.2.4   Ratio Cut

The Kernighan-Lin algorithm yields partitions of comparable sizes, but these
sizes are predefined before partitioning. Since, there are natural clustering
structures in the circuit, predefining the partition size may not be well suited
for partitioning circuits, since there is no way to know the cluster size in circuits
before partitioning. To remedy this situation, Wei and Cheng proposed the
ratio cut as a new metric in order to locate natural clusters in the circuit and
at the same time force the partitions to be of equal sizes [WC89]. Given a
hypergraph $G = (V, E)$, let c_{ij} be the capacity of an edge connecting node
i and node j. Let (V_1, V_2) be a cut that separates a set of nodes V_1 from its
complement V_2 where $V_2 = V - V_1$. The capacity of this cut is equal to $C_{V_1 V_2} =
\sum_{i \in V_1} \sum_{j \in V_2} c_{ij}$. The ratio of this cut $R_{V_1 V_2}$ is defined as $R_{V_1 V_2} = \frac{C_{V_1 V_2}}{|V_1| \times |V_2|}$,
where $|V_1|$ and $|V_2|$ denote the cardinalities of subsets V_1 and V_2 respectively.

Circuit Type	Maximum Circuit Expansion	Net Cutset Reduction	Component Replication
Combinational	10%	41%	3.2%
	30%	43%	3.9%
Industrial	10%	15%	0.7%
	30%	9%	0.3%

Table 5.2: Comparison of different design styles.

The ratio cut is the cut that generates the minimum ratio. The maximum flow minimum cut method [FF62] prefers very uneven subsets which naturally give the lowest cost. Instead of minimizing the cost $C_{V_1 V_2}$, the ratio cut based approach minimizes the ratio $R_{V_1 V_2}$ to alleviate this hidden size effect. Cuts that go through weakly connected groups and groups of similar sizes correspond to smaller ratios. In this way, the minimization of all cuts according to their corresponding ratios balances the effect of minimizing the cost and the effect of keeping the resulting partitions of similar sizes.

Like many other partitioning problems, finding the ratio cut in a hypergraph belongs to the class of NP-complete problems [MS86]. Therefore, a good and fast heuristic algorithm is needed. A heuristic based on Fiduccia and Mattheyses algorithm was proposed in [WC89].

5.4 Simulated Annealing and Evolution

Simulated annealing and evolution belong to the probabilistic and iterative class of algorithms. The simulated annealing algorithm for partitioning is the simulation of the annealing process used for metals. As in the actual annealing process, the value of temperature is decreased slowly till it approaches the freezing point. The simulated evolution algorithm, simulates the biological process of evolution. Each solution is called a *generation*. The generations are improved in each iteration by using operators which simulate the biological events in the evolution process.

5.4.1 Simulated Annealing

Simulated Annealing is a special class of randomized local search algorithms. The optimization of a circuit partitioning with a very large number of components is analogous to the process of annealing, in which a material is melted and cooled down so that it will crystallize into highly ordered state. The energy within the material corresponds to the partitioning score. In an annealing process, the solid-state material is heated to a high temperature until it reaches an amorphous liquid state. It is then cooled very slowly according to a specific schedule. If the initial temperature is high enough to ensure a sufficiently

Algorithm SA
begin
   $t = t_0$;
   $cur\_part = ini\_part$;
   $cur\_score = \text{SCORE}(cur\_part)$;
   repeat
      repeat
        $comp1 = \text{SELECT}(part1)$;
        $comp2 = \text{SELECT}(part2)$;
        $trial\_part = \text{EXCHANGE}(comp1, comp2, cur\_part)$;
        $trial\_score = \text{SCORE}(trial\_part)$;
        $\delta s = trial\_score - cur\_score$;
        if $(\delta s < 0)$ **then**
           $cur\_score = trial\_score$;
           $cur\_part = \text{MOVE}(comp1, comp2)$;
        else
           $r = \text{RANDOM}(0, 1)$;
           if $(r < e^{-\frac{\delta s}{t}})$ **then**
              $cur\_score = trial\_score$;
              $cur\_part = \text{MOVE}(comp1, comp2)$;
      until (equilibrium at t is reached)
      $t = \alpha t$ (* $0 < \alpha < 1$ *)
   until (freezing point is reached)
end.

Figure 5.11: Algorithm SA.

random state, and if the cooling is slow enough to ensure that thermal equilibrium is reached at each temperature, then the atoms will arrange themselves in a pattern that closely resembles the global energy minimum of the perfect crystal.

Early work on simulated annealing used Metropolis algorithm [MRR53]. Since then, much work has been done in this field [CH90, GS84, KGV83, RVS84]. Simulated annealing process starts with a random initial partitioning. An altered partitioning is generated by exchanging some elements between partition. The resulting change in score, δs, is calculated. If $\delta s < 0$ (representing lower energy), then the move is accepted. If $\delta s \geq 0$ then the move is accepted with probability $e^{-\frac{\delta s}{t}}$. The probability of accepting an increased score decreases with the increase in temperature t. This allows the simulated annealing algorithm to climb out of local optimums in search for a global minimum. This idea is presented as a formal algorithm given by Figure 5.11.

The SELECT function is used to select two random components, one from each partition. These components are considered for exchange between the two partitions. The EXCHANGE function is used to generate a trial partioning and does not actually move the components. The SCORE function calculates the cost for the new partitioning generated. If the cost is reduced, this move is accepted and the components are actually moved using the MOVE function. The cost evaluated by the SCORE function can be either the cutsize or a combination of cutsize and other factors which need to be optimized. If the cost is greater than the cost for the partitioning before the component was considered for the move, the probability to accept this move is calculated using the RANDOM function. If the move is accepted, the MOVE function is used to actually move the components in between the partitions.

Simulated annealing is an important algorithm in the class of iterative, probabilistic algorithms. The quality of the solution generated by the simulated annealing algorithm depends on the initial value of temperature used and the cooling schedule. Temperature decrement, defined above as αt, is a geometric progression where α is typically 0.95. Performance can be improved by using the temperature decrement function, $t = te^{-0.7t}$. However, initial temperature and cooling schedule are parameters that are experimentally determined. The higher the initial temperature and the slower the cooling schedule the better is the result but time required to generate this solution is proportional to the steps in which the temperature is decreased.

5.4.2 Simulated Evolution

Simulated Evolution is in a class of iterative probabilistic methods for combinatorial optimization that exploits an analogy between biological evolution and combinatorial optimization.

In biological processes, species become better as they evolve from one generation to the next generation. The evolution process generally eliminates the "bad" *genes* and maintains the "good" genes of the old generation to produce "better" new generation. This concept has been exploited in iterative improvement techniques for some combinatorial optimization problems [CP86, KB89, SR90, SR89]. In this kind of approach, each feasible solution to the problem is considered as a generation. The bad genes of the solution are identified and eliminated to generate a new feasible solution.

In the following discussion, we present a simulated evolution method, Stochastic Evolution (SE) developed by Saab and Rao [SR90]. SE is introduced as a general-purpose iterative stochastic algorithm that can be used to solve any combinatorial optimization problems whose states fit the certain state model given below.

The state model is defined as follows. Given a finite set M of movable elements and a finite set L of locations, a state is defined as a function $S : M \to L$ satisfying certain state-constraints. Also, each state S has an associated cost given by COST(S). The SE algorithm retains the state of lowest cost among those produced by a procedure called PERTURB, thereby generating a new generation. Each time a state is found which has a lower cost than the best state so far, SE decrements the counter by R, thereby increasing the number of its iterations before termination. The general outline of the SE algorithm is given in Figure 5.12.

PERTURB Procedure: In the biological processes, each gene of a specie in the current generation has to prove its suitability under the existing environmental conditions in order to remain unchanged in the next generation. The PERTURB procedure implements this feature by requiring that each movable element $m \in M$ in the current state S has to prove that its location $S(m)$ is suitable to remain unchanged in the next state of the algorithm. Using the state function model described above, the moves are described as follows. Given S and $m \in M$, a move from S with respect to m is just a change in the value of $S(m)$, i.e., a move generates a new function $S' : M \to L$ such that $S'(m) \neq S(m)$ while $S'(m') = S(m')$ for all $m \neq m' \in M$. A move from a state S generates a function $S' : M \to L$ which may not be a state since it may violate certain state-constraints. This function has to be converted into a state before next iteration begins. The cost function should be suitably extended to include such functions. During each call to PERTURB, the elements of the set M of movable elements are scanned in some ordering. The choice of this ordering is problem-specific.

When element $m \in M$ is being scanned, we assume $S : M \to L$ be the existing function that may or may not satisfy the state-constraints. A unique sub-move, which is a move from S, is associated with m that generates a new function $S' : M \to L$ such that $S'(m) \neq S(m)$. The details of the sub-move associated with m will be given in below for the partitioning problem. Define $Gain(m) = \text{COST}(S) - \text{COST}(S')$ as the reduction in cost after the sub-move is performed. The procedure PERTURB decides whether or not to accept the sub-move associated with the element m. This decision is made stochastically by using a non-positive control parameter p as follows. The value of $Gain(m)$ is compared to a integer r randomly generated in the interval $[p, 0]$. If $Gain(m) > r$, then the sub-move to S' is accepted; otherwise, the sub-move is rejected. Since $r \leq 0$, sub-moves with positive gains are always accepted. The algorithm then scans the next element in M. The final function S generated after scanning all elements of M may not satisfy the state-constraints of the problem. In such a case, a function MAKE-STATE(S) is called to reverse the fewest number of latest sub-moves accepted so that all the state-constraints are satisfied. The outline of PERTURB procedure can be outlined as given by Figure 5.12.

Some modifications to the above structure of PERTURB are possible. For example, only a subset M' of M may be scanned in order to save computation time.

The UPDATE procedure: This procedure is used for updating the value of the control parameter p. Initially, p is set to a negative value close to zero so that only moves with small negative gains are performed. It has been observed that moves with large negative gains tend to upset the optimization process and only increase the running time of the algorithm. Hence, the value of p is

Algorithm SE
begin
    $S = S_0$;   (* initial state *)
    $S_{BEST} = S$;   (* save initial state *)
    $p = p_0$;   (* initialize control parameter *)
    $\gamma = 0$;   (* initialize counter *)
    repeat
        $C_{pre} = \text{COST}(S)$;
        $S = \text{PERTURB}(S, p)$;
        $C_{cur} = \text{COST}(S)$;
        $\text{UPDATE}(p, C_{pre}, C_{cur})$;
        if $(\text{COST}(S) < \text{COST}(S_{BEST}))$ **then**
            $S_{BEST} = S$;   (* save best state *)
            $\gamma = \gamma - R$;   (* decrement counter *)
        else
            $\gamma = \gamma + 1$;   (* increment counter *)
    until $(\gamma > R)$;   (* stopping criterion *)
    return (S_{BEST});   (* report best state *)
end.

Procedure PERTURB(S)
begin
    for(each $m \in M$) **do**
        $S' = \text{SUB-MOVE}(S, m)$;
        $Gain(m) = \text{COST}(S) - \text{COST}(S')$;
        if $Gain(m) > \text{RANDOM}(p, 0)$
            $S = S'$;
    $S = \text{MAKE-STATE}(S)$;
    return S;
end.

Procedure UPDATE(p, C_{pre}, C_{cur})
begin
    if $C_{pre} = C_{cur}$ **then**
        $p = p - 1$;
    else
        $p = p_0$;
end.

Figure 5.12: Algorithm SE.

reduced only when necessary. During each iteration, the cost C_{cur} of the new state is compared to the cost C_{pre} of the previous state. If both costs are same, p is decremented. Otherwise, p is reset to its initial value. The parameter p is decremented to give the algorithm a chance to escape a local minimum via an uphill climb. The procedure UPDATE is given in Figure 5.12.

Choice of R: The stopping criterion parameter R acts as the expected number of iterations the SE algorithm needs to achieve the objective. The quality of the final state obtained increases with the increase of R. If R is too large, then SE wastes time during the last iterations because it cannot find better states. On the other hand, if R is too small, then SE might not have enough time to improve the initial state.

Let us now discuss the application of SE to partitioning. Using the state model described above, movable elements of a state is the set of vertices, that is $M = V$, and locations of states are the two partitions, that is, $L = 1, 2$. A partition therefore is a function $S : V \rightarrow \{1, 2\}$, where the two partitions of the vertex set are the subsets V_1 and V_2. A state (or a bisection) is a partition which satisfy the state-constraint $|V_1| = |V_2|$. Then, the PERTURB procedure scans the vertex set V in some order, i.e., if u and v are two vertices and $u < v$, then u is scanned before v. The sub-move in PERTURB from S that is associated with a vertex $v \in V$ is a move that transfers v from its current partition to the other partition. More precisely, $S' = $ SUBMOVE(S, v) is an onto function such that $S'(v) = 3 - S(v)$ representing that vertex v is transferred from one partition to another partition and $S'(u) = S(u)$ for all $u \neq v$ representing that all the location of other vertices remain unchanged. Note that S', in general, represents a partition which may or may not be a bisection. After all the vertices have been scanned and the decisions to make the corresponding sub-moves have been made, the resulting function S may not be a state, which means that it may not represent a bisection. Suppose $|V_1| - |V_2| = k > 0$, then MAKE-STATE(S) generates a state from S by reversing the last appropriate $k/2$ sub-moves performed.

The time complexity of SE is proportional to the time required for the computation of the sub-moves and gains associated with each movable element of M. Suppose c is an upper bound on the time required for each sub-move and gain computation, then each iteration of SE runs in $O(c \times |M|)$ time. Since c is either a constant or is linear in the problem size, the SE algorithm for these problems requires either linear or quadratic time per iteration.

The simulated evolution and simulated annealing algorithms are computation intensive. The key difference between these two kinds of algorithms is that the simulated evolution uses the history of previous trial partitionings. Therefore, it is more efficient than simulated annealing. However, it takes more space to store the history of the previous partitioning than the simulated annealing.

5.5 Other Partitioning Algorithms

Besides the group migration and simulated annealing/evolution methods, there are other partitioning methods. In this section, we will present the metric allocation algorithm. The references to other algorithms are provided at the end of the chapter.

5.5.1 Metric Allocation Method

Initial work on measuring the connectivity with a metric was carried out by Charney and Plato [CP68]. They showed that using electrical analog of the network minimizes the distance squared between the components. Partitioning starts after all the values of the metric have been computed; these values are calculated from eigenvalues of the network. This method is described by Cullum, Donath, and Wolfe [CDW75].

The basic metric allocation partitioning algorithm starts with a set V of the nodes and a set $S = S_1, S_2, ..., S_N$ of the nets. A metric value over $V \times V$ is computed. Nodes in V are then partitioned into subsets $V_1, V_2, ..., V_k$ such that sum of the areas in V_i is less than or equal to A for all i and the number of nets with members both internal to V_i and external to V_i is less than T for all i, where, A and T represent the area and terminal constraint for each partition, respectively. The algorithm given in Figure 5.13 determines if a k-way partition can be done to satisfy the requirements.

The function CONSTRUCT-ST is used to construct the spanning trees for each net in the netlist. All the edges of these spanning trees are added to a set L by using the function ADD-EDGES. The procedure SORT-ASCENDING sorts L in an ascending order on the metric used. Each vertex v_i is assigned to a individual group G_i by the function INITIALIZE_GROUPS. The groups to which vertices v_i and v_j, joined by edge e_{ij}, belong are collapsed to form a single group if the area and terminal count restriction is not violated. The merging process is carried out by MERGE-GROUPS. This routine also keeps track of the order in which the groups are merged. The function AREA is used to calculate area of a group while function COUNT gives the number of terminals in a group. If such mergings of the groups reduce the number of groups to K or less, the set of groups is returned by the algorithm. If after merging all possible groups, if the number of groups is greater than K, then the smallest group is selected by using function SELECT_SMALL. An attempt is made to merge this group with another group which causes the least increase in area and terminal count of the resulting group. If such a group is found the flag *merge_success* is set to TRUE. The function STORE-MIN is used to store the group which causes the smallest increase in area and terminal count. The function RESTORE-MIN returns the group which is stored by STORE-MIN. If the smallest group consists of only a single component and *merge_success* is FALSE, the algorithm returns a null set indicating failure. If the smallest group consists of more than one component and the *merge_success* flag is set to FALSE, function SELECT_LARGE is used identify the largest group among

Algorithm METRIC-PARTITION
begin
    for($i = 1$ to N) **do**
        CONSTRUCT-ST(S_i);
        ADD-EDGES(S_i, L);
    SORT-ASCENDING($L, metric$);
    no_groups = INITIALIZE-GROUPS(V);
    while($L \neq \phi$) **do**
        e_{ij} = SELECT-EDGE(L);
        if($(G_i \neq G_j)$ **and**
            $(\text{AREA}(G_i) + \text{AREA}(G_j) \leq A)$ **and**
            $(\text{COUNT}(G_i) + \text{COUNT}(G_j) \leq T)$)
            MERGE-GROUPS(G_i, G_j);
            $no\_groups = no\_groups - 1$;
            if($no\_groups \leq K$)
                return(G);
        else
            continue;
    while($no\_groups > K$) **do**
        G_i = SELECT $\_$ SMALL();
        for($j = 1$ to no_groups) **do**
            if($i \neq j$) **and**
                $(\text{AREA}(G_i) + \text{AREA}(G_j) \leq A)$ **and**
                $(\text{COUNT}(G_i) + \text{COUNT}(G_j) \leq T)$)
                STORE-MIN(G_j);
                $merge\_success$ = TRUE;
        MERGE-GROUPS(G_i, RESTORE-MIN(G_j));
        if($no\_groups \leq K$)
            return(G);
        if($merge\_success$ = FALSE)
            if(SIZE(G_i) = 1)
                return(ϕ);
        else
            G_j = SELECT $\_$ LARGE();
            DECOMPOSE(G_j, G_k, G_l);
end.

Figure 5.13: Algorithm METRIC-PARTITION.

all groups. This group is decomposed into two subgroups by using function DECOMPOSE and procedure is repeated.

5.6 Performance Driven Partitioning

In recent years, with the advent of the high performance chips, the on-chip delay has been greatly reduced. Typically on-chip delay is in the order of a few nanoseconds while on-board delay is in the order of a few milliseconds. The on-board delay is three orders of magnitude larger than on-chip delay. If a critical path is cut many times by the partition, the delay in the path may be too large to meet the goals of the high performance systems. The design of a high performance system requires partitioning algorithms to reduce the cutsize as well as to minimize the delay in critical paths. The partitioning algorithms, which deal with high performance circuits, are called as *timing (performance) driven* partitioning algorithms and the process of partitioning for such circuits is called timing (performance) driven partitioning.

For timing driven partitioning algorithms, in addition to all the other constraints, timing constraints have to be satisfied. Discussion on these types of partitioning problems for FPGAs can be found in Chapter 11. Timing driven partitioning plays a key role in MCM design and will be discussed in Chapter 12.

The partitioning problem for high performance circuits can be modeled using directed graphs. Let $G = (V, E)$ be a weighted directed graph. Each vertex $v_i \in V$ represents a component (gate) in the circuit and each edge represents a connection between two gates. Each vertex v_i has a weight $GD(v_i)$, specifying the gate delay associated with the gate corresponding to v_i. Each edge (v_i, v_j) has a delay associated with it, which depends on the partitions to which v_i and v_j belong. The edge delay, $ED_{ij} = (d_1, d_2, d_3)$ specifies the delay between v_i and v_j. The delay associated with edge (v_i, v_j) is d_1 if the edge is cut at chip level. If the edge is cut at board level the delay associated with the edge is d_2 and it is d_3 if the edge gets cut at system level. This problem is very general and is still a topic of intensive research.

A timing driven partitioning addresses the problem of clustering a circuit for minimizing its delay, subject to capacity constraints on the clusters. The early work on this problem was done by Stone [Sto66]. When the delay inside a cluster is assumed to be negligible compared to the delay across the clusters, then the following algorithm by Lawler, Levitt and Turner [LLT69], which uses a unit delay model, can be used. The circuit components are represented by a group of vertices or nodes and the nets are represented as directed edges. Each vertex, v_i, has a weight, $W(v_i)$, attached to it indicating the area of the component. A label, $L(v_i)$, is given to each node, v_i, to identify the cluster to which the node belongs. The labeling is done as follows: All the input nodes are labeled 0. A node, v_i, all of whose predecessors have been labeled, is identified. Let k be the largest predecessor label, $WP_i(k)$ be the total weight of all the k-predecessors, and M be the largest weight that can be accommodated in a

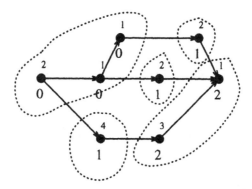

Figure 5.14: Labeling Sequence and clusters formed.

cluster. If $WP_i(k) + W(v_i) \leq M$, label of vertex v_i is set to k, i.e., $L(v_i) = k$, otherwise, vertex v_i gets the label, $k + 1$. After all the vertices are labeled, a vertex, v_j, is identified such that none of the successors of v_j have the same label as v_j. The vertex v_j and all its $k-$predecessors form a cluster. The vertex v_j is called the root of the cluster. Similar procedure is carried out till all the vertices are clustered. This clustering mechanism may cause a vertex to be in more than one cluster in which case it has to be replicated appropriately. The label for any vertex v_i,as defined above, is the maximum delay of the signal when the signal reaches the vertex v_i after assuming the delay inside a cluster is zero. Thus the above model represents the minimization of the maximum delay of signal under the area constraints when the delay inside a cluster is assumed zero. Figure 5.14 shows a digraph representing a circuit. The number above a vertex indicates the weight of the vertex while the number below a vertex denotes the label of the vertex. M is set equal to 4. Clusters formed are also shown in Figure 5.14.

The clusters (e.g., chips) have large capacities, and very likely, the critical path inside a cluster will be comparable to the total delay of the circuit. Therefore, to be more general, it is better to use more realistic delay model. In a general delay model, each gate of the combinational circuit has a delay associated with it. Considering this problem, Murgai, Brayton and Sangiovanni-Vincentelli [MBV91] proposed an algorithm to reduce this delay. The key idea is to label the vertices (gates) according to the clusters' internal delay. Then number of clusters is minimized without increasing the maximum delay. Minimizing the number of clusters and vertices reduces the number of components and hence the cost of the design. The number of clusters is minimized by merging, subject to a capacity constraint.

5.7 Summary

Partitioning divides a large circuit into a group of smaller subcircuits. This process can be carried out hierarchically until each subcircuit is small enough to be designed efficiently. These subcircuits can be designed independently and simultaneously to speed up the design process. However, the quality of the design may suffer due to partitioning. The partitioning of a circuit has to be done carefully to minimize its ill effects. One of the most common objectives of partitioning is to minimize the cutsize which simplifies the task of routing the nets. The size of the partitions should be balanced. For high performance circuits, the number of times a critical path crosses the partition boundary has to be minimized to reduce delay. Partitioning for high performance circuits is an area of current research, especially so with the advent of high performance chips, and packaging technologies.

Several factors should be considered in the selection of a partitioning algorithm. These factors include difficulty of implementation, performance on common partitioning problems, and time complexity of the algorithm. Group migration method is faster and easier to implement. Metric allocation method is more costly in computing time than group migration method, and hardest to implement since it requires numerical programming. The results show that simulated annealing usually takes much more time than the Kernighan-Lin algorithm does. On a random graph, however, the partitions obtained by simulated annealing are likely to have a smaller cutsize than those produced by the Kernighan-Lin algorithm. Simulated evolution may produce better partition than simulated annealing, but it has larger space complexity. The algorithms for bipartitioning presented in this chapter are practical methods. They are mainly used for bipartitioning, but can be extended to multiway partitioning.

5.8 Exercises

1. Partition the graph shown in Figure 5.15, using Kernighan-Lin algorithm.

† 2. Extend Kernighan-Lin algorithm to multiway partitioning of graph.

3. Apply Fiduccia-Mattheyses algorithm for the graph in Figure 5.15 by considering the weights for the vertices, which represent the areas of the modules. Areas associated with the vertices are: $v_1 = 10$, $v_2 = 12$, $v_3 = 8$, $v_4 = 15$, $v_5 = 13$, $v_6 = 20$, $v_7 = 9$, $v_8 = 7$, $v_9 = 14$ and $v_{10} = 9$. The areas of the two partitions should be as equal as possible. Is it possible to apply the Kernighan-Lin algorithm in this problem?

† 4. For the graph in Figure 5.16, let the delay for the edges going across the partition be 20 nsec. Each vertex has a delay which is given below. Consider vertex v_1 as the input node and vertex v_8 as the output node. Partition the graph such that the delay between the input node and

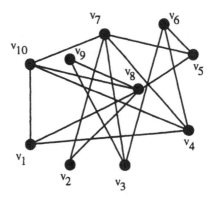

Figure 5.15: A graph partitioning problem.

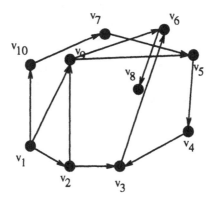

Figure 5.16: A delay minimization problem in graph partitioning.

output node is minimum and the partitions have the same size. The delays for the vertices are $d(v_1) = 3$ nsec, $d(v_2) = 2$ nsec, $d(v_3) = 1$ nsec, $d(v_4) = 2$ nsec, $d(v_5) = 3$ nsec, $d(v_6) = 4$ nsec, $d(v_7) = 3$ nsec, $d(v_8) = 8$ nsec, $d(v_9) = 7$ nsec and $d(v_1 0) = 5$ nsec.

5. Apply the vertex replication algorithm to the graph given in Figure 5.16.

† 6. Implement Fiduccia-Mattheyses and the Kernighan-Lin algorithms for any randomly generated instance, and compare the cutsize.

† 7. Implement the Simulated Annealing and Simulated Evolution algorithms described in the text. Compare the efficiency of these two algorithms on a randomly generated example. In what aspects do these algorithms differ?

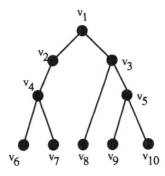

Figure 5.17: A problem instance showing critical net.

8. Compare the performance of the Simulated Annealing algorithm for different values of α.

† 9. Consider the tree shown in Figure 5.17, which represents a critical net. Partition the tree into four partitions, two of which will be on one chip and the other two partitions on another chip. Let the delay values of each vertex be the same as that in problem 3. Let the interchip delay be 20 nsec and the delay between the two partitions on the same chip be 10nsec. The objective is to partition the tree into four partitions such that the longest delay path from the root of the tree to any of its leaves is minimized and the number of vertices on each partition is as equal as possible.

‡ 10. Suggest modifications to the Kernighan-Lin algorithm to speed up the algorithm.

† 11. In the application of vertex replication technique, as the vertex replication percentage increases, the cutsize decreases. However, at the same time, the layout area increases as well. Is it possible to get a graph showing the relation between cutsize and circuit area as the component replication percentage varies? Use a randomly generated example. From your result, can you obtain an optimal strategy such that the trade off between cutsize and layout area is compromised?

† 12. Implement the ratio cut algorithm. Is it possible to use the vertex replication technique in the ratio cut model?

Bibliographic Notes
Many other partitioning approaches have been proposed in solving circuit partitioning problems, such as network flow [FF62], and eigenvector decomposition [FK86, Hal70] etc. The maximum flow minimum cut algorithm presented by

Ford and Fulkerson [FF62] is an exact algorithm for finding a minimum cost bipartitions for a network without considering the area constraints for the partitions. In many cases, e.g., system level partitioning, the partitioning problem with objective to reduce the number of interconnections, does not represent the actual problem because the area constraints are not considered in this model. Certain algorithms simplify the partitioning problem by restricting the range of the circuits that can be partitioned e.g., partitioning algorithms for planar graphs [Dji82, LT79, Mil84]. Partitioning problem in planar graphs has been discussed in [LT79]. But clearly all circuits cannot be represented as planar graphs. Hence, planar graph algorithms are not very practical in the partitioning of VLSI circuits.

There is an interesting trend in which an interactive man-machine approach is used in solving partitioning problems. Interested readers should read [BKM+66, HMW74, Res86]. A 'Functional Partitioning' which takes into account certain structural qualities of logic circuits, namely loops and reconverging fan-out subnets, can be found in [Gro75]. A new objective function to reduce the number of pins was presented in [Hit70].

A partitioning which intends to form partitions with equal complexity, e.g., similar in terms of area, yield and speed performance, was introduced in [YKR87]. A partitioning model was formulated in which components are assigned probabilities of being placed in bins, separated by partitions. The expected number of nets crossing partitions is a quadratic function of these probabilities. Minimization of this expected value forces condensation of the probabilities into a 'definite' state representing a very good partitioning [Bia89].

A neural network model was proposed in [YM90] for circuit bipartitioning. The massive parallelism of neural nets has been successfully exploited to balance the partitions of a circuit and to reduce the external wiring between the partitions. A constructive partitioning method based on resistive network optimization was developed in [CK84]. Another partitioning technique called clustering was presented in [CB87, Joh67, McF83, McF86, Raj89, RT85]. The simulated annealing algorithm described in this chapter, generates one move at random, evaluates the cost of the move, and then accepts it or rejects it. Greene and Supowit [GS84] proposed an algorithm whereby a list of moves is generated and the moves are taken from the list by a random selection process. In [CLB94] J. Cong, Z. Li, and R. Bagrodia present two algorithms in the acyclic multi-way partitioning approach.

Chapter 6

Floorplanning and Pin Assignment

After the circuit partitioning phase, the area occupied by each block (sub-circuit) can be estimated, possible shapes of the blocks can be ascertained and the number of terminals (pins) required by each block is known. In addition, the netlist specifying the connections between the blocks is also available. In order to complete the layout, we need to assign a specific shape to a block and arrange the blocks on the layout surface and interconnect their pins according to the netlist. The arrangement of blocks is done in two phases; Floorplanning phase, which consists of planning and sizing of blocks and interconnect and the Placement phase, which assign a specific location to blocks. The interconnection is completed in the routing phase. In the placement phase, blocks are positioned on a layout surface, in a such a fashion that no two blocks are overlapping and enough space is left on the layout surface to complete the interconnections. The blocks are positioned so as to minimize the total area of the layout. In addition, the locations of pins on each block are also determined.

The input to the Floorplanning phase is a set of blocks, the area of each block, possible shapes of each block and the number of terminals for each block and the netlist. If the layout of the circuit within a block has been completed then the dimensions(shape) of the block are also known. The blocks for which the dimensions are known are called *fixed* blocks and the blocks for which dimensions are yet to be determined are called *flexible* blocks. Thus we need to determine an appropriate shape for each block (if shape is not known), location of each block on the layout surface, and determine the locations of pins on the boundary of the blocks. The problem of assigning locations to fixed blocks on a layout surface is called the *Placement* problem. If some or all of the blocks are flexible then the problem is called the *Floorplanning* problem. Hence, the placement problem is a restricted version of the floorplanning problem. If one asks for planning of the interconnect in addition to floorplanning, then it is referred to as the chip planning problem . Thus floorplanning is a restricted version of chip planning problem. The terminology is slightly confusing as floorplanning problems are placement problems as well but these terminologies have been widely used and accepted. It is desirable that the pin locations are identified at the same time when the block locations are fixed. However, due to the complexity of the placement problem, the problem of identifying the pin locations for the blocks is solved after the locations of all the blocks are known. This process of identifying pin locations is called *pin assignment*.

Chip planning, Floorplanning and Placement phases are very crucial in overall physical design cycle. It is due to the fact, that an ill-floorplanned layout cannot be improved by high quality routing. In other words, the overall quality of the layout, in terms of area and performance is mainly determined in the chip planning, floorplanning and placement phases. In this chapter we will review Floorplanning and pin assignment algorithms.Algorithms for placement will be discussed in the subsequent chapter.

There are several factors that are considered by the chip planning, floorplanning, pin assignment and placement algorithms. These factors are discussed below:

1. **Shape of the blocks:** In order to simplify the problem, the blocks are assumed to be rectangular. The shapes resulting from the floorplanning algorithms are mostly rectangular for the same reason. The floorplanning algorithms use *aspect ratios* for determining the shape of a block. The aspect ratio of a block is the ratio between its height and its width. Usually there is an upper and a lower bound on the aspect ratios, restricting the dimensions that the block can have. More recently, other shapes such as L-shapes have been considered, however dealing with such shapes is computationally intensive.

2. **Routing considerations:** In chip planning, it is required that routing is considered as an integral part of the problem. In placement and floorplanning algorithms it maybe sufficient to estimate the area required for routing. The blocks are placed in a manner such that there is sufficient routing area between the blocks, so that routing algorithms can complete the task of routing of nets between the blocks. If complete routing is not possible, placement phase has to be repeated.

3. **Floorplanning and Placement for high performance circuits:** For high performance circuits the blocks are to be placed such that all critical nets can be routed within their timing budgets. In other words, the length of critical paths must be minimized. The floorplanning(placement) process for high performance circuits is also called as *performance driven* floorplanning(placement).

4. **Packaging considerations:** All of these blocks generate heat when the circuit is operational. The heat dissipated should be uniform over the entire surface of the group of blocks placed by the placement algorithms. Hence, the chip planning, floorplanning and placement algorithms must place the blocks, which generate a large amount of heat, further apart

from each other. This might conflict with the objective for high performance circuits and some trade off has to be made.

5. **Pre-placed blocks:** In some cases, the locations of some of the blocks may be fixed, or a region may be specified for their placement. For example, in high performance chips, the clock buffer may have to be located in the center of the chip. This is done with the intention to reduce the time difference between arrival time of the clock signal at different blocks. In some cases, a designer may specify a region for a block, within which the block must be placed.

In this chapter, we will discuss floorplanning and pin assignment problems in different design styles. Section 6.1 discusses the Floorplanning problem and algorithms for the floorplanning problems. Section 6.2 presents a brief introduction to chip planning, while pin assignment is discussed in Section 6.3. In Section 6.4, we discuss integrated approach to these problems.

6.1 Floorplanning

As stated earlier, Floorplanning is the placement of flexible blocks, that is, blocks with fixed area but unknown dimensions. It is a much more difficult problem as compared to the placement problem (discussed in Chapter 7). In floorplanning, several layout alternatives for each block are considered. Usually, the blocks are assumed to be rectangular and the lengths and widths of these blocks are determined in addition to their locations. The blocks are assigned dimensions by making use of the aspect ratios. The aspect ratio of a block is the ratio of the width of the block to its height. Usually, there is an upper and a lower bound on the aspect ratio a block can have as the blocks cannot take shapes which are too long and very thin. Initial estimate on the set of feasible alternatives for a block can be made by statistical means, i.e., by estimating the expected area requirement of the block. Many techniques of general block placement have been adapted to floorplanning. The only difference between floorplanning and general block placement is the freedom of cells' interface characteristic. Like placement, inaccurate data partly affects floorplanning. In addition to the inaccuracy of the cost function that we optimize, the area requirements for the blocks may be inaccurate.

Floorplanning algorithms are typically used in hierarchical design. This is due to the fact that, although the dimensions of each leaf of the hierarchical tree may be known, the blocks at the node level in the tree are *flexible*, i.e., they can take any dimension. Hence, the floorplanning algorithms are used at each of the nodes in the tree so that the area of the layout is minimum and the position of all the blocks are identified.

6.1.1 Problem Formulation

The input consists of $B_1, B_2, \ldots, B_n$ circuit blocks, with area $a_1, a_2, \ldots, a_n$ respectively. Associated with each block are two aspect ratios A_i^l and A_i^h, which give the lower and the upper bound on the aspect ratio for that block. The floorplanning algorithm has to determine the width w_i and height, h_i of each block B_i such that $A_i^l \leq \frac{h_i}{w_i} \leq A_i^h$. In addition to finding the shapes of the blocks, the floorplanning algorithm has to generate a valid placement such that the area of the layout is minimized.

A *slicing floorplan* is a floorplan which can be obtained by recursively partitioning a rectangle into two parts either by a vertical line or a horizontal line. The cut tree obtained by min-cut algorithm is known as *slicing tree*. A slicing tree is a binary tree in which each leaf represents a partition and each internal node represents a cut. Consider the floorplan as shown in Figure 6.1. Partitions are labeled with letters and cutlines are labeled with numbers. Figure 6.1(b) shows the slicing tree for the floorplan in Figure 6.1(a). Figure 6.1(c) is the slicing tree indicating the cut direction. Figure 6.1(d) shows a floorplan for which there is no valid slicing tree.

A floorplan is said to be *hierarchical* of *order k*, if it can be obtained by recursively partitioning a rectangle into at most k parts. The hierarchy of a hierarchical floorplan can be represented by a *floorplan tree*. Figure 6.2 shows a hierarchical floorplan of order 5 and its floorplan tree. Each leaf in the tree corresponds to a basic rectangle and each internal node corresponds to a composite rectangle in the floorplan. An important class of hierarchical floorplans is the set of all slicing floorplans.

6.1.1.1   Design Style Specific Floorplanning Problems

Floorplanning is not carried out for some design styles. This is due to the fixed dimensions of blocks in some design styles.

1. **Full custom design style:**   Floorplanning for general cells is the same as discussed above.

2. **Standard cell design style:**   In standard cell design style, the dimensions of cells are fixed, and floorplanning problem is simply the placement problem. For large standard cell design, circuit is partitioned into several regions, which are floorplanned, before cells are placed in regions.

3. **Gate array design style:**   Like standard cells, the floorplanning problem is same as placement problem.

6.1.2   Classification of Floorplanning Algorithms

Floorplanning methods can be classified as follows:

1. Constraint based methods.

2. Integer programming based methods.

3. Rectangular dualization based methods.

4. Hierarchical tree based methods.

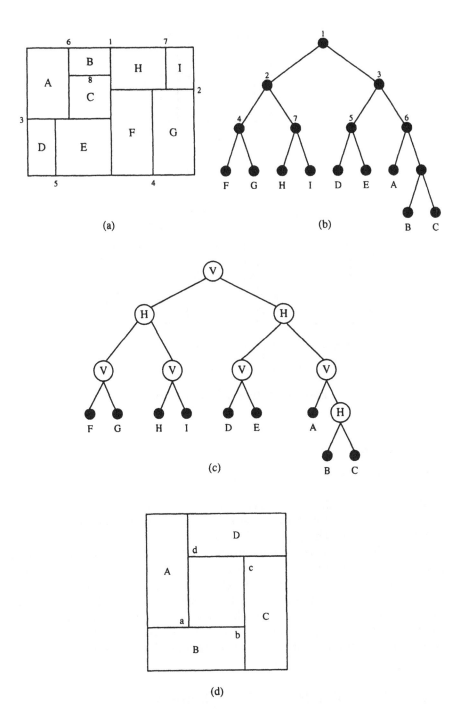

Figure 6.1: A floorplan with slicing tree and a non-slicing floorplan.

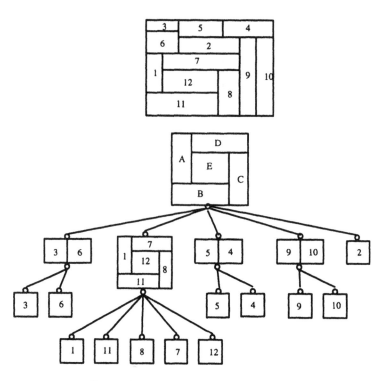

Figure 6.2: Hierarchical Floorplan.

5. Simulated Evolution algorithms

6. Timing Driven Floorplanning Algorithms

These methods will be discussed in the following subsections.

Beside the methods stated above, several simple methods may be used, such as min-cut method. The process of min-cut can be used to construct a sized floorplan. The first phase of min-cut method i.e., bipartition of a weighted graph, helps in constructing the floorplan. The weight of the vertex roughly estimates the area taken up by the block. This weight may represent the area of the corresponding cell in general-cell placement. The initial sized floorplan represents an empty base rectangle whose area is the total of all weights of the vertices of the weighted graph and each node in the tree represents a rectangular room in the layout area. All the floorplans that can be generated with min-cut bipartitioning are slicing floorplans.

6.1.3   Constraint Based Floorplanning

This method, proposed by Vijayan and Tsay [VT91], constructs a floorplan of optimal area that satisfies (respects) a given set of constraints. A set of horizontal and vertical topological (i.e., ordering) constraints is derived from the relative placement of blocks. Given a constraint set, it is usually the case that there is no reason to satisfy all the constraints in the set. This is especially true when a majority of the blocks have flexible shapes. A floorplan is said to respect a constraint, if for each pair of blocks, the floorplan satisfies at least one constraint (horizontal or vertical). A constraint set is said to be *overconstrained* if it has many redundant constraints. It is desirable to derive a complete constraint set from the input relative placement and then to remove those redundant constraints that result in reduction of floorplan area.

A topological constraint set of a set of blocks is given by two directed acyclic graphs (G_H, G_V): G_H is the horizontal constraint graph and G_V is the vertical constraint graph. In order to reduce the floorplan area, the heuristic iteratively removes a redundant constraint from the critical path of either G_H or G_V and also iteratively reshapes the blocks on the critical paths of the two graphs. Critical path is the longest path in G_H or G_V. The *input* to the algorithm is a constraint set (G_H, G_V) of the set of blocks. To minimize the floorplan area, repeat steps 1 and 2 until there is no improvement in the floorplan area.

1. **Step 1:** Repeat the following steps until no strongly redundant edges on P_H or P_V exist. G_H and G_V are topologically sorted and swept. Either P_H or P_V, whichever is more critical is selected, where P_H and P_V are the critical paths of G_H and G_V respectively. The strongly redundant edge on the selected critical path is eliminated.

2. **Step 2:** The current shapes of the blocks are stored and a path, either P_V or P_H is selected depending on which of the two is more critical. All the flexible blocks on the selected path are reshaped. G_H and G_V are scanned again to construct the new floorplan. If the newly generated floorplan is better than the previous one the stored block shapes are updated. All the steps described above are repeated, a specified number of times.

Each pass of the algorithm constitutes one execution of two steps. Constraint reduction takes place in step 1 and step 2 does the reshaping of the blocks. If the chip dimensions are fixed, the passes are repeated until the target dimensions are reached. Otherwise the passes can be repeated until there is improvement in the floorplan area. Typically three or four passes are required.

The purpose of removing a redundant edge on the critical path is to break the path into two smaller paths. A good choice for such a redundant edge is the one which is nearest to the center point of the path. The above heuristic removes only one redundant constraint from a critical path at each iteration, and thus seeks to minimize the number of constraints removed. An edge can be checked for strong redundancy in constant time if we maintain the adjacency matrix of G_H and G_V. It takes $O(n^2)$ time to set up adjacency matrices. A topological sort of a directed acyclic graph with n nodes and m edges takes $O(m + n)$ time. The number of topological sorts executed depends on the number of redundant edges removed, the user-specified value for the number of reshaping iterations, and the number of passes.

6.1.4 Integer Programming Based Floorplanning

In this section an integer programming formulation for generating the floorplan developed by Sutanthavibul, Shragowitz and Rosen [SSR91] is presented.

The floorplanning problem is modeled as a set of linear equations using 0/1 integer variables. Two types of constraints are considered: the overlap constraints and the routability constraints. The overlap constraints prevent any two blocks from overlapping whereas the routability constraints estimate the routing area required between the blocks. For the critical nets, net lengths are specified which should not be exceeded. The length of the net depends on the timing budget of that net. The critical net constraints ensure that the length of the critical nets does not exceed this specified value. We now describe how the constraints can be developed.

1. **Block overlap constraints for fixed blocks:** Given two fixed (rigid) blocks, B_{ri} and B_{rj} which should not overlap, we have four possible ways to position the two blocks so as to avoid overlap. Let $\{x_i, y_i, w_i, h_i\}$ and $\{x_j, y_j, w_j, h_j\}$ be the 4-tuples associated with blocks B_{ri} and B_{rj} respectively, where (x_i, y_i) gives the location of the block, w_i is the width of the block and h_i is the height of the block. The block B_{rj} can be positioned to the right, left, above or below block B_{ri}. These conditions transformed into equations given below:

$$
\begin{aligned}
x_i + w_i \le x_j &\quad (B_{rj} \text{ is to the right of } B_{ri}), &\quad \text{or} \\
x_i - w_j \ge x_j &\quad (B_{rj} \text{ is to the left of } B_{ri}), &\quad \text{or} \\
y_i + h_i \le y_j &\quad (B_{rj} \text{ is to the above of } B_{ri}), &\quad \text{or} \\
y_i - h_j \ge y_j &\quad (B_{rj} \text{ is to the below of } B_{ri}) &\quad (1)
\end{aligned}
$$

   To satisfy one of these equations, two 0-1 integer variables x_{ij} and y_{ij} are used for each pair of blocks. Two bounding functions W and H are defined such that, $|x_i - x_j| \le W$ and $|y_i - y_j| \le H$. W can be equal to $W_{\max}$ which is the maximal allowed width of the chip or $W = \sum_{i=1}^{p+q+r} w_i$. Similarly, $H = H_{\max}$, the maximal allowed height of the chip or $H = \sum_{i=1}^{p+q+r} h_i$. Equation set (1) can be rewritten with the introduction of the integer variables to generate the 'or' condition as,

$$
\begin{aligned}
x_i + w_i &\le x_j + W(x_{ij} + y_{ij}) \\
x_i - w_j &\ge x_j + W(1 - x_{ij} + y_{ij}) \\
y_i + h_i &\le y_j + W(1 + x_{ij} - y_{ij}) \\
y_i - h_j &\ge y_j + W(2 - x_{ij} - y_{ij}) \qquad (2)
\end{aligned}
$$

   As the integer variables x_{ij} and y_{ij} can take either 0 or 1 values, only one of the above equations in (2) will be active and other equations will be true depending on the value of x_{ij} and y_{ij}. For example, when $x_{ij} = y_{ij} = 1$, the first equation in (2) becomes active and all other equations are true.

$$
\begin{aligned}
x_i &\ge 0, \qquad y_i \ge 0 \\
x_i + w_i &\le W, \\
y^* &\ge y_i + h_i \qquad (3)
\end{aligned}
$$

where y^* is the height to be minimized. To allow rotation of the blocks so as to optimize the solution, another integer variable z_i is used for each block. z_i is 0 when the block is in its initial orientation and 1 when the block is rotated by $90°$. The constraints for the fixed blocks can be rewritten as:

$$x_i + z_i h_i + (1 - z_i) w_i \leq x_j + M(x_{ij} + y_{ij})$$
$$x_i - z_i h_j - (1 - z_j) w_j \geq x_j + M(1 - x_{ij} + y_{ij})$$
$$y_i + z_i w_i + (1 - z_i) h_i \leq y_j + M(1 + x_{ij} - y_{ij})$$
$$y_i - z_j w_j - (1 - z_j) h_j \geq y_j + M(2 - x_{ij} - y_{ij}) \tag{4}$$

where, $M = \max(W, H)$. Constraints (3) are rewritten as:

$$x_i \geq 0, \quad y_i \geq 0,$$
$$x_i + (1 - z_i) w_i + z_i h_i \leq W,$$
$$y^* \geq y_i + (1 - z_i) w_i + z_i h_i \tag{5}$$

where y^* is the height to be minimized. The floorplanning problem, for fixed blocks without taking into consideration either routing areas or critical nets can be solved by finding the minimum y^* subject to constraints (4) and (5).

2. **Block overlap constraints for flexible blocks:** So far we discussed about fixed blocks. We can now see how constraints for flexible blocks can be developed. The flexible blocks can take rectangular shapes within a limited aspect ratio range i.e. its width and height can be varied keeping the area fixed. The non-linear area relation is linearized about the point of maximum allowable width by applying the first two members of the Taylor series giving,

$$h_i = h_{i0} + \Delta w_i \lambda_i$$

where,

$$h_{i0} = \frac{A_i}{w_{jmax}},$$
$$\lambda_i = \frac{A_i}{w_{imax}^2},$$
$$\Delta w_i = w_{imax} - w_i$$

where Δw_i is a continuous variable for block B_{fi}. The overlap constraints for a flexible block B_{fi} and a fixed block B_{rj} can be written as:

$$
\begin{array}{lll}
x_i + w_{imax} - \Delta w_i \leq x_j, & (B_{fj} \text{ is to the right of } B_{ri}), & \text{or} \\
y_i + h_{i0} + \Delta w_i \lambda_i \leq y_j, & (B_{fj} \text{ is above } B_{ri}), & \text{or} \\
x_i - w_j \geq x_j, & (B_{fj} \text{ is to the left of } B_{ri}), & \text{or} \\
y_i - h_j \geq y_j, & (B_{fj} \text{ is below } B_{ri}) & (6)
\end{array}
$$

Using two integer variables x_{ij} and y_{ij} per block pair as was done for fixed blocks, the 'or' condition between the equations can be satisfied.

The same set of equations can be extended to get overlap constraints between two flexible blocks. Using the same technique, the interconnection length constraints and routing area constraints can be developed. This set of equations are the input to any standard linear programming software package such as LINDO. The locations of the blocks and their dimensions are variables, the values of which are calculated by the software depending on the constraints and the objective function.

In [CF98], authors present a new convex programming formulation of the area minimization with a lesser numbers of variables and constraints than previous papers.

6.1.5 Rectangular Dualization

The partitioning process generates a group of subcircuits and their interconnections. This output from a partitioning algorithm can be represented as a graph $G = (V, E)$ where the vertices of the graph correspond to the subcircuits and the edges represent the interconnections between the subcircuit. The floorplan can be obtained by converting this graph into its *rectangular dual* and this approach to floorplanning is called *rectangular dualization*. A rectangular dual of graph $G = (V, E)$ consists of non-overlapping rectangles which satisfy the following properties:

1. Each vertex $v_i \in V$ corresponds to a distinct rectangle R_i, $1 \le i \le |V|$.

2. For every edge $(v_i, v_j) \in E$, the corresponding rectangles R_i and R_j are adjacent in the rectangular dual.

When this method is directly applied to the graph generated by partitioning, it may not be possible to satisfy the second property for generating the rectangular dual.

The problem of finding a suitable rectangular dual is a hard problem. In addition, there are many graphs which do not have rectangular duals. A further complication arises due to areas and aspect ratios of the blocks. In rectangular dualization, areas and aspect ratios are ignored to simplify the problem. As a result, the output cannot be directly used for floorplanning. Kozminski and Kinnen [KK84] have presented an algorithm for finding a rectangular dual of a planar triangulated graph. Usually, the graph is processed only if a rectangular dual for the graph exists. Bhasker and Sahni [BS86] have extended the approach in [KK84] to present a linear time algorithm for finding a rectangular dual of a planar triangulated graph.

A planar triangular graph (PTG) G is a connected planar graph that satisfies the following properties:

1. every face (except the exterior) is a triangle.

2. all the internal vertices have a degree ≥ 4.

3. all cycles that are not faces have length ≥ 4.

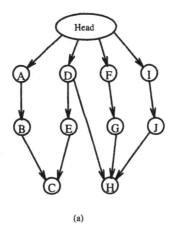

(a)

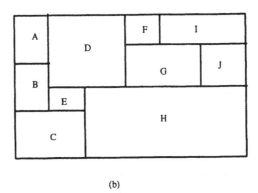

(b)

Figure 6.3: Conversion of planar digraph to a floorplan.

Given a PTG, a planar digraph is constructed which is a directed graph. Once a planar digraph is constructed, it can be converted into a floorplan as shown in Figure 6.3. Lokanathan and Kinnen [LK89] presented a procedure for floor-planning that minimizes routing parasitics using rectangular dualization. The use of rectangular dualization maximizes adjacency of blocks that are heavily connected or connected by critical nets.

6.1.6   Hierarchical Tree Based Methods

Hierarchical tree based methods represent a floorplan as a tree. Each leaf in the tree corresponds to a block and each internal node corresponds to a composite block in the floorplan. A floorplan is said to be *hierarchical* of *order k*, if it can be obtained by recursively partitioning a rectangle into at most k parts. Physical hierarchy can be generated in two ways: top-down partitioning or bottom-up clustering. Partitioning assumes that the relative areas (or number of nodes) Within partitions, at a given level of hierachy, may be fixed during a top down construction of a decomposition tree (or partitioning tree). There is no justification, except convenience, for this assumption. The optimal choice of relative areas varies from problem instance to problem instance, but there is no way to determine a desirable ratio, in top-down construction. Placements performed by min-cut method, a popular partitioning algorithm, often creates lot of vacant space or white space. Clustering on the other hand is a bottom-up algorithm for constructing a decomposition tree (or cluster tree).

In [DEKP89] a hierarchical floorplanner for arbitrary size rectangular blocks using the clustering approach has been proposed. At each level of the hierarchy, highly connected blocks (or clusters of blocks) are grouped together into larger clusters. At each level, the number of blocks is limited to five so that simple pattern enumeration and exhaustive search algorithms can be used later. Blocks (or block clusters) which are connected by edges of greater than average edge weight are grouped into a single cluster, if the resulting cluster has less than five members.

After forming the hierarchical clustering tree, a floorplanner and a global router together perform a top-down traversal of the hierarchy. Given an overall aspect ratio goal and I/O pin goal, at each level of the hierarchy, the floorplanner searches a simple library of floorplan templates and considers all possible room assignments which meet the combined goals of aspect ratios and I/O pins. At each level, the global routing problem is formulated as a series of minimum steiner tree problems in partial 3-trees. The global routing solution at the current level is used as the I/O pin goal for the floorplan evaluation, and as base for the global routing refinement at the next level. This floorplanning and global routing create constraints on the aspect ratio of the rooms, and gives assignments of I/O pins on the walls of the rooms, which are recursively transmitted downward as sub-goals to the floorplanner and global router. While evaluating the cost of a given floorplan template and room assignment, both chip area and net path length are considered. When undesirable block shapes and pin positions are detected, alternate floorplan templates and room assignments are tried by backtracking and using automatic module generators. This algorithm performs better than other well-known deterministic algorithms and generates solutions comparable to random-based algorithms.

Ting-Chi Wang and D. F. Wong [WW90] have presented an optimal algorithm for a special class of floorplans called *hierarchical floorplans of order 5*. Two types of blocks have been considered; L-shaped and rectangular. The algorithm takes a set of implementations for each block as input and identifies the best implementation for each block so that the resulting floorplan has minimum area.

6.1.7 Floorplanning Algorithms for Mixed Block and Cell Designs

All the algorithms discussed in the previous section can be used for floorplanning of Mixed Block and Cell (MBC) designs. These designs can be viewed as a set of blocks in a sea of cells. This is a popular ASIC layout design style. However, these algorithms were not implemented as a part of any tool which can generate floorplans for MBC designs. In this section, we describe some of the algorithms that were developed as a part of a tool specifically designed for floorplanning of MBC designs.

In [AK90], a heuristic algorithm has been developed for MBC designs. The algorithm employs a combined floorplanning, partitioning and global routing strategy. The main focus of the algorithm is in reducing the white space costs and the wiring cost. In [Sec88], the simulated annealing approach is used to solve the floorplanning problem for MBC designs. In [UKH85], "CHAMP", a floorplanning tool for MBC designs using the hierarchical approach has been presented. In [USS90, PSS+88, cL93], the floorplanning problem for MBC designs has been considered. All existing floorplanning algorithms, except [cL93], for the MBC designs restrict the block shapes to rectangular in order to simplify the problem at hand. Even in [cL93], only the pre-designed block shapes are considered to be rectilinear and the shapes generated for soft modules are always rectangular with varying aspect ratios. None of the existing floorplanning algorithms for MBC designs take advantage of the flexibility of the standard cell regions.

6.1.8 Simulated Evolution Algorithms

[RR96] describes a Simulated Evolution (Genetic) Algorithm for the Floorplan Area Optimization problem. The algorithm is based on suitable techniques for solution encoding and evaluation function definition, effective crossover and mutation operators, and heuristic operators which further improve the method's effectiveness. An adaptive approach automatically provides the optimal values for the activation probabilities of the operators. Experimental results show that the proposed method is competitive with the most effective ones as far as the CPU time requirements and the result accuracy is considered, but it also presents some advantages. It requires a limited amount of memory, it is not sensible to special structures which are critical for other methods, and has a complexity which grows linearly with the number of implementations. Finally, it is demonstrated that the method is able to handle floorplans much larger (in terms of number of basic rectangles) than any benchmark previously considered in the literature.

In [FSZ+97] a Multi-Selection-Multi-Evolution (MSME) scheme for parallelizing a genetic algorithm for floorplan optimization is presented and its implementation with MPI and its experimental results are discussed. The experimental results on a 16 node IBM SP2 scalable parallel computer have shown that the scheme is effective in improving performance of floorplanning over that of a sequential implementation. The parallel version could obtain better results with more than 90parallel program could reduce both chip area and maximum path delay by more than 8also speed up the evolution process so that there could be higher probability of obtaining a better solution within a given time interval.

The genetic algorithm (GA) paradigm is a search procedure for combinatorial optimization problems. Unlike most of other optimization techniques, GA searches the solution space using a population of solutions. Although GA has an excellent global search ability, it is not effective for searching the solution space locally due to crossover-based search, and the diversity of the population sometimes decreases rapidly. In order to overcome these drawbacks, the paper [TKH96] proposes a new algorithm called immunity based GA (IGA) combining features of the immune system (IS) with GA. The proposed method is expected to have local search ability and prevent premature convergence. IGA is applied to the floorplan design problem of VLSI layout. Experimental results show that IGA performs better than GA.

6.1.9 Timing Driven Floorplanning

With increasing chip complexities and the requirement to reduce design time, early analysis is becoming increasingly important in the design of performance critical CMOS chips. As clock rates increase rapidly, interconnect delay consumes an appreciable portion of the chip cycle time, and the floorplan of the chip significantly affects its performance.

[SYTB95] presents a timing-influenced floorplanner for general cell IC design. The floorplanner works in two phases. In the first phase the modules are restricted to be rigid and the floorplan to be slicing. The second phase of floorplanner allows modification to the aspect ratios of individual modules to further reduce the area of the overall bounding box. The first phase is implemented using genetic algorithm while in the second phase, a constraint graph based approach is adopted.

In [YSAF95] a timing driven floorplanning program for general cell layouts is presented. The approach used combines quality of force directed approach with that of constraint graph approach. A floorplan solution is produced in two steps. First a timing and connectivity driven topological arrangement is obtained using a force directed approach. In the second step, the topological arrangement is transformed into a legal floorplan. The objective of the second step is to minimize the overall area of the floorplan. The floorplanner is validated with circuits of sizes varying from 7 to 125 blocks.

[NLGV95] describes a system for early floorplan analysis of large designs. The floorplanner is designed to be used in the early stages of system design, to optimize performance, area and wireability targets before detailed implementation decisions are made. Unlike most floorplanners which optimize timing by considering only a subset of paths this floorplanner performs static timing analysis during the floorplan optimization process, instead of working on a subset of the paths. The floorplanner incorporates various interactive and automatic floorplanning capabilities.

6.1.10   Theoretical advancements in Floorplanning

In [PL95] P. Pan and C. L. Liu propose two area minimization methods for general floorplans with respect to the floorplan sizing problem.

The traditional algorithm for area minimization of slicing floorplans due to Stockmeyer has time and space complexity $O(n^2)$ in the worst case. For more than a decade, it has been considered the best possible. [Shi] presents a new algorithm of worst-case time and space complexity $O(n \log n)$, where n is the total number of realizations for the basic blocks, regardless whether the slicing is balanced or not. It has also been shown that $\theta(n \log n)$ is the lower bound on the time complexity of any area minimization algorithm. Therefore, the new algorithm not only finds the optimal realization, but also has the optimal running time.

In [PSL96], the complexity of the area minimization problem for hierarchical floorplans has been shown to be NP-complete (even for balanced hierarchical floorplans). A new algorithm has been presented for determining the nonredundant realizations of a wheel. The algorithm has time cost $O(k^2 \log k)$ and space cost $O(k^2)$ if each block in a wheel has at most k realizations. Based on the new algorithm for wheel, the authors have designed a new pseudo-polynomial area minimization algorithm for hierarchical floorplans of order-5. The time and space costs of the algorithm are $O((nM)^2 \log(nM))$ and $O(n^2 M)$, respectively, where n is the number of basic blocks and M is an upper-bound on the dimensions of the realizations of the basic blocks. The area minimization algorithm was implemented. Experimental results show that it is very fast.

In [CT], the authors have found an $\Omega(k^2)$ lower bound for area optimization of spiral floorplans. Let F be a spiral floorplan where each of its five basic rectangles has k implementations. It has been shown that there can be as many as $\Omega(k^2)$ useful implementations generated for F, in the worst case. This implies that the previously known $O(k^{\frac{2}{\log k}})$-time algorithm is almost optimal.

In [PL], the authors have presented two area minimization methods for general floorplans, which can be viewed as generalizations of the classical algorithm for slicing floorplans of Otten (1982) and Stockmeyer (1983) in the sense that they reduce naturally to their algorithm for slicing floorplans. Compared with the branch-and-bound algorithm of Wimer et al (1989), which does not have a nontrivial performance bound, these methods are provably better than an exhaustive method examined for many other examples.

[HL97] presents a formal algebraic specification (in SETS notation) that is appropriate for VLSI physical design layout and capable of representing both the floorplan topology and the modules' dimensions. The specification proposed allows a concise and rigorous representation of arbitrarily complex composite floorplans. This algebraic description unifies-under a rotation-invariant single-expression formalism-slicing and non-slicing generalized wheels floorplans. As needed by specific floorplan algorithms, it supports either a topology-dimensionless description or the introduction of module dimensions. Finally, it

allows an eightfold reduction-over previous representations-of the total number
of floorplan solutions considered in floorplanning problem algorithms.

In [KD97] a new method of non-slicing floorplanning is proposed, which
is based on the new representation for non-slicing floorplans, called bounded
slicing grid (BSG) structure. The authors have developed a new greedy algorithm based on the BSG structure, running in linear time, to select the
alternative shape for each soft block so as to minimize the overall area for general floorplan, including non-slicing structures. A new stochastic optimization
method, named genetic simulated annealing (GSA) for general floorplanning
is proposed. Based on BSG structure, SA-based local search and GA-based
global crossover is extended to L-shaped, T-shaped blocks and high density
packing of rectilinear blocks is obtained.

In [DSKB95], it is shown that for any rectangularly dualizable graph, a
feasible topology can be obtained by using only either straight or Z-cutlines
recursively within a bounding rectangle. Given an adjacency graph, a potential
topology, which may be nonslicible and is likely to yield an optimally sized
floorplan, is produced first in a top-dozen fashion using heuristic search in AND-
OR graphs. The advantage of this technique is four-fold: (i) accelerates top-
down search phase, (ii) generates a floorplan with minimal number of nonslice
cores, (iii) ensures safe routing order without addition of pseudo-modules, and
(iv) solves the bottom-up algorithm efficiently for optimal sizing of general
floorplans in the second phase.

[TY95] addresses the problem of minimizing wiring space in an existing slicing floorplan. Wiring space is measured in terms of net density, and the existing
floorplan is adjusted only by interchanging sibling rectangles and by mirroring
circuit modules. An exact branch and bound algorithm and a heuristic are
given for this problem. Experiments show that both algorithms are effective in
reducing wiring space in routed layouts.

6.1.11 Recent Trends

Several new trends are emerging in floorplanning, we discuss a few of them.
Interactive floorplanning can improve productivity, improve performance and
reduce die size. In [EK96a] an interactive floorplanner based on the genetic
algorithm is presented. Layout area, aspect ratio, routing congestion and maximum path delay are optimized simultaneously. The design requirements are
refined interactively as knowledge of the obtainable cost tradeoffs is gained
and a set of feasible solutions representing alternative and good tradeoffs is
generated. Experimental results illustrate the special features of the approach.

In [YTK95], a hybrid floorplanning methodology is proposed. Two hierarchical strategies for avoiding local optima during iterative improvement are
proposed: (1) Partial Clustering, and (2) Module Restructuring. These strategies work for localizing nets connecting small modules in small regions, and
conceal such small modules and their nets during the iterative improvement
phase. This method is successful in reducing both area and wire length in
addition to reducing the computational time required for optimization. Al-

though the method only searches slicing floorplans, the results are superior to the results obtained even with non-slicing floorplans.

In [WC95], a new approach to solve a general floorplan area optimization problem is proposed. By using the analogy between a floorplan and a resistive network, it has been shown that a class of zero wasted area floorplan can be achieved under the shape constraint of continuous aspect ratio. However, in many practical designs, each module may have constraints on its dimensions such as minimum length or width. In this paper, the authors have defined the floorplan area minimization problem under the constrained aspect ratio and give necessary conditions for the realization of zero wasted area floorplan under the shape constraints. A set of optimization methods is developed to minimize the wasted area if no zero wasted area floorplan is achievable.

6.2 Chip planning

Both floorplanning and placement problems either ignore the interconnect or consider it as a secondary objective. Chip planning is an attempt to integrate floorplanning and interconnect planning. The basic idea is to comprehend impact of interconnect as early as possible.

6.2.1 Problem Formulation

The input consists of $B_1, B_2, \ldots, B_n$ circuit blocks, with area $a_1, a_2, \ldots, a_n$ respectively. Associated with each block are two aspect ratios A_i^l and A_i^h, which give the lower and the upper bound on the aspect ratio for that block. In addition, we have $S_1, S_2, \ldots, S_m$ signals. For each signal we have criticality, width, source and sink. The chip planning algorithm has to determine the width w_i and height, h_i of each block B_i and layout of each signal such that $A_i^l \leq \frac{h_i}{w_i} \leq A_i^h$. In addition to finding the shapes of the blocks, the chip planning algorithm has to generate a valid placement for blocks and interconnect such that the area of the layout is minimized.

6.3 Pin Assignment

The purpose of pin assignment is to define the signal that each pin will receive. Pin assignment may be done during floorplanning, placement or after placement is fixed. If the blocks are not designed then good assignment of nets to pins can improve the placement. If the blocks are already designed, it may be possible to exchange a few pins. This is because some pins are *functionally equivalent* and some are *equipotential*. Two pins are called functionally equivalent, if exchanging the signals does not effect the circuit. For example, exchanging two inputs of a gate does not effect the output of the gate. Two pins are equipotential if both are internally connected and hence represent the same net. The output of the gate may be available on both sides, so the out-

functionally equivalent pins   equipotential pins

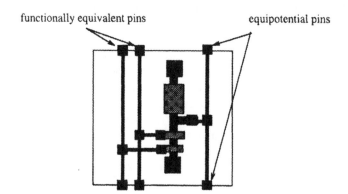

Figure 6.4: Functionally equivalent and equipotential pins.

put signal can be connected on any side. Figure 6.4 shows both functionally equivalent pins and equipotential pins.

6.3.1 Problem Formulation

The purpose of pin assignment is to optimize the assignment of nets within a functionally equivalent pin groups or assignment of nets within an equipotential pin group. The objective of pin assignment is to reduce congestion or reduce the number of crossovers. Figure 6.5 illustrates the effectiveness of pin assignment. Note that a net can be assigned to any equipotential pin within a set of functionally equivalent pins. The pin assignment problem can be formally stated as follows: Given a set of terminals $T_1, T_2, \ldots, T_n$ and a set of pins $P_1, P_2, \ldots, P_m$, $m > n$. Each T_i is assigned to pin P_i, $i = 1, 2, \ldots, n$. Let $\mathcal{E}_{P_i}$ be the set of pins which are equipotential and equivalent to P_i, the objective of pin assignment is to assign each T_i to a pin in $\mathcal{E}_{p_i}$ such that a specific objective function is minimized. The objective functions are typically routing congestions. For standard cell design, it may be the channel density.

6.3.1.1 Design Style Specific Pin Assignment Problems

Pin assignment problems in different design styles have different objectives.

1. **Full custom design style:** In full custom, we have two types of pin assignment problems. At floorplanning level, the pin location along the boundary of the block can be changed as the block is assigned a shape. This assignment of pins can reduce routing congestions. Thus, not only we can change pin assignment of pins, we can also change the location of pins along the boundary. At placement level, the options are limited to assigning the nets to pins. Notice that in terms of problem formulation,

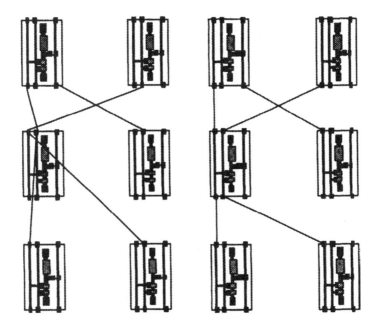

Figure 6.5: Impact of pin assignment.

we can declare all pins of a flexible block as functionally equivalent to achieve pin assignment in floorplanning.

2. **Standard cell design style:** The pin assignment problem for standard cells is essentially that of permuting net assignment for functionally equivalent pins or switching equipotential pins for a net.

3. **Gate array design style:** The pin assignment problem for gate array design style is the same as that of standard cells.

Assignment problems mostly occur in semi custom design styles such as gate arrays or standard cells.

In gate array design, the cells are pre-fabricated and are arranged on the master. Pin assignment problem in this type of design style is to assign to each terminal a functionally equivalent slot such that wiring cost is minimized. Slots in this case are the pin locations on pre-designed (library) cells. In standard cells, however, equipotential pins appear as feedthroughs. Since no wiring around the cell is needed, the wire length decreases with the use of feedthroughs.

6.3.2 Classification of Pin Assignment Algorithms

The pin assignment techniques are classified into general techniques and special pin assignment techniques. General techniques are applicable for pin

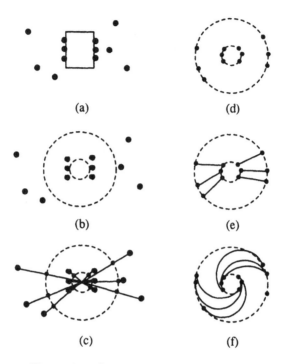

Figure 6.6: Concentric circle mapping.

assignment at any level and any region within a chip. Such techniques are applied at board level as well as chip level. On the other hand, the special pin assignment technique can be used for assignment of pins within a specific region such as channel or a switchbox.

6.3.3 General Pin Assignment

There are several methods in this category as discussed below:

1. **Concentric Circle Mapping:** To planarize the interconnections, this method models the pin assignment problem by using two concentric circles [Kor72]. The pins on the component being considered are represented as points on the inner circle whereas the points on the outer circle represent the interconnections to be made with other components. The concentric circle mapping technique solves the pin assignment problem by breaking it into two parts. The first part is the assignment of pins to points on the two circles and in the second part the points on the inner and outer circles are mapped to give the interconnections.

   For example, consider the component and the pins shown in Figure 6.6(a). The two circles are drawn so that the inner circle is inside all the pins

on the component being considered while the outer circle is just inside
the pins that are to be connected with the pins of this component. This
is shown in Figure 6.6(b). Lines are drawn from the component center
to all these pins as shown in Figure 6.6(c). The points on the inner and
outer circle are defined by the intersection of these lines with the circles
(Figure 6.6(d)). The pin assignment is completed by mapping the points
on the outer circle to those on the inner circle in a cyclic fashion. The
worst and the best case assignment are shown in Figure 6.6(e) and (f).

2. **Topological Pin Assignment:** Brady [Bra84] developed a technique
   which is similar to concentric circle mapping and has certain advantages
   over the concentric circle mapping method. With this method it is easier
   to complete pin assignment when there is interference from other components and barriers and for nets connected to more than two pins. If a net
   has been assigned to more two pins than the pin closest to the center of
   the primary component is chosen and all other pins are not considered.
   Hence in this case only one pin external to the primary component is
   chosen. The pins of the primary component are mapped onto a circle as
   in the concentric circle method. Then beginning at the bottom of the
   circle and moving clockwise the pins are assigned to nets and hence they
   get assigned in the order in which the external pins are encountered. For
   nets with two pins the result is the same as that for concentric circle
   mapping.

3. **Nine Zone Method:** The nine zone method, developed by Mory-Rauch,
   is a pin assignment technique based on zones in a Cartesian coordinate
   system [MR78]. The center of the coordinate system is located inside a
   group of interchangeable pins on a component. This component is called
   pin class. A net rectangle is defined by each of the nets connected to the
   pin class. There are nine zones in which this rectangle can be positioned
   as shown in Figure 6.7. The positions of these net rectangles are defined
   relative to the coordinate system defined by the current pin class.

6.3.4 Channel Pin Assignment

In design of VLSI circuits, a significant portion of the chip area is used for
channel routing. Usually, after the placement phase, the positions of the terminals on the boundaries of the blocks are not completely fixed and they still
have some degree of freedom to move before the routing phase begins. Figure 6.8 shows how channel density could be reduced by moving the terminals.
Figure 6.8(a) shows a channel which needs three tracks. By moving the pins,
the routing can be improved such that it requires one track as shown in Figure 6.8 (b). The channel pin assignment problem is the problem of assigning
positions for the terminals, subject to constraints imposed by design rules and
the designs of the previous phases, so as to minimize the density of the channel. The problem has various versions depending on how the pin assignment

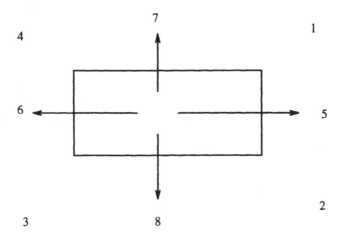

Figure 6.7: The nine pin zones.

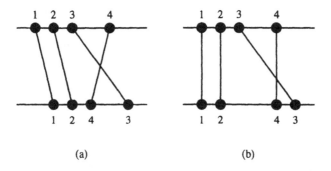

Figure 6.8: Reducing channel density by moving terminals.

constraints are specified. Many special cases of this problem have been investigated. In [GCW83], Gopal, Coppersmith, and Wong considered the channel routing problem with movable terminals.

In [YW91] the channel pin assignment problem in which assignment of terminals is subject to linear order position constraints is solved using a dynamic programming formulation by Yang and Wong. Their method is described briefly below. Except minor changes for clarity, the discussion is essentially the same as it appears in [YW91].

Since the terminals are linearly ordered we have a set of terminals at the top given by TOP in which the terminals $t_1 < t_2 < \ldots < t_p$. Similarly the terminal set for terminals at the bottom is given by BOT in which the terminals are ordered $b_1 < b_2 < \ldots < b_q$. Each terminal t_i on the top and b_i on the bottom have a corresponding set given by T_i and B_i which indicate the possible

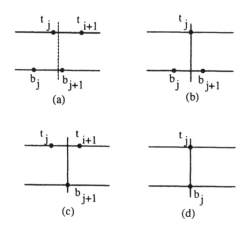

Figure 6.9: Four types of (i, j, k) solutions.

positions these terminals can occupy. A solution to this problem is called an (i,j,k)-*solution* if it assigns exactly $t_0, t_1, t_2, \ldots, t_i, b_0, b_1, b_2, \ldots, b_j$ to the first k columns of the channel where t_0 and b_0 correspond to a auxiliary column and two trivial nets which consist of only one terminal are introduced. The main idea used by the algorithm is to first compute a density function using dynamic programming and then use backtracking to reconstruct an optimal solution.

Let L be the length of the channel and N be the number of nets to be routed. The set of terminals on the top boundary of the channel is denoted as TOP, and the set of terminals on the bottom boundary of the channel is denoted as BOT. In this implementation, separation constraints and position constraints are not considered and but the terminals are definitely allowed to be within a certain position only i.e. the length L of the channel. The (i, j, k) solutions can be classified into the following four types, according to pin assignment at column k as illustrated in Figure 6.9.

- Type 0: No terminal is assigned to either endpoints of column k as shown in Figure 6.9(a).

- Type 1: Only t_i is assigned to the top endpoint of column k as shown in Figure 6.9(b).

- Type 2: Only b_j is assigned to the bottom endpoint of column k as shown in Figure 6.9(c).

- Type 3: Both t_i and b_j are assigned to column k as shown in Figure 6.9(d).

Let $d(i, j, k)$ be the density of the channel considering i terminals at the top and j terminals at the bottom after consideration of k columns. Let $x(i, j, k)$, $y(i, j, k)$ and $z(i, j, k)$ be the local densities (*crossing numbers*) at column k considering i nets at the top and j nets at the bottom for Type 1, Type 2 and Type 3 solutions respectively. Let $R_1(i,j)$ denote the set of nets with one terminal in $\{t_1, t_2, \ldots, t_{i-1}, b_1, b_2, \ldots, b_j\}$ and one terminal in TOP $\cup$ BOT - $\{t_1, t_2, \ldots, t_i, b_1, b_2, \ldots, b_j\}$, and the net containing t_i, if it is not trivial. Let $R_2(i,j)$ denote the set of nets with one terminal in $\{t_1, t_2, \ldots, t_i, b_1, b_2, \ldots, b_{j-1}\}$, and one terminal in TOP $\cup$ BOT - $\{t_1, t_2, \ldots, t_i, b_1, b_2, \ldots, b_j\}$, and the net containing b_j, if it is not trivial. Let $R_3(i,j)$ denote the set of nets with one terminal in $\{t_1, t_2, \ldots, t_{i-1}, b_1, b_2, \ldots, b_{j-1}\}$, and one terminal in TOP $\cup$ BOT -$\{t_1, t_2, \ldots, t_i, b_1, b_2, \ldots, b_j\}$, and the net containing t_i or b_j, if it is not trivial and if they do not belong to the same net. Hence we have

$$x(i,j,k) = \begin{cases} +\infty & \text{if } K \notin T_i \\ |R_1(i,j)| & \text{otherwise} \end{cases}$$

$$y(i,j,k) = \begin{cases} +\infty & \text{if } K \notin B_j \\ |R_2(i,j)| & \text{otherwise} \end{cases}$$

$$z(i,j,k) = \begin{cases} +\infty & \text{if } K \notin T_i \cap B_j \\ |R_1(i,j)| & \text{otherwise} \end{cases}$$

The algorithm for optimal channel pin assignment is shown in Figure 6.10. It is easy to see that the time complexity of the algorithm Linear-CPA is $O(pql)$.

6.4   Integrated Approach

The various stages in the physical design cycle evolved as the entire problem is extremely complex to be solved altogether at once. But over the years, with better understanding of the problems, attempts are made to merge some steps of the design cycle. For example, floorplanning was considered as a problem of just finding the shapes of the blocks without considering routing areas. Over the years, the floorplanning problem has been combined with the placement problem [DEKP89, SSR91, WL86]. The placement problem is sometimes combined with the routing problem giving rise to the 'place and route' algorithms [Esc88, FHR85, SSV85, Sze86].

In this section, the approach used by Dai, Eschermann, Kuh and Pedram in BEAR [DEKP89] is described briefly.   BEAR is a macrocell-based layout system. The process of floorplanning is carried out in the following three steps:

1. **Clustering:** In this step, a hierarchical tree is constructed. Blocks that are strongly connected are grouped together in a cluster. Each cluster can have a limited number of blocks within it. The clustering process considers the shapes of the blocks to avoid a mismatch within the cluster. This step is repeated to build the cluster tree.

2. **Placement:** In this step, the tree is traversed top-down. The target shape and terminal goals for the root of the cluster tree is specified. This information is used to identify the topological possibility for the

Algorithm LINEAR-CPA()
begin
   (* initialize *)
   for $i = 0$ to p **do**
      for $j = 0$ to q **do**
         $d(i, j, 0) = +\infty$;
   for $k = 0$ to L **do**
      $d(0, 0, k) = 0$;
   COMPUTE-CROSS();
   for $k = 1$ to L **do**
      for $i = 0$ to p **do**
         for $j = 0$ to q **do**
            (* type 1 solution *)
            $D_1 = \max\{d(i - 1, j, k - 1), x(i, j, k)\}$;
            (* type 2 solution *)
            $D_2 = \max\{d(i, j - 1, k - 1), y(i, j, k)\}$;
            (* type 3 solution *)
            $D_3 = \max\{d(i - 1, j - 1, k - 1), z(i, j, k)\}$;

   if $d(p, q, l) = +\infty$ **then**
      return ϕ is not feasible;
   else
      (* backtracking for constructing optimal solution *)
      $i = p$; $j = q$;
      for $k = L$ down to 1 **do**
         if $d(i, j, k) = D_1$ **then**
            $f(i) = k$; $i = i - 1$;
         else if $d(i, j, k) = D_2$ **then**
            $g(j) = k$; $j = j - 1$;
         else if $d(i, j, k) = D_3$ **then**
            $f(i) = k$; $g(j) = k$;
            $i = i - 1; j = j - 1$;
      return $\pi = (f, g)$;
end.

Figure 6.10: The optimal channel pin assignment algorithm

Procedure COMPUTE-CROSS()
begin
    for $(i \ = \ 0 \ to \ p)$**do**
        for $(j \ = \ 0 \ to \ q)$ **do**
            COMPUTE($R_1(i,j)$);   (* Compute $|R_1(i,j)|$ *)
            COMPUTE($R_2(i,j)$);   (* Compute $|R_2(i,j)|$ *)
            COMPUTE($R_3(i,j)$);   (* Compute $|R_3(i,j)|$ *)
            for $(k \ = \ 0 \ to \ L)$ **do**
               $x(i,j,k) = +\infty;$
               $y(i,j,k) = +\infty;$
               $z(i,j,k) = +\infty;$
    for$(i \ = \ 0 \ to \ p)$ **do**
        for$(j \ = \ 0 \ to \ q)$**do**
            for$(k \in T_i)$ **do**
               $x(i,j,k) = |R_1(i,j)|;$
            for$(k \in B_j)$ **do**
               $y(i,j,k) = |R_2(i,j)|;$
            for$(k \in T_i \cap B_j)$ **do**
               $z(i,j,k) = |R_3(i,j)|;$
    end.

Figure 6.11: The COMPUTE-CROSS procedure for the channel pin assignment algorithm

clusters at the level below. This in turn sets the shape and terminal goals for the immediate lower levels in the hierarchy till at the leaf level the orientations of the blocks are determined. For each of the topologies, the routing space is determined. The selection of a particular topology is based on the area and the shape of the resulting topology and the connection cost. The system is developed so as to allow the user to control the trade off between the shape, the area and the connection costs. This strategy works well in case the blocks at the leaf level are flexible so that the shapes of these blocks can be adjusted to the shape of the cluster. On the other hand, if the leaf level blocks are fixed then this top-down approach can give unfavorable results. This is due to the fact that the information of the shape of these blocks at the leaf level are not considered by the objective function when determining the cluster shapes at higher levels of the cluster tree. This is rectified by passing the shape information from the leaves towards the root of the tree during the clustering phase. In addition, during the top-down placement step, a look-ahead is added so that the objective function can examine the shapes generated during clustering, at a level below the immediate level for which the shape is being determined.

3. **Floorplan optimization:** This an improvement step that resizes selected blocks iteratively. The blocks to be resized are identified by computing the longest path through the layout surface using the routing estimates done in the previous step.

6.5 Summary

Floorplanning and Pin assignment are key steps in physical design cycle. The pin assignment is usually carried out after the blocks have been placed to reduce the complexity of the overall problem.

Several placement algorithms have been presented. Simulated annealing and simulated evolution are two most successful placement algorithm. Although these algorithms are computationally intensive, they do produce good placements. Integer programming based algorithms for floorplanning have been also been successful. Several algorithms have been presented for pin assignment, including optimal pin assignment for channel pin assignment problems. The output of the placement phase must be routable, otherwise placement has to be repeated.

6.6 Exercises

1. Given the following 14 rectangles with their dimensions specified, write a program that will arrange all these rectangles within 5000 sq. units of area, if possible, or otherwise minimize the area required. The dimensions (width × height) of the rectangles are $R_1 = 15 \times 15, R_2 = 25 \times 15, R_3 = 10 \times 30, R_4 = 30 \times 20, R_5 = 10 \times 15, R_6 = 20 \times 5, R_7 = 10 \times 25, R_8 = 30 \times 15, R_9 = 10 \times 65, R_{10} = 10 \times 25, R_{11} = 20 \times 20, R_{12} = 10 \times 20, R_{13} = 30 \times 15, R_{14} = 40 \times 15$.

2. Recall that the aspect ratio of a block is the ratio of its height and width. If each rectangle in problem 1 can have three different aspect ratios, find the appropriate aspect ratio for each rectangle so that the area occupied by the rectangles is minimized. The set of aspect ratios for R_i, is R_i^a, is, $R_1^a = \{1.0, 1.2, 2.0\}$, $R_2^a = \{0.8, 1.5, 1.9\}$, $R_3^a = \{0.6, 2.0, 3.0\}$, $R_4^a = \{0.8, 1.2, 1.5\}$, $R_5^a = \{0.75, 1.2, 1.5\}$, $R_6^a = \{0.3, 0.9, 2.0\}$, $R_7^a = \{0.4, 1.2, 1.5\}$, $R_8^a = \{0.5, 1.0, 1.5\}$, $R_9^a = \{0.8, 3.0, 4.0\}$, $R_{10}^a = \{0.4, 1.2, 1.8\}$, $R_{11}^a = \{0.5, 1.0, 1.2\}$, $R_{12}^a = \{0.5, 0.9, 1.2\}$, $R_{13}^a = \{0.5, 1.0, 1, 5\}$, $R_{14}^a = \{0.4, 1.6, 2, 5\}$.

3. Use the lowest and highest aspect ratios for each rectangle in problem 2 as lower and upper bounds respectively and generate a placement which occupies minimum area.

† 4. Apply Simulated Annealing algorithm for pin assignment problem. In each iteration, pins of each block are permuted and routing congestion is estimated.

† 5. Develop an algorithm for pin assignment of a full custom layout based on concentric circle mapping.

‡ 6. Implement the channel pin assignment algorithm. Discuss the constraints, based on functionally equivalent and equipotential pins.

Bibliographic Notes

A floorplanning system designed to work within a hierarchical design environment supporting multiple design styles has been discussed in [MTDL90]. A technique for floorplanning and pin assignment of general cell layouts has been developed in [PMSK90]. A global floorplanning approach has been discussed in [PD86]. The approach is based on a combined min-cut and slicing paradigm. A pin assignment algorithm for improving the performance in standard cell design by improving the longest delay has been discussed in [SL90]. A pin assignment problem for macro-cells is discussed in [YYL88]. An approach which combines pin assignment and global routing has been developed in [Con89]. In [KK95], yield issues are considered. Authors demonstrate that for large area VLSI chips, especially those that incorporate some fault tolerance, changes in the floorplan can affect the projected yield. In [KK97], the authors have demonstrated that the floorplan of a chip can affect its projected yield in a nonnegligible way, for large area chips with or without fault-tolerance.

In [MAC98], a floorplanner for RF circuits based on a genetic algorithm that supports simultaneous placement and routing has been developed. In [MK98], Sequence-pair based placement method for hard/soft/pre-placed modules has been discussed (also discussed in chapter 7). In [ITK98], a new approach for the minimum area floorplanning is proposed where the shape of every module can vary under the constraint of area and floorplan topology. Simulating the air-pressure mechanics, the algorithms iterate to improve the layout to decide the shapes and positions of modules. In [CF98], a convex formulation of the floorplan area minimization problem is presented.

Chapter 7

Placement

Placement is a key step in physical design cycle. A poor placement consumes larger areas, and results in performance degradation. It generally leads to a difficult or sometimes impossible routing task. The input to the placement phase is a set of blocks, the number of terminals for each block and the netlist. If the layout of the circuit within a block has been completed then the dimensions of the block are also known. Placement phase is very crucial in overall physical design cycle. It is due to the fact, that an ill-placed layout cannot be improved by high quality routing. In other words, the overall quality of the layout, in terms of area and performance is mainly determined in the placement phase.

The placement of block occurs at three different levels.

1. **System level placement:** At system level, the placement problem is to place all the PCBs together so that the area occupied is minimum. At the same time, the heat generated by each of the PCBs should be dissipated properly so that the system does not malfunction due to overheating of some component.

2. **Board level placement:** At board level, all the chips on a board along with other solid state devices have to be placed within a fixed area of the PCB. All blocks are fixed and rectangular in shape. In addition, some blocks may be pre-placed. The PCB technology allows mounting of components on both sides. There is essentially no restriction on the number of routing layers in PCB. Therefore in general, the nets can always be routed irrespective of the quality of components placement. The objective of the board-level placement algorithms is twofold: minimization of the number of routing layers; and satisfaction of the system performance requirements. For high performance circuits, the critical nets should have lengths which are less than a specified value and hence the placement algorithms should place the critical components closer together. Another key placement problem is the temperature profile of the board. The heat dissipation on a PCB should be uniform, i.e., the chips which generate maximum heat should not be placed closer to each other. If MCMs are

used instead of PCBs, then the heat dissipation problem is even more critical, since chips are placed closer together on a MCM.

3. **Chip level placement:** At chip level, the problem can be either chip planning, placement or floorplanning along with pin assignment. The blocks are either flexible or fixed, and some of them may be pre-placed. The key difference between the board level placement problem and the chip level placement is the limited number of layers that can be used for routing in a chip. In addition, the circuit is fabricated only on one side of the substrate. This implies that some 'bad' placements maybe unroutable. However, the fact that a given placement is unroutable will not be discovered until routing is attempted. This leads to very costly delays in completion of the design. Therefore, it is very important to accurately determine the routing areas in the chip-level placement problems. Usually, two to four layers are used for routing, however, chips with four or more layers routings are more expensive to fabricate. The objective of a chip-level placement or floorplanning algorithm is to find a minimum area routable placement of the blocks. In some cases, a mixture of macro blocks and standard cells may have to be placed together. These problems are referred to as *Mixed block and cell* placement and floorplanning problems. At chip level, if the design is hierarchical then the placement and floorplanning is also carried out in a hierarchical manner. The hierarchical approach can greatly simplify the overall placement process.

In the following sections, we will discuss the chip-level placement. The placement problem for MCMs, which is essentially a performance driven Board level placement problem, will be discussed in Chapter 14.

In this chapter, we will discuss placement problems in different design styles. Section 7.1 discusses the problem formulation. Section 7.2 presents the classification of placement algorithms. Remaining sections present various algorithms for the placement problem.

7.1   Problem Formulation

The placement problem can be stated as follows: Given an electrical circuit consisting of fixed blocks, and a netlist interconnecting terminals on the periphery of these blocks and on the periphery of the circuit itself, construct a layout indicating the positions of each block such that all the nets can be routed and the total layout area is minimized. The objective for high performance systems is to minimize the total delay of the system, by minimizing the lengths of the critical paths. It is usually approximated by minimization of the length of the longest net. This problem is known as the *performance (timing) driven placement problem*. The associated algorithms are called *performance (timing) driven placement algorithms*.

The quality of a placement is based on several factors:

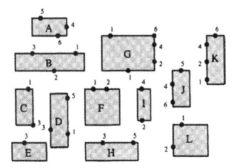

Figure 7.1: Blocks and set of interconnecting nets.

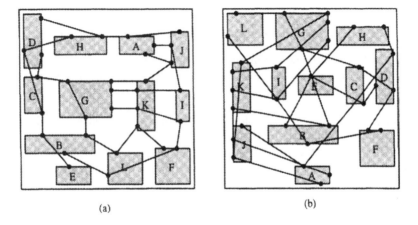

         (a)   (b)

Figure 7.2: Two different placements of the same problem.

1. layout area.

2. completion of routing, and

3. circuit performance.

The layout area and the routability of the layout are usually approximated by the topological congestion, known as *rat's nest*, of interconnecting wires. Consider the simple example in Figure 7.1. Two different placements for this example are shown in Figure 7.2. The topological congestion in Figure 7.2(a) is much less than that of in Figure 7.2(b). Thus, the placement given in Figure 7.2(a) can be considered more easily routable than the placement given in Figure 7.2(b). In many cases, several objectives may contradict each other. For example, minimizing layout area may lead to increased maximum wire length and vice versa.

Let us formally state the placement problem. Let $B_1, B_2, \ldots, B_n$ be the blocks to be placed on the chip. Each $B_i, 1 \leq i \leq n$, has associated with it a height h_i and a width w_i. Let $\mathcal{N} = \{N_1, N_2, N_3, \ldots, N_m\}$ be the set of nets representing the interconnection between different blocks. Let $Q = \{Q_1, Q_2, \ldots, Q_k\}$ represent rectangular empty areas allocated for routing between blocks. Let L_i denote the estimated length of net N_i, $1 \leq i \leq m$. The placement problem is to find iso-oriented rectangles for each of these blocks on the plane denoted by $\mathcal{R} = \{R_1, R_2, \ldots, R_n\}$ such that

1. Each block can be placed in its corresponding rectangle, that is, R_i has width w_i and height h_i,

2. No two rectangles overlap, that is, $R_i \cap R_j = \phi$, $1 \leq i, j \leq n$,

3. Placement is routable, that is, $Q_j, 1 \leq j \leq k$, is sufficient to route all the nets.

4. The total area of the rectangle bounding $\mathcal{R}$ and Q is minimized.

5. The total wirelength is minimized, that is, $\sum_{i=1}^{m} L_i$ is minimized. In the case of high performance circuits, the length of longest net $\max\{L_i \mid i = 1, \ldots, m\}$ is minimized.

The general placement problem is NP-complete and hence, the algorithms used are generally heuristic in nature.

Although the actual wiring paths are not known at the time of placement, however, a placement algorithm needs to model the topology of the interconnection nets. An interconnection graph structure which interconnects each net is used for this purpose. The interconnection structure for two terminal trees is simply an edge between the two vertices corresponding to the terminals. In order to model a net with more than two terminals, rectilinear steiner trees are used as shown in Figure 7.3(a) to estimate optimal wiring paths for a net. This method is usually not used by routers, because of the NP-completeness of steiner tree problem. As a result, minimum spanning tree representations are the most commonly used structures to connect a net in the placement phase. Minimum spanning tree connections (shown in Figure 7.3(b)) allow branching only at the pin locations. Hence, the pins are connected in the form of minimum spanning tree of a graph. *Complete graph* interconnection is shown in (Figure 7.3(c)). It is easy to implement such structures. However, this method causes many redundant interconnections, and results in longer wire length.

The large number of objective functions can be classified into two categories, net metrics and congestion metric. The net metrics deal with the assumption that all the nets can be routed without interfering with other nets or with the components. Usually the length of a net is important as the interconnection delays depend on the length of the wire. The net metrics only quantify the amount of wiring and do not account for the actual location of these wires. The examples of this kind of objective functions are the total length of all nets and the length of the longest net. The congestion metric is used to avoid the

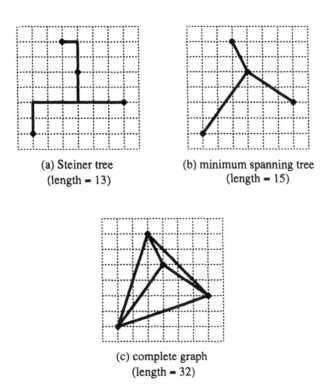

(a) Steiner tree
(length = 13)

(b) minimum spanning tree
(length = 15)

(c) complete graph
(length = 32)

Figure 7.3: Interconnection topologies.

buildup of many nets in a particular area leading to congestion. Example of congestion metric is the number of nets that intersect with a routing channel.

The layout surface on which the circuit is to be placed is modeled into either geometric or topological models. For the geometric model, the placement algorithms tend to accept the layout area as a fixed constraint and tend to optimize the interconnections. The geometric models are appropriate for design styles where placement aspects such as size, shape and public pin positions do not change during the layout process such as PCB design. On the other hand, the placement systems which model the layout surface as a topological model assume the constraint to be the completion of interconnections and optimize the layout area. Topological models are appropriate for more flexible design styles such as full custom designs.

7.1.1   Design Style Specific Placement Problems

Different design styles impose different restrictions on the layout and have different objectives in placement problems.

1. **Full custom:** In full custom design style, the placement problem is the packing problem concerned with placing a number of blocks of different sizes and shapes tightly within a rectangular area. There is no restriction on how the blocks can be placed within the rectangle except that no two blocks may overlap. The primary objective is to minimize the total layout area. The irregularity of the block shapes is usually the main cause of unused areas. Since unused area increases the total area, the blocks must be placed so as to minimize unused areas. The objective of minimizing the layout area sometimes conflicts with the objective of minimizing the maximum length of a net. Therefore, in high performance circuit design, additional constraints on net lengths must also be considered.

2. **Standard cells:** The standard cell placement problem is somewhat simpler than the full custom placement problem, as all the cells have the same height. Cells are placed in rows and minimizing layout area is equivalent to minimizing the summation of channel heights and minimizing the width of the widest row. In order to reduce overall area, all rows should have equal widths. The total area consists of area required for the cell rows and the area required for routing or the channel area. The routing area estimates, which determine the channel height, play a key role in determining the overall area of the design. With the advent of over-the-cell routing, in which the empty spaces over the standard cell rows are used for routing, the channels in standard cells have almost disappeared giving rise to *channelless* standard cell designs. Standard cells are designed so the power and ground nets run horizontally through the top and bottom of cells.

3. **Gate arrays:** As mentioned in the previous chapter, in case of gate arrays, the partitioning of a circuit maps the circuit onto the gates of the gate array. Hence the problem of partitioning and placement is essentially the same in this design style. If partitioning does not actually assign gate locations, then a placement algorithm has to be used to assign subcircuits or gates to the slots on the gate array. Given a set of blocks $B_1, B_2, \ldots, B_n$, and set of slots $S_1, S_2, \ldots, S_r$, $r \geq n$, assign each block B_i to a slot S_j such that no two blocks are assigned to the same slot and the placement is routable. For high performance designs, additional constraints on net lengths have to be added.

Another situation, where gate array partitioning and placement may be different, is when each 'gate' in the gate array is a complex circuit. In this case, the circuit is partitioned such that each subcircuit is equivalent to a 'gate'. The placement algorithm is then used to find the actual assignment. This happens to be the case in FPGAs and is discussed in Chapter 13.

7.2 Classification of Placement Algorithms

The placement algorithms can be classified on the basis of :

1. the input to the algorithms,

2. the nature of output generated by the algorithms, and

3. the process used by the algorithms.

Depending on the input, the placement algorithms can be classified into two major groups: *constructive placement* and *iterative improvement* methods. The input to the constructive placement algorithms consists of a set of blocks along with the netlist. The algorithm finds the locations of blocks. On the other hand iterative improvement algorithms start with an initial placement. These algorithms modify the initial placement in search of a better placement. These algorithms are typically used in an iterative manner until no improvement is possible.

The nature of output produced by an algorithm is another way of classifying the placement algorithms. Some algorithms generate the same solution when presented with the same problem, i.e., the solution produced is repeatable. These algorithms are called *deterministic* placement algorithms. Algorithms that function on the basis of fixed connectivity rules (or formulae) or determine the placement by solving simultaneous equations are deterministic and always produce the same result for a particular placement problem. Some algorithms, on the other hand, work by randomly examining configurations and may produce a different result each time they are presented with the same problem. Such algorithms are called as *probabilistic* placement algorithms.

The classification based on the process used by the placement algorithms is perhaps the best way of classifying these algorithms. There are two important class of algorithms under this classification: simulation based algorithms and partitioning based algorithms. Simulation based algorithms simulate some natural phenomenon while partitioning based algorithms use partitioning for generating the placement. The algorithms which use clustering and other approaches are classified under 'other' placement algorithms.

7.3 Simulation Based Placement Algorithms

There are many problems in the natural world which resemble placement and packaging problems. Molecules and atoms arrange themselves in crystals, such that these crystals have minimum size and no residual strain. Herds of animals move around, until each herd has enough space and it can maintain its predator-prey relationships with other animals in other herds. The simulation based placement algorithms simulate some of such natural processes or phenomena. There are three major algorithms in this class: simulated annealing, simulated evolution and force directed placement. The simulated annealing algorithm simulates the annealing process which is used to temper metals. Simulated evolution simulates the biological process of evolution while the force directed placement simulates a system of bodies attached by springs. These algorithms are described in the following subsections.

7.3.1   Simulated Annealing

Simulated annealing is one of the most well developed placement methods available [BJ86, GS84, Gro87, Haj88, HRSV86, LD88, Oht86, RSV85, RVS84, SL87, SSV85]. The simulated annealing technique has been successfully used in many phases of VLSI physical design, e.g., circuit partitioning. The detailed description of the application of simulated annealing method to partitioning may be found in Chapter 5. Simulated annealing is used in placement as an iterative improvement algorithm. Given a placement configuration, a change to that configuration is made by moving a component or interchanging locations of two components. In case of the simple pairwise interchange algorithm, it is possible that a configuration achieved has a cost higher than that of the optimum but no interchange can cause a further cost reduction. In such a situation the algorithm is trapped at a local optimum and cannot proceed further. Actually this happens quite often when this algorithm is used on real life examples. Simulated annealing avoids getting stuck at a local optimum by occasionally accepting moves that result in a cost increase.

In simulated annealing, all moves that result in a decrease in cost are accepted. Moves that result in an increase in cost are accepted with a probability that decreases over the iterations. The analogy to the actual annealing process is heightened with the use of a parameter called *temperature T*. This parameter controls the probability of accepting moves which result in an increased cost. More of such moves are accepted at higher values of temperature than at lower values. The acceptance probability can be given by $e^{\frac{-\Delta C}{T}}$, where ΔC is the increase in cost. The algorithm starts with a very high value of temperature which gradually decreases so that moves that increase cost have lower probability of being accepted. Finally, the temperature reduces to a very low value which causes only moves that reduce cost to be accepted. In this way, the algorithm converges to a optimal or near optimal configuration.

In each stage, the configuration is shuffled randomly to get a new configuration. This random shuffling could be achieved by displacing a block to a random location, an interchange of two blocks, or any other move which can change the wire length. After the shuffle, the change in cost is evaluated. If there is a decrease in cost, the configuration is accepted, otherwise, the new configuration is accepted with a probability that depends on the temperature. The temperature is then lowered using some function which, for example, could be exponential in nature. The process is stopped when the temperature has dropped to a certain level. The outline of the simulated annealing algorithm is shown in Figure 7.4.

The parameters and functions used in a simulated annealing algorithm determine the quality of the placement produced. These parameters and functions include the cooling schedule consisting of initial temperature (*init_temp*),

Algorithm SIMULATED-ANNEALING
begin
    $temp$ = INIT-TEMP;
    $place$ = INIT-PLACEMENT;
    while ($temp$ > FINAL-TEMP) **do**
        while ($inner\_loop\_criterion$ = FALSE) **do**
            new_place = PERTURB($place$);
            ΔC = COST(new_place) - COST($place$);
            if ($\Delta C < 0$) **then**
                $place$ = new_place;
            else if (RANDOM$(0,1) > e^{\frac{\Delta C}{temp}}$) **then**
                $place$ = new_place;
        $temp$ = SCHEDULE($temp$);
  end.

Figure 7.4: The Simulated Annealing algorithm.

final temperature ($final\_temp$) and the function used for changing the temperature (SCHEDULE), $inner\_loop\_criterion$ which is the number of trials at each temperature, the process used for shuffling a configuration (PERTURB), acceptance probability (F), and the cost function (COST). A good choice of these parameters and functions can result in a good placement in a relatively short time.

Sechen and Sangiovanni-Vincentelli developed TimberWolf 3.2, which is a standard cell placement algorithm based on Simulated Annealing [SSV85]. TimberWolf is one of the most successful placement algorithms. In this algorithm, the parameters and functions are taken as follows. For the cooling schedule, $init\_temp$ = 4000000, $final\_temp$ = 0.1, and SCHEDULE $(T) = \alpha(T) \times T$ where $\alpha(T)$ is a cooling rate depending on the current temperature T. $\alpha(T)$ is taken relatively low when T is high, e.g. $\alpha(T) = 0.8$ when the cooling process just starts, which means the temperature is decremented rapidly. Then, in the medium range of temperature, $\alpha(T)$ is taken 0.95, which means that the temperature changes more slowly. When the temperature is in low range, $\alpha(T)$ is again taken 0.8, the cooling procedure go fast again. In this way, there are a total of 117 temperature steps. The graph for the cooling schedule is shown in Figure 7.5. The value of $inner\_loop\_criterion$ is taken according to the size of the circuit, e.g., 100 moves per cell for a 200-cell circuit and 700 moves per cell for a 3000-cell circuit are recommended in [SSV85]. The new configuration is generated by making a weighted random selection from one of the following:

1. the displacement of a block to a new location,

2. the interchange of locations between two blocks,

3. an orientation change for a block.

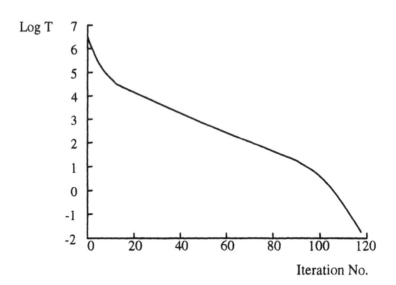

Figure 7.5: Cooling schedule in TimberWolf.

The alternative 3 is used only when the new configuration generated by using alternative 1 is rejected. The ratio r of single block displacement to pairwise interchange should be carefully chosen to give a best overall result. An orientation change of a block is simply a mirror image of that block's x-coordinate. The cost function is taken as:

$$COST = cost1 + cost2 + cost3$$

where *cost1* is the weighted sum of estimate length of all nets, *cost2* is the penalty cost for overlapping, and *cost3* is the penalty cost for uneven length among standard cell rows.

$$cost1 = \sum_{i \in nets}[xspan(i) \times HWeight(i) + yspan(i) \times VWeight(i)]$$
$$cost2 = \sum_{i \neq j \in blocks}[overlap(i,j)]^2$$
$$cost3 = \sum_{i \in rows}|ActRowLength(i) - DesRowLength(i)| \times factor$$

Where, $xspan(i)$ and $yspan(i)$ are the horizontal and vertical spans of the minimum bounding rectangle of net i. Horizontal and vertical weights ($HWeight$ and $VWeight$) are introduced so that each net can have different priority to be optimized, e.g. critical nets can have higher priority, and one direction can be favored over another direction. The quadratic function in *cost2* is used to penalize more heavily on large overlaps than small ones. Actually overlap is not allowed in the placement. However, it takes large amount of computer time to remove all overlapping. So, it is more efficient to allow overlapping during intermediate placement and use a cost function to penalize the overlapping.

ActRowLength(i) and *DesRowLength(i)* are the actual row length and desired row length for the *i*th row, respectively. The *factor* is used so that the minimum penalty for the difference in length of rows is *factor*.

The simulated annealing is one of the most established algorithms for placement problems. It produces good quality placement. However, Simulated Annealing is computationally expensive and can lead to longer run times. Therefore, it is only suitable for small to medium sized circuits.

7.3.2 Simulated Evolution

Simulated evolution (genetic algorithm) is analogous to the natural process of mutation of species as they evolve to better adapt to their environment. It has been recently applied to various fields. Readers are referred to the chapter 5 for the description of simulated evolution algorithm used in partitioning.

We use the example of gate array placement problem to explain the simulated evolution algorithm. In gate array placement problem, the layout plane is divided into $S = \{S_1, S_2, \ldots, S_r\}$ slots. The problem of placing cells $B = \{B_1, B_2, \ldots, B_n\}$, where $n \leq r$, is to assign some S_j to each B_i, such that no two cells are assigned to the same slot. The algorithm starts with an initial set of placement configurations, which is called the *population*. This initial placement can be generated randomly. The individuals in this population represent a feasible placement to the optimization problem and are actually represented by a string of symbols. The symbols used in the solution string are called *genes*. A solution string made up of genes is called a *chromosome*. A *schema* is a set of genes that make up a partial solution. Simulated evolution algorithm is iterative, and each iteration is called a *generation*. During each iteration the individuals of the population are evaluated on the basis of certain *fitness* tests which can determine the quality of each placement. Two individuals (corresponding to two possible placement configurations) among the population are selected as *parents* with probabilities based on their fitness. The better fitness an individual has, the higher the probability that it will be chosen. The operators called crossover, mutation and inversion, which are analogous to the counterparts in the evolution process, are then applied on the parents to combine 'genes' from each parent to generate a new individual called the *offspring*. The offsprings are then evaluated and a new generation is then formed by including some of the parents and the offsprings on the basis of their fitness in a manner that the size of population remains the same. As the tendency is to select high fitness individuals to generate offsprings and the weak individuals are deleted, the next generation tends to have individuals that have good fitness. The fitness of the entire population improves over the generations. That means the overall placement quality improves over iterations. At the same time, some 'bad' genes are inherited from previous generation even though the probability of doing so is quite low. In this way, it is assured that the algorithm does not get stuck at some local optimum. This is the basic mechanism of the algorithm which results in a good placement. The three genetic operators that are used for creating offsprings are discussed below.

1. **Crossover** : Crossover generates offsprings by combining schemata of two individuals at a time. This could be achieved by choosing a random cut point and generating the offspring by combining the left segment of one parent with the right segment of the other. However, after doing so, some blocks may be repeated while some other blocks may get deleted. This problem has been dealt with in many different ways. The amount of crossover is controlled by the crossover rate which is defined as the ratio of the number of offspring produced in each generation to the population size. The crossover rate determines the ratio of the number of searches in regions of high average fitness to the number of searches in other regions.

2. **Mutation:** This operator is not directly responsible for producing new offsprings but it causes incremental random changes in the offspring produced by crossover. The most commonly used mutation is pair-wise interchange. This is the process by which new genes which did not exist in the original generation can be generated. The mutation rate is defined as the percentage of the total number of genes in the population, which are mutated in each generation. It should be carefully chosen so that it can introduce more useful genes, and at the same time do not destroy the resemblance of *offsprings* to their *parents.*

3. **Selection:** After the offspring is generated, individuals for the next generation are chosen based on some criteria. There are many such selection functions used by various researchers. In *competitive selection* all the parents and offsprings compete with each other and the fittest individuals are selected so that the population remains constant. In *random selection* the individuals for the next generation are randomly selected so that the population remains constant. This could be advantageous considering the fact that by selecting the fittest individuals the population converges to individuals that share the same genes and the search might not converge to a optimum. However, if the individuals are chosen randomly, there is no way to gain improvements from older generation to new generation. By compromising both methods, *stochastic selection* makes selections with probabilities based on the fitness of each individual.

An algorithm developed by Cohoon and Paris [CP86] is shown in Figure 7.6. The scoring function is chosen to account for total net lengths and to penalize the placement with high wiring density in the routing channels. The score is given by:

$$\sigma = \frac{1}{2} \sum_{i \in nets} length(i) + \sum_{i \in HChannels} h_i'^2 + \sum_{i \in VChannels} v_i'^2$$

where

$$h_i' = \begin{cases} h_i - h_{avg} - h_{sd} & \text{if } h_i > h_{avg} - h_{sd} \\ 0 & \text{otherwise} \end{cases}$$

Algorithm GENIE
begin
   no_pop = SIZE-POP;
   $no\_offspring = no\_pop \times P_\psi$;
   (* P_ψ stands for the crossover rate. *)
   pop = CONSTRUCT-POP(no_pop);
   for ($i = 1$ to no_pop) **do**
   (* Evaluate score of each individual in
     the population on the basis of its fitness. *)
     SCORE($pop(i)$);
   for ($i = 1$ to $no\_generation$) **do**
   (* Generate $no\_generation$ generations *)
     for ($j = 1$ to $no\_offspring$) **do**
       (x, y) = CHOOSE-PARENT(pop)
       $offspring(j)$ = GENERATE(x, y);
       SCORE($offspring(j)$);
     pop = SELECT($pop, offspring, no\_pop$);
     for ($j = 1$ to no_pop) **do**
       MUTATE($pop(j)$);
   return highest scoring configuration in population;
end.

Figure 7.6: The Simulated Evolution algorithm.

$$v_i' = \begin{cases} v_i - v_{avg} - v_{sd} & \text{if } v_i > v_{avg} - v_{sd} \\ 0 & \text{otherwise} \end{cases}$$

where, h_i (v_i) is the number of nets intersecting horizontal (vertical) channel i. h_{avg} (v_{avg}) is the mean of h_i (v_i). h_{sd} (v_{sd}) is the standard deviation of h_i (v_i).

The parent choosing function is performed alternatively as either selecting parents with probabilities proportional to their fitness or selecting parents with probabilities proportional to their fitness and an additional constraint such that they have above average fitness. Two crossover operators can be used. One selects a random cell C_s and brings the four closest neighbors in parent 1 into neighboring slots in parent 2. At the same time, the cells in these slots in parent 2 are pushed outward until vacant slots are found. The other one selects a square of $k \times k$ cells from parent 1 where k is a random number with mean of 3 and variance of 1, and copy the square into parent 2. The result of this copying would result in the loss of some cells. So, before copying, the cells in parent 2 that are not part of square are being pushed outward into some vacant slots.

One possible mutation function is to use a greedy technique to improve the

placement cost. It selects a cell C_i on a net N_j and searches the cell C_k on the
same net that is farthest from cell C_i. C_k is then brought close to the cell C_i.
The cell which needs to be removed from that slot is pushed outward until a
vacancy is found.

Besides the implementation described above, there are other implementations, e.g. the genetic approach developed by Chan, Mazumder and Shahookar,
which uses a two-dimensional bitmap chromosome to handle the placement of
macro cells and gate arrays [CMS91]. In addition, the simulated evolution was
investigated in [CM89, KB87, Kli87, SM90a, SM90b].

7.3.3   Force Directed Placement

Force directed placement explores the similarity between placement problem
and classical mechanics problem of a system of bodies attached to springs.
In this method, the blocks connected to each other by nets are supposed to
exert attractive forces on each other. The magnitude of this force is directly
proportional to the distance between the blocks. According to Hooke's law, the
force exerted due to stretching of the springs is proportional to the distance
between the bodies connected to the spring. If the bodies were allowed to
move freely, they would move in the direction of the force until the system
achieved equilibrium. The same idea is used for placing the blocks. The final
configuration of the placement of blocks is the one in which the system achieves
equilibrium.

In [Qui75], Quinn developed a placement algorithm using force directed
method. In this algorithm, all the blocks to be placed are considered to
be rectangles. These blocks are classified as movable or fixed. Let $\mathcal{B} =
\{B_1, B_2, \ldots, B_n\}$ be the blocks to be placed, and (x_i, y_i) be the Cartesian coor-
dinates for B_i. Let $\Delta x_{ij} = |x_i - x_j|$, $\Delta y_{ij} = |y_i - y_j|$. $\Delta d_{ij} = \sqrt{(\Delta x_{ij})^2 + (\Delta y_{ij})^2}$.
Let F_x^i (F_y^i) be the total force enacted upon B_i by all the other blocks in the
x-direction (y-direction). Then, the force equations can be expressed as:

$$F_x^i = \sum_{j=1}^{n}[-k_{ij} \times \Delta x_{ij}]$$
$$F_y^i = \sum_{j=1}^{h}[-k_{ij} \times \Delta y_{ij}]$$

where $i = 1, 2, \ldots, n$, k_{ij} = attractive constant between blocks B_i and B_j and
$k_{ij} = 0$ if $i = j$. The blocks connected by nets tend to move toward each other,
and the force between them is directly proportional to the distance between
them. On the other hand, the force model does not reflect the relationship
between unconnected blocks. In fact, the unconnected blocks tend to repel
each other. So, the above model should be modified to include these repulsion
effects. Since the formulation of the force equation in x-direction is the same
as in y-direction, in the following, only the formulation in x-direction will be
discussed.

$$F_x^i = \sum_{j=1}^{n}[-k_{ij} \times \Delta x_{ij} + \delta k_{ij} \times R \times \Delta x_{ij}/\Delta d_{ij}], \quad i = 1, 2, \ldots, n$$

where $\delta k_{ij} = 1$ when $k_{ij} = 0$, and $\delta k_{ij} = 0$ when $k_{ij} = 1$. R is the repulsion constant directly proportional to the maximum of k_{ij} and inversely proportional to n.

In addition, it is also desirable to locate the center of all movable blocks in some predetermined physical location (usually the geometric center of the layout plane) so that the placement of blocks is balanced. Physically, it is equivalent to have the forces acted upon the set of all movable blocks being removed. Suppose there are m movable blocks. Then, the force equations become:

$$F_x^i = \sum_{j=1}^{n} [-k_{ij} \times \Delta x_{ij} + \delta k_{ij} \times R \times \Delta x_{ij} / \Delta d_{ij}] - F_{ext}, \quad i = 1, 2, \ldots, n$$

where F_{ext} is the total external force acted upon the set of all movable blocks by the fixed blocks and $F_{ext} = \{\sum_{i=1}^{m} \sum_{j=m+1}^{n} [-k_{ij} \times \Delta x_{ij} + \delta k_{ij} \times R \times \Delta x_{ij} / \Delta d_{ij}]\}/m$.

The placement problem now becomes a problem in classical mechanics and the variety of methods used in classical mechanics can be applied. To solve for the set of force equations, one of the methods is to set the potential energy equal to $\sum_{i=1}^{n} [F_x^{i2} + F_y^{i2}]$, and apply the unconstrained minimization method, i.e., Fletcher-Reeves method [FR64], since the solution of the force equations correspond to the state of zero potential energy of the system.

Besides the implementation presented by Quinn [Qui75], there are various implementations [AJK82, Got81, HK72, HWA78, Oht86, QB79].

7.3.4 Sequence-Pair Technique

A packing of set of rectangles is nothing but non-overlapping placement of rectangles. Sequence-pair is a representation of such a packing in terms of a pair of module name sequences. In algorithms like simulated annealing solution space is infinite and thus the algorithms stops the search for optimal solution half-way and outputs the result. A finite solution space which includes an optimal solution is the key for successful optimization. Sequence-pair technique generates such a finite solution space. Murata et. al. proved in [MFNK96] that searching the solution space generated by sequence-pair technique using simulated annealing placement algorithm where move is change of the sequence-pair, gives efficient rectangular packing.

A procedure called **Gridding** is used to encode a placement on a chip to a sequence-pair. Let $\mathbf{P}$ be a packing of m modules on chip $\mathbf{C}$. In *gridding* procedure m non-intersecting, non-overlapping lines (lines doesn't cross boundaries of modules also) are drawn from south-west corner to north-east corner of the chip and each line passes through one module diagonally. These lines can be linearly ordered and this is S_1 of sequence-pair (S_1, S_2). Second order S_2 of sequence-pair can be obtained by drawing similar kind of lines from south-east corner of the chip to north-west corner of the chip.

Given a sequence-pair (S_1, S_2) one of the optimal solution under the constraint can be obtained in $O(m^2)$ time by applying the longet path algorithm

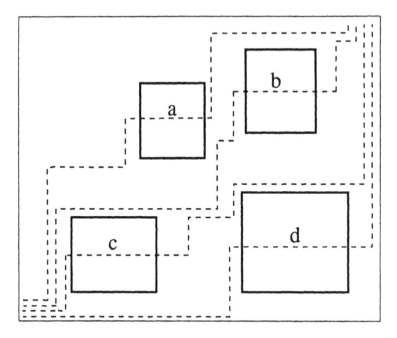

Figure 7.7: Sequence pair for the given placement is a, b, c, d.

for vertex weighted directed acyclic graphs. A relation ($LeftOf, RightOf,$
$BelowOf, AboveOf$) between a pair of modules can be determined based on
the location of modules relative to each other. From the given sequence-pair
(S_1, S_2) it is easy to generate such relations between modules of the chip. If
$leftof(m_1) = (m_2)$ means m_2 is left side of m_1. m_2 will be left side of m_1 if
m_2 is before m_1 in both the orders of sequence-pair.

In Figure 7.7 sequence-pair for the given placement is $abcd, cdab$.
LeftOf(a) = () , Modules that are after a in both the orders of sequence-pair.
RightOf(a) = (b),Modules that are before a in both the orders of sequence-pair.
AboveOf(a) = (), Modules that are before a in first order and after 'a' in second
order of sequence-pair.
BelowOf(a) = (c,d), Modules that are after a in first order and before 'a' in
second order of sequence-pair.

In Figure 7.8 horizontal and vertical constraint graphs are generated for the
sequence-pair $abcd, cdab$.

A directed and vertex-weighted graph called "horizontal-constraint graph"
G_h, can be constructed using modules as verticies, module widths as weight of
vetices and $leftof$ relation as edges of graph. Similarly using "below" relation
and height of the block vertical-constraint graph G_v, can be generated. For
both the graphs source and sink vertices are out side the chip boundary with

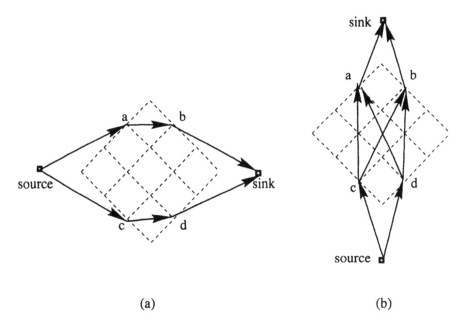

(a)   (b)

Figure 7.8: (a) Horizontal and (b) Vertical Constraint Graphs from a given sequence-pair.

weight of zero. Neither of these graphs contains any directed cycle. Module pairs that have horizontal edges in G_h do not overlap horizontally and similarly module pairs that have vertical edges in G_v do not overlap vertically. Thus no two modules overlap each other in the resultant placement because any pair of modules are either in horizontal or vertical relation. The width and height of the chip is determined by the longest path length between the source and the sink in G_h and G_v. Since the width and height of the chip is independently minimum, the resultant packing is the best of all the packings under the constraint. The longest path length calculation on each graph can be done in $O(m2)$ time, proportional to the number of edges in the graph.

For a given chip $\mathbf{C}$ of m modules $O(m!)^2$ sequence-pairs are possible and each sequence-pair can be mapped to a packing in $O(m^2)$ time, and atleast one of the sequence-pair corresponds to the optimal packaging solution. When the orientation of the block is not fixed then the size of the solution space increases to $(m!)^2 2^m$.

Authors of [MFNK96] applied this technique in a simulated annealing algorithm where move is a change of the sequence-pair. They have used three kinds of pair-interchanges.

1. Two module names in S_1, for placement optimization.

2. Two module names in both S_1 and S_2 for placement optimization.

3. Width and Height of a module for orientation optimization.

The initial sequence-pair is made as $S_1 = S_2$, which corresponds to a linear horizontal arrangement of modules. The temparature was decreased exponentially. Operation 1 was performed with higher probability in higher temperatures and operation 3 was performed with higher probability in lower temperatures to achieve better results.

The above technique can be extended to consider wire lengths also.

7.3.5 Comparison of Simulation Based Algorithms

Both the simulated annealing and simulated evolution are iterative and probabilistic methods. They can both produce optimal or near-optimal placements, and they are both computation intensive. However, the simulated evolution has an advantage over the simulated annealing by using the history of previous trial placements. The simulated annealing can only deal with one placement configuration at a time. In simulated annealing it is possible that a good configuration maybe obtained and then lost when a bad configuration is introduced later. On the other hand, the good configuration has much better chance to survive during each iteration in simulated evolution since there are more than one configurations being kept during each iteration. Any new configuration is generated by using several configurations in simulated evolution. Thus, history of previous placements can be used. However, the genetic method has to use much more storage space than the simulated annealing since it has to memorize all individual configurations in the population. Unlike simulated annealing and simulated evolution, force directed placement is applicable to general designs, such as full custom designs. The force-directed methods are relatively faster compared to the simulated annealing and genetic approaches, and can produce good placement.

7.4   Partitioning Based Placement Algorithms

This is an important class of algorithms in which the given circuit is repeatedly partitioned into two subcircuits. At the same time, at each level of partitioning, the available layout area is partitioned into horizontal and vertical subsections alternately. Each of the subcircuits so partitioned is assigned to a subsection. This process is carried out till each subcircuit consists of a single gate and has a unique place on the layout area. During partitioning, the number of nets that are cut by the partition is usually minimized. In this case, the group migration method can be used.

7.4.1   Breuer's Algorithm

The main idea for Breuer's algorithm [Bre77a, Bre77b] is to reduce the number of nets being cut when the circuit is partitioned. Various objective functions have been developed for this method. These objective functions are as given below.

1. **Total net-cut objective function:** All the nets that are cut by the partitioning are taken into account. This sum includes all nets cut by both horizontal and vertical partitioning cut lines. Minimizing this value is shown to be equivalent to minimizing the semi-perimeter wire-length [Bre77a, Bre77b].

2. **Min-max cut value objective function:** In the case of standard cells and gate arrays, the channel width depends on the number of nets that are routed through the channel. The more the number of nets the larger is the channel width and therefore the chip size. In this case the objective function is to reduce the number of nets cut by the cut line across the channel. This will reduce the congestion in channels having a large number of nets but at the expense of routing them through other channels that have a fewer number of nets or through the sparser areas of the channel.

3. **Sequential cut line objective function:** A third objective function is introduced to ease the computation of net cuts. Even though the above two objective functions represent a placement problem more accurately, it is very difficult to compute the minimum net cuts. This objective function reduces the number of nets cut in a sequential manner. After each partition, the number of nets cut is minimized. This greedy approach is easier to implement, however, it may not minimize the total number of nets cut.

In addition to the different objective functions, Breuer also presented several placement procedures in which different sequence of cut lines are used.

1. **Cut Oriented Min-Cut Placement:** Starting with the entire chip, the chip is first cut by a partition into two blocks. The circuit is also partitioned into two subcircuits so that the net cut is minimized. All the blocks formed by the partition are further partitioned by the second cut line and this process is carried out for all the cut lines. This partitioning procedure is sequential and easy to implement but it does not always yield good results because of the following two reasons. Firstly, while processing a cut line, the blocks created by the previous cut lines have to be partitioned simultaneously. Secondly, when a cut line partitions a block into two, the blocks to be placed in one of the partition might not fit in the partition created by the cut line as it might require more space than the block to be placed in the other partition (see Figure 7.9(a)).

2. **Quadrature Placement Procedure:** In this procedure, each region is partitioned into four regions of equal sizes by using horizontal and vertical cut lines alternatively. During each partitioning, the cutsize of the partition is minimized. As it cuts through the center and reduces the

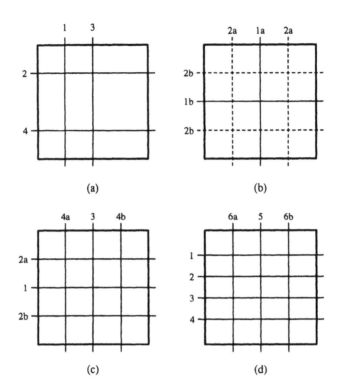

Figure 7.9: Different sequences of cut lines.

cutsize, this process reduces the routing density in the center. Currently, this is the most popular sequence of cut lines for min-cut algorithms (see Figure 7.9(b)).

3. **Bisection Placement Procedure:**   The layout area is repeatedly bisected (partitioned into two equal parts) by horizontal cut lines until each subregion consists of one row. Each of these rows is then repeatedly bisected by vertical cut lines till each resulting subregion contains only one slot thus fixing the positions of all blocks. This method is usually used for standard cell placement and does not guarantee the minimization of the maximum net cut per channel (see Figure 7.9(c)).

4. **Slice Bisection Placement Procedure:**   In this method, a suitable number of blocks are partitioned from the rest of the circuit and assigned to a row, which is called a slicing, by horizontal cut lines. This process is repeated till each block is assigned to a row. The blocks in each row are then assigned to columns by bisecting using vertical cut lines. This technique is most suitable for circuits which have a high degree of inter-connection at the periphery since this procedure tends to reduce the wire

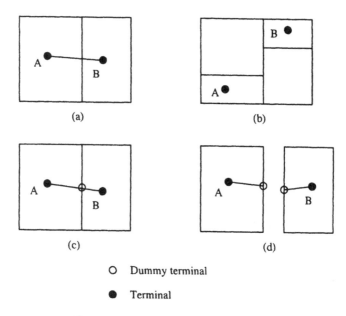

○ Dummy terminal

● Terminal

Figure 7.10: Terminal propagation.

congestion at the periphery (see Figure 7.9(d)).

In any procedure described above, if the partitioning is to minimize the number of nets cut by the partition, a group migration method can be used in the partitioning process.

7.4.2 Terminal Propagation Algorithm

The partitioning algorithms partitioned the circuit merely to reduce the net cut. Therefore, the partitioning algorithms cannot be directly used for placement. This is illustrated in Figure 7.10. If the partitioning algorithm were to be used directly, terminals A and B may move away from each other as a result of partitioning, as shown in Figure 7.10(b). This not only increases the net length but increases the congestion in the channels as well. Hence unlike partitioning algorithms, placement algorithms which are based on partitioning need to preserve the information regarding the terminals which are connected and fall into two different partitions because of the cut. This can be done by propagating a dummy terminal to the nearest point on the boundary, when a net connecting two terminals is cut, as shown in Figure 7.10(c). When this dummy terminal is generated, the partitioning algorithm will not assign the two terminals in each partition, as shown in Figure 7.10(b), into different partitions as this would not result in a minimum cut. This method called the *terminal propagation* method was developed by Dunlop and Kernighan [DK85].

Algorithm CLUSTER-GROWTH
begin
    let B be the set of blocks to be placed;
    select a seed block B block from B;
    place B in the layout;
    $B = B - B$;
    while $(B \neq \phi)$ **do**
        select a block B from B;
        place B in the layout;
        $B = B - B$;
end.

Figure 7.11: The Cluster growth algorithm.

7.5 Other Placement Algorithms

In this section, different kind of placement algorithms are described, which are neither simulation based nor partition based. These include cluster growth, quadratic assignment, resistive network optimization, and branch-and-bound algorithms.

7.5.1 Cluster Growth

In this constructive placement algorithm, the bottom-up approach is used. Blocks are placed sequentially in a partially completed layout. The seed or the first block is usually selected and placed by the user. After the seed block is placed, other blocks are selected and placed one by one to complete the layout. The selection and placement techniques differentiate in various cluster growth techniques.

In cluster growth algorithm, the block that is highly connected (have the most connections) to the already placed blocks is selected to be placed. Then, this block is placed either close to the block that it is highly connected to or a exhaustive search is carried out for the best possible location for the block. The outline of the cluster growth algorithm is shown in Figure 7.11.

The random constructive placement is a degenerate form of cluster growth. In this case, the selection of blocks is made randomly and its position is also fixed randomly. As this method does not take into account the interconnections and other circuit features, in most of the cases, it does not produce a good layout. This method is sometimes utilized to generate a basic layout for an iterative placement algorithm.

7.5.2 Quadratic Assignment

This method solves an abstract version of the gate array placement problem. It assumes that the blocks are points and have zero area. The cost of connecting two blocks B_i and B_j given by c_{ij} is stored in a connection matrix. The distance between slot k and slot l, given by d_{kl} is stored in a distance matrix. The objective is to map the blocks onto slots such that the product of connectivity and distance between the slots to which the blocks have been mapped (which gives the net length), for all the blocks, is minimized. This objective is equivalent to minimizing the total wire length for the circuit. This placement problem has been formulated as a quadratic assignment problem by Hall [Hal70]. If C is the connection matrix and c_i is the sum of all elements in the ith row of C, then a diagonal matrix D can be defined as,

$$d_{ij} = \begin{cases} 0, & \text{if i} \neq \text{j}, \\ c_i, & \text{if } i = j \end{cases}$$

Let a matrix E be defined as $E = D - C$ and $X^T = [x_1, x_2, \ldots, x_n]$ and $Y^T = [y_1, y_2, \ldots, y_n]$ be row vectors representing the $x-$ and $y-$ coordinates of the desired solution. Hall proved that a nontrivial solution is obtained and the objective function is minimized if the smallest eigenvalues of the matrix E are chosen. The corresponding eigenvectors X and Y then give the coordinates of all the blocks.

7.5.3 Resistive Network Optimization

The placement problem has been transformed into the problem of minimizing the power dissipation in a resistive network by Cheng and Kuh [CK84]. The objective function, which is the squared Euclidean wire length, is written in a matrix form. This representation is similar to the matrix representation of resistive networks. This method can include fixed blocks in the formulation. Also, blocks with irregular sizes are allowed within cell rows. The algorithm comprises of subprograms which are used for optimization, scaling, relaxation, partitioning and assignment. The efficiency of the method comes from the fact that it takes advantage of the sparsity of the netlist. Slot constraints are used which guarantee the placement of blocks to be legal and each block is allocated to one slot. There are upto n constraints, where n is the number of blocks. The slot constraints are given by the equation

$$\sum_{i=1}^{n} x_i^j = \sum_{i=1}^{n} p_i^j, \qquad 1 \leq j \leq n$$

The algorithm maps the given circuit to a resistive network in which the pads and fixed blocks are represented as fixed voltage sources. Using the slot constraints the power dissipation in the circuit is minimized which causes the blocks to cluster around the center of the chip. The higher order slot constraints when applied cause the blocks to spread out. This step is called the *scaling step*. A repeated partitioning and relaxation then aligns the blocks with the slot locations.

7.5.4 Branch-and-Bound Technique

The general branch-and-bound algorithm can be applied to the placement problem. This method can be used for small circuits as it is a computationally intensive method. The method assumes that all the feasible solutions and the scores of these solutions are known. All these solutions make up a set called the *solution set*. The solution can be systematically searched. The search can be actually represented by a tree structure. The leaves of the tree are all the solutions. The selection of a solution is equivalent to traversing a branch of the tree and this step is called the *branch* step. If at any node in this tree a solution yields a score which is greater than the currently known lowest, then the search continues in another part of the decision tree. This step is the *bound* step. Hence the algorithm actually prunes the decision tree which results in reduced computation.

Consider a gate array with three slots S_1, S_2, S_3 and three blocks B_1, B_2, B_3. At the first level of the tree, the root has three branches, each corresponding to a different placement of B_1 in three different slots. All the child nodes of the root will have two branches, each specifying two positions of B_2 in the remaining two slots. Finally, all grand children of root will have exactly one branch, specifying the slot for B_3.

The branch-and-bound algorithm traverses the tree and computes the cost of the solution at any given node. The cost can simply be the total wire length due to the placement of blocks upto that node. If this cost is higher than another known placement, this subtree need not be explored.

7.6 Performance Driven Placement

The delay at chip level, which depends on interconnecting wires plays a major role in determining the performance of the chip. As the blocks in a circuit become smaller and smaller, the size of the chip decreases. As a result, the delay due to the connecting wire becomes a major factor for high performance chips. The placement algorithms for high performance chips have to generate placements which will allow routers to route nets within the timing requirements. Such problems are called performance driven placement and the algorithms are called performance driven algorithms. The performance driven placement algorithms can be classified into two major categories, one which use the net-based approach and the other which use the path-based approach. In path-based approach [Don90, JK89], the critical paths in the circuit are considered and the placement algorithms try to place the blocks in a manner that the path length is within its timing constraint. On the other hand, the net-based approach [DEKP89, Dun84, Oga86, HNY87, MSL89], tries to route the nets to meet the timing constraints on the individual nets instead of considering the paths. In this case, the timing requirement for each net has to be decided by the algorithm. Usually a pre-timing analysis generates the bounds on the netlengths which the placement algorithms have to satisfy while placing the blocks. Gao, Vaidya and Liu [GVL91] presented a algorithm for high performance placement. The algorithm consists of the following steps:

1. Upper bounds for the netlengths are deduced from the timing requirements which is a part of the input to the algorithm. Each net has a set of
   such upper bounds. This provides the algorithm with maximum flexibility. The timing requirements are expressed by a set of linear constraints
   which are solved using convex programming techniques. A new convex
   programming algorithm is used for which the computational complexity
   depends only on the number of variables rather than the number of linear
   constraints.

2. A modified min-cut placement algorithm is used to obtain the placement
   of the blocks. The upper bounds calculated in the previous step guide
   the min-cut algorithm in placing the blocks. The min-cut algorithm,
   is a modified version of the Fiduccia's min-cut algorithm which tries to
   minimize the number of nets whose lengths exceed their corresponding
   upper bounds in addition to minimizing the size of the cutset.

3. The next step is to check whether all timing requirements are satisfied in
   the placement generated by the modified min-cut placement algorithm.

4. In case all the timing requirements are met, the placement is valid and
   is accepted. Otherwise the set of upper-bounds obtained in step 1 is
   modified and the steps 2 and 3 are repeated. Most other algorithms
   could not handle situations where the placement generated did not meet
   the timing specifications.

7.7 Recent Trends

In Very Deep Sub-Micron(VDSM) designs, Placement problem is considered
much more than simply achieving the routability of the design and minimizing
the chip die area . Several other critical issues such as timing, zero clock-
skew, even power distribution are increasing the complexity of the placement
problem exponentially. Since placement phase is one of the early phases of the
IC physical design, lot of attention is paid to placement phase in IC design
cycle.

Timing driven placement is very critical to IC design and some of the
techniques to perform timing driven placement are discussed in [DNA+90],
[RMNP97], [SS95], and [SKT97].

Algorithms to estimate the wire lengths are becoming part of placement
algorithms because accurate estimation of wire lengths help to fix the problems
in placement phase itself rather than in routing phase. Estimation of wirelength
during the placement stage helps to understand the routability of the design.
One of the techniques to estimate wire length is discussed in [CKM+98].

1. Early Placement to obtain better Wire Load Models(WLM) for synthesis.
   WLM is a parameter for the delay estimate for logic synthesis algorithm.

2. Placement for Cross talk avoidance.

3. Placement for Minimizing clock skew.

4. Placement for even power distribution.

7.8   Summary

Placement is a key step in physical design cycle. Several placement algorithms have been presented. Simulated annealing and simulated evolution are two most successful placement algorithm. Although these algorithms are computationally intensive, they do produce good placements. Integer programming based algorithms for floorplanning have been also been successful. Several algorithms have been presented for pin assignment, including optimal pin assignment for channel pin assignment problems. The output of the placement phase must be routable, otherwise placement has to be repeated.

7.9   Exercises

1. Consider the following blocks. $R_1 = 15 \times 15, R_2 = 25 \times 15, R_3 = 10 \times 30, R_4 = 30 \times 20, R_5 = 10 \times 15, R_6 = 20 \times 5, R_7 = 10 \times 25, R_8 = 30 \times 15, R_9 = 10 \times 65, R_{10} = 10 \times 25$. If D_{ij} represents the center to center distance between blocks i and j then determine if these blocks can be placed together so that the distances between the blocks are within their specified values. The distances that are to be maintained are $D_{12} = 30, D_{23} = 35, D_{14} = 18, D_{34} = 40, D_{24} = 30, D_{13} = 45$.

† 2. Implement the Simulated Annealing algorithm. Consider the graph shown in Figure 7.12. Each vertex represents rectangle R_i whose dimensions are specified in problem 1, the edges of the graph represent the connectivity of the blocks. Use the Simulated Annealing algorithm to generate a placement.

† 3. For the placement obtained in problem 5, implement a pin assignment algorithm which will reduce the total net length and minimize the maximum length of a net. Generate a complete routing for this placement. The routing can be generated on an uniform grid and two nets can intersect only when they are perpendicular to each other.

‡ 4. Implement the Simulated annealing algorithm for the general cell placement problem.
    Hint: Instead of exchanging a big block with a small block, a big block can be exchanged with a cluster of small blocks.

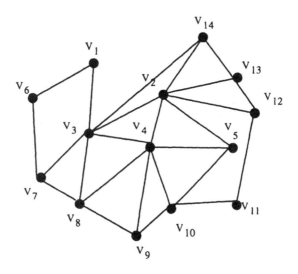

Figure 7.12: Example of a connectivity graph.

‡ 5. For a given placement, implement a pin assignment algorithm which will rotate the pins on the blocks either in clockwise or anticlockwise direction till the total net length is reduced. While rotating the pins on the blocks, the order of these pins must be maintained.

6. For the blocks specified in problem 1 generate the integer program constraints to solve the placement problem. Consider some of the blocks as flexible and solve the floorplanning problem using the integer program.

† 7. Implement a placement algorithm for high performance circuits which takes into account path delays instead of net delays.

‡ 8. Modify the min-cut algorithm to incorporate the terminal propagation scheme.

‡ 9. Develop Simulated Evolution algorithm for standard cells design.

‡ 10. Develop Simulated Evolution algorithm for full custom designs.

‡ 11. Several industrial libraries allow cells with different cell heights. This leads to irregular shape channels. Suggest modifications required for applying the Simulated annealing algorithm to standard cells of uneven heights.

‡ 12. Implement force directed placement algorithm for gate array design style.

Bibliographic Notes
A linear assignment algorithm for the placement problem has been discussed in [Ake81]. Two partition/interchange processes are described in [Pat81] for

solving the placement problem. The graph is partitioned into several smaller graphs for initial placement in both the methods and finally interchange optimization is carried out. The simulated annealing optimization method has been adapted to the placement of macros on chips for full custom design in [JJ83].

A hierarchical placement procedure incorporating detailed routing and timing information has been discussed in [GKP+90]. The procedure is based on the min-cut method. Global routing and timing analysis is carried out after every cut which guides the subsequent cell partitioning. [Leb83] discusses an interactive program to get a good floorplan. It includes graphical output, block and pad manipulation and a cost function for estimation of total wire length. In [MM93] S. Mohan and P. Mazumder present a placement algorithm in the distributed computing environment. In [SSL93] S. Sutanthavibul, E. Shragowitz, and R. Lin present a timing-driven placement algorithms for high performance VLSI chips. In [SDS94] Shanbhag, Danda, and Sherwani presented an algorithm for mixed macro block and standard cell designs. Algorithm for mixed macro-cell and standard-cell placement to minimize the chip size and interconnection wire length is presented in [XGC97]. Quadratic placement technique is revisited in [ACHY97].

Chapter 8

Global Routing

In the placement phase, the exact locations of circuit blocks and pins are determined. A netlist is also generated which specifies the required inter-connections. Space not occupied by the blocks can be viewed as a collection of regions. These regions are used for routing and are called as *routing regions*. The process of finding the geometric layouts of all the nets is called *routing*. Nets must be routed within the routing regions. In addition, nets must not short-circuit, that is, nets must not intersect each other.

The input to the general routing problem is:

1. Netlist,

2. Timing budget for nets, typically for critical nets only,

3. Placement information including location of blocks, locations of pins on the block boundary as well as on top due to ATM model (sea-of-pins model), location of I/O pins on the chip boundary as well as on top due to C4 solder bumps,

4. RC delay per unit length on each metal layer, as well as RC delay for each type of via.

The objective of the routing problem is dependent on the nature of the chip. For general purpose chips, it is sufficient to minimize the total wire length, while completing all the connections. For high performance chips, it is important to route each net such that it meets its timing budget. Usually routing involves special treatment of such nets as clock nets, power and ground nets. In fact, these nets are routed separately by special routers.

A VLSI chip may contain several million transistors. As a result, tens of thousands of nets have to be routed to complete the layout. In addition, there may be several hundreds of possible routes for each net. This makes the routing problem computationally hard.

One approach to the general routing problem is called *Area Routing*, which is a single phase routing technique. This technique routes one net at a time

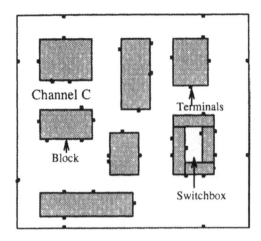

Figure 8.1: Layout of circuit blocks and pins after placement.

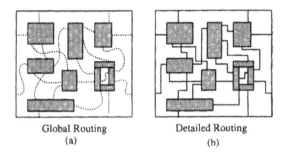

Figure 8.2: Two phases of Routing.

considering all the routing regions. However, this technique is computationally infeasible for an entire VLSI chip and is typically used for specialized problems, and smaller routing regions.

The traditional approach to routing, however, divides the routing into two phases. The first phase is called *global routing* and generates a 'loose' route for each net. In fact it assigns a list of routing regions to each net without specifying the actual geometric layout of wires (see Figure 8.2(a)). The second phase, which is called *detailed routing*, finds the actual geometric layout of each net within the assigned routing regions (see Figure 8.2(b)). Unlike global routing, which considers the entire layout, a detailed router considers just one region at a time. The exact layout is produced for each wire segment assigned to a region, and vias are inserted to complete the layout. In fact, even when the routing problem is restricted to a routing region, such as channels (see definition below), it cannot be solved in polynomial time, i.e., the channel routing problem is NP-complete [Szy85].

In this book, we will briefly describe area routing techniques. Basically, we will follow the two phase approach to routing. In the following, we will discuss global and detailed routing in more detail. Figure8.3 shows a typical two phase routing approach.

The global routing consists of three distinct phases; Region definition, Region Assignment, and Pin assignment. The first phase of global routing is to partition the entire routing space into routing regions. This includes spaces between blocks and above blocks, that is, OTC areas. Between blocks there are two types of routing regions: channels and 2D-switchboxes. Above blocks, the entire routing space is available, however, we partition it into smaller regions called 3D-switchboxes. Each routing region has a capacity, which is the maximum number of nets that can pass through that region. The capacity of a region is a function of the design rules and dimensions of the routing regions and wires. A *channel* is a rectangular area bounded by two opposite sides by the blocks. Capacity of a channel is a function of the number of layers(l), height(h) of the channel, wire width(w) and wire separation(s), i.e., $Capacity = \frac{l \times h}{w+s}$. For example, if for channel C shown in Figure 8.1, $l = 2$, $h = 18\lambda$, $w = 3\lambda$, $s = 3\lambda$, then the capacity is $\frac{2 \times 18}{3+3} = 6$. In a five layer process, only M1, M2 and M3 are used for channel routing. Note that channel may also have pins in th middle. The pins in the middle are actually used to make connections to nets routed in 3D-switchboxes. A *2D-switchbox* is a rectangular area bounded on all sides by blocks. It has pins on all four sides as well as pins in the middle. The pins in the middle are actually used to make connections to nets routed in 3D-switchboxes. A *3D-switchbox* is a rectangular area with pins on all six sides. The pins on the bottom are the pins which allow for connections to nets in channels, 2D-switchboxes and nets using ATM (sea-of-pins) on top of blocks. The pins on the top may be required to connect to C4 solder bumps.

Consider the five metal layer process and assume that blocks use upto third metal layer for internal routing. In this case, channel and 2D-switchboxes will be used in M1, M2 and M3 to route regions between the blocks. Furthermore, the M4 and M5 routing space will be partitioned into several smaller routing regions. The three different routing regions are shown in Figure ref3dswitchbox-6. Another approach to region definition is to partition the M4 and M5 along block boundaries. In this case, channels and 2D-switchboxes will be routed in five metal layers. In addition the regions on top of blocks will be 3D-switchboxes and need to be routed in M4 and M5.

The second phase of global routing can be called region assignment. The purpose of this phase to identify the sequence of regions through which a net will be routed. This phase must take into account the timing budget of each net and routing congestion of each routing region. After the region assignment, each net is assigned a pin on region boundaries. This phase of global routing is called pin assignment. The region boundaries can be between two channels, channel and 3D-switchbox, 2D-switchbox and a 3D-switchbox among others. The pin assignment phase allows the regions to be somewhat independent.

After global routing is complete, the output is pin locations for each net on

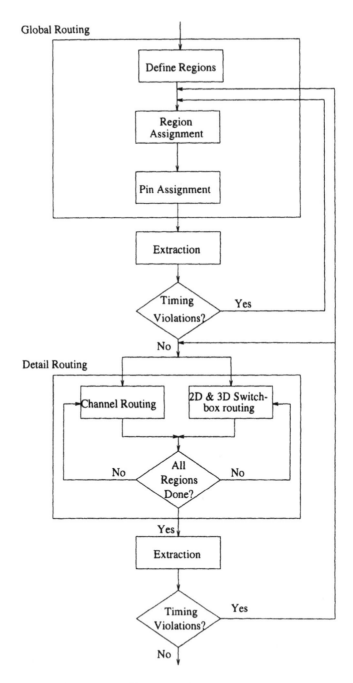

Figure 8.3: The two phase routing flow.

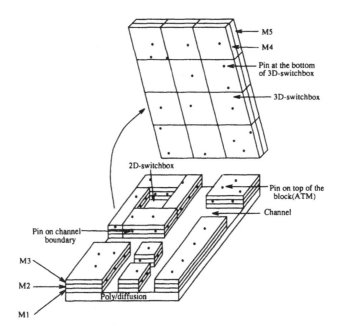

Figure 8.4: Three different routing regions.

the all the region boundaries it crosses. Using this information, we can extract the length of the net and estimate its delay. If some net fails to meet its timing budget, it needs to be ripped-up or global routing phase needs to be repeated.

Detailed routing includes channel routing, 2D-switchbox and 3D-switchbox routing. Typically channels and 2D-switchboxes should be routed first, since channels may expand. After channels and 2D-switchboxes have been routed, the pin locations for 3D-switchboxes are fixed and then their routing can be completed. Channels are routed in a specific order to minimize the impact of channel expansion on the floorplan.

After detailed routing is completed, exact wire geometry can be extracted and used to compute RC delays. The delay model not only considers the geometry (length, width, layer assignments and vias) of a net, but also the relationship of this net with other nets. If some nets fail to meet their timing constraints, they need to be ripped-up or detailed routing of the specific routing region needs to be repeated.

In this chapter, we discuss techniques for global routing. We will also discuss some techniques that can be used for area routing. In Chapter 7 we will discuss the detailed routing techniques. Chapter 8 is dedicated to routing techniques on top of blocks. Chapter 9 discusses the routing of special nets, such as clock and power nets.

Global routing has to deal with two types of nets. Critical nets, which must be routed in high performance layers and other nets. Fir very critical

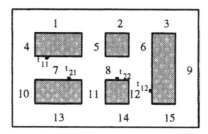

Figure 8.5: The horizontal and vertical channels.

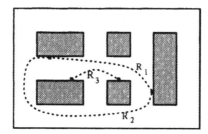

Figure 8.6: Two possible connections between source and target.

nets, global router must use a path which takes the nets from its channel (or 2D-switchbox) through 3D-switchboxes to the termination point in a channel or a 2D-switchbox or C4 bump. Other nets which may not need use of M4 and M5 can be routed through a sequence of channels. Global router must not allocate more nets to a routing region than the region capacity. Let us illustrate this concept of global routing by an example. Suppose that each channel in Figure 8.5 has unit capacity. We consider routing of two nets $N_1 = \{t_{11}, t_{12}\}$ and $N_2 = \{t_{21}, t_{22}\}$. There are several possible routes for net N_1. Two such routes R_1 and R_2 are shown in Figure 8.6. If the objective is to route just N_1, obviously R_1 is a better choice. However, if both N_1 and N_2 are to be routed, it is not possible to use R_1 for N_1 since it would make N_2 unroutable. Thus global routing is computationally hard since it involves trade-offs between routability of all nets and minimization of the objective function. In fact, we will see that global routing of even a single multi-terminal net is NP-complete. In order to simplify presentation, in the rest of the chapter, we will consider global routing with channels and 2D-switchboxes. We will note exceptions for 3D-switchboxes, as and when appropriate. In addition, we will assume that timing constraints are translated into length constraints, hence the objective is to route each net within its length budget.

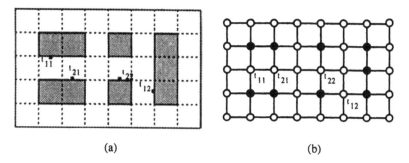

Figure 8.7: Grid Graph Model.

8.1   Problem Formulation

The global routing problem is typically studied as a graph problem. The routing regions and their relationships and capacities are modeled as graphs. However, the design style strongly effects the graph models used and as a result, there are several graph models. Before presenting the problem formulation of global routing, we discuss three different graph models which are commonly used.

The graph models for area routing capture the complete layout information and are used for finding exact route for each net. On the other hand, graph models for global routing capture the adjacencies and routing capacities of routing regions. We discuss three graph models *viz;* grid graph model, checker board model and the channel intersection graph model. Grid graphs are most suitable for area routing while the channel intersection graphs are most suitable for global routing.

1. **Grid Graph Model:** The simplest model for routing is a *grid graph.* The grid graph $G_1 = (V_1, E_1)$ is a representation of a layout. In this model, a layout is considered to be a collection of unit side square cells arranged in a $h \times w$ array. Each cell c_i is represented by a vertex v_i, and there is an edge between two vertices v_i and v_j, if cells c_i and c_j are adjacent. A terminal in cell c_i is assigned to the corresponding vertex v_i. The capacity and length of each edge is set equal to one, i.e., $c(e) = 1, l(e) = 1$. It is quite natural to represent blocked cells by setting the capacity of the edges incident on the corresponding vertex to zero. Figure 8.7(b) shows a grid graph model for a layout in Figure 8.7(a).

   Given a grid graph, and a two terminal net, the routing problem is simply to find a path connecting the vertices, corresponding to the terminals, in the grid graph. Whereas, for a multi-terminal net, the problem is to find a Steiner tree in the grid graph.

   The more general routing problems may consider k-dimensional grid graphs, however, the general techniques for routing essentially remain

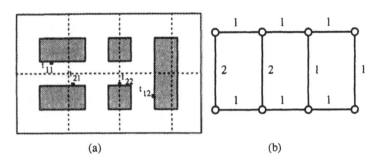

(a)   (b)

Figure 8.8: Checker Board Graph.

the same in all grids. In fact, routing in grids, should be considered as area routing, since the actual detailed route of the net is determined.

2. **Checker Board Model:** Checker board model is a more general model than the grid model. It approximates the entire layout area as a 'coarse grid' and all terminals located inside a coarse grid cell are assigned that cell number. The checker board graph $G_2 = (V_2, E_2)$ is constructed in a manner analogous to grid graph. The edge capacities are computed based on the actual area available for routing on the cell boundary. Figure 8.8(b) shows a checker board graph model of a layout in Figure 8.8(a). Note that the partially blocked edges have unit capacity, whereas, the unblocked edges have a capacity of 2. Given the cell numbers of all terminals of a net, the global routing routing problem is to find a routing in the coarse grid graph.

   A checker board graph can also be formed from a cut tree of floorplan. A block b_i in a floorplan is represented by a vertex v_i and there is an edge between vertices v_i and v_j if the corresponding blocks b_i and b_j are adjacent to each other. Note that, unlike the cells in a grid, two adjacent modules in a cut tree of a floorplan may not entirely share a boundary with each other. Figure 8.9(b) shows an example of a checker board graph for a cut tree of a floorplan in Figure 8.9(a).

3. **Channel Intersection Graph Model:** The most general and accurate model for global routing is the channel intersection model. Given a layout, we can define a channel intersection graph $G_3 = (V_3, E_3)$, where each vertex $v_i \in V_3$ represents a channel intersection CI_i. Two vertices v_i and v_j are adjacent in G_3 if there exists a channel between CI_i and CI_j. In other words, the channels appear as edges in G_3. Figure 8.10(b) shows a channel intersection graph for a layout in Figure 8.10(a). Let $c(e)$ and $l(e)$ be the capacity and length of a channel associated with edge $e \in E_3$. The channel intersection graph should be extended to include the pins as vertices so that the connections between the pins can be considered in this graph. For example, the extended channel intersection graph in

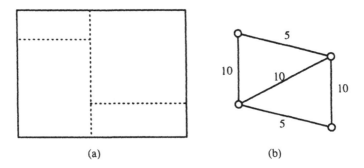

(a)   (b)

Figure 8.9: Checker Board Graph of a Floorplan.

Figure 8.11(b) is obtained by adding vertices representing terminals to the channel intersection graph in Figure 8.10(b).

In the rest of the chapter, the type of routing graph will be clear from the context and will be denoted as $G = (V, E)$.

The global routing problem of two terminal nets is to find path for each net in the routing graph such that the desired objective function is optimized. In addition, the number of nets using each edge (traffic through the corresponding channel) should not violate the capacity of that edge. For example, the global routes for nets N_1 and N_2 are shown as the paths P_1 and P_2 in Figure 8.11(b). It is obvious from the example that routing of one net at a time causes ordering problem for nets. It is important to note that the overall optimal solution may consist of suboptimal solutions of individual nets.

For a net with more than two terminals, the path model discussed above is not appropriate. In fact, global routing of multi-terminal nets can be formulated as a Steiner tree problem. As defined in Chapter 3, a Steiner tree is a tree interconnecting a set of specified points called *demand points* and some other points called *Steiner points*. The number of Steiner points is arbitrary. The global routing problem can be viewed as a problem of finding a Steiner tree for each net in the routing graph such that the desired objective function is optimized. In addition, the capacity of the edges must not be violated. As discussed earlier, a typical objective function is to minimize the total length of selected Steiner trees. In high-performance circuits, the objective function is to minimize the maximum wire length of selected Steiner trees. A more precise objective function for high-performance circuits is to minimize the maximum diameter of selected Steiner trees. The *diameter* of a Steiner tree is defined as the maximum length of a path between any two vertices in the Steiner tree. If there is no feasible solution to an instance of a global routing problem, then the netlist is not routable as the capacity constraints of some edges can not be satisfied. In such cases, the placement phase has to be carried out again.

The formal statement of global routing problem is as follows: Given, a netlist $N = \{N_1, N_2, \ldots, N_n\}$, the routing graph $G = (V, E)$, find a Steiner tree T_i

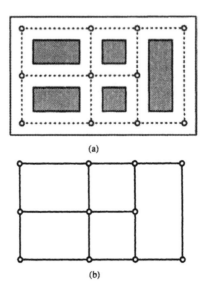

(a)

(b)

Figure 8.10: Channel intersection graph.

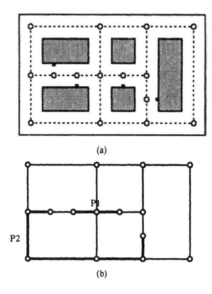

(a)

(b)

Figure 8.11: Extended channel intersection graph.

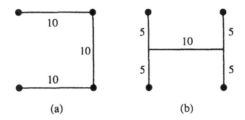

Figure 8.12: Difference between diameter and length of a net.

for each net $N_i, 1 \leq i \leq n$, such that, the capacity constraints are not violated, i.e., $U(e_j) \leq c(e_j)$ for all $e_j \in E$, where $U(e_j) = \sum_{i=1}^{n} x_{ij}$ is the number of wires that pass through the channel corresponding to edge e_j ($x_{ij} = 1$ if e_j is in T_i, it is 0 otherwise). A typical objective function is to minimize the total wire length ($\sum_{i=1}^{n} L(T_i)$), where $L(T_i)$ is the length of Steiner tree T_i.

In the case of high-performance chips the objective function is to minimize the maximum wire length ($\max_{i=1}^{n} L(T_i)$). Note that minimization of maximum wire length may not directly reduce the diameter of the Steiner trees. Consider the example shown in Figure 8.12. The two Steiner trees are both of length 30, but the Steiner tree shown in Figure 8.12(b) has diameter equal to 20, which is much smaller that the diameter of the tree shown in Figure 8.12(a).

8.1.1   Design Style Specific Global Routing Problems

The objective of global routing in each design style is different. We will discuss the global routing problem for full custom, standard cell and gate array. Global routing problem for FPGA and MCM is discussed in Chapters 11 and 12 respectively.

1. **Full custom:** The global routing problem formulation for full custom design style is similar to the general formulation described above. The only difference is how capacity constraints guide the global routing solution. In the general formulation the edge capacities cannot be violated. In full custom, since channels can be expanded, some violation of capacity constraints is allowed. However, major violation of capacities which leads to significant changes in placement are not allowed. In such case, it may be necessary to carry out the placement again.

2. **Standard cell:** In the standard cell design style, at the end of the placement phase, the location of each cell in a row is fixed. In addition, the capacity and location of each feedthrough is fixed. However, the channel heights are not fixed. They can be changed by varying the distance between adjacent cell rows to accommodate wires assigned by a global router. As a result, they do not have a predetermined capacity. On the other hand, feedthroughs have predetermined capacity. The area of a

standard cell layout is determined by the total cell row height and the total channel height, where the total cell row height is the summation of all cell row heights and the total channel height is the summation of all channel heights. As the total cell row height is fixed, the layout area could only be minimized by minimizing the total channel height. As a result, standard cell global routers attempt to minimize the total channel height. Other optimization functions include the minimization of the total wire length and the minimization of the maximum wire length.

The edge set of $G = (V, E)$ are partitioned into two disjoint sets E^v and E^h, i.e., $E = E^v \cup E^h$. Edges in E^v represent feedthroughs, whereas, edges in E^h represent channels. Capacity of each edge $e_j \in E^v$ is equal to the number of wires that can pass through the corresponding feedthrough. Whereas, the capacity of an edge $e \in E^h$ is set to infinity. Let E_{ij}^h represent a jth edge in i^{th} channel and let $E_i^h = \cup_{\forall j} E_{ij}^h$ for all $i = 1, 2, \ldots, p$, where p is the total number of channels in the layout.

Thus, the global routing problem is to find a Steiner tree T_i for each net $N_i, 1 \leq i \leq n$, such that, the capacity constraints are not violated, i.e., $U(e_j) \leq c(e_j)$ for all $e_j \in E^v$, where $U(e_j) = \sum_{i=1}^n x_{ij}$ is the number of wires that go through the feedthrough corresponding to edge e_j ($x_{ij} = 1$ if e_j is in T_i, it is 0 otherwise). The optimization function is either to minimize the total wire length ($\sum_{i=1}^n L(T_i)$) or to minimize the maximum wire length ($\max_{i=1}^n L(T_i)$) or to minimize the total channel height $\sum_{i=1}^p \max\{U(e)|e \in E_i^h\}$. $L(T_i)$ is the length of Steiner tree T_i.

If there is no feasible solution for a global routing problem, feedthrough capacities are not sufficient. (see Figure 8.13.) Additional feedthroughs should be inserted in order to allow global routing.

Recently, a new approach, called over-the-cell routing, has been presented for standard cell design, in which, in addition to the channels and feedthroughs the over-the-cell areas are available for routing. Availability of over-the-cell areas changes the global routing problem. In Chapter 8, this approach is discussed in detail.

3. **Gate array:** In gate array design style, the size and location of all cells and the routing channels and their capacities are fixed by the architecture. This is the key difference between gate array and other design styles. Unlike the full custom design style and standard cell design style the primary objective of the global routing in gate arrays is to guarantee *routability*. The secondary objective may be to minimize the total wire length or to minimize the maximum wire length. Other than these objectives, the formulation of global routing problem in gate array design style is same as the general global routing formulation. If there is no feasible solution to a given instance of global routing problem, the netlist can not be routed (see Figure 8.14). In this case, the placement phase has to be carried out again as the capacity of routing channels is fixed in gate array design style.

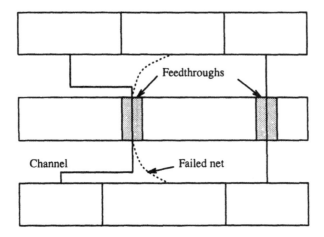

Figure 8.13: Not enough feedthroughs in standard cell design style.

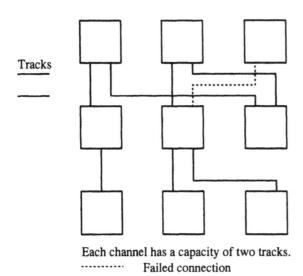

Each channel has a capacity of two tracks.
··········· Failed connection

Figure 8.14: Not all nets are routable in gate array design style.

8.2   Classification of Global Routing Algorithms

Basically, there are two kinds of approaches to solve global routing problem; the sequential and the concurrent.

1. **Sequential Approach:** In this approach, as the name suggests, nets are routed one by one. However, once a net has been routed it may block other nets which are yet to be routed. As a result, this approach is very sensitive to the order in which the nets are considered for routing. Usually, the nets are sequenced according to their criticality, half perimeter of the bounding rectangle and number of terminals. The criticality of a net is determined by the importance of the net. For example, clock net may determine the performance of the circuit and therefore it is considered to be a very important net. As a result, it is assigned a high criticality number. The nets on the critical paths are assigned high criticality numbers since they also play a key role in determining the performance of the circuit. The criticality number and other factors can be used to sequence nets. However, sequencing techniques do not solve the net ordering problem satisfactorily. In a practical router, in addition to a net ordering scheme an improvement phase is used to remove blockages when further routing of nets is not possible. However, this also may not overcome the shortcoming of sequential approach. One such improvement phase involves 'rip-up and reroute' [Bol79, DK82] technique, while other involves 'shove-aside' technique. In 'rip-up and reroute', the interfering wires are ripped up, and rerouted to allow routing of the affected nets. Whereas, in 'Shove-Aside' technique, wires that will allow completion of failed connections are moved aside without breaking the existing connections. Another approach [De86] is to first route simple nets consisting of only two or three terminals since there are few choices for routing such nets. Usually such nets comprise a large portion of the nets (up to 75%) in a typical design. After the simple nets have been routed, a Steiner tree algorithm is used to route intermediate nets. Finally, a maze routing algorithm is used to route the remaining multi-terminal nets (such as power, ground, clock etc.) which are not too numerous.

   The sequential approach includes:

   (a) Two-terminal algorithms:
       i. Maze routing algorithms
       ii. Line-probe algorithms
       iii. Shortest path based algorithms
   (b) Multi-terminal algorithms:
       i. Steiner tree based algorithms

2. **Concurrent Approach:** This approach avoids the ordering problem by considering routing of all the nets simultaneously. The concurrent

approach is computationally hard and no efficient polynomial algorithms are known even for two-terminal nets. As a result, integer programming methods have been suggested. The corresponding integer program is usually too large to be employed efficiently. Hence, hierarchical methods that work top down are employed to partition the problem into smaller subproblems, which can be solved by integer programming. The integer programming based concurrent approach will be presented in this chapter.

8.3 Maze Routing Algorithms

Lee [Lee61] introduced an algorithm for routing a two terminal net on a grid in 1961. Since then, the basic algorithm has been improved for both speed and memory requirements. Lee's algorithm and its various improved versions form the class of maze routing algorithms.

Maze routing algorithms are used to find a path between a pair of points, called the source(s) and the target(t) respectively, in a planar rectangular grid graph. The geometric regularity in the standard cell and gate array design style lead us to model the whole plane as a grid. The areas available for routing are represented as unblocked vertices, whereas, the obstacles are represented as blocked vertices. The objective of a maze routing algorithm is to find a path between the source and the target vertex without using any blocked vertex. The process of finding a path begins with the exploration phase, in which several paths start at the source, and are expanded until one of them reaches the target. Once the target is reached, the vertices need to be retraced to the source to identify the path. The retrace phase can be easily implemented as long as the information about the parentage of each vertex is kept during the exploration phase. Several methods of path exploration have been developed.

8.3.1 Lee's Algorithm

This algorithm, which was developed by Lee in 1961 [Lee61], is the most widely used algorithm for finding a path between any two vertices on a planar rectangular grid. The key to the popularity of Lee's maze router is its simplicity and and its guarantee of finding an optimal solution if one exists.

The exploration phase of Lee's algorithm is an improved version of the breadth-first search. The search can be visualized as a wave propagating from the source. The source is labeled '0' and the wavefront propagates to all the unblocked vertices adjacent to the source. Every unblocked vertex adjacent to the source is marked with a label '1'. Then, every unblocked vertex adjacent to vertices with a label '1' is marked with a label '2', and so on. This process continues until the target vertex is reached or no further expansion of the wave can be carried out. An example of the algorithm is shown in Figure 8.15. Due to the breadth-first nature of the search, Lee's maze router is guaranteed to find a path between the source and target, if one exists. In addition, it is guaranteed to be the shortest path between the vertices.

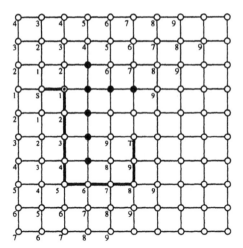

Figure 8.15: A net routed by Lee's algorithm.

The input to the Lee's Algorithm is an array B, the source(s) and target(t) vertex. $B[v]$, denotes if a vertex v is blocked or unblocked. The algorithm uses an array L, where $L[v]$ denotes the distance from the source to the vertex v. This array will be used in the procedure $RETRACE$ that retraces the vertices to form a path P, which is the output of the Lee's Algorithm. Two linked lists *plist* (Propagation list) and *nlist* (Neighbor list) are used to keep track of the vertices on the wavefront and their neighbor vertices respectively. These two lists are always retrieved from tail to head. We also assume that the neighbors of a vertex are visited in counter-clockwise order, that is top, left, bottom and then right.

The formal description of the Lee's Algorithm appears in Figure 8.16. The time and space complexity of Lee's algorithm is $O(h \times w)$ for a grid of dimension $h \times w$.

The Lee's routing algorithm requires a large amount of storage space and its performance degrades rapidly when the size of the grid increases. There have been numerous attempts to modify the algorithm to improve its performance and reduce its memory requirements.

Lee's algorithm requires up to $k+1$ bits per vertex, where k bits are used to label the vertex during the exploration phase and an additional bit is needed to indicate whether the vertex is blocked. For an $h \times w$ grid, $k = \log_2(h \times w)$. Acker [Ake67] noticed that, in the retrace phase of Lee's algorithm, only two types of neighbors of a vertex need to be distinguished; vertices toward the target and vertices toward the source. This information can be coded in a single bit for each vertex. The vertices in wavefront L are always adjacent to the vertices in wavefront $L - 1$ and $L + 1$. Thus, during wave propagation, instead of using a sequence $1, 2, 3, \ldots$, the wavefronts are labeled by a sequence

```
Algorithm LEE-ROUTER (B, s, t, P)
   input: B, s, t
   output: P
begin
   plist = s;
   nlist = φ;
   temp = 1;
   path_exists = FALSE;
   while plist ≠ φ do
       for each vertex vᵢ in plist do
           for each vertex vⱼ neighboring vᵢ do
           if B[vⱼ] = UNBLOCKED then
               L[vⱼ] = temp;
               INSERT(vⱼ,nlist);
               if vⱼ = t then
                   path_exists = TRUE;
                   exit while;
           temp=temp + 1;
           plist = nlist;
           nlist = φ;
       if path_exists = TRUE then RETRACE (L, P);
       else path does not exist;
end.
```

Figure 8.16: Algorithm LEE-ROUTER.

like $0, 0, 1, 1, 0, 0, \ldots$. The predecessor of any wavefront is labeled differently from its successor. Thus, each scanned vertex is either labeled '0' or '1'. Besides these two states, additional states ('block' and 'unblocked') are needed for each vertex. These four states of each vertex can be represented by using exactly two bits, regardless of the problem size. Compared to Acker's scheme, Lee's algorithm requires at least 12 bits per vertex for a grid size of 2000×2000.

It is important to note that Acker's coding scheme only reduces the memory requirement per vertex. It inherits the search space of Lee's original routing algorithm, which is $O(h \times w)$ in the worst case.

8.3.2   Soukup's Algorithm

Lee's algorithm explores the grid symmetrically, searching equally in the directions away from target as well as in the directions towards it. Thus, Lee's algorithm requires a large search time. In order to overcome this limitation, Soukup proposed an iterative algorithm in 1978 [Sou78]. During each iteration, the algorithm explores in the direction toward the target without changing the direction until it reaches the target or an obstacle, otherwise it goes away from

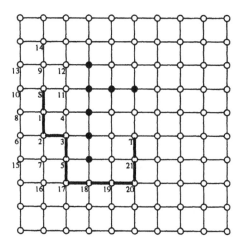

Figure 8.17: A net routed by Soukup's router.

the target. If the target is reached, the exploration phase ends. If the target is not reached, the search is conducted iteratively. If the search goes away from the target, the algorithm simply changes the direction so that it goes towards the target and a new iteration begins. However, if an obstacle is reached, the breadth-first search is employed until a vertex is found which can be used to continue the search in the direction toward the target. Then, a new iteration begins. Figure 8.17 illustrates the Soukup's algorithm with an example. In Figure 8.17, the number near a vertex indicates the order in which that vertex was visited.

Figure 8.18 contains the formal description of Soukup's Algorithm. The notation used in the algorithm is similar to that used in the Lee's algorithm except for the array L. We use $L[v]$ to denote the order in which the vertex v is visited during the exploration phase in this algorithm. Function $DIR(v_1, v_2)$ returns the direction from v_1 to v_2. Function $NGHBR\text{-}IN\text{-}DIR(v_1, v_2)$ returns the neighbor of v_2 which is in the direction from v_1 to v_2.

The Soukup's Algorithm improves the speed of Lee's algorithm by a factor of 10 to 50. It guarantees finding a path if a path between source and target exits. However, this path may not be the shortest one. The search method for this algorithm is a combined breadth-first and depth-first search. The worst case time and space complexities for this algorithm are both $O(h \times w)$, for a grid of size $h \times w$.

8.3.3   Hadlock's Algorithm

An alternative approach to improve upon the speed was suggested by Hadlock in 1977 [Had75]. The algorithm is called Hadlock's minimum detour algorithm. This algorithm uses A^* search method.

Algorithm SOUKUP-ROUTER (B, s, t, P)
    input: B, s, t
    output: P
begin
    $plist = s$;
    $nlist = \phi$;
    $temp = 1$;
    $path_exists = $ FALSE;
    while $plist \neq \phi$ **do**
        for each vertex v_i in $plist$ **do**
            for each vertex v_j neighboring v_i **do**
                if $v_j = t$ **then**
                    $L[v_j] = temp$;
                    $path_exists = $ TRUE;
                    exit while;
                if $B[v_j] = $ UNBLOCKED **then**
                (* If the direction of the search is toward the
                target, the search continues in this direction *)
                    if $DIR(v_i, v_j) = $ TO-TARGET
                    then $L[v_j] = temp$;
                      $temp = temp + 1$;
                      INSERT $(v_j, plist)$;
                      while $B[$NGHBR-IN-DIR$(v_i, v_j)]=$
                      UNBLOCKED **do**
                      $v_j = $ NGHBR-IN-DIR(v_i, v_j);
                      $L[v_j] = temp$;
                      $temp = temp + 1$;
                      INSERT $(v_j, plist)$;
                else
                  $L[v_j] = temp$;
                  $temp = temp + 1$;
                  INSERT $(v_j, nlist)$;
        $plist = nlist$;
        $nlist = \phi$;
    if $path_exists = $ TRUE **then** RETRACE (L, P);
    else path does not exist;
end.

Figure 8.18: Algorithm SOUKUP-ROUTER.

Algorithm HADLOCK-ROUTER(B, s, t, P)
    input: B, s, t
    output: P
begin
    $plist = s$;
    $nlist = \phi$;
    $detour = 0$;
    $path\_exists$ = FALSE;
    while $plist \neq \phi$ **do**
        for each vertex v_i in $plist$ **do**
            for all vertices v_j neighboring v_i **do**
                if $B[v_j]$ = UNBLOCKED **then**
                    $D[v_j]$ = DETOUR-NUMBER(v_j);
                    INSERT (v_j,$nlist$);
                    if $v_j = t$ **then**
                        $path\_exists$ = TRUE;
                        exit while;
        if $nlist = \phi$ **then**
            $path\_exists$ = FALSE;
            exit while;
        $detour$ = MINIMUM-DETOUR($nlist$);
        for each vertex v_k in $nlist$ **do**
            if $D[v_k] = detour$ **then** INSERT($v_k, plist$);
        DELETE ($nlist, plist$);
    if $path\_exists$ = TRUE **then** RETRACE (L, P);
    else path does not exist;
  end.

Figure 8.19: Algorithm HADLOCK-ROUTER.

Hadlock observed that the length of a path(P) connecting source and target can be given by $M(s,t) + 2d(P)$, where $M(s,t)$ is Manhattan distance between source and target and $d(P)$ is the number of vertices on path P that are directed away from the target. The length of P is minimized if and only if d is minimized as $M(s,t)$ is constant for given pair of source and target. This is the essence of Hadlock's algorithm. The exploration phase, instead of labeling the wavefront by a number corresponding to the distance from the source, uses the detour number. The detour number of a path is the number of times that the path has turned away from the target. Figure 8.20 illustrates the Hadlock's algorithm with an example. In Figure 8.20, the number near a vertex indicates the order in which that vertex was visited.

A formal description of Hadlock's Algorithm is given in Figure 8.19. Function DETOUR-NUMBER(v) returns detour number of a vertex v. Procedure DELETE($nlist, plist$) deletes the vertices which are in $plist$ from $nlist$. Func-

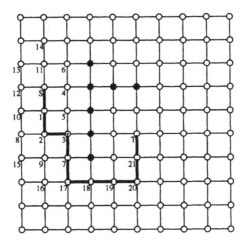

Figure 8.20: A net routed by Hadlock's Algorithm.

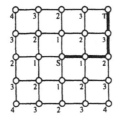

Figure 8.21: Lee's algorithm in the worst case.

tion MINIMUM-DETOUR($nlist$) returns the minimum detour number among all vertices in the list $nlist$.

The worst case time and space complexity of Hadlock's algorithm is $O(h \times w)$ for a grid of size $h \times w$.

8.3.4 Comparison of Maze Routing Algorithms

Maze routing algorithms are grid based methods. The time and space required by these algorithms depend linearly on their search space.

The search in Lee's algorithm is conducted by using a wave propagating from the source. The algorithm searches symmetrically in every direction, using the breath-first search technique. Thus, it guarantees finding a shortest path between any two vertices if such a path exists. However, the worst case happens when the source is located at the center and the target is located at a corner of routing area, in which all the vertices have to be scanned before the

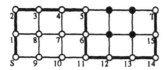

Figure 8.22: Soukup's algorithm does not find the shortest path.

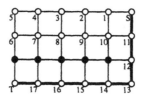

Figure 8.23: Soukup's algorithm in the worst case.

target is reached. (see Figure 8.21.)

The Soukup's algorithm remedies the shortcoming of the breadth-first search method by using a depth-first search until an obstacle is encountered. If an obstacle is encountered, a breadth-first search method is used to get around the obstacle. The search time in Soukup's algorithm is usually smaller than the Lee's algorithm due to the nature of depth-first search method. However, this algorithm may not find a shortest path between the source and target. In Figure 8.22, the Soukup's algorithm explores all the vertices and does not find the shortest path between s and t. The worst case of Soukup's algorithm occurs when the search goes in the direction of the target, which is opposite the direction of the passageway through the obstacle. Figure 8.23 shows an example in which Soukup's algorithm scans all vertices while finding a path between s and t.

The Hadlock's algorithm aims at both reducing the search time and finding an optimal path between given two vertices. Basically, the Hadlock's algorithm

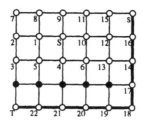

Figure 8.24: Hadlock's algorithm in the worst case.

is a breadth-first search method. As a result, it finds a shortest path if one exists. The difference between the Hadlock's algorithm and Lee's algorithm is the way in which the wavefront is labeled. The Hadlock's algorithm label the wavefront by the detour number instead of the distance from the source used in the Lee's algorithm. In this way, the search can prefer the direction toward the target to the direction away from the target. This search time is shorter than the Lee's algorithm. When the direction of the search goes toward the target and opposite the passageway through the obstacle, the worst case happens (see Figure 8.24).

All the maze routers and many of their variations are grid based methods. Information must be kept for each grid node. Thus, a very large memory space is needed to implement these algorithms for a large grid. To give an approximate estimate, a chip of size $10000\lambda \times 10000\lambda$ requires as much as 350 MBytes of memory and 66 seconds to route one net on a 15MIPS workstation. There may be 5000 to 10000 nets in a typical chip. Such numbers make these maze routing algorithms infeasible for large chips. In order to reduce the large memory requirements and run times, line-probe algorithms were developed.

8.4 Line-Probe Algorithms

The line-probe algorithms were developed independently by Mikami and Tabuchi in 1968 [MT68], and Hightower in 1969 [Hig69]. The basic idea of a line probe algorithm is to reduce the size of memory requirement by using line segments instead of grid nodes in the search. The time and space complexities of these line-probe algorithms is $O(L)$, where L is the number of line segments produced by these algorithms.

The basic operations of these algorithms are as follows. Initially, lists *slist* and *tlist* contain the line segments generated from the source and target respectively. The generated line segments do not pass through any obstacle. If a line segment from *slist* intersects with a line segment in *tlist*, the exploration phase ends; otherwise, the exploration phase proceeds iteratively. During each iteration, new line segments are generated. These segments originate from 'escape' points on existing line segments in *slist* and *tlist*. The new line segments generated from *slist* are appended to *slist*. Similarly, segments generated from a segment in *tlist* are appended to *tlist*. If a line segment from *slist* intersects with a line segment from *tlist*, then the exploration phase ends. The path can be formed by retracing the line segments in set *tlist*, starting from the target, and then going through the intersection, and finally retracing the line segments in set *slist* until the source is reached.

The data structures used to implement these algorithms play an important role in the efficiency considerations of the search for obstructions to probes. Typically two lists, one for the horizontal lines and one for the vertical lines are used. The use of two separate lists allows lines parallel to the direction of the probe to be ignored, thus expediting the search.

The Mikami and the Hightower algorithms differ only in the process of

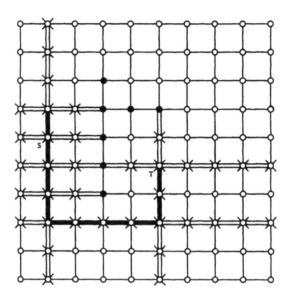

Figure 8.25: A net routed by Mikami-Tabuchi's Algorithm.

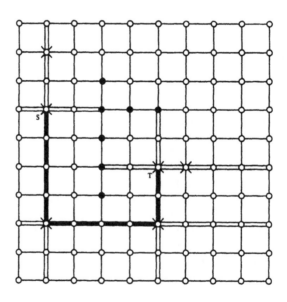

Figure 8.26: A net routed by Hightower's Algorithm.

Algorithm LINE-PROBE-ROUTER(s, t, P)
   input: s, t
   output: P
begin
   new_slist = line segments generated from s;
   new_tlist = line segments generated from t;
   while $new\_slist \neq \phi$ **and** $tlist \neq \phi$ **do**
      $slist = new\_slist$;
      $tlist = new\_tlist$;
      for each line segment l_i in $slist$ **do**
         for each line segment l_j in $tlist$ **do**
            if INTERSECT(l_i, l_j)=TRUE **then**
               $path\_exists$ = TRUE;
               exit while;
      $new\_slist = \phi$;
      for each line segment l_i in $slist$ **do**
         for each escape point e on l_i **do**
            GENERATE(l_k, e);
            INSERT(l_k, new_slist);
      $new\_tlist = \phi$;
      for each line segment l_i in $tlist$ **do**
         for each escape point e on l_i **do**
            GENERATE(l_k, e);
            INSERT(l_k, new_tlist);
   if $path\_exists$=TRUE **then** RETRACE;
   else a path can not be found;
end.

Figure 8.27: Algorithm LINE-PROBE-ROUTER.

choosing escape points. In Mikami's algorithm, every grid node on the line segment is an 'escape' point, which generates new perpendicular line segments. This search is similar to the breadth-first search, and is guaranteed to find a path if one exists. However, the path may not be the shortest one. Figure 8.25 shows a path generated by Mikami's algorithm. On the other hand, Hightower's algorithm makes use of only a single 'escape' point on each line segment. In the simple case of a probe parallel to the blocked vertices, the escape point is placed just past the endpoint of the segment. Figure 8.26 shows a path generated by Hightower's algorithm. Hightower has described three such processes, designed to help the router find a path around different types of obstacles. The disadvantage in generating fewer escape points in Hightower's algorithm essentially means that it may not be able to find a path joining two points even when such a path exists.

A formal description of these two algorithm is given in Figure 8.27. (As

	Algorithms				
	Maze Routing			Line-Probe	
	Lee	Soukup	Hadlock	Mikami	Hightower
Time complexity	$h \times w$	$h \times w$	$h \times w$	L	L
Space complexity	$h \times w$	$h \times w$	$h \times w$	L	L
Finds path if one exists?	yes	yes	yes	yes	no
Is the path shortest?	yes	no	yes	yes	no

Table 8.1: Comparison of different algorithms.

these two algorithms basically are the same, we just use one description for both of them.) Procedure GENERATE(l, e) generates a line-probe l from an escape point e, whereas, INSERT($l, list$) adds a line-probe l to the *list*. Function INTERSECT(l_i, l_j) returns TRUE if line-probes l_i and l_j intersect, it returns FALSE otherwise.

Maze routers and many of their variations are grid based methods. Information must be kept for each grid node. Thus, a very large memory space is needed to implement these algorithms for a large grid. The line-probe algorithms, however, require the information to be kept for each line segment. Since the number of line segments is very small compared to the nodes in a grid, the required memory is greatly reduced. The key difference between the two line probe algorithms is that, the Mikami's algorithm can find a path between any two vertices if one exists. This path may not be the shortest path. Hightower's algorithm may not be able to find a path joining two vertices even if such a path exists. A comparison of the maze routing algorithms and line-probe algorithms in their worst cases is given in Table 8.1. ($h \times w$ denotes the size of grid and L denotes the number of line segments generated in line-probe algorithms).

8.5 Shortest Path Based Algorithms

A simple approach to route a two-terminal net uses Dikjstra's shortest algorithm [Dij59]. Given, a routing graph $G = (V, E)$, a source vertex $s \in V$ and a target vertex $t \in V$ a shortest path in G joining s and t can be found in $O(|V|^2)$ time. The algorithm in Figure 8.28 gives formal description of an algorithm based on Dijkstra's shortest path algorithm for global routing a set $\mathcal{N}$ of two-terminal nets in a routing graph G. The output of the algorithm is a set $\mathcal{P}$ of paths for the nets in $\mathcal{N}$. A path $P_i \in \mathcal{P}$ gives a path for net $N_i \in \mathcal{N}$. The time complexity of the algorithm SHORT-PATH-GLOBAL-ROUTER is $O(\mathcal{N}|V|^2)$.

Note that the length of an edge is increased by a factor $\alpha > 1$ whenever a congested edge is utilized in the path of a net. This algorithm is suitable

Algorithm SHORT-PATH-GLOBAL-ROUTER($G, \mathcal{N}, \mathcal{P}$)
    input: $G, \mathcal{N}$
    output: $\mathcal{P}$
begin
    for each N_i in $\mathcal{N}$ **do**
        $P_i = $ DIJKSTRA-SHORT-PATH(G, N_i);
        for each e_j in P_i **do**
            $c(e_j) = c(e_j) - 1$;
            if $c(e_j) < 0$ **then** $l(e_j) = \alpha \times l(e_j)$;
end.

Figure 8.28: Algorithm SHORT-PATH-GLOBAL-ROUTER.

for channel intersection graph, since it assumes that congested channels can be expanded. If the edge congestions are strict, the algorithm can be modified to use 'rip-up and reroute' or 'shove aside' techniques [Bol79, DK82].

8.6   Steiner Tree based Algorithms

Global routing algorithms presented so far are not suitable for global routing of multi-terminal nets. Several approaches have been proposed to extend maze routing and line-probe algorithms for routing multi-terminal nets. In one approach, the multi-terminal nets are decomposed into several two-terminal nets and the resulting two-terminal nets are routed by using a maze routing or line-probe algorithm. The quality of routing, in this approach, is dependent on how the nets are decomposed. This approach produces suboptimal results as there is hardly any interaction between the decomposition and the actual routing phase. In another approach, the exploration can be carried out from several terminals at a time. It allows the expansion process to determine which pairs of pins to connect, rather than forcing a predetermined net decomposition. However, the maze routing and line-probe algorithms cannot optimally connect the pins. In addition, these approaches inherit the large time and space complexities of maze routing and line-probe algorithms.

The natural approach for routing multi-terminal nets is Steiner tree approach. Usually *Rectilinear Steiner Trees* (RST) are used. A rectilinear Steiner tree is a Steiner tree with only rectilinear edges. The length of a tree is the sum of lengths of all the edges in the tree. It is also called the cost of the tree. The problem of finding a minimum cost RST is NP-hard [GJ77]. In view of NP-hardness of the problem, several heuristic algorithms have been developed. Most of the heuristic algorithms depend on minimum cost spanning tree. This is due to a special relationship between Steiner trees and minimum cost spanning trees. Hwang [Hwa76a, Hwa79] has shown that the ratio of the cost of a minimum spanning tree (MST) to that of an optimal RST is no greater than

$\frac{3}{2}$. Let S be a net to be routed. We define an underlying grid $G(S)$ of S (on an oriented plane) as the grid obtained by drawing horizontal and vertical lines through each point of S (see Figure 8.29). Let $G_c(S)$ be a complete graph for S. An MST for net S is a minimum spanning tree of $G_c(S)$ (see Figure 8.29.) Note that, there may be several MST's for a given net and they can be found easily. Using Hwang's result, an approximation of the optimal RST can be obtained by rectilinearizing each edge of an MST. Different ways of rectilinearizing the edges of T give different approximations. If an edge (i,j) of T is rectilinearized as a shortest path between i and j on the underlying grid $G(S)$, then it is called as a staircase edge layout. For example, all the edge layouts in Figure 8.29 are staircase layouts. A staircase layout with exactly one turn on the grid G(S) is called as an L-shaped layout. A staircase layout having exactly two turns on the grid $G(S)$ is called as a Z-shaped layout. For example, the edge layout of P_3 and P_1 in Figure 8.29 are L-shaped and Z-shaped layouts respectively. An RST obtained from an MST T of a net S, by rectilinearizing each edge of T using staircase layouts on $G(S)$ is called S-RST. An S-RST of T, in which the layout of each MST edge is a L-shaped layout is called an L-RST of T. An S-RST of T, in which the layout of each MST edge is a Z-shaped layout is called a Z-RST of T. An optimal S-RST (Z-RST, L-RST) is an S-RST (Z-RST, L-RST) of the least cost among all S-RST's (Z-RST's, L-RST's). It is easy to see that an optimal L-RST may have a cost larger than an optimal S-RST (see Figure 8.30), which in turn may have a cost larger than the optimal RST. Obviously, least restriction on the edge layout gives best approximation. However, as the number of steps allowed per edge is increased it becomes more difficult to design an efficient algorithm to find the optimal solution.

The organization of the rest of this section is as follows: First, we discuss a separability based algorithm to find an optimal S-RST from a separable MST. This is followed by a discussion on non-rectilinear Steiner trees. We also discuss MIN-MAX Steiner tree that are used for minimizing the traffic in the densest channels. These three approaches do not consider the presence of obstacles while finding approximate rectilinear Steiner tree for a net. At the end of this section, we discuss a weighted Steiner tree approach that works in presence of obstacles and simultaneously minimizes wire lengths and density of the routing regions.

8.6.1   Separability Based Algorithm

In [HVW85], Ho, Vijayan, and Wong presented an approach to obtain an optimal S-RST from an MST, if the MST satisfies a special property called *separability*. A pair of nonadjacent edges is called *separable* if staircase layouts of the two edges does not intersect or overlap. An MST is called as a *separable MST* (SMST) if all pairs of non-adjacent edges satisfy this property. In other words, such an MST is called to have *separability* property. If an edge is deleted from an SMST, the staircase layouts of the two resulting subtrees do not intersect or overlap each other. Overlaps can occur only between edges that are incident on a common vertex. This property enables the use of dynamic

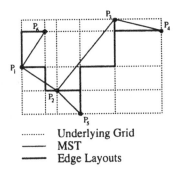

```
........  Underlying Grid
————      MST
————      Edge Layouts
```

Figure 8.29: Grid, MST, and edge layouts.

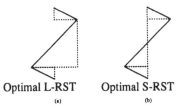

Optimal L-RST	Optimal S-RST
(a)	(b)

Figure 8.30: Different edge layouts for an MST.

programming techniques to obtain an optimal S-RST.

The algorithm works in two steps. In the first step, an SMST T is constructed for a given net N_p by using a modified Prim's algorithm [Pri57] in $O(|N_p|^2)$ time. In the second step, an optimal Z-RST is obtained by using the SMST obtained in the first step in $O(|N_p| \times t_{max}^6)$ time, where t_{max} is the maximum of $t(e)$ over all edges e and $t(e)$ denote the number of edges on the underlying grid $G(N_p)$ traversed by any staircase layout of an edge e of the MST T of net N_p. According to the Z-sufficiency Theorem in [HVW85], an optimal Z-RST is an optimal S-RST. The optimal S-RST is used as an approximation of the minimum cost RST. In the rest of this section, we discuss these two steps in detail.

1. **Algorithm SMST:** Let $G_c(N_p)$ denote the complete graph for net N_p. For any vertex i and j, $x(i)$ and $y(i)$ denote the x, y-coordinates of vertex i on a Cartesian plane, and $dist(i,j)$ denotes the total length of shortest path between i and j. Function PRIM(G_c, W, T), takes the complete graph(G_c) and an array(W) containing weights of edges in G_c as input. It generates a separable MST T using a modified Prim's algorithm for MST, which has a time complexity of $O(|N_p|^2)$. The formal description of algorithm SMST appears in Figure 8.31. The time complexity of algorithm SMST is $O(|N_p|^2)$.

Algorithm SMST(N_p, T)
   input:N_p
   output:T
begin
   Construct the complete graph $G_c(N_p)$ of the net N_p.
   for each edge (i,j) of $G_c(N_p)$ **do**
      $3tuple(i,j) = (dist(i,j), -|y(i) - y(j)|,$
         $-\max(x(i), x(j)));$
   PRIM($G_c(N_p), 3tuple(i,j), T$);
end.

Figure 8.31: Algorithm SMST.

2. **Algorithm Z-RST:** The input to the algorithm is an SMST T of a net N_p. By hanging the input separable MST T by any leaf edge r, a rooted tree T_r can be obtained. For each edge e in T, let T_e denote the subtree of T_r that hangs by the edge e. Given a Z-shaped layout z of an edge e, we let $M_z[e]$ denote the Z-RST of the subtree T_e, which has the minimum cost among all Z-RST's of T_e, in which the layout of the edge e is constrained to be the Z-shape z. $M_z[e]$ can be computed recursively as follows: Let $e_i, i = 1, 2, \ldots, d$ be the d child edges of e in the rooted tree T_r. For each child edge e_i and for each possible Z-shaped layout z_{ij} of the edge e_i, recursively compute the constrained optimal Z-RST's $M_{z_{ij}}[e_i]$ of the subtrees T_{e_i}. Let the number of such constrained Z-RST's for a subtree T_{e_i} be denoted as $t(e_i)$. Taking one such Z-RST for each subtree T_{e_i}, and merging these subtree Z-RST's with the layout z of the edge e, results in a Z-RST of T_e. Since the tree T has the separability property, the only new overlaps that can occur during this merging are among the edges $e, e_1, \ldots, e_d$, which are all incident on a common point. Therefore, the total amount of overlap in the resulting Z-RST of T_e is the sum of the overlaps among the layouts of the edges $e, e_1, \ldots, e_d$, added to the sum of overlaps in the selected Z-RST's of the subtrees T_{e_i}. Enumerate all combinations of selecting one of the Z-RST's $M_{z_{ij}}[e_i]$ for each subtree T_{e_i}, and for each such combination compute the resulting Z-RST of T_e. The constrained optimal Z-RST $M_z[e]$ of the subtree T_e is simply the one with the least cost. To compute the optimal Z-RST of the entire rooted tree T_r, recursively compute (as explained above) the constrained optimal Z-RST's $M_z[r]$ for each Z-shaped layout z of the root edge r, and select that Z-RST of the smallest cost. (see Figure 8.32)

A recursive definition of Function LEAST-COST is given Figure 8.33. Function LEAST-COST($z, T_e, cost, M_z[e]$) takes a Z-shaped layout z of an edge e, and a subtree T_e as input. The output of function LEAST-COST is the optimal Z-RST(denoted as $M_z[e]$) of T_e for the z layout of edge e and the cost(denoted as $CostM_z[e]$) of $M_z[e]$. Function CHILD-

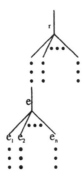

Figure 8.32: Structure used in the algorithm Z-RST.

EDGES-NUM(e) returns the number of child edges of an edge e.

Let r be the leaf edge that is used to hang the SMST T, and T_r be the tree obtained, then the output of the algorithm in Figure 8.34 is the optimal Z-RST (M) of T and its cost $CostM$.

The fact that the algorithm Z-RST constructs the optimal S-RST follows from the separability of the input MST and from the Z-sufficiency theorem stated below.

Theorem 6 *(Z-Sufficiency Theorem): Given an SMST T of a point set S of cardinality n, there exists a Z-RST of T whose cost is equal to the cost of an optimal S-RST of T.*

The worst case time complexity of algorithm Z-RST is $O(|N_p| \times t_{max}^6)$, where t_{max} is the maximum of $t(e)$ over all edges e.

8.6.2   Non-Rectilinear Steiner Tree Based Algorithm

Burman, Chen, and Sherwani [BCS91] studied the problem of global routing of multi-terminal nets in a generalized geometry called δ-geometry in order to improve the layout and consequently enhance the performance. The restriction of layout to rectilinear geometry, and thus only rectilinear Steiner trees, in the previous Steiner tree based global routing algorithms was necessary to account for restricted computing capabilities. Recently, because of enhanced computing capabilities and the need for design of high performance circuit, non-rectilinear geometry has gained ground. In order to obtain smaller length Steiner trees, the concept of separable MST's in δ-geometry was introduced. In δ-geometry, edges with angles $i\pi/\delta$, for all i, are allowed, where δ (≥ 2) is a positive integer. $\delta = 2$, 4 and ∞ correspond to rectilinear, 45^0 and Euclidean geometries respectively. Obviously, we can see that δ-geometry always includes rectilinear edges and is a useful with respect to the fabrication technologies. It has been proved [BCS91]

Function LEAST-COST($z, T_e, CostM_z[e], M_z[e]$)
   input: z, T_e
   output: $CostM_z[e], M_z[e]$
begin
   if CHILD-EDGES-NUM(e) $\neq 0$ **then**
      for each child edge e_i of e **do**
         for each layout z_{ij} of e_i **do**
            LEAST-COST($z_{ij}, T_{e_i}, CostM_{z_{ij}[e_i]}, M_{z_{ij}}[e_i]$);
      for (each combination of the layouts containing one
         optimal Z-RST layout for each T_{e_i}) **do**
         merge layouts in the combination with the layout z
         of edge e;
         calculate the resulting cost of merged layout;
      $CostM_z[e]$ = minimum cost among all merged layouts;
      $M_z[e]$ = the layout corresponding to the minimum cost;
   else (* The bottom edge is reached *)
      $M_z[e] = z$;
      $CostM_z[e] =$ cost of z;
end.

Figure 8.33: Function LEAST-COST.

Algorithm Z-RST ($r, T_r, CostM, M$)
   input: r, T_r
   output: $CostM, M$
begin
   $CostM = \infty$;
   for each z-shaped layout z of r **do**
      LEAST-COST ($z, T_r, CostTempM, TempM$);
      if $CostTempM < CostM$ **then**
         $M = TempM$;
         $CostM = CostTempM$;
end.

Figure 8.34: Algorithm Z-RST.

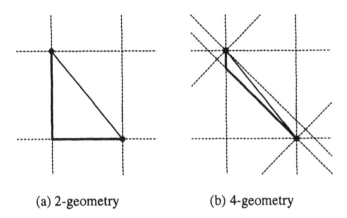

(a) 2-geometry   (b) 4-geometry

Figure 8.35: Comparison of the rectilinearization in 2-geometry and in 4-geometry.

that for an even $\delta \geq 4$, all minimum cost spanning trees in δ-geometry satisfy the separability property.

Theorem 7 *Any minimum spanning tree for a given point set in the plane is δ-separable for any even $\delta \geq 4$.*

Therefore, there exists a polynomial time algorithm to find an optimal Steiner tree in δ-geometry, which is derivable from the separable minimum spanning tree. The experiments have shown that tree length can be reduced up to 10-12% by using 4-geometry as compared to rectilinear geometry (2-geometry). Moreover, length reduction is quite marginal for higher geometries. As a consequence, it is sufficient and effective to consider layouts in 4-geometry in the consideration of global routing problem. An example of the derivation of a Steiner tree for a simple two-terminal net in 4-geometry is shown in Figure 8.35(b) while Figure 8.35(a) shows the derivation of a Steiner tree for the same net in rectilinear geometry. Clearly, the tree length in 4-geometry is shorter than the one in the rectilinear geometry.

8.6.3 Steiner Min-Max Tree based Algorithm

The approach in [CSW89] uses a restricted case of Steiner tree in global routing problem, called *Steiner Min-Max Tree* (SMMT) in which the maximum weight edge is minimized (real vertices represent channels containing terminals of a net, Steiner vertices represent intermediate channels, weights correspond to densities). They give an $O(\min\{|E| \log\log|E|, |V|^2\})$ time algorithm for obtaining a Steiner min-max tree in a weighted coarse grid graph $G = (V, E)$. The weight of an edge in E is a function of current density, capacity, and measures crowdedness of a border. Each vertex in V is labeled with demand or

Algorithm SMMT(G, d, T)
    input:G, d
    output:T
begin
    T = An MST of G;
    while EXIST-ODSV(T, d)=TRUE **do**
        v = GET-ODSV(T, d); (* a Steiner leaf*)
        REMOVE(v, T);
    output T;
end.

Figure 8.36: Algorithm SMMT.

(potential) Steiner depending on whether it is respectively a terminal of net N_i
or not. A Steiner min-max tree of G dictates a global routing that minimizes
traffic in the densest channel. While the Steiner min-max tree method tends to
route nets through less crowded channels, it is also desirable to have nets with
short length. Therefore, among all Steiner min-max trees of the given net, we
are interested in those with minimum length. The problem of finding a Steiner
min-max tree whose total length is minimized is NP-hard.

Given, a weighted coarse grid graph $G = (V, E)$ and a boolean array d
such that $d[v]$ is true if the vertex $v \in V$ corresponds to a terminals of N_i.
An SMMT T of N_i can be obtained using algorithm in Figure 8.36. Function
EXIST-ODSV(T, d) returns TRUE if there exists a one-degree Steiner vertex
in T. Function GET-ODSV(T, d) returns a one-degree Steiner vertex from T.
REMOVE(v, T) removes vertex v and edges incident on it from T.

Theorem 8 *Algorithm SMMT correctly computes a Steiner min-max tree of
net N_i in weighted grid graph $G = (V, E)$ in $O(\min\{|E| \log \log |E|, |V|^2\})$ time.*

A number of heuristics have been incorporated in the global router based on
the min-max Steiner trees. The nets are ordered first according to their priority,
length and multiplicity numbers. The global routing is then performed in two
phases: the SMMT-phase and the SP-phase. The SP-phase is essentially a
minimum-spanning tree algorithm. The SMMT-phase consists of J_1 steps and
the SP-phase consists of J_2 steps, where J_1 and J_2 are heuristic parameters
based on the importance of density and length minimization in a problem,
respectively.

In the SMMT-phase, the nets are routed one by one, using the algorithm
SMMT. At the j-th step of the SMMT-phase, if the length of routing of N_i is
within a constant factor, c_j of its minimum length then it is accepted, otherwise,
the routing is rejected. Once a net is routed during SMMT-phase, it will not
be routed again.

Algorithm LAYOUT-WRST $(\mathcal{R}, N_i)$
begin
    $T =$ an MST of N_i;
    $L_0 = \phi$;
    for $j = 1$ **to** $n - 1$ **do**
      $min\_cost = \infty$;
      for $i = 1$ **to** k **do**
        FIND-P$(i, e_j, \mathcal{R})$;
        $Q_{i,j} =$ MERGE$(L_{j-1}, P_i(e_j))$;
        CLEANUP$(Q_{i,j})$;
        if WT$(Q_{i,j}) < min\_cost$ **then** $L_j = Q_{i,j}$;
end.

Figure 8.37: Algorithm LAYOUT-WRST.

In the SP-phase, the nets are routed one by one by employing a shortest-path heuristic and utilizing the results from the SMMT-phase. At the j-th step we accept a routing only if it is better than the best routing obtained so far.

8.6.4 Weighted Steiner Tree based Algorithm

Several global routing algorithms have been developed that consider minimizing the length of Steiner tree as the primary objective and minimizing the traffic through the routing areas as the secondary objective and vice versa. In [CSW92], Chiang, Sarrafzadeh, and Wong proposed a global router that simultaneously minimizes length and density by using a weighted Steiner tree. Consider a set $\mathcal{R} = \{R_1, R_2, \ldots, R_m\}$ of weighted regions in an arbitrary-style layout, where weight of a region is proportional to its density and area. The regions with blockages are assigned infinite weights. A weighted Steiner tree is a Steiner tree with weighted lengths, i.e., an edge with length l in a region with weight w has weighted length lw. A *weighted rectilinear Steiner tree* (WRST) is a weighted Steiner tree with rectilinear edges. A minimum-weight WRST is a WRST with minimum total weight.

The 2-approximate algorithm to find an approximation of minimum-weight WRST is as discussed below: First step of this algorithm is to find an MST T for a given net N_i using Prim's algorithm. Let $e_1, e_2, \ldots, e_{n-1}$ be the edges of T. In the second step, the edges of T are rectilinearized one by one. In general, there are more than one possible staircase layouts for an edge e_j of T. Let $\{P_1(e_j), P_2(e_j), \ldots, P_k(e_j)\}$ be a subset of all possible staircase layouts for edge e_j. Let L_{j-1} denotes the staircase layout of edges $e_1, e_2, \ldots, e_{j-1}$. Let $Q_{i,j}$ be the layout obtained by merging L_{j-1} and $P_i(e_j)$. L_j is selected to be the minimum cost layout among all $Q_{i,j}$.

The formal description of the algorithm is given in Figure 8.37. Function FIND-P$(i, e_j, \mathcal{R})$ finds $P_i(e_j)$ and function CLEANUP$(Q_{i,j})$ removes over-

lapped layouts. Function $WT((Q_{i,j})$ gives the total weighted length of $Q_{i,j}$.
The time complexity of algorithm LAYOUT-WRST is $O(|N_i|^2)$.

8.7 Integer Programming Based Approach

The problem of concurrently routing all the nets is computationally hard.
The only known technique uses integer programming. In fact, the general
global routing problem formulation can be easily modified to a 0/1 integer programming formulation. Given a set of Steiner trees for each net and a routing
graph, the objective of such an integer programming formulation is to select
a Steiner tree for each net from its set of Steiner trees without violating the
channel capacities while minimizing the total wire length. This approach is well
suited when there is a preferred set of Steiner trees for each net. However, as
the size of input increases the time required to solve corresponding integer program increases exponentially. Thus it is necessary to break down the problem
into several small subproblems, solve them independently and combine their
solutions in order to solve the original problem.

8.7.1 Hierarchical Approach

In this section, we discuss the hierarchical based integer program for global
routing, presented by Heisterman and Lengaur [HL91]. Let $S = \{S_i | 1 \leq i \leq n\}$
denote a set of sets of vertices in the routing graph $G = (V, E)$. Let $T = \{T_{ij}\}, j = 1, 2, \ldots, l_i$, denote a set of Steiner trees for $S_i, i = 1, 2, \ldots, n$. Then,
the global routing problem can be formulated as an integer program by taking
an integer variable x_{ij} to denote the number of nets which are routed using T_{ij}.
S_i is called a net type and T_{ij} a route for S_i. Let n_i denote the number of nets
corresponding to the net type S_i for $i = 1, 2, \ldots, n$. The following constraints
have to be met:

$$\sum_{j=1}^{l_i} x_{ij} = n_i, i = 1, \ldots, n \text{ (completeness constraints)}$$
$$\sum_{(ij), e \in T_{ij}} x_{ij} + x_e = c(e), e \in E \text{(capacity constraints)}$$

The variable x_e is a slack variable for edge e which denotes the free capacity
of e. Technology constraints may have to be added to this system. The cost
function to be minimized is

$$\sum_{i=1}^{n} \sum_{j=1}^{l_i} l(T_{ij}) \times x_{ij}$$

where $l(T_{ij})$ is the length of the Steiner tree T_{ij}.

The resulting integer program is denoted by R. It cannot be solved efficiently
because of its size and NP-hardness of integer programming. Hierarchical global
routing methods break down the integer program into pieces small enough to
be solved exactly. The solutions of these pieces are then combined by a variety

Figure 8.38: A floorplan and its preprocessed cut-tree.

of methods. This results in an approximate solution of the global routing problem.

Hierarchical methods that work top down are especially effective because they can take into account global knowledge about the circuit. Top-down methods start with a cut-tree for the circuit emerging from a floorplanning phase that uses circuit partitioning methods. The cut-tree is preprocessed so that each interior node of the tree corresponds to a simple routing graph as shown in Figure 8.38.

The cut-tree is then traversed top down. At each node a global routing problem is solved on the corresponding routing graph. The solutions for all nodes in a level of the cut-tree are combined. The resulting routing influences the definition of the routing problems for the nodes in the next lower level.

The small integer programs corresponding to the routing problems at each interior node of the cut-tree can be solved by general integer programming techniques. However, this solution may be computationally infeasible. In the best case the linear relaxation of the integer program, i.e., the linear program obtained by eliminating the integrality constraint has to be solved. Since a large number of integer programs have to be solved during the course of global routing, speeding up the computations is necessary. One possibility is to round off the solution of the linear relaxation deterministically or by random methods. This may not lead to an optimal solution. So, it is desirable to exactly solve the integer program corresponding to the routing problem at interior nodes of the cut-tree. Because integer programs corresponding to small routing graphs are quite structured, appropriate preprocessing can substantially reduce the size of the integer programs, and sometimes eliminate them altogether. For example, there exists a greedy algorithm [HL91] to solve the corresponding integer programming problem for a small routing graph H_4 in Figure 8.39. This algorithm will be described in the remainder of the section.

We assume that the length of each edge is the distance between the centers of vertices. In Figure 8.39, a specific floorplan pattern is depicted that is dual to H_4. Figure 8.40 depicts all possible net types and routes for H_4. The size of the integer program R_4 that corresponds to H_4 is reduced by combinatorial arguments on the patterns.

A simple greedy preprocessing strategy can be used for reducing the size of the integer program R_4. This strategy is the first phase of the greedy routing

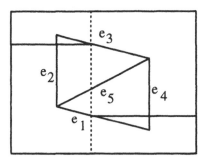

Figure 8.39: The routing graph H_4 and its floorplan.

algorithm. It constructs a smaller integer program R'_4 and is followed by two
more phases. The second phase further reduces the size of R'_4 and constructs a
small mixed integer program R''_4. The third phase solves the integer program
R''_4.

During the first phase, for $i = 1, \ldots, 5$, the algorithm routes $\min\{c(e_i), n_i$
$\}$ nets by using T_{i1}. It delete the routed nets from the problem instance and
reduces the capacity of each edge e_i by the number of routed nets crossing
e_i. Now if there are still nets left in type S_i, then e_i is saturated. This fact
eliminates all routing patterns in Figure 8.40 that include edge e_i. After the
deletion, the pattern set for $i = 1, 4$ is identical with the pattern set for $i = 9$,
and the pattern set for $i = 2, 3$ is identical with the pattern set for $i = 10$.
Thus the net types S_1 and S_4 are merged with net type S_9 and the net types
S_2 and S_3 are merged with net type S_{10}. Net type S_5 cannot be eliminated.
As a result, the original problem R_4 has been reduced into a problem R'_4 with
fewer variables and constraints.

The second phase further reduces R'_4. The result is a very small mixed
integer program R''_4 that can be solved with traditional integer programming
techniques. Two cases have to be distinguished.

1. **There are no more nets of Type S_5 to be routed:** In this case,
   the integer program only contains nets of types $S_6 - S_{11}$. An inspection
   of Figure 8.40 shows that for $i = 6, \ldots, 10$, the long routing patterns for
   net type S_i also occur as routing patterns for net type S_{11}. This suggests
   elimination of the variables corresponding to the long routing patterns
   for net types S_6 to S_{10} from R'_4. The variables $x_{ij}, i = 6, \ldots, 10$ then
   count the nets of these types that are routed with the short routes. All
   other nets of these types should be counted by the variables for net type
   S_{11}. This can be achieved by introducing slack variables x_i to denote the
   number of nets of type S_i that can not be routed by the short routing
   patterns of type S_i for $i = 6, 7, \ldots, 10$. Thus, the completeness constraints
   can now be given as:

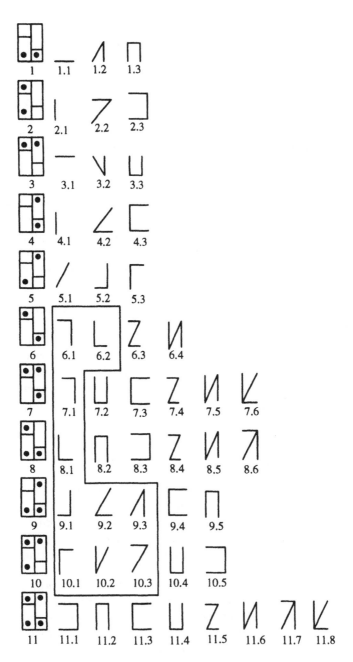

Figure 8.40: All net types and routing patterns for routing graph H_4.

$$x_i + \sum_{j=1}^{l_i} x_{ij} = n_i, i = 6, \ldots, 10$$

$$\sum_{j=1}^{l_{11}} x_{11,j} - \sum_{i=6}^{10} x_i = n_{11}$$

All the other equations remain the same. This yields the integer program R_4''.

The integrality constraints in the resulting integer program can be eliminated for some variables. Specifically, since all coefficients are integers, the integrality constraint can be omitted for one variable per constraint. The number of integer variables is thus reduced and the solution of the integer program by such techniques as branch-and-bound becomes more efficient.

2. **There are more nets of Type S_5 to be routed:** In this case, edge e_5 is saturated after the first phase. This saturation eliminates all routing patterns containing this edge. The resulting integer program is then subjected to an analogous reduction procedure as in the first case. Nets of types $S_7 - S_{10}$ that cannot be routed with short routes are subsumed in type S_{11}.

The third phase of the algorithm solves the small mixed integer program R_4'' obtained in Phase 2 and interprets the solution.

In addition to the formulation of the global routing problem as finding a set of Steiner trees described above, the global routing problem can also be formulated as finding the optimal spanning forest (a generalization of optimal spanning trees) on a graph that contains all of the interconnection information. Cong and Preas presented a concurrent approach based on this formulation [CP88].

8.8   Performance Driven Routing

With the advent of deep submicron technology, interconnect delay has become an important concern in high performance circuit design. Interconnect delay is now a significant part of the total net delay. The reduction in feature sizes has resulted in increased wire resistance and net delay. The increased proximity between the devices and interconnection wires resulting in increased cross-talk noise.

The routers should now model the cross-talk noise between adjacent nets during topology generation. Buffer Insertion, wire sizing, and high performance topology constructions are some of the techniques adopted to reduce generate routing for high performance circuits. Zhou and Wong [Won98] considered crosstalk avoidance during global routing.

Lillis, Cheng, Lin and Ho [CH96] presented techniques for performance driven routing techniques with explicit area-delay trade-off and simultaneous wire sizing. In [Buc98] Lillis and Buch present table-lookup methods for improved performance driven routing.

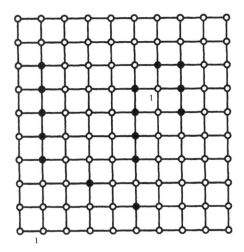

Figure 8.41: A grid, blocked vertices and a two-terminal net.

8.9 Summary

Global routing assigns a sequence of routing channels to each net without violating the capacity of channels. In addition, it typically optimizes the total wire length. In high performance circuits, the optimization function is to minimize the critical RC delay of the nets. Different design style have different objective functions. In standard cell design style, the optimization function is to minimize the total channel height. Whereas, in gate array design style the objective is to guarantee routability.

The global routing algorithms fall roughly into two categories: One is the sequential approach and the other is the concurrent approach. In sequential approach, the nets are routed one by one. However, the nets which have been already routed may block the nets to be routed later. Thus, the order in which the nets are routed is very important. Maze routing algorithms, line-probe algorithms and Steiner tree based algorithms are important classes of algorithms in this approach. The first two class of algorithms are used for two-terminal nets, whereas, the Steiner tree algorithms are used for the multi-terminal nets. The general rectilinear Steiner tree problem is NP-hard, however, approximate algorithms have been developed for this problem. The concurrent approach takes a global view of all nets to be routed at the same time. This approach requires use of computationally expensive methods. One such method uses integer programming. Integer program for an overall problem is normally too large to be handled efficiently. Thus, hierarchical top down methods are used to break the problem into smaller sub-problems. These smaller sub-problems can be solved efficiently. The solutions are then combined to obtain the solution of original global routing problem.

8.10   Exercises

1. Design and implement an algorithm to find the extended channel intersection graph if the size and location of all cells are known.

2. Assume that several nets have been assigned feed-throughs in a standard cell layout with K cell rows. A two-terminal net N that starts at a terminal on cell row i and ends at a terminal on cell row j has to be added to this layout, where $1 \leq i \leq j \leq K$. Design an optimal algorithm to assign feed-throughs to N such that increase in the overall channel height of the layout is minimized.

3. Figure 8.41 shows a grid graph with several blocked vertices. It also shows terminals of a two-terminal net N_1 marked by '1'. Use the Lee's algorithm to find:

   (a) the path for N_1.
   (b) the number of nodes explored in (a).

   Use the Soukup's algorithm to find (a) and (b). Use the Hadlock's algorithm to find (a) and (b).

†4. Extend Lee's maze router so that it generates a shortest path from source to target with the least number of bends.

†5. Design an efficient heuristic algorithm based on maze routing to simultaneously route two 2-terminal nets on a grid graph. Compare the routing produced by this algorithm with that produced by Lee's maze router by routing one net at a time.

6. Give an example for which the Hightower line-probe algorithm does not find a path even when a path exists between the source and the target.

†7. In Mikami's line-probe router, every grid node on the line segment is an escape point on each line segment. Whereas, Hightower's algorithm makes use of only single escape point on each line segment. As a result, Hightower's algorithm runs faster than Mikami's algorithm. Also, Hightower's algorithm may not be able to find a path even when one exists. On the other hand, Mikami's algorithm always finds a path if one exists. The number and location of escape points plays very important role in the performance of the router.

   Implement a line-probe router which can use k number of escape points, where k is a user specified parameter. Use an efficient heuristic for the location of the escape points.

8. In Figure 8.42, terminals of two nets N_1 and N_2 are shown on a grid graph. Terminals of net N_1 are marked by '1' and that of N_2 are marked by '2'. Find an MRST for N_1.

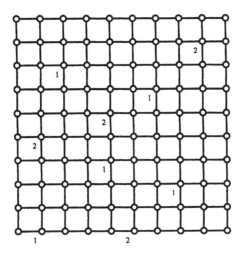

Figure 8.42: A grid and net terminals.

9. For the example in Figure 8.42, find an RST for each net N_1 and N_2 such that they do not intersect with each other and

    (a) the summation of the cost of these two RST's is minimum.

    (b) the maximum of the costs of these two RST's is minimum.

10. Design an algorithm to determine an MRST of a multi-terminal net in a $2 \times n$ grid graph.

11. Compute the number of intersection points in an underlying grid of a set of points in δ-geometry. Is it sufficient to consider just the edges of underlying grid graph to construct a Steiner tree in δ-geometry?

†12. Why does the algorithm Z-RST gives an optimal S-RST for a separable MST? In other words, prove Theorem 6.

†13. Implement the algorithm to find an optimal S-RST for any given net.

†14. Prove Theorem 7 and modify the algorithm Z-RST to use δ-geometry.

†15. Prove Theorem 8.

†16. The problem of finding a Steiner tree for a K-terminal net in a grid graph is known to be NP-complete. Design an efficient heuristic algorithm based on maze routing for this problem.

Bibliographic Notes
Besides the classes of global routing algorithms described above, there are other global routing algorithms that use different approaches and have different optimization functions. Shragowitz and Keel proposed a global router based on

a multicommodity flow model [SK87]. Vecchi and Kirkpatrick discussed the global wiring by simulated annealing [VK83]. A practical global router for row-based layout such as sea-of-gate, gate array and standard cell was developed by Lee and Sechen in 1988 [LS88]. Karp and Leighton discuss the problem of global routing in two-dimensional array [KLR+87]. An interior point method (Karmarkar's Algorithm) can be applied to solve the linear programming model of global routing problem [HS85, AKRV89, Van91]. A path selection global router is developed by Hsu, Pan, and Kubitz [HPK87]. A novel feature of the algorithm is that the *active vertices* (vertices in the net which are not yet connected) are modeled as magnets during the path search process. Several global routing algorithms, including the one based on wave propagation and diffraction, a heuristic minimum tree algorithm using "common edge" analysis, an overflow control method, and global rerouting treatment are discussed in [Xio86]. A simple but effective global routing technique was proposed by Nair, which iterates to improve the quality of wiring by rerouting around congested areas [Nai87]. A global routing algorithm in a cell synthesis system was proposed by Hill and Shugard [HS90], which includes detailed geometric information specific to the cell synthesis problem. The system models diffusion strips, congestion and existing feedthroughs as a cost function associated with regions on the routing plane.

The placement and routing can be combined together so that every placement can be judged on the basis of the routing cost. Researchers have produced some useful results in this direction. Burstein and Hong presented an algorithm to interleave routing with placement in a gate array layout system [BH83]. Dai and Kuh presented an algorithm for simultaneous floorplanning and global routing based on hierarchical decomposition [DK87a]. Suaris and Kedem presented an algorithm for integrated placement and routing based on quadri-section hierarchical refinement [SK89]. An algorithm which combines the pin assignment step and the global routing step in the physical design of VLSI circuits is presented by Cong [Con89]. The sequential algorithms for routing require large execution time. Jonathan Rose [Ros90] developed a parallel global routing algorithm which route multiple nets in parallel by relaxing data dependencies. The speedup is achieved at expense of losing some quality of the routing. The global routing problem is formulated at each level of hierarchy as a series of the minimum cost Steiner tree problem in a special class of partial 3-trees, which can be solved optimally in linear time. In [CH94] Chao and Hsu present a new algorithm for constructing a rectilinear Steiner tree for a given set of points. In [HXK+93] Hong, Xue, Kuh, Cheng, and Huang present two performance-driven Steiner tree algorithms for global routing which consider the minimization of timing delay during the tree construction as the goal. In [HHCK93] Huang, Hong, Cheng, Kuh propose an efficient timing-driven global routing algorithm where interconnection delays are modeled and included during routing and rerouting process in order to minimize the routing area as well as to satisfy timing constraint.

Chapter 9

Detailed Routing

In a two-phase routing approach, detailed routing follows the global routing phase. During the global routing phase, wire paths are constructed through a subset of the routing regions, connecting the terminals of each net. Global routers do not define the wires, instead, they use the original net information and define a set of restricted routing problems. The detailed router places the actual wire segments within the region indicated by the global router, thus completing the required connections between the terminals.

The detailed routing problem is usually solved incrementally, in other words, the detailed routing problem is solved by routing one region at a time in a predefined order. The ordering of the regions is determined by several factors including the criticality of routing certain nets and the total number of nets passing through a region. A routing region may be channel, 2D-switchbox or a 3D-switchbox. Channels can expand in Y direction and their area can be determined exactly only after the routing is completed. If this area is different than the area estimated by the placement algorithm, the placement has to be adjusted to account for this difference in area. If the floorplan is slicing then a left to right sweep of the channels can be done such that no routed channel has to be ripped up to account for the change of areas. Consider the example shown in Figure 9.1(a). In this floorplan, if channel 1 is routed first followed by routing of channel 2 and channel 3, no rerouting would be necessary. In fact, complete routing without rip-up of an already routed channel is possible if the channels are routed in the reverse partitioning order. If the floorplan is non-slicing, it may not be possible to order the channels such that no channel has to be ripped up. Consider the example shown in Figure 9.1(b). In order to route channel 2, channel 1 has to be routed so as to define all the terminals for channel 2. Channel 2 has to be routed before channel 3 and channel 3 before channel 4. Channel 4 requires routing of channel 1 giving rise to a cyclic constraint for ordering the channels. This situation is resolved by the use of L-channels or 2D-switchboxes. L-channels are not simple to route and are usually decomposed. Figure 9.1(c) shows decomposition of an L-channel into two 3-sided channels while Figure 9.1(d) shows decomposition of an L-channel into

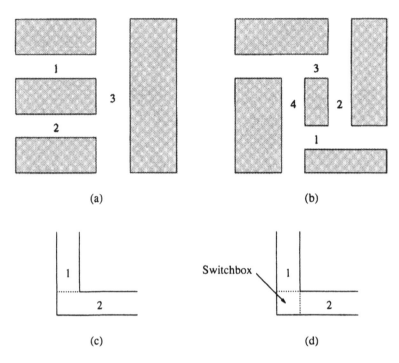

Figure 9.1: Channels and Switchboxes.

two 3-sided channels and a 2D-switchbox. The area of switchboxes (both 2D
and 3D) is fixed and the main issue is routability. That is, given a specific area
and pin locations for all the nets, is this switchbox routable ?. If a switchbox
is unroutable, then the design must be re-global routed. In terms of routing
complexity, channels are easy to route, 2D-switchboxes are harder and 3D-
switchboxes are hardest to route.

Characteristics of a routing problem largely depend upon the topology of
the routing region. Routing regions consist of one or more layers. In the
general case, even single-layer routing problems are NP-complete [Ric84]. In
multi-layer routing problems, the wires can switch adjacent layers at certain
locations using *vias*. A via is an electrical connection (contact) between wire
segments on adjacent layers. In many multi-layer models, the layers are restricted to contain either horizontal or vertical *segments* (a straight piece of
wire placed on a single layer) of a wire. This type of model is known as a
restricted layer model or *reserved layer model*. Multilayer routing problems are
also NP-complete [Szy85], even when the routing region has a simple shape. For
this reason many of the algorithms for multi-layer routing problems are heuristic in nature. Different detailed routing strategies have been developed with
a variety of objectives, but all the detailed routing problems share some common characteristics. These characteristics deal with routing constraints. For

example, wires must satisfy some geometric restrictions which often concern wire thickness, separation, and path features. One obvious restriction present in all routing problems is intersection; that is, no two wires from different nets are allowed to cross each other on the same layer.

A primary objective function of a router is to meet timing constraints for each net and complete the routing of all the nets. Channel routers attempt to minimize the total routing area. Various secondary objective functions have also been considered, such as, improve manufacturability by minimizing the number of vias and jogs, improve performance by minimizing crosstalk between nets and delay for critical nets, among others. Minimizing vias is important, since vias are difficult to fabricate due to the mask alignment problem. In addition, via's increase delay and are therefore undesirable in high-performance applications. Other objective functions include minimization of the average or total length of a net, and minimization of the number of vias per net.

In this chapter, we discuss the routing problem and various algorithms proposed to solve different versions of the routing problem. In the next section, we first formulate the routing problem and classify different routing problems.

9.1 Problem Formulation

As mentioned earlier, the detailed routing problem is solved by solving one routing region at a time. The routing area is first partitioned into smaller regions. Since, the global router only assigns wires to different regions, the detailed routing problem is to find the actual geometric path for each wire in a region. The complexity of the routing problems varies due to many factors including shape of the routing region, number of layers available, and number of nets. However, the shape of the region is perhaps the most important factor. Before presenting the routing problem formally, we describe important considerations and models used in routing.

9.1.1 Routing Considerations

In general, the routing problem has many parameters. These parameters are usually dictated by the design rules and the routing strategy.

1. **Number of terminals:** Majority of nets are two terminal nets, however, the number of terminals in a net may be very large. This is especially true for global nets such as clock nets. In order to simplify the routing problem, traditionally, routing algorithms assume all nets to be two terminal nets. Each multi-terminal net is decomposed into several two terminal nets. More recently, algorithms which can directly handle multi-terminal nets have also been developed.

2. **Net width:** The width of a net depends on the layer it is assigned and its current carrying capacity. Usually, power and ground nets have different widths and routers must allow for such width variations.

3. **Pin locations:**   In channels, pins are located on the top and bottom boundaries. In addition, pin may be located on the sides as well as in the middle of the channel to connect to 3D-switchboxes. The pins on the sides are assigned by the global router. In 2D-switchboxes, the pin are located on all four sides as well as in the middle. The most general form of routing region is a 3D-switchbox, which has pins on all six sides. The pins of the bottom are assigned by the global router so that nets are pass from channels and 2D-switchboxes to 3D-switchboxes and vice versa. The pins on the sides allow nets to pass from one 3D-switchbox to another. The pins on the top allow nets to connect to C4 solder bumps.

4. **Via restrictions:** The final layout of a chip is specified by means of masks. The chip is fabricated one layer at a time, and the masks for the various layers must align perfectly to fabricate the features such as vias which exist in two layers. Perfect alignment of masks is difficult, and thus vias were normally only allowed between adjacent layers. Even between two layers, minimization of vias reduces mask alignment problems. Improvements in the chip manufacturing technology have reduced mask alignment problems, and today stacked vias (vias passing through more than two layers) can be fabricated. However, vias still remain a concern in routing problems and must be minimized to improve yield, performance and area.

5. **Boundary type:** A boundary is the border of the routing region which contains the terminals. Most detailed routers assume that the boundaries are regular (straight). Even simple routing problems which can be solved in polynomial time for regular boundaries become NP-hard for the irregular boundary routing problem. Some recent routers [Che86, CK86, VCW89] have the capability of routing within irregular boundaries.

6. **Number of layers:** Almost all fabrication processes allow three or four layers of metal for routing. Recently, a fifth metal layer has also become available; however, its usage is restricted due to its cost. Six and seven layer processes are expected to be available within two to three years. Most existing detailed routers assume that there are two or three layers available for routing. Recently, several n-layer routers have also been developed. Each layer is sometimes restricted to hold either vertical or horizontal segments of the nets. It is expected that as the fabrication technology improves, more and more layers will be available for routing. In our formulation of five metal process, channel and 2D-switchbox routers must route in M1, M2 and M3. While, 3D-switchbox router must route in M4 and M5.

7. **Net types:** Some nets are considered critical nets. Power, ground, and clock nets fall in this category. Power, and ground wires need special consideration since they are normally wider than signal wires. Clock nets require very careful routing preference, since the delay of the entire chip may depend on clock routing. Due to this type of restriction placed on

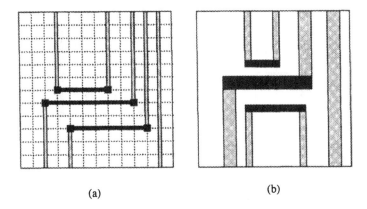

(a)   (b)

Figure 9.2: (a) Grid-based. (b) Gridless.

critical nets, they need to be routed before signal nets using specialized routers or often routed by hand.

9.1.2 Routing Models

For ease of discussion and implementation of net-routing problems, it is often necessary to work at a more abstract level than the actual layout. In many cases, it is sufficient to use a mathematical wiring model for the nets and the rules that they must obey. For instance, wires are usually represented as paths without any thickness, but the spacing between these wires is increased to allow for the actual wire thickness and spacing in the layout. The most common model used is known as the *grid-based* model. In this model, a rectilinear (or possibly octilinear) grid is super-imposed on the routing region and the wires are restricted to follow paths along the grid lines. A horizontal grid line is called a *track* and a vertical grid line is called a *column*. Any model that does not follow this 'gridded' approach is referred to as a *gridless model*.

In the grid-based approach, terminals, wires and vias are required to conform to the grid. The presence of a grid makes computation easy but there are several disadvantages associated with this approach, including the large amount of memory required to maintain the grid and restricted wire width. The gridless approach, on the other hand, allows arbitrary location of terminals, nets, and vias. Moreover, nets are allowed arbitrary wire widths. Due to these advantages, the gridless approach is gaining more popularity than the grid-based approach [Che86, CK86]. Figure 9.2 illustrates some of the differences in grid-based and gridless routing.

Routing problems can also be modeled based on the layer assignments of horizontal and vertical segments of nets. This model is applicable only in multi-layer routing problems. If any net segment is allowed to be placed in any layer, then the model is called an *unreserved layer model*. When certain type of seg-

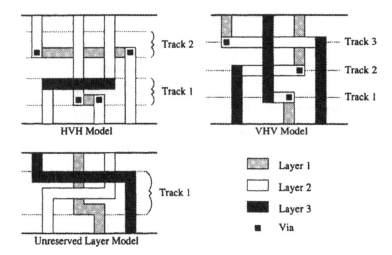

Figure 9.3: A comparison between HVH, VHV, and unreserved layer models.

ments are restricted to particular layer(s), then the model is called a *reserved layer model*. Most of the existing routers use reserved layer models. In a two-layer routing problem, if the layer 1 is reserved for vertical segments and layer 2 is reserved for horizontal segments, then the model is called a VH model. Similarly, a HV model allows horizontal segments in layer 1 and vertical segments in layer 2. Two-layer models can be extended to three-layer routing models: VHV (Vertical-Horizontal-Vertical) or HVH (Horizontal-Vertical-Horizontal). In the VHV model the first and third layers are reserved for routing the vertical segments of nets and the second layer is reserved for routing the horizontal segments. On the other hand, in the HVH model, the first and third layers are reserved for routing the horizontal segments of nets and the second layer is reserved for routing the vertical segments. The HVH model is preferred to the VHV model in channel routing because, in contrast with the VHV model, the HVH model offers a potential 50% reduction in channel height.

Figure 9.3 shows an example of the HVH model using two tracks, the VHV model using three tracks, and the unreserved model using only one. The HVH model and unreserved layer models show more than one trunk per track in Figure 9.3. This is done because the horizontal segments were placed on different layers, the figure offsets them slightly for a clearer perspective of the routing. An unreserved layer model has several other advantages over the reserved layer model. This model uses less number of vias and in fact, in most cases, can lead to an optimal solution, i.e., a solution with minimum channel height. The unreserved routing model also has it disadvantages, such as, routing complexity, blocking of nets, among others. Generally speaking, reserved layer and gridded routers are much faster than gridless and unreserved layer routers.

Another unreserved layer model based on use of *knock-knees* has also been

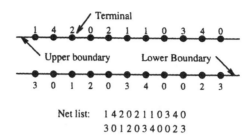

Figure 9.4: A channel and its associated net list.

proposed. The knock-knee model allows two nets to share a grid point if they are in different layers. This model has the advantage of avoiding undesirable electrical properties caused due to overlap of wire segments, such as capacitive coupling.

We now discuss the problem formulation for both channel and switchbox routing problems.

9.1.3 Channel Routing Problems

A *channel* is a routing region bounded by two parallel rows of terminals. Without loss of generality, it is assumed that the two rows are horizontal. The top and the bottom rows are also called *top boundary* and *bottom boundary*, respectively. Each terminal is assigned a number which represents the net to which that terminal belongs to (see Figure 9.4). Terminals numbered zero are called *vacant terminals*. A vacant terminal does not belong to any net and therefore requires no electrical connection. The net list of a channel is the primary input to most of the routing algorithms.

The horizontal dimension of the routed channel is called the *channel length* and the vertical dimension of the routed channel is called the *channel height*. The horizontal segment of a net is called a *trunk* and the vertical segments that connect the trunk to the terminals are called its *branches*. The horizontal line along which a trunk is placed is called a *track*. A *dogleg* is a vertical segment that is used to maintain the connectivity of the two trunks of a net on two different tracks. A pictorial representation of the terms mentioned above is shown in Figure 9.5.

A channel routing problem (CRP) is specified by four parameters: Channel length, Top (Bottom) terminal list, Left (Right) connection list, and the number of layers. The channel length is specified in terms of number of columns in grid based models, while in gridless models it is specified in terms of λ. The Top and the Bottom lists specify the terminals in the channel. The Top list is denoted by $T = (T_1, T_2, ..., T_m)$ and the bottom list by $B = (B_1, B_2, ..., B_m)$. In grid based models, T_i (B_i) is the net number for the terminal at the top (bottom) of the ith column, or is 0 if the terminal does not belong to any

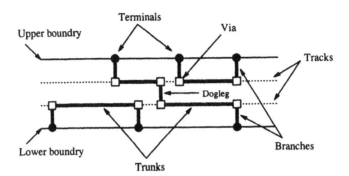

Figure 9.5: Terminology for channel routing problems.

net. In gridless model, each terminal, T_i (B_i), indicates the net number to
which the ith terminal. The Left (Right) Connection list, consist of nets that
enter the channel from the left (right) end of the channel. It is an ordered list
if the channel to the left (right) of the given channel has already been routed.

Given the above specifications, the problem is to find the interconnections
of all the nets in the channel including the connection sets so that the channel
uses minimum possible area. A solution to a channel routing problem is a set
of horizontal and vertical segments for each net. This set of segments must
make all terminals of the net electrically equivalent. In the grid based model,
the solution specifies the channel height in terms of the total number of tracks
required for routing. In gridless models, the channel height is specified in terms
of λ.

The main objective of the channel routing is to minimize the channel height.
Additional objectives functions, such as, minimizing the total number of vias
used in a multilayer routing solution, and minimizing the length of any particular net are also used. In practical designs, each channel is assigned a height
by the floorplanner and the channel router's task is to complete the routing
within the assigned height. If channel router cannot complete the routing in
the assigned height, channel has to expand, which changes the floorplan. This
requires routing the channels in a predefined order, so that such expansions
can be accommodated, without major impact on the floorplan.

In grid based models, the channel routing problem is essentially assignment
of horizontal segments of nets to tracks. Vertical segments are used to connect
horizontal segments of the same net in different tracks and to connect the terminals to the horizontal segments. In gridless models, the problem is somewhat
similar except the assignment of horizontal segments is to specific locations in
the channel rather than tracks. There are two key constraints which must be
satisfied while assigning the horizontal and vertical segments.

1. **Horizontal Constraints:** There is a horizontal constraint between two
   nets if the trunks of these two nets overlap each other when placed on the

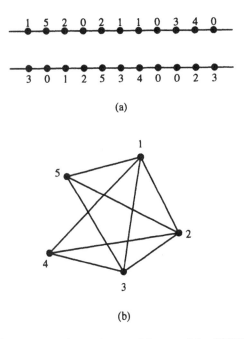

Figure 9.6: A routing problem and its HCG.

same track. For a net N_i, the interval spanned by the net, denoted by I_i is defined by (r_i, l_i), where r_i is the right most terminal of the net and l_i is the leftmost terminal of the net. Given a channel routing problem, a *horizontal constraint graph* (HCG) is a undirected graph $G_h = (V, E_h)$ where

$$V = \{v_i | v_i \text{ represents } I_i \text{ corresponding to } N_i\}$$

$$E_h = \{(v_i, v_j) | I_i \text{ and } I_j \text{ have non-empty intersection}\}$$

Note that HCG is in fact an interval graph as defined in chapter 3. Figure 9.6(a) shows a channel routing problem and the associated horizontal constraint graph is shown in Figure 9.6(b).

The HCG plays a major role in determining the channel height. In a grid based two-layer model, no two nets which have a horizontal constraint maybe assigned to the same track. As a result, the maximum clique in HCG forms a lower bound for channel height. In the two-layer gridless model, the summation of widths of nets involved in the maximum clique determine the lower bound.

2. **Vertical Constraints:** A net N_i, in a grid based model, has a vertical constraint with net N_j if there exists a column such that the top terminal of the column belongs to N_i and the bottom terminal belongs to N_j and $i \neq j$. In case of the gridless model, the definition of vertical constraint is

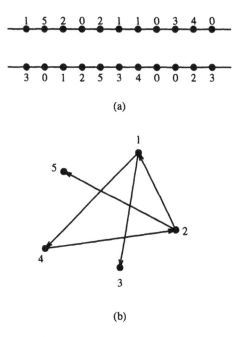

Figure 9.7: A simple routing problem and its VCG.

somewhat similar except that the overlap is between the actual vertical
segments rather than terminals in a column. Given a channel routing
problem, a *vertical constraint graph* (VCG) is a directed graph $G_v = (V, E_v)$, where,

$$E_v = \{(v_i, v_j) | N_i \text{ has vertical constraint with } N_j\}$$

It is easy to see that a vertical constraint, implies a horizontal constraint,
however, the converse is not true. Figure 9.7(b) shows the vertical constraint graph for the channel routing problem in Figure 9.7(a).

   Consider the effect of a directed path in the vertical constraint graph on
the channel height. If doglegs are not allowed then the length of the longest
path in VCG forms a lower bound on the channel height in the grid based
model. This is due to the fact that no two nets in a directed path may be
routed on the same track. Note that if VCG is not acyclic than some nets must
be doglegged. Figure 9.8(a) shows a channel routing problem with a vertical
constraint cycle while Figure 9.8(b) shows how a dogleg can be used to break a
vertical constraint cycle. Figure 9.8(c) shows vertical constraint cycle involving
four nets. In Figure 9.8(d), we show one possible routing for the example in
Figure 9.8(c).

   The two constraint graphs can be combined to form a mixed graph called
the Combined Constraint Graph (CCG) which has the same vertex set as the

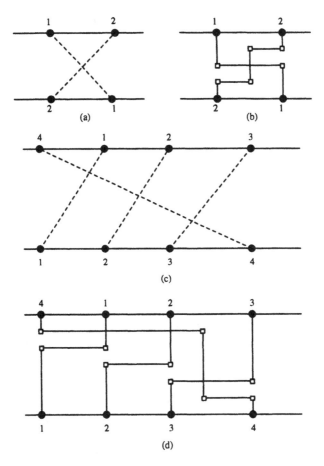

Figure 9.8: A cyclic vertical constraint.

HCG and VCG while the edge set is the union of E_h and E_v. The combined constraint graph for Figure 9.6(a) is shown in Figure 9.9.

Two interesting graphs related to channel routing problem are the permutation graph and the circle graph. The permutation graph can only be defined for channel routing problem for two terminal nets and no net has both of its terminal on one boundary(see Chapter 3). These graphs allow us to consider the channel routing problem as a graph theoretic problem.

Note that, we do not address the channel routing problem with pins in the middle of the channel in this book. This problem is largely a research topic and it is currently solved by using area routers.

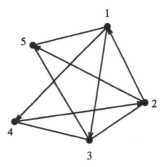

Figure 9.9: A combined constraint graph.

9.1.4 Switchbox Routing Problems

Switchbox routing problem is a generalization of the channel routing problem, where terminals are located on all four sides. Switchboxes are formed in two ways. There maybe be a four sided enclosed region within which the routing must be completed or a four sided region maybe formed due to the intersection of two channels. A switchbox is formally defined as a rectangular region R ($h \times w$) where h and w are positive integers. Each pair (i, j) in R is a grid point. The ith column and jth row or track are the sets of grid points. The 0th and hth columns are the LEFT and RIGHT boundaries respectively. Similarly, the 0th and wth rows are TOP and BOTTOM boundaries respectively. The connectivity and location of each terminal are represented as LEFT$(i) = k$, RIGHT$(i) = k$, TOP$(i) = k$, or BOTTOM$(i) = k$ depending upon the side of the switchbox it lies on, where i stands for the coordinate of the terminal along the edge and k is a positive integer specifying the net to which the ith terminal belongs to.

Since it is assumed that the terminals are fixed on the boundaries, the routing area in a switchbox is fixed. Therefore, the objective of switchbox routing is not to minimize the routing area but to complete the routing within the routing area. In other words, the switchbox routing problem is a routability problem, i.e., to decide the existence of a routing solution. Unlike the channel routing problem, switchbox routing problem is typically represented by its circle graph (see Chapter 3).

Note that we do not address the 3D-switchbox routing in this book. This problem is solved by using area routing approaches. Some concepts and algorithms related to 3D-switchbox and OTC routing will be discussed in Chapter 8.

9.1.5 Design Style Specific Detailed Routing Problems

In this section, we discuss the detailed routing problem with respect to different design styles.

1. **Full custom:** The full custom design has both channels and switchboxes. As explained earlier, depending on the design, the order in which the channels and 2D-switchboxes are routed is important. A 3D-switchbox can only be routed after all the channels and 2D-switches under it have been routed. The objective of a detailed routing algorithm is to complete the routing in a manner that each net meets it timing constraint and minimum area is utilized for routing. Other constraints such as manufacturability, reliability and performance constraints are also used.

2. **Standard cells:** The standard cell design style has channels of uniform lengths which are interleaved with cell rows. Hence the detailed routing problem is reduced to routing channels. Unlike in the full-custom design, the order in which the channels are routed is not important. This is possible since global router assigns pins in the feedthroughs. Typically, regions on top of cells can be used for 3D-switchbox routing. This will be explained in more detail in Chapter 8. The objective is to route all the nets in the channel so that the height of the channel is minimized. Additional constraints such as minimizing the length of the longest net and restricting length of critical nets within some prespecified limits are used for high performance standard cell designs.

3. **Gate arrays:** The gate arrays have channels of fixed size and hence the detailed routing algorithms have to route all the nets within the available routing regions. If the detailed router cannot route all the nets, the partitioning process may have to be repeated till the detailed routers can route all the nets. For high performance routing net length constraints must added.

9.2   Classification of Routing Algorithms

There could be many possible ways for classifying the detailed routing algorithms. The algorithms could be classified on the basis of the routing models used. Some routing algorithms use grid based models while some other algorithms use the gridless model. The gridless model is more flexible as all the wires in a design need not have the same widths. Another possible classification scheme could be to classify the algorithms based on the strategy they use. Thus we could have greedy routers, hierarchical routers, etc. to name a few. We classify the algorithms based on the number of layers used for routing. Single layer routing problems frequently appear as sub-problems in other routing problems which deal with more than one layers. Two and three layer routing problems have been thoroughly investigated. Recently, due to improvements in the fabrication process, fourth and fifth metal layers have also been allowed but this process is expensive compared to three-layer metal process. Several multi-layer routing algorithms have also been developed recently, which can be used for routing MCMs which have up to 32 layers.

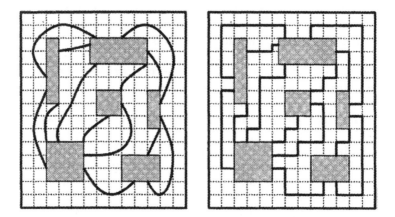

Figure 9.10: Single layer routing problem.

9.3   Single-Layer Routing Algorithms

A general single-layer routing problem can be stated as follows.   Given
a routing region, a netlist $\mathcal{N} = \{\ N_1,\ N_2,\ \ldots,\ N_n\ \}$, a set of terminals
$\{T_{i,j}, i = 1, \ldots, n, j = 1, \ldots, n_i\}$ where, $T_{i,j}$ specifies the jth terminal of net N_i,
a set of blocks $\mathcal{B} = \{B_1, B_2, \ldots, B_l, B_{l+1}, B_{l+2}, \ldots, B_m\}$, where $B_1, B_2, \ldots, B_l$
are flippable and $B_{l+1}, B_{l+2}, \ldots, M_m$ are not flippable (a block is flippable if
orientation of its terminals is not determined). Also given is a set of design
rule parameters which specify the necessary widths of wire segments and the
minimum spacing between wires. The single-layer routing problem is to find a
set of wire segments inside the routing region which complete the connections
required by the netlist without violating any design rule. Figure 9.10 shows
an instance of a single-layer routing problem. Figure 9.10(a) gives the global
routing of the instance of the problem and Figure 9.10(b) gives the detailed
routing of wires on a single layer.

Although the general single layer routing problem is conceptually easier
than the multi-layer routing problem, it is still computationally hard. In single-
layer routing, the fundamental problem is to determine whether all the nets can
be routed. This problem is called *single-layer routability problem* and is known
to be NP-complete [Ric84]. Figure 9.11 shows an instance of a single-layer
routing problem that is unroutable.

There are many practical restricted versions of the single-layer routing
problem which are easier to handle than the general single-layer routing problem [MST83]. For example, consider the following:

1. There are no flippable blocks, i.e., $l = 0$.

2. All the blocks are flippable, i.e., $m = l$.

3. All the nets are two-terminal nets with no flippable blocks.

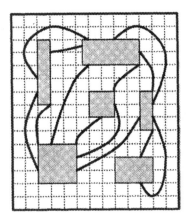

Figure 9.11: A unroutable single layer example.

4. All the nets are two-terminals nets with all flippable blocks.

5. There are no blocks inside the routing region and all nets are two-terminal nets.

6. There are no blocks inside the routing region, the nets are two-terminal and the terminals lie on a single row.

Problems 3, 4, and 5 are commonly known as the variations of *river routing*. Problem 6 is known as *single row routing problem*.

Several special cases of the single layer routing problem can be solved in polynomial time [BP83, DKS$^+$87, LP83, Mal90, SD81, Tom81].

Although, single layer routing problem appears restricted when one considers that fabrication technology allows three layers for routing. However, single layer routing still can be used for power and ground routing, bus routing, over-the-cell routing and some clock routing problems. During floorplanning, the sequence of the input and output busses is determined for each block. Since a bus may have a very large number of nets, it is advisable to pre-route the buses. Buses are routed such that the output bus of a block is in the same sequence as that of the input bus of a receiving block. Since the input and output busses of the blocks have the same sequence, it may be possible to make the interconnections between blocks on a single layer. This also minimizes vias and area required for bus routing. Power and ground nets are sometimes also routed in a single layer due to electrical considerations. The power and ground routing problems will be considered in Chapter 9. In the three layer environment, in certain regions, the two underlying metal layers may be blocked and only the top layer is available for routing. In this case, additional nets may be routed using single layer techniques on the third layer. This is a typical situation in over the cell routing. We will discuss over-the-cell routing in Chapter 8. Finally, for high performance circuits, clock nets may be routed in a single layer,

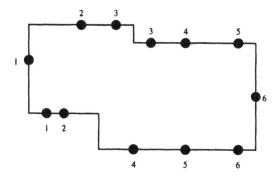

Figure 9.12: An example of the general river routing problem.

as electrical effects of vias are undesirable due to performance considerations. In this section, we discuss two special cases of single layer routing problem, the river routing problem and the single row routing problem.

9.3.1   General River Routing Problem

River routing is a special case of the single layer routing under following assumptions. All the terminals lie on the boundary of the region. Every net consists of exactly two terminals and there are no blocks in the region. Terminals are located in such a way that no crossover between nets is necessary for a solution to exist, that is, nets are planar. Figure 9.12 shows an example of a general river routing problem.

A special case of general river routing problem, which has attracted a lot of attention is simply called the *river routing* problem [Hsu83a, JP89, LP83, MST83, TH90]. It is essentially a single layer channel routing problem for two terminal nets, such that each net has one terminal on each boundary. Figure 9.13 shows an example of a river routing problem. We will concentrate on the general river routing problem and present an algorithm for an arbitrary shaped rectilinear routing region.

9.3.1.1   General River Routing Algorithm

In this section, we discuss the general river routing algorithm presented by Hsu [Hsu83a]. This algorithm is capable of routing in arbitrary shaped rectilinear routing regions and guarantees that a solution will be found if one exists. The algorithm is gridless and allows arbitrary net widths and wire separations. Although, the algorithm is developed for two terminal nets, it can be easily extended for multi-terminal nets. We start by defining some terminology.

Let a path be a alternating sequence of horizontal and vertical segments connecting two terminals of a net. A terminal is called starting terminal if it is connected to the first segment of a path. Similarly, the terminal connected

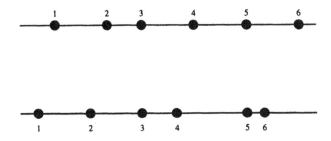

Figure 9.13: A simple river routing problem.

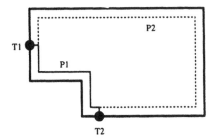

Figure 9.14: Two possible paths of a net along the boundary.

with the last segment is called an ending terminal. Without loss of generality, it will be assumed that every path is counter-clockwise along the boundary. Every net has two possible paths along the boundary and therefore there are the two possible choices of starting terminal for the net. Figure 9.14 shows two possible paths along the boundary for a net $N_i = T_1, T_2$. Path P_1 has T_1 as its starting terminal and path P_2 has T_2 as its starting terminal. The general river routing algorithm routes one net at a time and consists of four phases. In the first phase, the starting terminal of each net is determined. In the next phase, net order is determined by the sequence of terminals on the boundaries. Based on the net order, path searching is done by routing each net, in order, as close to the boundaries as possible. Unnecessary corners are then removed by flipping the corners in the last phase. We will now briefly discuss these phases.

Starting Terminal Assignment: As stated earlier, each net has two possible paths along the boundary. The starting terminal for a net is chosen independent of all other nets, such that the shorter path is selected. In order to select a starting terminal for a net, the length of the path in the counter-clockwise direction is computed and compared to the half of the total length of the boundary of the routing region. In figure 9.14, terminal $T_{i,1}$ is assigned to be the starting terminal, since path P_1 is shorter than path P_2. Figure 9.16

shows an example of the starting terminal assignment for a netlist.

Net Ordering: Every path is counter-clockwise and begins at the starting terminal, as a result, the order in which the nets are routed is very important. A net can only be routed after all the nets 'contained' by the net are already routed. A net N_i is contained by another net N_j, if all the terminals of N_i are on the boundary between the starting and ending terminals of the net N_j. Note that only the counter-clockwise boundary is considered.

To determine the net order, a circular list of all terminals ordered in counter-clockwise direction according to their positions on the boundaries is generated. A planarity check is performed to determine if the given instance is routable. If the given instance is routable, the nets are ordered by NET-ORDERING algorithm as given below. The basic idea is to just push the starting terminal of the nets on the stack as they are encountered. A number is assigned to a net N_i, when the algorithm encounters the ending terminal of net N_i and the top item on the stack is the starting terminal of net N_i. This ensures that all the nets contained in net N_i are assigned a number, before assigning a number to net N_i. Net N_i is then deleted from further consideration and algorithm continues until all nets have been numbered. The formal description of the algorithm appears in Figure 9.15.

Consider the example shown in Figure 9.16. Starting at terminal 1, terminals are considered in counter-clockwise order. The net N_1 is assigned first, as its ending terminal is encountered, while the top of the stack has the starting terminal of N_1. The starting terminal of N_2 is pushed onto the stack, followed by pushing of starting terminal of net N_3. The net N_3 is number second as its ending terminal is encountered next. The final net ordering for the example in Figure 9.16 is $\{1, 3, 8, 7, 6, 5, 2, 4\}$. The 's' next to a terminal indicates that the terminal is a starting terminal.

Path Searching: Based on the net order, each net is routed as close to the pseudo-boundary as possible. For the first net, the pseudo-boundary is the boundary of the region. For the second net, the segments of the first net and the segments of the boundary not covered by the first net form the pseudo-boundary. In other words, each time a net is routed, the region available for routing is modified and the boundary of this region is referred to as *pseudo-boundary*. The path of the net is checked for design rule violations by checking the distances between the counter-clockwise path of the net and the pseudo-boundary not covered by the net. If a violation occurs, it implies that the given problem is unroutable. Figure 9.17 shows the pseudo-boundary 'abcdihgf' for net N_i and the path is created by routing as close to 'abcd' as possible. The path is then checked against the remaining segments of the pseudo-boundary, i.e., 'ihgf' for design rule violations.

Corner Minimization: Once the path searching for all nets has been completed without design rule violation, a feasible solution has been found. However, the routing technique described above pushes all the paths outward

Algorithm NET-ORDERING
begin
    for $i = 1$ to $2n$ **do**
        if END-TERMINAL(T_i) **then**
            MATCHED(T_i) = 0;
        else
            MARKED(T_i) = 0;
    $stack = \phi$;
    $i = 1$;
    T = any terminal in the circular list;
    while $i \leq n$ **do**
        if START-TERMINAL(T) **and** MARKED(T) = 0
        then PUSH($T, stack$);
            MARKED(T) = 1;
        else if END-TERMINAL(T) **and** MATCHED(T) = 0
        then $T1$ = POP($stack$);
            if $T = T1$ **then**
                MATCHED(T) = 1;
                ASSIGN-NUMBER(i, NET(T));
                $i = i + 1$;
            else exit;
        T = next terminal in the circular list;
end.

Figure 9.15: Algorithm NET-ORDERING

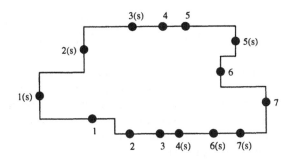

Figure 9.16: The assignment of starting terminals

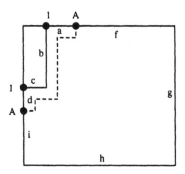

Figure 9.17: Pseudo-boundary and path creation.

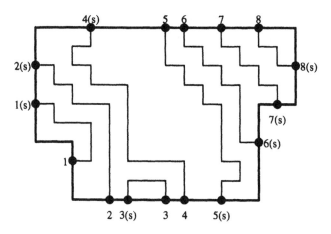

Figure 9.18: A boundary-packed solution after path searching.

against the boundaries and the excess space is left vacant in the center of the routing region. Figure 9.18 shows an example of the routing after path searching.

The corner minimization is a systematic method of flipping corners toward the inside of the routing region. Corners are minimized one net at a time. The order of the nets for this operation is precisely the reverse of the order determined by the previous net ordering step. That is, the corners of the paths are minimized starting with nets from the center of the routing region towards the boundary of the routing region.

Every corner of a path belongs to one of the eight possible cases as shown in Figure 9.19. Since every path is routed in the counter-clockwise direction, in four cases the corners can be flipped towards the inside of the routing region. Figure 9.19 shows the four cases a, b, c, d which can be respectively transformed

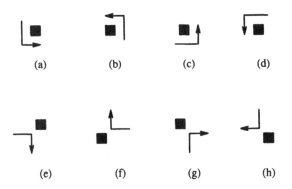

Figure 9.19: Eight possible cases of a corner.

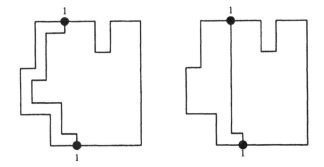

Figure 9.20: Length minimization by flipping corners.

to cases e, f, g, h by flipping towards the inside. The inside of the region is indicated by a filled dot. A pseudo-boundary is generated in the same way as in the path searching step and design violation checks are performed before all corner flips. If the intended corner flip does not create any design rule violation, the corner is flipped and two corners are eliminated from the path. Otherwise, this corner is skipped and the next corner of the path is checked. Figure 9.20 shows an example of a path before and after flipping of corners.

9.3.2 Single Row Routing Problem

Given a set of two-terminal or multi-terminal nets defined on a set of evenly spaced terminals on a real line, called the *node axis*, the single row routing problem (SRRP) is to realize the interconnection of the nets by means of non-crossing paths. Each path consists of horizontal and vertical line segments on a single layer, so that no two paths cross each other. Moreover, no path is allowed to intersect a vertical line more than once, i.e., backward moves of nets

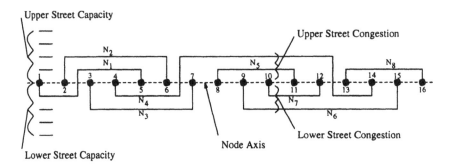

Figure 9.21: Basic terminology and a single row realization of the net list L_1.

are not allowed.

For an example consider the net list $\mathcal{N} = \{N_1, N_2, \ldots, N_8\}$ where $N_1 = \{1, 5\}$, $N_2 = \{2, 6\}$, $N_3 = \{4, 14\}$, $N_4 = \{3, 7\}$, $N_5 = \{8, 11\}$, $N_6 = \{9, 15\}$, $N_7 = \{10, 12\}$, $N_8 = \{13, 16\}$. A single row realization of $\mathcal{N}$ is shown in Figure 9.21.

The area above the node axis is called the *upper street* while the area below the node axis is called the *lower street*. The number of horizontal tracks available for routing in the upper street is called *upper street capacity*. Similarly the number of horizontal tracks available in the lower street is called the *lower street capacity*. Due to symmetry in single row routing, the upper street capacity is usually equal to the lower street capacity. For a given realization, the number of the horizontal tracks needed in the upper street is called the *upper street congestion* (C_{us}) and the number of horizontal tracks needed in the lower street is called the *lower street congestion* (C_{ls}). The term *dogleg* is used to describe a bend in a net, when it makes an interstreet crossing. The *between-nodes congestion* C_B of a realization is the maximum number of interstreet crossings between a pair of adjacent terminals. For the realization shown in Figure 9.21, $C_{us} = 2$, $C_{ls} = 2$, $C_B = 1$. The net N_1 is doglegged once, while the net N_3 is doglegged twice.

The objective function considered most often is to minimize the maximum of upper and lower street congestions, i.e., minimize Q_0, where $Q_0 = \max\{C_{us}, C_{ls}\}$. To minimize the separation between the two adjacent terminals it is sometimes necessary to minimize C_B. In practical problems, $Q_0 \leq 3$ and $C_B \leq 2$. Other objective functions include minimizing the total number of doglegs in a realization or to minimize number of doglegs in a wire.

9.3.2.1 Origin of Single Row Routing

The SRRP was introduced by So in the layout design of multilayer circuit boards [So74]. It has received considerable attention [HS84a, KKF79, RS83, RS84, TKS76, TMSK84, TKS82]. So proposed a systematic approach to the routing of large multi-layer printed circuit board problem(MPCBP). This approach consists of a well defined decomposition of the MPCBP into several independent single layer single row routing problems. The scheme decomposes the MPCBP routing problem into five phases:

1. via assignment,

2. placement of via columns,

3. layering,

4. single row routing, and

5. via elimination,

In the via assignment phase, each multi-terminal net is decomposed into several two terminal nets. A net whose terminals lies on the same row or same column is connected by a wire. A net is not directly decomposable if it contains two terminals not in the same row or same column. In this case vias are introduced to facilitate decomposition of the net. In the second phase of decomposition the via columns are permuted to minimize the wire lengths. Obviously this change is meaningful only if vias appear column-wise. In particular, in this step, the locations of two via columns are exchanged without violating any of the net connections.

In the third phase of decomposition, a single row routing problem is decomposed into several single row routing problems so that each subproblem can be routed to satisfy the upper and lower street constraints in a different layer. Usually, half the available layers are used for realization of the row problems, and the other half of the available layers is used for the column problems.

Sufficient conditions for a realization with minimum congestion along with a routing algorithm were presented by Ting, Kuh, and Shirakawa [TKS76]. It was shown that an arbitrary set of nets can be realized if upper and lower street capacities are unbounded. Kuh, Kashiwabara, and Fujisawa [KKF79], presented an interval diagram representation of the single row routing problem. This representation played an important role in the research and development of several algorithms for the single row routing problem. The interval diagram representation of an example is given in Figure 9.22. The broken line shown in Figure 9.22(b) is called the *reference line*. The layout is obtained by stretching out the reference line and setting it on top of the node axis. The interval lines for each net are mapped topologically onto vertical and horizontal paths. The nets and its segments above the reference line are mapped onto paths in the upper street, while the nets and its segments below the reference line are mapped onto paths in the lower street. This process defines a unique realization as shown in Figure 9.22(c). An important implication of the interval diagram representation is that it reduces the single row routing problem to finding an optimal permutation of nets and thus greatly enhances the understanding of the problem.

Kuh, Kashiwabara, and Fujisawa [KKF79] also proposed an algorithm for minimizing street congestion. It was based on the number of possible orderings

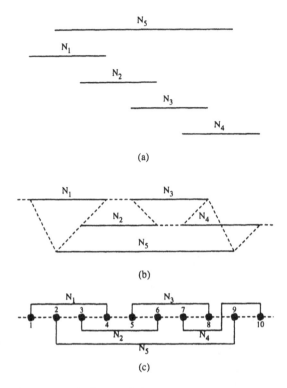

Figure 9.22: Interval diagram representation

(permutations) of nets that can be considered for routing. Another important contribution of Kuh, Kashiwabara, and Fujisawa [KKF79] is the development of necessary and sufficient conditions for the optimal realization of the single row routing problem. These conditions were based on the idea of the *cut number* of a net. The cut number of a terminal is the number of nets passing over that terminal. The cut number of a net is maximum among all cut numbers of its terminals. Example in Figure 9.23 shows the concept of cut number. Let q_i be the cut number of net N_i. Then in the Figure 9.23, $q_1 = 2$, $q_2 = 3$, $q_3 = 4$, $q_4 = 3$, $q_5 = 4$, and $q_6 = 2$. Let q_{max} and q_{min} be the maximum and minimum over the cut numbers of all nets, respectively. Then in the Figure 9.23, $q_{max} = 4$ and $q_{min} = 2$.

The main idea behind the necessary and sufficient condition is the optimal partitioning of nets at each terminal. A realization is optimal with congestion equal to $q_t = \frac{q_{max}}{2}$ if at each terminal with cut number c there are at least $k = c - q_t$ nets above and k nets passing below that terminal, not counting the net to which the terminal belongs. In other words, if c nets cover a terminal, then a realization with congestion equal to q_t is optimal only if the nets covering this terminal can be partitioned into two sets, each containing at least k nets.

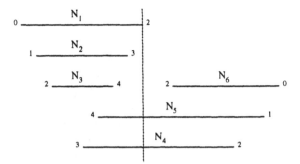

Figure 9.23: The concept of cut numbers.

Although this condition can be used to verify whether a given realization is optimal, it does not lead to any optimal routing algorithm. In fact there is strong evidence that no such algorithm may exist. Arnold [Arn82] proved that the problem of finding a layout with minimum congestion is NP-Hard.

Based on the concept of cut-numbers, a lower bound is also presented in [KKF79].

Theorem 9 *For any feasible realization* $Q_0 \geq \max\{q_{min}, \lceil \frac{q_{max}}{2} \rceil\}$.

In [TMSK84] Trang, Marek-Sadowska, and Kuh proposed a heuristic algorithm for minimizing street congestion. The proposed algorithm is based on permuting the nets according to their cut number. It is observed that nets having larger cut numbers should be placed inside, i.e., near the middle of the permutation, while nets having lower cut numbers should be placed outside, i.e., at the ends of the permutation. In [DL87b] Du and Liu showed that the algorithm in [TMSK84] does indeed produce optimal results if all the nets belong to one 'group'. However, if the net list has more than one group than the algorithm given in [TMSK84] may not produce optimal results. The set of nets that covers at least one common node is said to form a *group*. Du and Liu proposed an algorithm that takes the group structure into account. They use the idea of *local cut number* that is, the cut number of a net with respect to a group. The algorithm routes the largest group first and then tries to route the nets in the adjacent group while trying to satisfy the heuristic criterion of placing nets with larger local cut numbers inside and nets with smaller local cut numbers outside. This algorithm produced better results than the one reported in [TMSK84].

In [DIN87] Du et al. investigated the problem of minimizing the between-node congestion which is the congestion between two adjacent nodes. They developed a fast algorithm for the case when the number of horizontal tracks available as well as the number of vertical tracks available between adjacent nodes is fixed. Their algorithm is an extension of Han and Sahni's algorithm.

9.3.2.2   A Graph Theoretic Approach

In [SD89a, SD89b, SDR89, SDR90], Sherwani and Deogun developed a new graph-theoretic approach to single row routing problems. This approach models a single row routing problem with three graphs, an overlap graph, a containment graph and an interval graph. It was found that several relationships exist between the properties of an SRRP and the graph representation. In [SD89a], a new heuristic algorithm has been developed based on this approach. This algorithm achieves substantially better results than the existing algorithms. In [SD89b], new lower bounds for SRRP have been developed and in [SDR89] the problem of single row routing with a restricted number of doglegs is investigated.

We will briefly discuss the principle results obtained by the graph theoretic approach. Let R be a set of evenly spaced terminals on the node axis. Let $\mathcal{N} = \{N_1, N_2, \ldots, N_n\}$ be a set of two-terminal nets defined on R. Each net N_i can be uniquely specified by two distinct terminals l_i and r_i called the left touch point and the right touch point, respectively, of N_i. Abstractly, a net can be considered as an interval bounded by left and right touch points. Thus for a given set of nets, an interval diagram depicting each net as an interval can be easily constructed. Given an interval diagram corresponding to an SRRP, we can define an interval graph G_I containment graph G_C and overlap graph G_O. The definitions of these graphs may be found in chapter 3.

The approach presented in [SD89a] uses modified cut numbers. The cut number is a very important criterion in determining the position of a net in the final routing. Usually only cliques are considered in computation of cut-numbers. For improving the utility of cut-numbers, they consider not only cliques but also the clique intersections. Two cliques are said to have *high clique intersection* if the number of nets they have in common is at least equal to one-half of the maximum size of the two cliques, otherwise cliques have low clique intersection. If the clique intersection is relatively high between two cliques, these cliques are collapsed to form a bigger *pseudo-clique*. For routing purposes a pseudo-clique is treated as a clique. This operations of clique collapsing continues until all clique intersections are relatively low. The cut-number of a net with respect to its pseudo-clique is called the *modified cut-number*. It is easy to see that this approach behaves like Trang et al.'s algorithm if all clique intersections are high while behaving like Du et al.'s algorithm if all clique intersections are low. In addition, it even produces a good solution for problem sets for which some clique intersections are high and some clique intersections are low.

9.3.2.3   Algorithm for Street Congestion Minimization

In [SD89a], an algorithm to minimize the street congestion was developed. The basic function of the algorithm, denoted as SRRP_ROUTE, is to find maximal pseudo-cliques in the interval graph G_I representing the sub-problem under consideration. This goal is accomplished by finding and collapsing maximal cliques. If the clique intersection of the two neighboring (pseudo-)cliques is

Algorithm SRRP-ROUTE ( )
begin
    FIND-CLIQUES (L, C)
    COMBINE-CLIQUES (L, C, D)
        (* D contains super-cliques $SC_j, j = 1, \ldots, r$ *)
    MAX-PSEUDO-CLIQUE (D, SC_k)
    SOLVE (SC_k, M)
    for $j = 1$ **to** $k - 1$
        INSERT (SC_j)
    for $j = k + 1$ **to** r
        INSERT (SC_j)
end.

Figure 9.24: Algorithm SRRP-ROUTE

high then these are combined and a pseudo-clique is formed. All modified cut-numbers are computed according to the pseudo-cliques. First, the maximum pseudo-clique SC_{ik} is routed using a greedy approach similar to the approach used in [TMSK84]. Other pseudo-cliques to the left and right of this maximum pseudo-clique are then routed. An outline of the algorithm is in Figure 9.24.

In the following, we give a brief description of the main procedures of SRRP-ROUTE:

Procedure FIND-CLIQUES: This procedure decomposes the given problem into several smaller single-row-routing problems by identifying the linear ordering of cliques $C_i, 1 \leq i < n$ of the interval graph G_I.

Procedure COMBINE-CLIQUES: This procedure finds the clique intersections between adjacent cliques, and forms a pseudo-clique if the clique intersection is high. This process is carried out until all clique intersections between all pseudo cliques are low. The clique collapsing parameter can be changed.

Procedure SOLVE: This procedure returns a permutation of the nets of a sub-problem obtained by placing them according to the greedy heuristic based on the modified cut numbers. This procedure is used to route the maximum pseudo-clique.

Procedure INSERT: This procedure combines solutions of two adjacent sub-problems to produce a solution for the larger problem defined by the combination of the two sub-problems. It inserts the new nets belonging to the new clique into the existing solution so that nets with higher modified cut numbers are assigned to inner tracks, while nets lower modified cut numbers are assigned to outer tracks.

The number of nets in any sub-problem cannot be greater than n. Moreover, procedure SOLVE has a time complexity of $O(n \log n)$. Procedure INSERT has a time complexity of $O(n)$. The first loop thus takes $O(n^2 \log n)$ time. Similarly, the second loop also takes $O(n^2 \log n)$. Therefore, the worst case time complexity of the algorithm is $O(n^2 \log n)$.

9.3.2.4   Algorithm for Minimizing Doglegs

The problem of finding a layout without doglegs is of interest because of the limited amount of inter-pin distance available in IC's. This problem has been considered before by Raghavan et al., [RS84] when an algorithm for checking feasibility of routing without doglegs was developed. The authors however did not present a characterization. Using the graph model, in [SD89a] a characterization of single row routing problems which can be solved without doglegs is presented.

Theorem 10 *An SRRP can be routed without any doglegs if and only if the corresponding overlap graph is bipartite.*

Similarly, a sufficient condition for routing with at most one dogleg per net was also established.

Theorem 11 *An SRRP can be realized with at most one dogleg per net if the corresponding containment graph G_C is null.*

Using this graph representation, three algorithms for minimum-bend single row routing problem have recently been reported [SWS92]. It was shown that the proposed algorithms have very tight performance bounds. In particular, it is proved that the maximum number of doglegs per net is bounded by $O(k)$, where k is the size of the maximum clique in certain graph representing the problem. Expected value of k is $\Theta(\sqrt{n})$ and in practical examples $k = O(1)$, where n is the number of nets.

We will briefly describe one of the algorithms, which is based on the decomposition of the given SRRP into several smaller SRRPs so that interval graph for each subproblem is null. This operation is called *independent set decomposition* of G_I. The motivation for this algorithm is derived from the fact that using the interval graph representing a SRRP, the problem can be decomposed into k subproblems and each one of these subproblems can be routed without any doglegs. The key therefore, is to combine the routing of these subproblems such that maximum number of doglegs per net is minimized. The independent set decomposition of G_I can be achieved by using the algorithm to find maximum clique in an interval graph described in Chapter 3. Using the k independent sets, an algorithm, denoted K-DOGLEG-I is presented, which combines the routing of these sets into a routing for the given SRRP. The formal description of the algorithm is in Figure 9.25.

Theorem 12 *The Algorithm K-DOGLEG-I routes a given net list L with at most $O(k)$ doglegs per net in $O(n \log n)$ time, where $k = C_I$ and n is the total number of nets.*

Algorithm K-DOGLEG-I()
begin
    Phase 1:
        (* Use Left_edge algorithm to decompose $\mathcal{N}$ into
        k independent net lists *)
        (* lists $(\mathcal{N}_1, \mathcal{N}_2,..., \mathcal{N}_k)$ of $\mathcal{N}$. *)
        $\mathcal{N}_i$=LEDGE($\mathcal{N}$); $(i = 1,.., k)$.
        (* Assign $\mathcal{N}_1$ to the upper street. *)
        for ($N_i \in \mathcal{N}_1$ $i = 1,.., m_1$) **do**
            $T_1^U = T_1^U \bigcup N_i$;
        (* Assign N_2 to the lower street. *)
        for ($N_i \in \mathcal{N}_2$ $i = 1,.., m_2$) **do**
            $T_1^B = T_1^B \bigcup N_i$;
    Phase 2:
        (* Insert the remaining independent net lists.*)
        $t = U; u = 1; l = 1$;
        for $(G_I^i$ $i = 3,.., k)$ **do**
            for ($N_j \in \mathcal{N}_i(j = 1,.., m_i)$) **do**
                (* Find the smallest track which contains N_j. *)
                $k = \min\{q|1 \leq q \leq p, N_j \in T_q^t\}$;
                if (N_j contained by previously routed net at T_k^t)
                    then
                    (* Insert N_j under T_k^t. *)
                    INSERT(N_j, T_k^t);
                else
                  (* Assign the new net to the outer track. *)
                  $T_p^t = T_p^t \bigcup N_j$;
        (* Switch street. *)
        if ($t = U$) **then**
            $t = B; l = l + 1; p = l$;
        else
            $t = U; u = u + 1; p = u$;
  end.

Figure 9.25: Algorithm K-DOGLEG-I

Proof: Given a net list L, the algorithm decomposes it into k independent net lists $L_i, i = 1, \ldots, k$, and routes the first independent net list L_1 on the upper street and the second independent set L_2 on the lower street. This operation can be completed without any doglegs. Then the algorithm inserts all the nets in the remaining $k - 2$ independent net lists into the existing layout. Since inserting one independent net list causes at most $O(1)$ more doglegs to each net in the layout, hence inserting all the remaining $k - 2$ independent net lists causes at most $O(k)$ more doglegs to each net in the layout. So the total dogleg number per net is $O(k)$. On the other hand, in an interval graph, k is equal to C_I. Therefore, the algorithm K-DOGLEG-I can route a given net list with at most $O(C_I)$ doglegs per net.

All operations of the K-DOGLEG-I algorithm, except the track finding operation can be carried out in constant time. Each find operation can be accomplished by a binary search in $O(\log n)$ time. Therefore the total time complexity is $O(n \log n)$. $\square$

9.4   Two-Layer Channel Routing Algorithms

Two-layer channel routing differs from single-layer routing in that two planar set of nets can be routed if vias are not allowed, and a non-planar set of nets can be routed if vias are allowed. For this reason, checking for routability is unnecessary, as all channel routing problems can be completed in two layers of routing if vias are allowed. Therefore, the key objective function is to minimize the height of the channel.

For a given grid-based channel routing problem, any solution to the problem requires at least a minimum number of tracks. This requirement is called the *lower bound* for that problem. Since the lower bound is the minimum number of tracks that is required, it is unnecessary to reduce the number of tracks beyond the lower bound and therefore, it is important to calculate the lower bound of the number of tracks before solving a particular routing instance. Following theorem presents the lower bounds for channel routing problems assuming two-layer reserved layer routing models with no doglegs allowed. Let $h_{\max}$ and $v_{\max}$ represent the maximum clique in the HCG and the longest path in VCG, respectively for a routing instance.

Theorem 13 *The lower bound on the number of tracks of a two-layer dogleg free routing problem is* $\max\{h_{\max}, v_{\max}\}$.

For grid-less channel routing problems, the width of nets must be taken into account while computing v_{max} and h_{max}.

9.4.1   Classification of Two-Layer Algorithms

One method of classifying two-layer channel routing algorithms would be to classify them based on the approach the algorithms use. Based on this classification scheme we have:

1. LEA based algorithms: LEA based algorithms start with sorting the trunks from left to right and assign the segments to a track so that no two segments overlap.

2. Constraint Graph based routing algorithms: The constraint based routing algorithms use the graph theoretic approach to solve the channel routing problem. The horizontal and vertical constraints are represented by graphs. The algorithms then apply different techniques on these graphs to generate the routing in the channel.

3. Greedy routing algorithm: The greedy routing algorithm uses a greedy strategy to route the nets in the channel. It starts with the leftmost column and works towards the right end of the channel by routing the nets one column at a time.

4. Hierarchical routing algorithm: The hierarchical router generates the routing in the channel by repeatedly bisecting the routing region and then routing each net within the smaller routing regions to generate the complete routing.

In the following subsection, we present a few routers from each category.

9.4.2   LEA based Algorithms

The Left-Edge algorithm (LEA), proposed by Hashimoto and Stevens [HS71], was the first algorithm developed for channel routing. The algorithm was initially designed to route array-based two-layer PCBs. The chips are placed in rows and the areas between the rows and underneath the boards are divided into rectangular channels. The basic LEA has been extended in many different directions. In this section, we present the basic LEA and some of its important variants.

9.4.2.1   Basic Left-Edge Algorithm

The basic LEA uses a reserved layer model and is applicable to channel routing problems which do not allow doglegs and any vertical constraints. Consequently, it does not allow cyclic vertical constraints.

The left-edge algorithm sorts the intervals, formed by the trunks of the nets, in ascending order, relative to the x coordinate of the left end points of intervals. It then allocates a track to each of the intervals, considering them one at a time (following their sorted order) using a greedy method. To allocate an interval to a track, LEA scans through the tracks from the top to the bottom and assigns the net to the first track that can accommodate the net. The allocation process is restricted to one layer since the other layer is used for the vertical segments (branches) of the nets. The detailed description of LEA is in Figure 9.26. Figure 9.27 shows a routing produced by LEA. Net N_1 is assigned to track 1. Net N_2 is assigned to track 2 since it intersects with N_1 and cannot be assigned to track 1. Net N_3 is similarly assigned to track 3. Net N_4 is

Algorithm LEFT-EDGE $(\mathcal{N}, \mathcal{I})$
begin
    FORM-INTERVAL$(\mathcal{N}, \mathcal{I})$;
    FORM-HCG$(\mathcal{I}, \text{HCG})$;
    $d = $ DENSITY(HCG);
    let $T = \{T_1 T_2, \ldots, T_d\}$ denote the set of routing
       tracks from top to bottom;
    SORT-INTERVAL$(\mathcal{I})$;
    for $i = 1$ to n **do**
       for $j = 1$ to d **do**
          if DOES-NOT-OVERLAP(I_i, T_j) **then**
            assign interval I_i to T_j;
    for $i = 1$ to n **do**
       (* connect the vertical segments of net N_i to its *)
       (* horizontal segment *)
       VERTICAL-SEGMENT(left(I_i), left(N_i));
       VERTICAL-SEGMENT(right(I_i), right(N_i));
end.

Figure 9.26: Algorithm LEFT-EDGE

assigned to track 1 since it does not intersect with N_1. The following theorem which establishes the optimality of LEA is easy to prove.

Theorem 14 *Given a two-layer channel routing problem with no vertical constraints, LEA produces a routing solution with minimum number of tracks.*

The input to the algorithm is a set of two-terminal nets $\mathcal{N} = \{N_1, N_2, \ldots, N_n\}$. Procedure FORM-INTERVAL forms interval set $\mathcal{I} = \{I_1, I_2, \ldots, I_n\}$ from N. Once the intervals are formed, FORM-HCG forms the horizontal constraint graph HCG from $\mathcal{I}$. Note that the HCG is a interval graph corresponding to interval set $\mathcal{I}$. Procedure DENSITY computes the maximum clique size in HCG. This maximum clique size is a lower bound on the given channel routing problem instance. SORT-INTERVAL sorts the intervals in $\mathcal{I}$ in the ascending order of their x-coordinate on their left edge. Procedure VERTICAL-SEGMENT connects the vertical segments with the corresponding horizontal segment. The time complexity of this algorithm is $O(n \log n)$, which is the time needed for sorting n intervals.

The assumption that no two nets share a common end point is too restrictive, and as a result LEA is not a practical router for most channel routing problems. The restrictions placed on the router in order to achieve optimal results are not practical for most channel routing problems. However, LEA can be used to route PCB routing problems with vertical constraints since there is sufficient space between the adjacent pins to create a *jog*. LEA is also useful as a initial router for routing of channels with vertical constraints. The basic

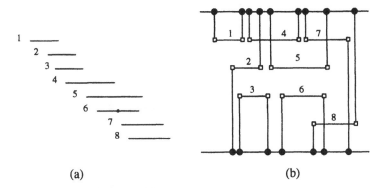

Figure 9.27: Left-edge channel routing.

idea is to create a layout with design rule violations and then use clean up procedures to remove the violations.

9.4.2.2 Dogleg Router

One of the drawbacks of LEA is that it places an entire net on a single track. It has been observed that this leads to routings with more tracks than necessary. Consider Figure 9.28(a), which shows a simple channel routing problem that has been routed using LEA and uses three tracks. On the other hand, if a dogleg is introduced in net N_2, the same problem can be routed using only two tracks. We recall that a dogleg is a vertical segment that is used to maintain the connectivity of two trunks (subnets) that are on two different tracks. The insertion of doglegs, may not necessarily reduce the channel density. A badly placed dogleg can lead to an increase in channel density. Finding the smallest number and locations of doglegs to minimize the channel density is shown to be NP-complete [Szy85].

Deutsch [Deu76] proposed an algorithm known as *dogleg router* by observing that the use of doglegs can reduce channel density. The dogleg router is that it allows multi-terminal nets and vertical constraints. Multi-terminal nets may have terminals on both sides of the channel and often form long horizontal constraint chains. In addition, there are several critical nets, such as clock nets, which pose problems because of their length and number of terminals. These type of nets can be broken into a series of two-terminal subnets using doglegs and each subnet can be routed on a different track. Like LEA, the Dogleg router uses a reserved layer model. Restricting the doglegs to the terminal positions reduces the number of unnecessary doglegs and consequently reduces the number of vias and the capacitance of the nets. The dogleg router cannot handle cyclic vertical constraints.

The dogleg router introduces two new parameters: *range* and *routing sequence*. Range is used to determine the number of consecutive two-terminal

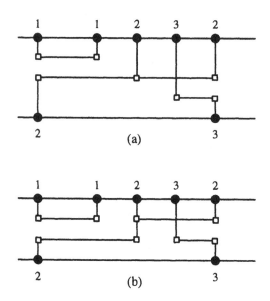

Figure 9.28: Using a dogleg to reduce channel density.

subnets of the same net which can be placed on the same track. Increasing the range parameter will result in fewer doglegs. The routing sequence specifies the starting position and the direction of routing along the channel. Unlike LEA, the routing can start from any end and work towards the opposite end. Different results can be obtained by starting at different corners: top-left, top-right, bottom-left, bottom-right. Furthermore, instead of starting from the top to the bottom or from the bottom to the top, the algorithm can alternate between topmost and bottommost tracks. This scheme results in eight different routing sequences: top-left $\rightarrow$ bottom-left, top-left $\rightarrow$ bottom-right, top-right $\rightarrow$ bottom-left, top-right $\rightarrow$ bottom-right, bottom-left $\rightarrow$ top-left, bottom-left $\rightarrow$ top-right, bottom-right $\rightarrow$ top-left, and bottom-right $\rightarrow$ top-right. (The left side of the arrow indicates the starting corner and the right side of the arrow indicates the alternate corner). Consider the example shown in Figure 9.29(a). If the range is set to 1 and we set the routing sequence to top-left $\rightarrow$ bottom-right, then Figure 9.29(b) shows routing steps in dogleg router. Notice that nets N_2 and N_3 use doglegs.

The complexity of the algorithm is dominated by the complexity of LEA. As a result, the complexity of the algorithm is $O(n \log n + nd)$, where n is the total number of two-terminal-nets after decomposition and d is the total number of tracks used. Note that the parameter's range and routing sequence can be changed to get different solutions of the same routing problem. A large value of range keeps the number of doglegs smaller. If the number of two-terminal subnets of a net is less than the value of a range, then that net is routed without any dogleg. Varying the routing sequence can also lead to a reduced

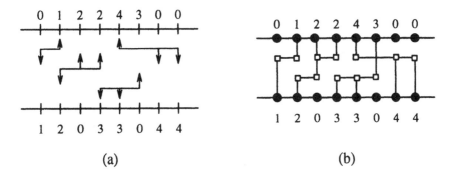

(a)

(b)

Figure 9.29: Example routed by dogleg router.

channel height. Dogleg router can easily be extended to gridless routing model. Experimentally dogleg routers achieve far superior results as compared to LEA, often requiring very few tracks beyond the channel density.

9.4.2.3 Symbolic Channel Router: YACR2

LEA does not allow vertical constraints thereby making it impractical for most of the channel routing problems. If a given channel is routed using LEA, then vertical constraint violations may be introduced by the router which need to be removed to get a legal routing solution. Note that a vertical constraint violation is a localized problem and may be resolved by anyone of the two methods:

1. local rip-up and reroute

2. localized maze routing.

In the second approach, vacant space surrounding the column in which vertical constraint violation occurred can be used to resolve the violation. Usually several horizontal segments of tracks as well as several vertical columns are not used for routing of any nets. Since the general maze routing technique is very time consuming and vertical constraint violations are local in nature, special maze routing techniques can used to remove vertical constraint violations. In case any vertical constraint violations cannot be resolved, new tracks can be added to resolve the constraints.

Based on these observations, Reed, Sangiovanni-Vincentalli, and Santamauro [RSVS85] proposed YACR2 (Yet Another Channel Router). In order to explain how vertical constraint violations are handled in YACR2 we define the concept of vertical overlap factor, which indicates the total number of tracks that a vertical constraint violation spans. Precisely stated, let us assume that column c_i has a vertical constraint violation between net N_t that has to be connected to the top boundary and net N_b that has to be connected to the bottom boundary. Also assume that N_t is assigned to track t_p and N_b is assigned to track t_q. Track t_p is above track t_q and tracks are numbered in increasing order from top to bottom boundary. Then vertical constraint $vof(c_i) = (p - q + 1)$. For any column c_j, if there is no vertical constraint violation in c_j, then $vof(c_j) = 0$. The basic idea of YACR2 is to select nets in an order and assign nets to tracks in a way such that $vof(c_i)$ is as minimum as possible for each column c_i. After assigning nets to tracks, specialized maze routing techniques are used to resolve the violations. If a vertical constraint violation cannot be resolved using maze routing technique, additional tracks are used to complete the routing.

The algorithm works in four different phases. First three phases are essentially for assigning nets to tracks with the objective of minimizing $vof(c_i)$ for each column c_i. In attempt to minimize $vof(c_i)$, the algorithm starts with the nets belonging to the maximum density column. After assigning tracks to the nets belonging to the maximum density column, it uses LEA to assign tracks to nets that are to the right of the maximum density column and then assigns to the nets that are to the left of the maximum density column. A modified LEA is used to assign tracks that are to the left of the maximum density column. It can be thought of as a right-edge algorithm, since it works from right to left.

As mentioned earlier that the goal of selecting and assigning nets to tracks is to minimize the total number of vertical constraint violations so that it is easy for the simplified maze routers to complete the routing. However, it is impossible to determine all the vertical constraint violations caused by the placement of a certain net. In fact some of the vertical constraint violations may occur between the net under consideration and nets yet to be routed. Since the nets are routed without doglegs, the vertical constraint graph can be used to estimate the possibility of an assignment giving rise to a violation, and the difficulty involved in removing the violation if it occurs. The techniques of selecting and assigning nets used in [RSVS85] are rather complicated and readers are referred to [RSVS85] for the details. It should be noted that any technique will work; however, vof may be very high for some column making vertical constraint violation resolution steps rather complicated.

After track assignments of horizontal segments, at the end of phase III, has been achieved, appropriate vertical segments are placed in the columns with $vof = 0$. In phase IV, the columns with vertical constraint violations are examined one at a time to search for legal connection between the nets and their terminals. Instead of applying the general purpose maze routing technique, three different maze routing techniques are applied to resolve the vertical constraint violations in this phase. These three techniques (strategies) are called maze1, maze2, and maze3. At each column with a vertical constraint violation, maze1 strategy is used first. If maze1 fails to resolve the violation, maze2 is applied. If maze2 fails, then maze3 is applied to resolve the violation. In case all three strategies fail, the channel is enlarged by adding one track and the process is repeated.

To explain how the maze routing techniques work, let us assume that column c_i has a vertical constraint violation between net N_t that has to be connected to the top boundary and net N_b that has to be connected to the bottom boundary. Also assume that N_b is assigned to track t_p and N_t is assigned to track t_q. Tack t_p is above track t_q.

The Maze1 technique checks for either one of the following:

1. No vertical segments exist between t_{p-1} and t_q on column c_{i-1} or c_{i+1}.

   In this case a jog is used in net N_t in track t_{p-1} to resolve the violation (see Figure 9.30(a) and (b));

2. No vertical segment exist between t_p and t_{q+1} on column c_{i-1} or c_{i+1}.

   In this case a jog is used in net N_b in track t_{q+1} to resolve the violation (see Figure 9.30(c) and (d));

3. No vertical segment exist between t_{p-1} and some t_s, between t_p and t_q, on column c_{i-1} and between t_{s-1} and t_{q+1} on column c_{i+1}, or vice versa.

   In this case net N_t uses jogs in tracks t_{p-1} and t_s and net N_b uses jogs in tracks t_{s-1} and t_{q+1} to resolve vertical constraint violation (see Figure 9.30(e) and (f)).

In case maze1 technique cannot resolve the vertical constraint violation, maze2 technique is used in attempt to resolve the violation. Maze2 checks for one of the following:

1. A track, column pair (t_r, c_j) such that: (a) there are no horizontal segments in track t_r between columns c_i and c_j; (b) there are no vertical segments in c_j between t_r and t_q; (c) the horizontal segment of net N_t in track t_q either crosses column c_j or can be extended to c_j without causing a horizontal constraint violation; and (d) t_r is above t_p.

   In this case, net N_t uses a dogleg in track t_r to resolve the violation as shown in Figure 9.31(a) and (b).

2. A track, column pair (t_r, c_j) such that: (a) there are no horizontal segments in track t_r between columns c_i and c_j; (b) there are no vertical segments in c_j between t_r and t_p; (c) the horizontal segment of net N_b in track t_p either crosses column c_j or can be extended to c_j without causing a horizontal constraint violation; and (d) t_r is below t_q.

   In this case, net N_b uses a dogleg in track t_r to resolve the violation as shown in Figure 9.31(c) and (d).

If none of the conditions in maze2 techniques are satisfied, maze3 technique is applied. As opposed to the local maze routing, the pattern based approach of YACR2 is efficient and avoids long routes; at the same time, it is limited in scope as opposed to local maze routing techniques. If none of the maze routing techniques can resolve vertical constraint violations, new tracks are added to complete the routing.

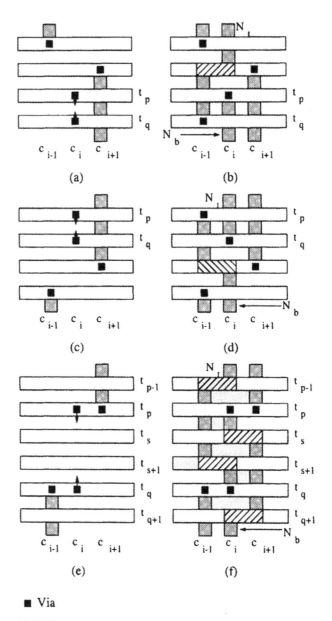

■ Via

▨ Metal 1 (vertical segments)

☐ Metal 2 (horizontal segments)

▨ Area where metal 1 passes horizontally under metal 2

Figure 9.30: Maze1 routing.

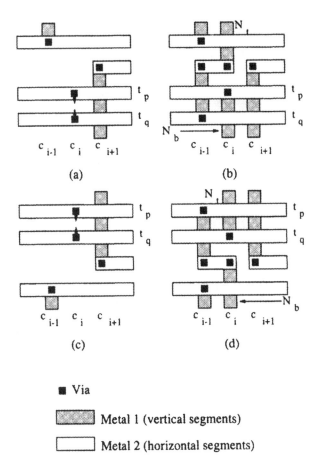

■ Via

▨ Metal 1 (vertical segments)

☐ Metal 2 (horizontal segments)

Figure 9.31: Maze2 routing.

9.4.3 Constraint Graph based Routing Algorithms

Consider a channel routing problem with no vertical constraints. Obviously, the number of tracks needed is determined by the maximum clique h_{max} in the horizontal constraint graph(HCG). In this case, LEA produces optimal results, if no doglegs are allowed. In presence of vertical constraints, the length of the longest path $v_{\max}$ in vertical constraint graph(VCG) also plays a key role in determining the channel height. In particular, the nets which lie on long paths in the vertical constraint graph, must be carefully assigned to tracks. In order to explain the effect of long vertical chains, let us define length of ancestor and descendent chains of a net N_i. Let v_i represent N_i in VCG. Let A_i denote the length of the longest path from a vertex of zero in-degree to v_i in VCG. Similarly, let D_i denote the length of longest path from v_i to a vertex of zero

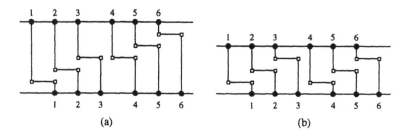

Figure 9.32: Effect of net merging on channel height.

out-degree in VCG. It is easy to see that

$$v_{\max} = \max_{i=1,n}(A_i + D_i) - 1$$

Consider the nets N_3 and N_4 as shown in Figure 9.32. If N_3 and N_4 are assigned to the same track then the channel height is given by

$$ch\_ht \geq \max\{A_3 + A_4, A_3 + D_4, A_4 + D_3, A_4 + D_4\}$$

In other words, if we consider N_3 and N_4 as a new net, then a new vertical constraint chain is created which consists of longer of two ancestor chains and longer of two descendent chains.

$$ch\_ht \geq \max\{\max\{A_3, A_4\} + \max\{D_3, D_4\}\}$$

Figure 9.32(a) shows the effect of assigning two nets to the same track without considering the constraint chains. The channel height for this solution is equal to 6. A better assignment resulting in a channel height of 4 is shown in Figure 9.32(b). Based on the above equation, we make the following observation. In order to minimize the effect of vertical constraint chains on channel height, two nets may be assigned to the same track only if both of them have small ancestor chains, or both of them have small descendent chains. Several algorithms have been developed which are based on this observation. In this section, we discuss the first constraint graph based algorithm and its grid-less variant.

9.4.3.1   Net Merge Channel Router

In 1982, Yoshimura and Kuh [YK82] presented a new channel routing algorithm (YK algorithm) for two-layer channel routing problems based on net merging. This work was the first attempt to analyze the graph theoretic structure of the channel routing problem. YK algorithm considers both the horizontal and vertical constraint graphs and assigns tracks to nets so as to minimize the effect of vertical constraint chains in the vertical constraint graph. It does not allow doglegs and cannot handle vertical constraint cycles. The YK algorithm partitions the routing channel into a number of regions called *zones* based on the horizontal segments of different nets and their constraints. The basic observation is that a column by column scan of the channel is not necessary as nets within a zone cannot be merged together and must be routed in a separate track. This observation improves the efficiency of the algorithm. The algorithm proceeds from left to right of the channel merges nets from adjacent zones. The nets that are merged are considered as one composite net and are routed on a single track. In each zone, new nets are combined with the nets in the previous zone. After all zones have been considered, the algorithm assigns each composite net to a track. The key steps in the algorithm are zone representation, net merging to minimize the vertical constraint chains, and track assignment. Throughout our discussion, we will use the example given in [YK82], since that example serves as a benchmark.

1. **Zone Representation of Horizontal Segments:** Zones are in fact maximal clique in the interval graph defined by the horizontal segments of the nets. The interval graph of the net list in Figure 9.33(a) is shown in Figure 9.33(e). In terms of an interval graph the clique number is the density of the channel routing problem.

   In order to determine zones, let us define $S(i)$ to be the set of nets whose horizontal segments intersects column i. Assign *zones* the sequential number to the columns at which $S(i)$ are maximal. These columns define zone 1, zone 2, etc., as shown in the table 9.33(c), for the example in Figure 9.33. The cardinality of $S(i)$ is called local density and the maximum among all local densities is called maximum density which is the lower bound on the channel density. In should be noted that a channel routing problem is completely characterized by the vertical constraint graph and its zone representation.

2. **Merging of Nets:** Let N_i and N_j be the nets for which the following two conditions are satisfied:

   1. There is no edge between v_i and v_j in HCG.

   2. There is no directed path between v_i and v_j in VCG.

   If these conditions are satisfied, net N_i and net N_j can be merged to form a new composite net.

   The operation of merging net N_i and net N_j modifies the VCG by shrinking node v_i and and node v_j into node $v_{i.j}$, and updates the zone representation by replacing net N_i and net N_j by net $N_{i.j}$ which occupies the consecutive zones including those of net N_i and net N_j.

   Let us consider the example shown in Figure 9.33(a). Net N_6 and net N_9 are merged and the modified VCG along with the zone representation is shown in Figure 9.34. The updated vertical constraint graph and the zone representation correspond to the net list in Figure 9.34, where N_6

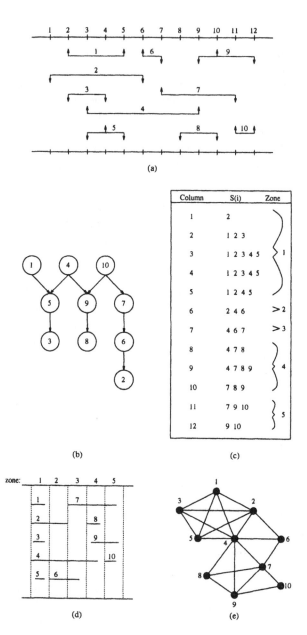

Figure 9.33: Example of zone representations.

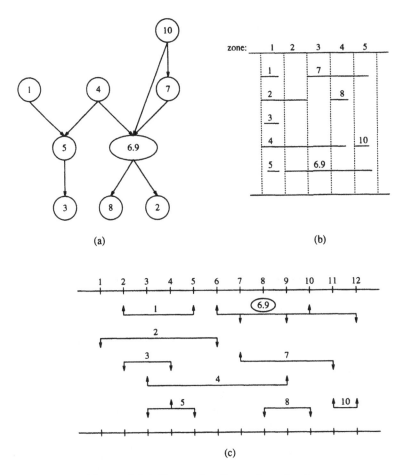

Figure 9.34: Example of merging of nets.

and N_9 are replaced by net $N_{6.9}$. The algorithm, given in Figure 9.35 merges nets as long as two nets from different zones can be merged.

In each iteration, the nets ending in zone z_i are added to the list L. While the nets starting in z_{i+1} are kept in list R. Function MERGE then merges two list L and R so as to minimize the increase in the longest path length in VCG. The list L' returned by function MERGE consists of all the nets merged by the function. These nets are not considered further.

In Figure 9.36 we illustrate how the vertical constraint graph is updated by the algorithm NET-MERGE. The length of the longest path in VCG is 4 and size of the maximum clique is 5, therefore any optimal solution takes at least 5 tracks. In first iteration, $L = \{N_1, N_3, N_5\}$ and $R = \{N_6\}$. There are three possible net mergings, $N_{1.6}$, $N_{3.6}$, and $N_{5.6}$. Merging N_1

Algorithm NET-MERGE
begin
    $L = \phi$;
    for $z = (z_1 \text{ to } z_{t-1})$ **do**
        $L = L + \{z_i - (z_i \cap z_{i+1})\}$;
        $R = \{z_{i+1} - (z_i \cap z_{i+1})\}$;
        $L' = $MERGE$(L, R)$;
        $L = L - L'$;
end.

Figure 9.35: Algorithm NET-MERGE

and N_6 creates a path of length 5, merging N_3 and N_6 creates a path of length 4, while merging N_5 and N_6 creates a path of length 4. Therefore either $N_{3.6}$ or $N_{5.6}$ may be formed. Let us merge N_5 and N_6. Similarly, in second iteration net N_1 and net N_7 are merged. In the fourth iteration N_{10} and N_4 are merged. The final graph is shown in Figure 9.36(e). The track assignment is straight forward. Each node in the final graph is assigned a separate track. For example, track 1 can be assigned to net $N_{10.4}$. Similarly, tracks 2 and 3 can be assigned to nets $N_{1.7}$ and net $N_{5.6.9}$, respectively. For net N_2 and net $N_{3.8}$, either track 4 or 5 can be assigned.

It should be noted that finding optimal net pairs for merging is a hard problem. This is due to the fact that the future effects of a net merge cannot be determined.

It is possible to improve the YK algorithm by allowing some look ahead or doing rip-up and re-merge operations. In 1982, YK algorithm represented a major step forward in channel routing algorithms. It formulated the problem and provided a basis for future development of three layer and multi-layer algorithms. It has been extended to three layer and gridless environment. These extended routers will be discussed later in the chapter.

9.4.3.2 Glitter: A Gridless Channel Router

All the algorithms presented thus far in the channel routing are grid-based. The main drawback of grid-based algorithms is that it is difficult to route nets with varying wire widths.

Chen and Kuh [CK86] first proposed a gridless variable-width channel router called *Glitter*. Glitter can utilize multiple layer technology and design rules. Terminals can be located at arbitrary positions and can be located on off-grid points. No columns or tracks are used in routing. Only the wire width, spacing, and via size are under consideration are used. Nets are allowed to have different wire widths to satisfy special design needs and improve the performance of the circuits. Glitter is a reserved-layer model routing algorithm.

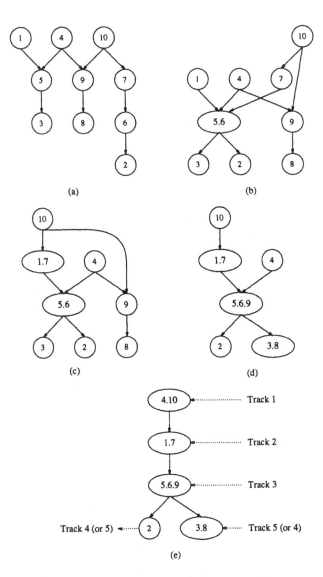

Figure 9.36: Illustration of algorithm NetMerge.

The basic idea of Glitter is somewhat similar to net merge algorithm. Instead of computing the longest vertical constraint chains in terms of tracks, the actual height of the vertical constraint chains is computed and used for assigning nets to locations in the channel.

Another key observation is that if each edge in the combined constraint graph is directed, then they specify a routing. Thus the routing problem is reduced to a problem of assigning directions to edges in the combined constraint graph.

Glitter uses vertical and horizontal constraint graphs to form a graph called *weighted constraint graph*. The weighted constraint graph combines all the vertical and horizontal constraints into the same graph, where each node represents a horizontal net or subnet, each directed edge represents a vertical constraint, and each undirected edge represents a horizontal constraint. The weight of the edge between node A and node B is the minimum vertical distance required between net A and net B. If net A needs to be placed above net B, then the edge should be directed from node A to node B.

To build the weighted constraint graph, vertical and horizontal constraints for each pair of nets (subnets) are checked. If there is more than one constraint, the larger weight will overrule the smaller, and the directed edge will overrule the undirected edge. If there are two contradictory directed edges (there is cycle in VCG), a dogleg must be introduced to break the cycle.

The upper boundary and the lower boundary are also represented by nodes in the weighted constraint graph. Since every net must be placed below the upper boundary, a directed edge will be generated from the upper boundary to each net. Similarly, there is a directed edge from each net to the lower boundary. The weight for a boundary constraint edge is the minimum distance required between the boundary and each net.

Figure 9.37(a) shows a simple example of the variable-width channel-routing problem. In the following analysis, we have assumed the following design rules for the example. The minimum wire spacing in layer 1 and 2 is assumed to be 3 and 2, respectively, the via size is assumed to be 2 × 2, and the minimum overlap width that each layer must extend beyond the outer boundary of the via is assumed to be 1. If the vertical wire width is 4 for every net and the horizontal wire width for each net is specified in Figure 9.37, then we can check the vertical and horizontal constraints for each pair of nets (subnets), and calculate the minimum distance required between them. For example, net 6 must be placed above net 2 by a minimum distance of 7, so there is a directed edge from node 6 to node 2 and the weight of this edge is 7. On the other hand, net 1 should be placed either above or below net 4 because their horizontal spans overlap each other. So there is an undirected edge between node 1 and node 4, and the minimum distance (edge weight) required is 6. The complete weighted constraint graph is shown in Figure 9.37(c).

After the weighted constraint graph is generated, the channel-routing problem can be formulated as follows. Given a weighted constraint graph, assign a direction to each undirected edge such that 1) no cycles are generated and 2) the total weight of the maximum weighted directed path (longest path) from

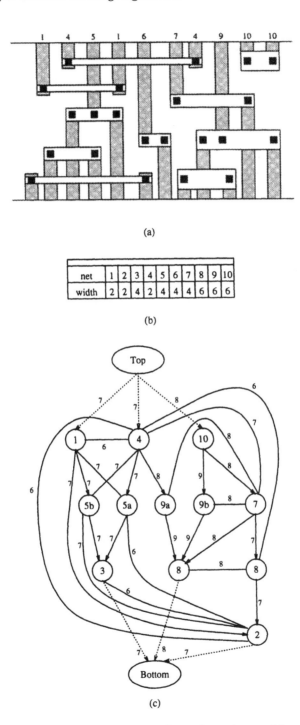

(a)

(b)

(c)

Figure 9.37: Variable-width channel routing problem.

the upper boundary to the lower boundary is minimized. In a graph which has m undirected edges, there are 2^m possible solutions.

Since the routing solution can always be obtained by assigning directions to undirected edges, the ordering of edge selection becomes very important. For each undirected edge in the weighted constraint graph, suppose the assignment of one direction will result in cycles. The only choice then is to assign the other direction. Those edges are called critical edges, and they should be assigned directions first.

The ancestor weight $ancw(i)$ of node i is the total weight of the maximum weighted ancestor chain of node i. Similarly, the descendant weight $desw(i)$ of node i is the total weight of the maximum weighted descendant chain of node i.

The label of each undirected edge(i, j) is defined as the maximum of $ancw(i)$ + $weight(i, j)$ + $desw(j)$ and $ancw(j) + weight(i, j) + desw(i)$. The label is a measure of the total weight of the maximum weighted directed path which passes through edge(i, j) if an improper direction is assigned to edge(i, j). The label is defined as ∞ if the undirected edge is a critical edge. If a label is larger than the maximum density of the channel, the meaning of this label becomes significant because it may be the new lower bound for minimum channel height. Undirected edge with a large label should therefore be assigned a proper direction as early as possible, so that the increase of the lower bound is less likely to occur.

Selection of nodes in the graph is based on the ancestor and descendent weights. If an unprocessed node has no ancestors (or all its ancestors have been processed), it can be placed close to the upper boundary (i.e., closer than other unprocessed nodes), and all the undirected edges connected to this node can be assigned outgoing directions. Similarly, if an unprocessed node does not have descendants (or all its descendants have been processed), it should be placed close to the lower boundary, and all the undirected edges connected to it can be assigned incoming directions. The unprocessed nodes with minimum $ancw(i)$ or $desw(i)$ are the candidates to be selected. A node is said to be *processed* if it has been placed close to the upper or lower boundary. An edge is said to be processed if it has been assigned a direction. The routing is complete when all the nodes (or edges) are processed.

The algorithm can be easily extended to accommodate irregularly-shaped channels by adding the boundary information to the weighted constraint graph. The Top node will represent the uppermost boundary, and the Bottom node will represent the lowermost boundary. The weight of boundary constraint edges should be modified to include the amount of indentation.

9.4.4   Greedy Channel Router

Assigning the complete trunk of a net or a two-terminal net segment of a multiterminal net severely restricts LEA and dogleg routers. Optimal channel routing can be obtained if for each column, it can be guaranteed that there is only one horizontal track per net. Based on this observation, one approach to

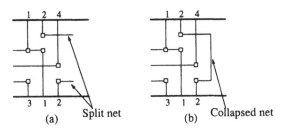

Figure 9.38: (a) A split net. (b) The collapsed split net.

reduce channel height could be to route nets column by column trying to join split horizontal tracks (if any) that belong to the same net as much as possible.

Based on the above observation and approach, Rivest and Fiduccia [RF82] developed *greedy channel router*. This makes fewer assumptions than LEA and dogleg router. The algorithm starts from the leftmost column and places all the net segments of a column before proceeding to the next right column. In each column, the router assigns net segments to tracks in a greedy manner. However, unlike the dogleg router, the greedy router allows the doglegs in any column of the channel, not necessarily where the terminal of the doglegged net occurs.

Given a channel routing problem with m columns, the algorithm uses several different steps while routing a column. In the first step, the algorithm connects any terminal to the trunk segment of the corresponding net. This connection is completed by using the first empty track, or the track that already contains the net. In other words, minimum vertical segment is used to connect a trunk to terminal. For example, net 1 in Figure 9.38(a) in column 3 is connected to the same net. The second step attempts to collapse any *split nets* (horizontal segments of the same net present on two different tracks) using a vertical segment as shown in Figure 9.38(b) A split net will occur when two terminals of the same net are located on different sides of the channel and cannot be immediately connected because of existing vertical constraints. This step also brings a terminal connection to the correct track if it has stopped on an earlier track. If there are two overlapping sets of split nets, the second step will only be able to collapse one of them.

In the third step, the algorithm tries to reduce the range or the distance between two tracks of the same net. This reduction is accomplished by using a dogleg as shown in Figure 9.39(a) and (b). The fourth step attempts to move the nets closer to the boundary which contains the next terminal of that net. If the next terminal of a net being considered is on the top(bottom) boundary of the channel then the algorithm tries to move the net to the upper(lower) track. In case there is no track available, the algorithm adds extra tracks and the terminal is connected to this new track. After all five steps have been completed, the trunks of each net are extended to the next column and the steps are repeated. The detailed description of the greedy channel routing

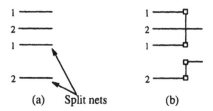

Figure 9.39: (a) Reducing the distance between split nets.

algorithm is in Figure 9.40.

The greedy router sometimes gives solutions which contain an excessive number of vias and doglegs. It has, however, the capability of providing solution even in presence of cyclic vertical constraints. The greedy router is more flexible in the placement of doglegs due to fewer assumptions about the topology of the connections. An example routed by the greedy channel router is shown in Figure 9.41.

9.4.5 Hierarchical Channel Router

In [BP83], Burstein and Pelavin presented a two layer channel router based on the reduction of the routing problem in $(m \times n)$ grid to the routing in $(2 \times n)$ grid and consistent utilization of 'divide and conquer' approach.

Let, $C(i, j)$ denote the cell in ith row and jth column in an $(m \times n)$ grid G. The terminals are assumed to be in the top and the bottom rows of the grid. Let, $h(i, j)$ denote the horizontal boundary shared by the cell $C(i, j)$ and $C(i + 1, j)$. Let, $v(i, j)$ denote the vertical boundary shared by the cell $C(i, j)$ and $C(i, j + 1)$. Each boundary in G has a capacity, which indicates the number of wires that can pass through that boundary.

In this approach, a large routing area is divided into two rows of routing tiles. The nets are routed globally in these rows using special types of steiner trees. The routing in each row is then further refined by recursively dividing and routing each of the rows. More specifically, the the $(m \times n)$ routing grid is partitioned into two subgrids; the top ($\lceil \frac{m}{2} \rceil \times n$) and the bottom ($\lfloor \frac{m}{2} \rfloor \times n$) subgrid. Each column in these subgrids is considered as a supercell. As a result, two rows of supercells are obtained, i.e., a $(2 \times n)$ grid is obtained. (see Figure 9.42.) Capacity of each vertical boundary in this grid will be the sum of corresponding boundary capacities in the original grid. The nets are routed one at a time in this $(2 \times n)$ grid. (see Figure 9.43.) Each row of the $(2 \times n)$ is then partitioned into a $(2 \times n)$ grid. The terminal positions for the routing in the new $(2 \times n)$ subproblems is defined by the routing in the previous level of hierarchy (see Figure 9.44). This divide and conquer approach is continued until single cell resolution is reached.

In the following, we discuss the algorithm for routing a net in the $(2 \times n)$ grid.

Algorithm GREEDY-CHANNEL-ROUTER ($\mathcal{N}$)
begin
    d = DENSITY($\mathcal{N}$);
(* calculate the lower bound of channel density *)
    insert d tracks to channel;
    for $i = 1$ to m **do**
        $T1$ = GET-EMPTY-TRACK;
        if $T1 = 0$ **then**
            ADD-TRACK($T1$);
            ADD-TRACK($T2$);
        else
            $T2$ = GET-EMPTY-TRACK;
            if $T2 = 0$ **then**
                ADD-TRACK($T2$);
        CONNECT($T_i, T1$);
        CONNECT($B_i, T2$);
        join split nets as much as possible;
        bring split nets closer by jogging;
        bring nets closer to either top or bottom boundary;
    while split nets exists **do**
        increase number of column by 1;
        join split nets as much as possible;
end.

Figure 9.40: Algorithm GREEDY-CHANNEL-ROUTER

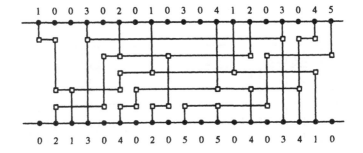

Figure 9.41: Channel routed using a greedy router.

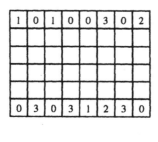

1	0	1	0	0	3	0	2
0	3	0	3	1	2	3	0

Figure 9.42: Reducing $(m \times n)$ grid to $(2 \times n)$ grid

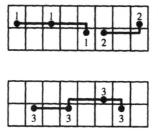

Figure 9.43: First level of hierarchy

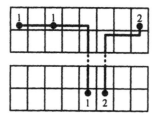 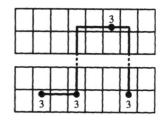

Figure 9.44: Second level of hierarchy

Wiring Within $(2 \times n)$ Grid: Let, p be a $(2 \times n)$ boolean matrix, such that, $p(i, j)$ is *true* if net N_p has a terminal in $C(i, j)$. Let, c_h an array of size n, such that, $c_h(j)$ indicates the cost that must be added to the cost of a net if it crosses the boundary $h(1, j)$. Let, c_v be a $(2 \times (n - 1))$ matrix such that $c_v(i, j)$ indicates the cost which must be added to the cost of a net if it crosses the boundary $v(i, j)$. Each element of c_h and c_v is a function of the capacity and utilization of the corresponding boundary; higher the ratio of utilization to the capacity, higher the cost of crossing of that boundary. Such a cost function reduces the probability of utilization of a congested boundary in the routes of the remaining nets. The algorithm to find a minimum cost tree for a net in $(2 \times n)$ grid is a recursive algorithm. This algorithms requires definitions of following terms:

1. $T^1(k)$: It is the minimum cost tree which interconnects the following set of cells:
$$\{C(i, j) : (j \leq k)\&(p(i, j) = true)\} \cup \{C(1, k)\}$$

   i.e., it is a minimum cost tree connecting the cells in first k columns that have terminals of N_p and cell $C(1, k)$.

2. $T^2(k)$: It is the minimum cost tree which interconnects the following set of cells:
$$\{C(i, j) : (j \leq k)\&(p(i, j) = true)\} \cup \{C(2, k)\}$$

   i.e., it is a minimum cost tree connecting the cells in first k columns that have terminals of N_p and cell $C(1, k)$.

3. $T^3(k)$: It is the minimum cost tree which interconnects the following set of cells:
$$\{C(i, j) : (j \leq k)\&(p(i, j) = true)\} \cup \{C(1, k), C(2, k)\}$$

   i.e., it is a minimum cost tree connecting the cells in first k columns that have terminals of N_p, cell $C(1, k)$ and cell $C(2, k)$.

4. $T^4(k)$: It denotes a minimum cost forest, containing two different trees $T_1^4(k)$ and $T_2^4(k)$: $T_1^4(k)$ uses cell $C(1, k)$, $T_2^4(k)$ uses $C(2, k)$ and the set
$$\{C(i, j) : (j \leq k)\&(p(i, j) = true)\}$$

Let, f and l denote the column number of leftmost and rightmost cells, i.e.,

$$f = \min\{k | p(1, k) \vee p(2, k) = true\},$$

$$l = \max\{k | p(1, k) \vee p(2, k) = true\}.$$

$T^i(k + 1), i = 1, 2, 3, 4$ is computed recursively from $T^i(k)$, as discussed below:

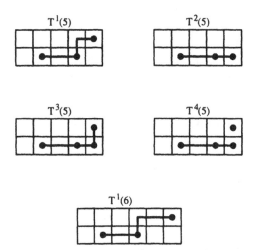

Figure 9.45: Recursively computing $T^1(6)$ from $T^i(5), i = 1, 2, 3, 4$

1. **Basis:** Trees $T^i(k)$ for $k = f$ can be computed trivially and they serve
   as basis for recursion. For $k \leq f$, $T^1(k)$ and $T^2(k)$ consists of a single
   vertex $C(1, k)$ and $C(2, k)$ respectively. $T^4(k)$ consists of the disjoint
   pair of vertices $C(1, k)$ and $C(2, k)$. Whereas, $T^3(1, k)$ consists of a path
   $C(1, k), \ldots, C(1, s), C(2, s), \ldots, C(2, k)$, where, $1 \leq s \leq k$ and

$$c_h(s) + \sum_{i=s}^{k-1} c_v(1, i) + c_v(2, i) \text{ is minimum.}$$

   Note that if the costs of all horizontal edges are same then $T^3(k)$ is
   $C(1, k), C(2, k)$.

2. **Recursive Step:** Suppose for $f \leq k \leq l$, $T^i(k), i = 1, 2, 3, 4$ are constructed. In order to construct $T^j(k+1)$, we simply enumerate all possible
   extensions from $T^i(k), i = 1, 2, 3, 4$ and select the cheapest one. An example in Figure 9.45 shows $T^i(5), i = 1, 2, 3, 4$ and computation of $T^1(6)$
   by extending $T^1(5)$ for net 3.

The above algorithm routes a net N_p in $O(n \log m)$ time.

In the higher levels of refinement the cost of crossing a boundary is higher
as the boundaries are partially utilized in the lower level of hierarchy. As a
result, the cost of later routes will be large. If the cost of a route is larger
than a user specified value LRG, the routes are allowed to detour outside the
channel, i.e., vertical tracks are added at the left or right ends of the channel.

Features	Algorithm						
	LEA	Dogleg	Y-K	Greedy	YACR2	Hierar- chical	Glitter
Model	grid- based	grid- based	grid- based	grid- based	grid- based	grid- based	grid- less
Dogleg	not allowed	allowed	allowed	allowed	allowed	allowed	not allowed
Layer assign- ment	reserved	reserved	reserved	reserved	reserved*	reserved	reserved
vertical constra- ints	not allowed	allowed	allowed	allowed	allowed	allowed	allowed
cyclic constra- ints	not allowed	not allowed	not allowed	allowed	allowed	not allowed	allowed

*with some exceptions

Table 9.1: Comparison of different features of two-layer routers.

9.4.6 Comparison of Two-Layer Channel Routers

Extensive research has been done on two layer channel routing. No algorithm is suitable for all problems and applications. Table 9.1 summarizes different features of two-layer routers discussed in this section.

As we have noted earlier, all algorithms are not equally good for all channel routing problems. Therefore, many benchmark channel routing examples have been proposed. The most famous benchmark is the *Deutsch difficult example*. Table 9.2 summarizes the routing results by different algorithms presented in this section on Deutsch difficult example. The table is reproduced from [PL88]. As it is clear from the table that YACR2 produces best routing for Deutsch difficult example both in terms of tracks and vias. However, YACR2 is considerably more complicated to implement as compared to Greedy router, which produces close to optimal results on most practical examples. In fact, it produces a solution within one or two tracks of the optimal.

9.5 Three-Layer Channel Routing Algorithms

The two-layer channel routing problem has been studied extensively in the past decade. In fact there are several two-layer channel routers that can produce solutions very close to the optimal solution (one or two tracks more than the minimum number of tracks required). About five years ago, a third metal layer became feasible. Most of the current gate-array technologies use three layers for routing. For example, the Motorola 2900ETL macrocell array is a bipolar gate array which uses three metal layers for routing. DEC's Alpha chip also

ROUTER	Tracks	Vias	Wire length
LEA	31	290	6526
Dogleg router	21	346	5331
Y-K router	20	403	5381
Greedy router	20	329	5078
Hierarchical router	19	336	5023
YACR2	19	287	5020

Table 9.2: Results on Deutsch difficult example.

uses three metal layer for routing. Intel's 486 chip used a three metal layer process and original Intel pentium was also fabricated on a similar process. As a consequence, a considerable amount of research has also been done on three-layer channel routing problem.

9.5.1 Classification of Three-Layer Algorithms

The three-layer routing algorithm can be classified into two main categories: The reserved layer and the unreserved layer model. The reserved layer model can further be classified into the VHV model and the HVH model.

Following theorems show the lower bounds of channel routing problem in three-layer reserved layer routing model, in terms of the maximum clique size in HCG and the longest path in VCG of the corresponding routing problem.

Theorem 15 *In the three layer VHV model, the lower bound on the number of tracks for a routing problem is* $h_{\max}$.

Theorem 16 *In the three layer HVH model, the lower bound on the number of tracks for a routing problem is* $\max\{v_{\max}, \frac{h_{\max}}{2}\}$.

Note that in VHV routing, the vertical constraints between nets no longer exist. Therefore, the channel height which is equal to the maximum density can always be realized using LEA. Almost all three-layer routers are extensions of two-layer routers. The net-merge algorithm by Yoshimura and Kuh [YK82] has been extended by Chen and Liu [CL84]. The gridless router Glitter [CK86] has been extended to Trigger by Chen [Che86].

Cong, Wong and Liu [CWL87] take an even general approach and obtain a three-layer solution from a two-layer solution. Finally, Pitchumani and Zhang [PZ87] partition the given problem into two subproblems and route them in VHV and HVH models. In this section, we discuss several three-layer channel routing algorithms.

9.5.2 Extended Net Merge Channel Router

In [CL84], Chen and Liu presented a three-layer channel router based on the net merging method and the left edge algorithm used in a two-layer channel routing algorithm by Yoshimura and Kuh [YK82].

As there are no vertical constraints in VHV and therefore the left edge algorithm is sufficient. In the HVH routing, vertical constraint still exist. As a result, there should not be a directed path in the VCG between nets that are placed in both first and third layers on the same track. The merging algorithm presented in [YK82] can be extended to the HVH problem. In three-layer routing, in addition to merging nets in the same layer between different zones the nets in the same zone between different layers can also be merged. Two types of merging are defined as follows:

1. **Serial merging:** If there is no horizontal and vertical constraints between N_i and N_j then they can be placed on the same layer and the same track. This operation is called as *serial merging*.

2. **Parallel merging:** If nets N_i and net N_j have horizontal constraints and if they do not have vertical constraints then they can be placed on the same track but in different layers. In case of HVH model, one of them is placed in the first layer, whereas the other is placed in the third layer. This operation is referred to as *parallel merging*

As in two-layer routing, the merging procedure is the essential element of the whole algorithm where two sets of nets are merged. Let $\mathcal{N}_P = \{N_1, N_2, \ldots, N_{n_p}\}$ and $\mathcal{N}_Q = \{M_1, M_2, \ldots, M_{n_q}\}$ be two sets of nets to be merged. Two nets $N_i \in \mathcal{N}_P$ and $M_j \in \mathcal{N}_Q$ are merged such that N_1 lies in the longest path in VCG before merging and farthest away from either the source node or the sink node, and M_j is neither ancestor nor descendent of N_i, and after merging N_i and M_j the increase of longest path in VCG is minimum.

If the merging is done between two adjacent zones i and $i + 1$, then $\mathcal{N}_P$ is the set of nets which terminates at zone i, $\mathcal{N}_Q$ is the set of nets that begin at zone $i + 1$.

Let us define the following:

1. Let $\mathcal{N}_B$ be the set of nets which begin at zone $i + 1$.

2. n_b is the number of nets in $\mathcal{N}_B$.

3. Let $\mathcal{N}_T$ be the set of nets which include: (a) nets which terminate at zone i. (b) nets which are placed on a track with no horizontal segment of nets on the other layer.

4. n_t is the number of nets in $\mathcal{N}_T$.

Before merging nets, all the nets in $\mathcal{N}_T$ have been placed on certain tracks, while nets in $\mathcal{N}_Q$ have not yet been placed on any track. Therefore, if net $M_2 \in \mathcal{N}_Q$ merges with any net in $\mathcal{N}_T$, no new track appears. Otherwise, if net $M_2 \in \mathcal{N}_Q$ merges with another net in $\mathcal{N}_Q$, then a new track appears. Therefore, if $n_q \leq n_t$ it is possible for the old tracks (where nets in T are placed) to contain all the nets in $\mathcal{N}_Q$. In such a case, in order to avoid the increase in the number of tracks, the parallel merging between nets in $\mathcal{N}_Q$ should be

Algorithm MERGE
begin
    $\mathcal{N}_Q = \mathcal{N}_B;\ n_q = n_b;$
    while $\mathcal{N}_Q \neq \phi$ **do**
        find $M_j \in Q;$
        $\mathcal{N}_Q = \mathcal{N}_Q - \{M_j\};$
        if $n_q \leq n_t$ **then** $\mathcal{N}_P = \mathcal{N}_T$
        else $\mathcal{N}_P = \mathcal{N}_T + \mathcal{N}_Q;$
        find $N_i \in \mathcal{N}_P;$
        if $N_i \in \mathcal{N}_T$ **then**
            (* serial (or parallel) merging (N_i, N_j) *)
            $\mathcal{N}_T = \mathcal{N}_T - N_i;$
            $n_t = n_t - 1;$
            $n_q = n_q - 1;$
        if $N_i \in \mathcal{N}_Q$ **then**
            (* parallel merging (N_i, N_j)
            $\mathcal{N}_Q = \mathcal{N}_Q - N_i;$
            $n_q = n_q - 2;$
        if N_i cannot be found **then**
            place N_j on a new track;
            $n_q = n_q - 1;$
            $n_t = n_t + 1;$
end.

Figure 9.46: Algorithm MERGE

avoided. Conversely, if $n_q > n_t$, the old tracks are not enough to contain all the nets in $\mathcal{N}_Q$, at least one new track appears. Parallel merging between nets in $\mathcal{N}_Q$ is allowed only in this case.

The details of the merging algorithm are given in Figure 9.46

As a special case, when merging starts, a parallel merging is made between all the nets which pass through the starting zone.

9.5.3   HVH Routing from HV Solution

In [CWL87], Cong, Wong, and Liu presented a general technique that systematically transforms, a two-layer routing solution into a three-layer routing solution. We will refer to this algorithm as CWL algorithm.

The focus of the CWL algorithm is very similar to the YK algorithm. In YK algorithm, nets are merged so that all merged nets forming a composite net are assigned to one track. The objective is to minimize the number of composite nets. In CWL algorithm, composite nets are merged together to form super-composite nets. The basic idea is to merge two composite nets such that the number of super-composite nets is minimized. Two composite nets in a super-composite net can then be assigned to two different layers on the *same* track. In order to find the optimal pair of composite nets that can be merged to form super-composite nets, a directed acyclic graph called *track ordering graph*, $TVCG = (V, E)$ is defined. The vertices in V represent the composite(tracks) in a given two layer solution. The directed edges in $G(S)$ represent the ordering restrictions on pairs of tracks. Composite interval t_i must be routed above composite interval t_j if there exists a net $N_p \in t_i$ and $N_q \in t_j$, such that N_p and N_q have a vertical constraint. Thus TVCG is in fact a vertical constraint graph between tracks or composite intervals. The objective of CWL algorithm is to find a track pairing which reduces the total number of such pairs. Obviously, we must have at least $\frac{|V|}{2}$ pairs. It is easy to see that the problem of finding an optimal track pairing of a given two layer solution S can be reduced to the problem of two processor scheduling in which tracks of V are tasks and $TCVG$ is the task precedence graph. Since the two processor scheduling problem can be optimally solved in linear time [Gab85, JG72], the optimal track permutation can also be found in linear time. Figure 9.47(b) shows the track ordering graph of the two-layer routing solution, shown in Figure 9.47(a), obtained by using a greedy router. Figure 9.47(c) shows an optimal scheduling solution for the corresponding two-processor scheduling problem.

The key problem is the number of tracks which are not paired. This happens due to adjacent vias. If adjacent vias can be removed between two non-paired tracks so that the tracks can be paired together, it would lead to saving of two empty tracks. The basic idea is to move the via aside and then a maze router can be used to connect the portion of the net containing x with the portion of the net containing the horizontal segment h_1 (see Figure 9.48).

In order to successfully merge non-paired tracks, we must minimize the number of adjacent vias between two tracks. This is accomplished by properly changing the processor (layer) assignment of tasks (tracks). It is easy to see that if tracks t_i and t_j are assigned to be routed in the pth track then the layer on which a particular track gets routed is still to be decided. In other words, for each track, we have two choices. However, the choice that we make for each track can affect the number of adjacent vias. This problem can be solved by creating a graph, which consists of vertices representing both the possible choices for the track. Thus, each track is represented by two vertices. Four vertices of two adjacent tracks are joined by edges. Thus each edge represents a possible configuration of two adjacent tracks. Each edge is assigned a weight which is equal to the number of adjacent vias if this configuration represented by the edge is used. It is easy to see that problem of finding optimal configuration can be reduced to a shortest path problem. Thus, the problem can be optimally solved in $O(n^2)$ time. Figure 9.47(d) shows the graph described above for the problem in Figure 9.47(c).

9.5.4 Hybrid HVH-VHV Router

In [PZ87], Pitchumani and Zhang developed a three-layer channel router that combines both a HVH and a VHV model based on the idea of partitioning

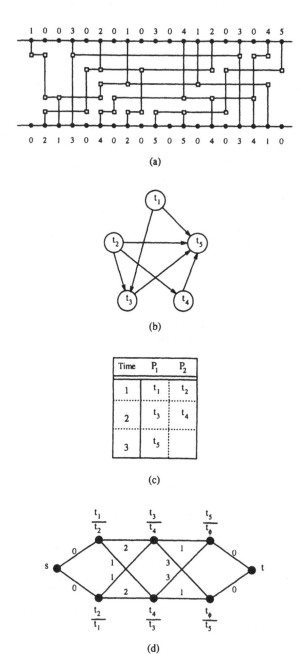

Figure 9.47: (a) A two-layer solution. (b) Track ordering graph. (c) An optimal scheduling solution.

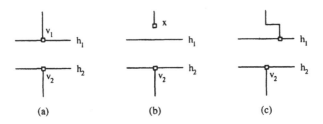

Figure 9.48: Local rerouting.

the channel. In this approach the channel can be thought of as two separate channels, not necessarily of the same size. One portion (upper or lower) is routed using the VHV and the other portion is routed using the HVH model. A transition track is usually needed between the two portions. The algorithm does not allow any dogleg. One of the important feature of this algorithm is that the pure HVH and VHV can be treated as special cases of this hybrid approach. This is due to the fact that, in the extreme case, one of the portions may constitute the whole channel and the other may be nonexistent. Obviously, no transition track is needed in this case. As a consequence, the result of this approach is the best between pure HVH and VHV approaches.

The height of a channel depends on two parameters, v_{max} and h_{max}. If $v_{max} >> h_{max}$, then VHV is best suited for that channel, since it use only h_{max} number of tracks. On the other, if $h_{max} >> v_{max}$, then HVH is best suited for that channel, since it use only $\frac{h_{max}}{2}$ number of tracks. Figure 9.49(a) and (b) shows the two cases when HVH and VHV routing models generate optimal solution. s and t are two dummy nodes signifying top and bottom of the channel. In practice, many channel routing problems are in fact a combination of both HVH and VHV, as shown in Figure 9.49(c).

The hybrid algorithm partitions the given netlist into two netlist, such that each netlist is best suited for either VHV or HVH style of routing. It then routes them separately thus obtaining two sub-solutions. The algorithm then inserts transition tracks to complete the connections between the two routed sub-solutions. The hybrid algorithm consists of the following steps:

1. choose VHV-HVH or HVH-VHV model;

2. partition the set of nets into two sets; the HVH-set, the set of nets to be routed in the HVH portion and the VHV-set, those to be routed in VHV portion of the channel;

3. route the nets.

The key problem is the partitioning of the channel. It is important to note that not all partitions are routable in the hybrid scheme. It is easy to show that for a net in a partition, all the nets that have vertical constraints with this net must also be in the same partition for a valid routable partition. Based on this

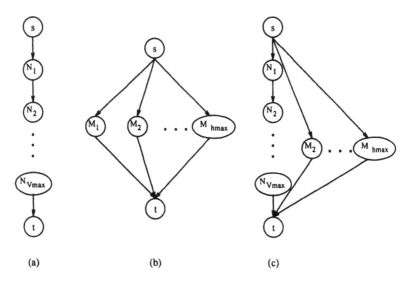

Figure 9.49: (a) CRP suited for VHV. (b) CRP suited for HVH. (c) CRP not suited for either HVH or VHV.

observation, any routable partition can be represented by a cut of the VCG with s and t on opposite sides of the cut, with all the nodes above the cut in upper-set and all the nodes below the cut in lower-set. Figure 9.50 shows an example of a cut inducing a routable partition; upper-set is $\{n_1, \ldots, n_{20}\}$ and lower-set is $\{m_1, \ldots, m_k\}$. The details of the partitioning algorithm based on the weighted cost function may be found in [PZ87].

To illustrate hybrid routing, we use the same example as in [PZ87]. Figure 9.51(a) gives the netlists, while Figure 9.51(b) shows a hybrid routing with VHV (two tracks) in the upper region and HVH (three tracks) for the lower region. The routing uses six tracks (including the transition track), while pure VHV requires eight tracks and pure HVH uses seven. In some cases, the terminal connections of nets may be such that vertical runs that cross the boundary between the regions can change layers on one of the regular tracks; in such cases, the transition track may be removed as shown in Figure 9.51.

9.6   Multi-Layer Channel Routing Algorithms

As the VLSI technology improves, more layers are available for routing. As a result, there is a need for developing multilayer routing algorithms. It should be noted that many standard cell designs can be completed without channel areas by using over-the-cell techniques (see Chapter 8). It may be noted that many over-the-cell routing problems are similar to channel routing problems. In case of full custom, perhaps four layers would be sufficient to obtain layouts without any routing areas on the real estate. However, new technologies such

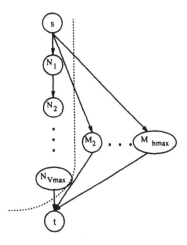

Figure 9.50: Partitioning for hybrid routing.

as MCM requires true multilayer capabilities since as many as 64 layers may be used.

In [Ham85], Hambrusch presented an algorithm for a n-layer channel router. The number of layers, the channel width, the amount of overlap and the number of contact points are four important factors for routing multi-terminal nets in multi-layer channels. An insight into the relationship between these four factors is also presented in [Ham85].

In [BBD+86], Braun developed a multi-layer channel router called *Chameleon*. Chameleon is based on YACR2. The main feature of Chameleon is that it uses a general approach for multilayer channel routing. Stacked vias can be included or excluded, and separate design rules for each layer can be specified. The Chameleon consists of two stages: a partitioner and a detailed router. The partitioner divides the problem into two and three-layer subproblems such that the global channel area is minimized. The detailed router then implements the connections using generalizations of the algorithms employed in YACR2.

In [ED86], Enbody and Du presented two algorithms for n-layer channel routing that guarantee successful routing of the channel for n greater that 3.

9.7 Switchbox Routing Algorithms

A switchbox is a generalization of a channel and it has terminals on all four sides. Switchbox routing problem is more difficult than a channel routing problem, because the main objective of channel routing is to minimize the channel height, whereas the objective of switchbox routing is to ensure that all the net are routed.

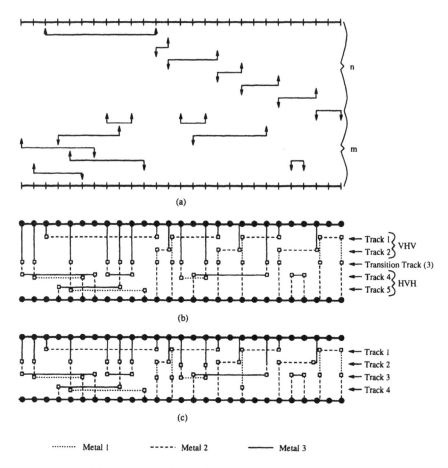

Figure 9.51: Example of hybrid routing.

9.7.1   Classification of switchbox routing algorithms

Switchbox routers can be classified as,

1. Greedy Routers,

2. Rip up and Reroute Routers and

3. Others.

Greedy routers are essentially extension of the Greedy channel router. Rip up and Reroute routers employ some algorithm for finding routes for nets and modifying the routes to accommodate additional nets. Several other techniques, such as computational geometry, simulated evolution have also been applied to switchbox routing. In this section, we review one algorithm from

each category. In this section we describe four different algorithms for switchbox routing. BEAVER [CH88] is an excellent router based on computational
geometry.

9.7.2   Greedy Router

In [Luk85], Luk presented a greedy switchbox router, which is an extension
of the greedy channel router [RF82]. As opposed to a channel, which is open on
the left and the right side, a switchbox is closed from all four sides. Moreover,
there are terminals on the left and right boundaries of a switchbox. Thus in
addition to the terminals on the upper and lower boundaries, the presence of
terminals on the left and right boundaries of the switchbox need to be considered while routing. This is achieved by the following heuristic:

1. **Bring in left terminals:**   The terminals on the left boundary are
   brought to the first column as horizontal tracks.

2. **Jog to right terminals:**   The step in greedy channel router in which
   the nets are jogged to their nearest top or bottom terminal is modified
   for the nets that have a terminal on the right boundary. Such nets are
   jogged to their *target rows*, in addition to jogging to the next top or
   bottom terminal. Jogging a net to the next top or bottom terminal is
   referred to as $JOG_{t/b}$, whereas, jogging a net to its target row is referred
   to as JOG_r. A target row of a net is a row of its terminal on the right
   boundary. The nets are jogged to their target rows according to the
   following priority.

   (a) First choice is a net whose right side of the target row and the
       vertical track between the net and target row is empty.

   (b) Second choice is a net whose right side of the target row is empty. In
       addition, the priority is also based on how close a net can be jogged
       to its target row.

   (c) Third choice is a net that can be brought closer to its target row.

Ties are resolved by giving higher priority to a net which is further from its
target row. The cyclic conditions at this step can be broken by allowing a net
to cross its target row. (see Figure 9.52.) Note that the optimal way to reach
the target row for a net is to jog once as each jogging wastes a vertical track,
too many joggings may result in running out of tracks. Excessive jogging can
be avoided by allowing a net to jog to its target row only if it can be brought
to or beyond half way between its initial position and the target row.

Several schemes of using $JOG_{t/b}$ and JOG_r have been suggested for a net
having terminals on the right and top/bottom boundaries.

1. **Scheme 1:** For a net that has a terminal on the right boundary JOG_r is
   performed until it reaches its target row. The top and bottom terminals
   are connected by branching out some net segments (see Figure 9.53(a)).

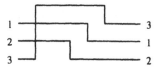

Figure 9.52: Cyclic constraints.

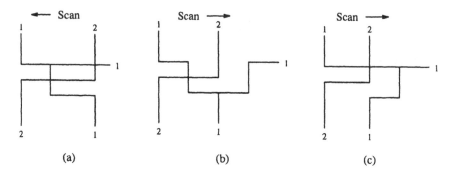

Figure 9.53: Routing schemes: (a) scheme 1,(b) scheme 2,(c) scheme 3

2. **Scheme 2:** $JOG_{t/b}$ is used until all top and bottom terminals are connected. JOG_r is used from the column where the last top/bottom terminal appears (see Figure 9.53(b)).

3. **Scheme 3:** In this scheme $JOG_{t/b}$ and JOG_r are used in parallel. Branching out is avoided by either using $JOG_{t/b}$ or JOG_r at each column (see Figure 9.53(c)).

4. **Scheme 4:** This scheme involves a combination of several schemes. If the rightmost top/bottom terminal of a net is in the rightmost $p\%$ of the switchbox, scheme 1 is used for that net. Otherwise, scheme 2 is used (see Figure 9.54).

Determining Scan Direction: Determining scan direction is equivalent to assigning *left* edge to one of the edges of the switchbox. Let $p_1 = (e_1, e_3)$ and $p_2 = (e_2, e_4)$ be the opposite pairs of edges of the switchbox. The objective of this step is to assign one of these pairs as the left-right pair and the other as the top-bottom pair. This operation is divided into two steps. In the first step, the top-bottom and the left-right pairs are assigned without identifying left or right edges in the left-right pair. In the second step the left and the right edges are identified.

The following measures are defined to achieve the first step. The *augmented density* of p_1 is defined as the overall minimum number of tracks required to maintain the connectivity of the terminals on opposite pair of edges in p_1 as in

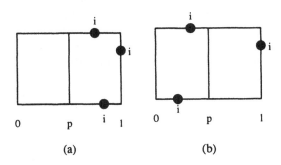

Figure 9.54: Routing scheme 4 (a) JOG_r for net N_i, (b) $JOG_{t/b}$ for net N_i

Algorithm GREEDY-SB-ROUTER
begin
    Determine the scan direction;
    Bring left terminals into column 1;
    for i = 1 to M **do**
        if no empty track exists **then**
            increase number of tracks;
        bring $T(i)$ and $B(i)$ to empty tracks;
        join split nets as much as possible;
        for each net with no right terminals **do**
            bring split nets closer by jogging;
        for each net with a right terminal **do**
            use scheme 4;
        if close to right edge **then**
            fanout to all target rows;
    while split net exist **do**
        join split nets as much as possible;
end.

Figure 9.55: Algorithm GREEDY-SB-ROUTER

the channel routing problem, plus the tracks required to connect the nets on the edges in p_2 with the nets on p_1. The *augmented channel density ratio* for $p_1 = (e_1, e_3)$ is defined as the ratio of the number of tracks available between e_1 and e_3 to the augmented density of p_1. In a similar manner the augmented channel density of p_2 can be defined. As the property of the greedy heuristic is to minimize the number of tracks perpendicular to the scan direction, it is obvious to assign the left-right pair to the pair that has smaller augmented channel density ratio.

Let $p_k = (e_i, e_j)$ be the left-right edge pair assigned in the last step. The edges in p_k are assigned the left and the right edges based on the following rules:

1. An edge in p_k which has more terminals and especially multiple terminals is selected as left edge. This reduces the burden on JOG_r and fanout operations.

2. An edge in p_k which is close to less congested region is selected as right edge. As a result, more free vertical tracks will be available to join split nets and fanout to target terminals.

The formal description of the GREEDY-SB-ROUTER appears in Figure 9.55.

9.7.3   Rip-up and Re-route Based Router

Shin and Sangiovanni-Vincentelli [SSV86] proposed a switchbox router based on an incremental routing strategy. It employs maze-running but has an additional feature of modifying already-routed nets. Some nets are even ripped-up and re-routed. It is this feature of MIGHTY that makes it suitable for channel and switch box routing. The cost function used for maze routing penalizes long paths and those requiring excessive vias. MIGHTY consists of two entities: a *path-finder* and a *path-conformer*. It is possible for the router to go into a loop in the modification phase. This can be avoided by using some sort of time-out mechanism. The overview of Algorithm MIGHTY is in Figure 9.56.

The worst case time complexity of the algorithm is more than $O(k^3 pnL)$, where p and k are the number of terminals and nets, respectively and L is the complexity of the maze routing algorithm.

9.7.4   Computational Geometry Based Router

In [CH88], Cohoon and Heck presented a switchbox routing algorithm called BEAVER, based on a delayed layering scheme with computational geometry techniques. The main objectives of BEAVER are the via and wire length minimization. BEAVER is an unreserved layer model routing algorithm. While routing a net, BEAVER delays the layer assignment as long as possible. One of the important features of BEAVER is that it uses priority queue to determine the order in which nets are interconnected. An overview of BEAVER is given in Figure 9.57.

Algorithm MIGHTY
begin
    1. Extend all pins on the boundaries of the region inside
           by one unit;
    $L \to \phi$; (* Initialize list *)
    2. (* path finder *)
    for each net **do**
        MAZE-ROUTE(net, L);
    3. sort L in increasing value of costs;
    4. **while** $L \neq \phi$ **do**
        Get next path p from L;
        if no grid cell in p is occupied **then**
            Implement p; goto step 5;
            else invoke the path-finder to find a new feasible
                minimum path connecting two unconnected
                subnets of the net;
            Let δ be the increase in cost for the new path p';
        if $\delta < MAXINCREASE$ **then**
            Implement p'; goto step 5;
            (* weak modification *)
            Push implemented nets around to
            obtain a 'good' connection for the given net;
            if weak modification fails **then**
                (* strong modification *)
                Remove an existing connection and
                try to obtain 'good' connection;
    5. Remove p from L;
end.

Figure 9.56: Algorithm MIGHTY

It can be seen from the algorithm that BEAVER uses up to three methods to find interconnections for nets: 1) *corner router*, 2) *line sweep router*, and 3) *thread router*.

Corner router: The corner router tries to connect terminals that form a *corner connection*. Such a connection is formed by two terminals if (1) they belong to the same net, (2) they lie on the adjacent sides of the switchbox, and (3) there are no terminals belonging to the net that lie between them on the adjacent sides. The corner router is also a preferred router since it is the fastest and because its connections tend to be part of the minimum rectilinear steiner tree for the nets.

During the initialization of the corner priority queue, each net is checked to see if it has a corner connection. A corner connection can be simply made by

Algorithm BEAVER
begin
    Initialize control information;
    Initialize corner-priority-queue;
    corner route;
    if there are nets to be routed **then**
        Initialize line-sweep-priority-queue;
        Line sweep route;
        if there are nets to be routed **then**
            Relax control constraints;
            Reinitialize line-sweep-priority-queue;
            Line sweep route;
            if there are nets to be routed **then**
                Initialize thread-priority-queue;
                Thread route;
    Perform layer assignment;
end.

Figure 9.57: Algorithm BEAVER

examining the control of the terminals that comprise the corner. If the section
of the control overlap, then the corner can be realized. Nets with one or two
corners need no further checks. However, straightforward connection of four
corner nets can introduce cycles. There are two types of cycles as shown in
Figure 9.58.

An overlap cycle is removed by routing only three of the corners. A four-
terminal cycle can be removed by routing only three of the corners. When the
corner router has to decide upon one of two such corners to route, the one with
least impact on the routability of other nets if preferred.

Linesweep router: The line sweep router is invoked when no more corner
connections can be made. However, if after current net's linesweep realization,
some other corner connection become realizable, then the linesweep router is
temporarily suspended until the corner priority queue gets emptied. For each
net five possibilities are considered: a wire with single bend, a single straight
wire, a dogleg connection with a unit-length cross piece, three wires arranged
as a horseshoe, and three wires in a stair arrangement. These are shown in
Figure 9.59.

To reduce the number of vias, straight line wires are preferred to dogleg
connections, dogleg to single-bend connections and single bend to two-bend
connections. In looking for its connections, the linesweep router uses the computational geometry technique of plane sweeping. One approach is to use scan-
lines to find straight line connections between disjoint subnets of the net in
question. It works in $O(n \log n + k)$ time. BEAVER uses three scan lines that

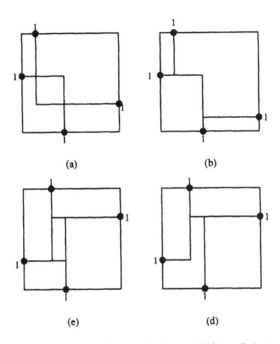

Figure 9.58: (a)Example of an Overlap Cycle and (b) its Solution (c) Example of a Four-Terminal Cycle and (d) its Solution

sweep the plane in tandem: one across the column, one across the row and a third to detect doglegs. Also, it employs bounding function to reduce the computational complexity. If some nets still remain unrealized, the control of existing nets is reduced and the process is repeated a second time. All remaining nets are then routed by a thread router.

Thread router: This router is invoked very sparingly. It is a maze-type router that seeks to find minimal length connections to realize the remaining nets. Since, this router does not restrict its connection to any preferred form, it will find a connection if one exists. Whenever a net N_i consists of more than one routable subnets, a maze expansion, on the lines of Soukup router is initiated. The expansion starts from a small subnet s that has not been used in an endeavor to minimize the wire-length and number of vias.

Layer Assignment: This phase primarily aims at minimizing the number of additional vias introduced. Since, at this stage, all grid points that any wire passes through are known, it is possible to optimally assign unlayered wires to a particular layer. BEAVER can also very easily extend this to achieve metal-maximization. A simple set of heuristics, based on coloring the grid points as red or black is presented in [CH88].

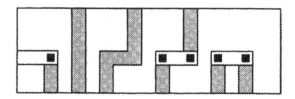

Figure 9.59: Prototype Linesweep Connections

Router	Vias	Wire length
GRS	36	529
MIGHTY	32	530
BEAVER	28	529

Table 9.3: Results on augmented dense switchbox.

9.7.5   Comparison of Switchbox Routers

Table 9.3 compares performance of the switchbox routers discussed in this chapter. It is apparent from the Table 9.3 that BEAVER routes the augmented dense switchbox with minimum number of vias, whereas, there is no significant difference in the wire lengths.

9.8   Summary

Detailed routing is one of the most fundamental steps in VLSI physical design cycle. The detailed routing problem is solved by routing the channels and switchboxes. Channel routing and switchbox routing problems have been studied extensively and several routing algorithms have been proposed by many researchers.

Routing results may differ based on the selection of routing models. A routing model can be grid-based where wires follow paths along the grid lines or gridless, where wires can be placed at any place as long as the design rules are not violated. Another model is be based on the layer assignments of different net segments. In reserved layer model, segments are allowed only to a particular layer(s). Most of the existing routers use reserved layer model. A model is unreserved if any segment is allowed to be placed in any layer.

The most widely considered objective function for routing a channel is the minimization of channel density. Other objectives are to minimize the length of routing nets and to minimize the number of vias. A routing algorithm must take into consideration the following: net types, net width, via restrictions, boundary types, number of layer available, degree of criticality of nets.

The main objective of (channel) routing is to minimize the total routing area. Most successful routers are simple in the approach. Greedy is one of the most efficient and easiest to implement channel routing algorithm. As the number of layers increase, the routing area used decreases. However, it should be noted that adding a layer is usually very expensive. In most of the cases, the routing area can virtually be eliminated using four layers by using advanced over-the-cell and over-the-block routing techniques discussed in [SBP95].

The objective of switchbox routing is to determine the routability. Several switchbox routing algorithms have been developed. Greedy and rip-up and reroute strategies have lead to successful routers.

9.9 Exercises

†1. Develop a river routing algorithm for a simple channel when the channel height is fixed. Note that a simple channel is the one with straight line boundaries. Given a channel routing problem, the router should first check if the given problem can be routed for a given channel height. In case it cannot be routed, the router should stop, otherwise the router should find a solution.

2. Prove that every single row routing problem is routable if no restrictions are placed on the number of doglegs and street congestions.

3. Given the net list in part (c) of Figure 9.61, find a single row routing such that the street congestion on both the streets is less than or equal to 3.

4. For the net list in Figure 9.61, does there exist a solution for which the between-node congestion is no more than 1 ?

†5. Prove Theorem 9.

†6. Prove Theorem 10. Extend this theorem to multi-terminal single row routing problems.

‡7. Prove Theorem 11. Does there exist a necessary and sufficient condition for SRRP with at most k ($k \geq 1$) doglegs per net ?

‡8. Does there exist a necessary and sufficient condition for SRRP for $C_B = 1$?

‡9. Consider the two row routing problem given in Figure 9.60. Given two rows of terminals separated by w tracks, the objective is to find a single layer routing with minimum congestions in the upper and lower streets. Note that the number of tracks in the middle street is fixed.

10. Give an instance of a channel routing problem in a two layer restricted (HV) model in which,

   (a) The channel height is greater than v_{max}.

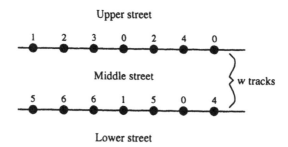

Figure 9.60: A two-row routing problem.

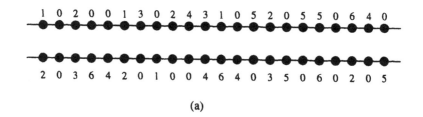

(a)

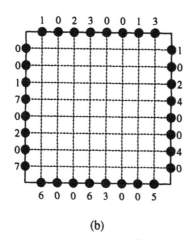

(b)

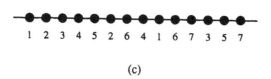

(c)

Figure 9.61: Routing problems.

(b) The channel height is greater than d_{max}.

(c) The channel height is greater than both v_{max} and d_{max}.

11. Give an instance of channel routing problem in two layer Manhattan model in which there are cyclic constraints.

12. Prove with an example that it is possible to get better results in channel routing by using an unrestricted model than the restricted model for a two layer channel routing problem using the Manhattan model.

13. Give an example of a channel routing problem for which the greedy router performs better than the hierarchical router.

14. Give an example of a channel routing problem for which the hierarchical router performs better than the greedy router.

15. Route the channel given in Figure 9.61(a) using the following routers and compare their results:

    (a) LEA.

    (b) Dogleg router.

    (c) Y-K algorithm.

    (d) Greedy channel router.

    (e) Hierarchical channel router.

    (f) YACR2.

16. In the greedy router, while joining split nets in a column, more than one nets can be joined. Formulate the problem of finding maximum number of nets that can be joined in a single column. Develop an $O(n \log n)$ time complexity algorithm for this problem, where n is the total number of tracks in the channel.

17. In the greedy channel router, while a column is being routed, segments of split nets are brought closer by using doglegs in case they cannot be joined. Develop an efficient strategy that will maximize the number of nets that can be brought closer.

18. Give an instance of channel routing problem in three-layer HVH model in which,

    (a) The channel height is greater than v_{max}.

    (b) The channel height is greater than $\frac{d_{max}}{2}$.

    (c) The channel height is greater than both $\frac{d_{max}}{2}$ and v_{max}.

19. Route the channel routing problem given in Figure 9.61(a) using extended net merge algorithm by Chen and Liu.

20. Show that in the hybrid HVH-VHV routing, the lower bound for the channel density is $d_{max}/2$, where d_{max} is the size of the maximum clique in the corresponding HCG.

21. Show that given a CRP, if $v_{max} = 1$ in the corresponding VCG, then every partition is routable in hybrid routing.

22. Develop a unreserved layer switchbox router when the terminals are located in any metal layer.

23. Develop a algorithms for routing in a three dimensional routing grid with planer upper surface and non-planer lower surface. Assume that the terminals are located at the lower surface.

24. Solve exercise 23 when the grid has non-planer upper surface.

Bibliographic Notes

The general single-layer routing problem was shown to NP-complete in [Ric84]. In [BP83, DKS+87, LP83, SD81, Tom81], several restricted single-layer routing problems have been solved optimally. River routing was defined in [DKS+87] and refined by Leiserson and Pinter [LP83]. Several extensions of the general river routing algorithm have been proposed [JP89, TH90]. In [JP89], river routing algorithm is extended to handle multiple, parallel channels. The river routing is called *feed-through river routing*, because wires must pass through gaps that are to be created between the components in each row. In [TH90], Tuan and Hakimi presented a variation of river routing that minimizes the number of jogs.

Tsukiyama et. al [TKS82], considered the restricted version of the via assignment problem where no net has more than one point in any given column. In [TK78], another restricted version in which each net is forced to use vias from the same via column is considered. Both [TKS82] and [TK78] have shown that, with their respective restrictions, deciding whether k via columns are sufficient to realize the netlist is an NP-Hard problem.

The layering problem was shown to be NP-Hard by Sahni, Bhatt, and Raghavan [SBR80]. As a consequence, heuristic algorithms have been studied for this problem. In practice $\max\{C_{ls}, C_{us}\} \leq 2$ and therefore heuristic algorithms for the restricted layering problem, in which $C_{ls} = C_{us} = 2$ has been considered by several researchers [GKG84, HS84a, TKS82]. The algorithm given in [TKS82] generates solutions with number of layers l, where, $l \leq 1.33 \times l^*$, where l^* is the optimal number of layers. In [HS84a], Han and Sahni presented two fast algorithms for layering. These algorithms consider one net at a time starting from the leftmost net. A net is assigned to the first layer in which the channel capacity allows its insertion. If this net cannot be assigned to any layer, a new layer is started. It was reported that these algorithms perform better than the algorithm proposed in [TKS82], that is, they use less layers. It was also reported that these algorithms are much faster than the one proposed in [TKS82]. Recently, Gonzalez et. al [GKG84] reported some results on layering; it is not, however, clear how their algorithm compares with earlier algorithms. It appears to be similar to the algorithm given in [HS84a].

Tsui and Smith [TI81] gave another formulation of single row routing problem. They considered only two terminal nets and obtained some necessary and sufficient conditions for the routability of a net list if upper and lower street capacities are known. Their idea is based on the number of blockages that a net would encounter in case all nets are laid out in the same street.

Han and Sahni [HS84a] proposed linear time algorithms for the special case of SRRP when the number of tracks available for routing is restricted to one, two or three. In 1983 they extended their work and presented simpler algorithms than that of [HS84b]. They introduced the notion of *incoming permutation* being the relative ordering of nets with respect to a certain terminal. The algorithm makes a left to right scan on the terminals. If a new net is starting at the terminal under consideration this net is inserted in all the incoming permutations to get a set of new permutations. The permutations which do not meet the street congestion requirement are deleted from further consideration. Thus, the main idea of the algorithm in [HS84a, HS84b] is to keep track of all legal permutations. Obviously, this algorithm is only practical for small values of upper and lower street capacities because of its exponential nature.

Raghavan and Sahni [RS84] investigated the complexity issues of single row routing problem and the decomposition process for the multi-layer circuit board problem. In [RS84] it is shown that the via assignment problem considered by [TK78] remains NP-Hard even for $k = 2$. They also prove that the problem of via column permutation is NP-Hard and remains so even if 2 via columns are allowed per net for decomposing. It is also shown that the problem of minimizing the total number of vias used is NP-Hard. The problem of finding a layout with minimum bends (doglegs) is also proved to be NP-Hard [RS84].

In [HS71], Hashimoto and Stevens first introduced the channel routing problem. Another column-by-column router has been proposed by Kawamoto and Kajitani [KK79] that guarantees routing with upper bound on the number of tracks equal to the density plus one, but additional columns are needed to complete the routing. Ho, Iyenger and Zhenq developed a simple but efficient heuristic channel routing algorithm [HIZ91]. The algorithm is greedy in nature and can be generalized to switchboxes and multi-layer problems.

Some other notable effort to solve switchbox problem have been reported by Hamachi and Ousterhout (Detour) [HO84]. This approach is an extension of the greedy approach for channel routing. In [Joo86], Joobbani proposed a knowledge based expert system called WEAVER for channel and switchbox routing. In [LHT89], Lin, Hsu, and Tsai presented a switchbox router based on the principle of evolution. In [GH89], Gerez and Herrmann presented a switchbox router called PACKER. This router is based on a stepwise reshaping technique. WEAVER [Joo86] is an elaborate rule-based expert system router that often produces excellent quality routing at the cost of excessive computation time. SILK [LHT89] a switchbox router based on the simulated evolution technique. A survey and comparison of switchbox routers has been presented by Marek-Sadowska [Sad92]. In [CPH94] Cho, Pyo, and Heath presented a parallel algorithm, using a conflict resolving method has been developed for the switchbox routing problem in a parallel processing environment.

Chapter 10

Over-the-Cell Routing and Via Minimization

The current trend in VLSI design is to develop high performance chips. The main objective of physical design is satisfy the performance needs while minimizing the die size. Historically, the gate delays limited the chip performance. The developments in fabrication process technology in the past two decades have resulted in a phenomenal decrease in feature sizes, and introduced additional metal layers for interconnections(routing). Deep sub-micron processes with five to seven metal layers for interconnections are now available for design of high performance and high density chips. The number of devices in a chip have increased from about a thousand devices in the early 70's to over twenty million devices now. The increase in the number of devices has led to a significant increase in number of interconnections. Interconnect delays, which were considered to be insignificant earlier, have now become comparable, if not more prominent than the gate delays.

With the availability of five to seven metal layers for interconnections, three dimensional routing techniques are necessary to satisfy the performance and density goals. The number of metal layers provide the third dimension. Therefore, for interconnect planning, routing *volume* needs to be considered, instead of just the routing *area*. The space in all metal layers across the entire die, both over active areas and in channels need to efficiently utilized for routing. This concept was first introduced in standard cell designs. Several existing channel routers can produce solutions only one or two tracks beyond optimum for most channels. Despite this fact, as much as 10% of the area in a typical layout was still consumed by routing. Considering a fixed placement, in view of this 'optimality' of channel routers, further reduction is only possible if some nets can be routed 'outside' the channel (as the area allocated for the standard cells is inherently fixed by the circuit design). In particular, the metal layers available over the cell rows can be used for the routing. This technique is called *over-the-cell* routing. The over-the-cell routing style for standard cell designs has become both practical and important as more and more metal layers are made available for routing.

Over-the-cell routing concept is used across the entire chip and it is not feasible to design complex high performance microprocessor chips (100Mhz to 1GHz) without adopting over-the-cell routing techniques. This book describes basic OTC routing algorithms for standard cell designs and advanced concepts can be found in [SBP95].

After a chip is completely routed, the layout is functionally complete and can be sent for fabrication. However, the layout is usually improved to reduce the possibility of fabrication errors, reduce the total chip area and therefore, improve performance. In most current technologies, two or more layers are available for routing. Most of the existing routing algorithms use a large number of vias to complete the routing. This is due to the fact that most routers use a reserved layer model. However, vias are undesirable from fabrication as well as circuit performance point of view and therefore, the number of vias should be kept as small as possible.

Significant volume of research exists on techniques for reduction of the number of vias in a completed detailed routing by re-assigning the wire segments to different layers. This kind of via minimization is called *Constrained Via Minimization* (CVM). Via minimization has also been considered without the restriction of completed routing. In this approach, the actual layout of wires can be changed and thus offers more flexibility as compared to the CVM approach. This via minimization approach is called *Unconstrained Via Minimization* (UVM) or *Topological Via Minimization* (TVM).

In this chapter, we discuss the problem of over-the-cell routing and via minimization to improve detailed routing solutions. In Section 10.1, we discuss the problem of over-the-cell routing. Both CVM and UVM problems have been considered in Section 10.2.

10.1 Over-the-cell Routing

The total layout area in the standard cell design style is equal to the sum of the total cell area and the total channel area. For a given layout, the total cell area is fixed. Thus, the total area of a layout can only be reduced by decreasing the total channel area. As several channel routers have been developed that complete channel routing with the number of tracks very close to the channel density, further improvement in the layout area is impossible if routing is done only in channels.

Internal routing of cells is typically completed using one metal layer. Therefore, the higher metal layers (M2 and M3) over-the-cell are un-utilized. The area in M2 and M3 can be utilized for routing of nets in order to reduce the channel height. As the number of layers allowed for routing increases, the over-the-cell routing problem becomes important. Since the conventional channel routing problem is known to be NP-hard [Szy85], and the over-the-cell channel routing problem is a generalization of the conventional channel routing problem, it is easy to see that the over-the-cell channel routing problem is also

NP-hard [GN87].

Several algorithms for over-the-cell routing have been presented, and the technique has proven to be very effective [CL88, CPL93, HSS93, LPHL91]. In the following, a review of algorithms for over-the-cell channel routing is presented. We start by describing the physical constraints for over-the-cell routing.

10.1.1 Cell Models

Based on the locations of the terminals there are four major classes of cell models : Boundary Terminal Models (BTM), and the Center Terminal Models (CTM), the Middle Terminal Model (MTM) and the Target Based Cell Model (TBC). Each of these classes contain several cell models based on the variations in other routing parameters, *i.e.*, the number of metal layers and permissibility of vias in over-the-cell areas.

- **Boundary Terminal Model(BTM):** This is the traditional cell model. This was introduced when only two metal layers were available for routing. In BTM, there are two parallel horizontal diffusion rows, one for the P-type transistors and the other for N-type transistors. The first metal layer (M1) is used to complete connections which are internal to the cells. The power and ground rails are in M2 layer, adjacent to each other, in the center of the cell row. Terminal rows are available in all layers and are located on the boundaries of the cells [HSS93]. This leaves a rectangular, over-the-cell routing area for each terminal row of the standard cells. The number of tracks available for over-the-cell routing is determined by the height of these rectangular areas and may vary depending on the cell library used. The entire over-the-cell area may be used for routing in the third metal (M3) layer. This model is used by most existing over-the-cell routers [CPL93, HSS93, HSS91]. This class of cell models is referred to as BTM or class of Boundary Terminal Models. (See Figure 10.1(a)). BTM contains, 2BTM (2 layer process), 3BTM-V (3 layer process when vias are not allowed in over-the-cell areas), and 3BTM+V (3 layer process when vias are allowed in over-the-cell areas).

- **Center Terminal Model(CTM):** This class of cell models is quite different than BTM in terms of terminal location. In CTM, the terminals are located in M2, in the middle of the cell. The power and ground rails are in M1 near the top and bottom cell boundaries respectively. Connections within the cell are completed in M1. Thus, M2 is only blocked by terminals, and M3 is completely unblocked (See Figure 10.1(b)). Over-the-cell routers may use two rectangular regions (about thirteen tracks wide) in M2 and M3.

- **Middle Terminal Model(MTM:** This model differs from the BTM and CTM in terms of terminal locations. In MTM, the terminals are located in two rows, one row is located k_1 tracks below the upper cell

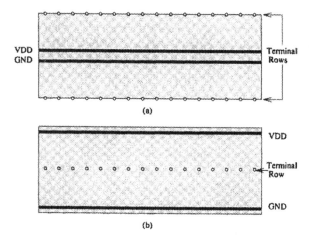

Figure 10.1: Existing Cell Models (a) BTM (b) CTM

boundary and another is located k_3 tracks above the lower cell boundary. As in CTM, in MTM, terminals are available only in M2 and the power and ground rails are in M1 near the top and bottom cell boundaries respectively (See Figure 10.2). Both terminals in a column of a cell are equi-potential. Intra-cell routing is completed in poly and M1, and does not block M2. As opposed to two rectangular regions in CTM, over-the-cell routers for MTM may use three rectangular regions in M2 and M3 as discussed below:

1. **T area:** k_1 track wide area between the upper cell boundary and the upper terminal row,

2. **C area:** k_2 track wide area between the lower terminal row and the upper terminal row,

3. **B area:** k_3 track wide area between the lower terminal row and the lower cell boundary.

- **Target Based Cell (TBC):** TBC is designed to effectively utilize the over-the-cell areas for routing. The terminals are in the form of long vertical strips in M1 layer, called *targets*. The exact location of the interconnection contacts on the targets is determined by the routing algorithm. The power and ground lines are located in M1 layer at the top and bottom cell boundaries, respectively. The TBC cells have targets of non-uniform heights and are placed arbitrarily, as shown in Figure 10.3. Since the power and ground lines and the targets are located in M1 layer, the over-the-cell areas in M2 and M3 areas are completely unblocked.

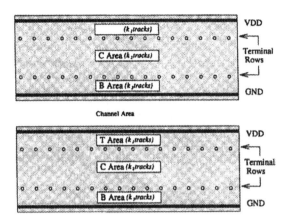

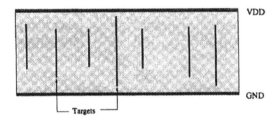

Figure 10.2: Middle Terminal Model (MTM)

Figure 10.3: Target Based Cell (TBC)

10.1.2 Two-Layer Over-the-Cell Routers

The two-layer routing problem essentially boils down to selection of two planar sets of segments. One of them is routed in the upper over-the-cell area and the other is routed in the lower over-the-cell region. The nets that are not selected are routed in the channel area. In the following, we discuss two algorithms for over-the-cell routing.

10.1.2.1 Basic OTC Routing Algorithm

In [CL90], Cong and Liu presented an algorithm for the over-the-cell channel routing. It divides the problem into the following three steps:

1. routing over the cells,

2. choosing net segments in the channel, and

Column: 1 2 3 4 5 6 7 8 9 10 11

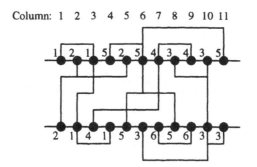

Figure 10.4: A valid over-the-cell routing solution.

3. routing in the channel.

The first step is formulated in a very natural way as the problem of finding a maximum independent set of a circle graph. Since the later problem
can be solved in quadratic time optimally, an efficient optimal algorithm is obtained for the first step. Also, the second step is formulated as the problem of
finding a minimum density spanning forest of a graph. The minimum density
spanning forest problem is shown to be NP-hard, so, an efficient heuristic algorithm is presented which produces very satisfactory results. A greedy channel
router [RF82] is used for the third step.

There are two routing layers in the channel, and there is a single routing
layer over-the-cells for inter-cell connections. Clearly, the over-the-cell routing
must be planar.

The first step of the over-the-cell channel routing problem is to connect terminals on each side of the channel using over-the-cell routing area on that side.
The same procedure is carried out for each side (upper or lower) of the channel
independently. Let t_{ij} denote the terminal of net N_i at column j. In a given
planar routing on one side of the channel, a hyperterminal of a net is defined to
be a maximal set of terminals which are connected by wires in the over-the-cell
routing area on that side. For example, for the terminals in the upper side of
the channel in Figure 10.4, $\{t_{5,4}, t_{5,6}, t_{5,11}\}$ is a hyperterminal of net 5. $\{t_{2,2}\}$
is also a hyperterminal. Obviously, when the routing within the channel step
(the third step) is to be done, all the hyperterminals of a net need to be connected instead of connecting all the terminals of the net, because the terminals
in each hyperterminal have already been connected in the over-the-cell routing
area. Intuitively, the fewer hyperterminals are obtained after routing over the
cells, the simpler the subsequent channel routing problem. Thus the first step
of the problem can be formulated as routing a row of terminals using a single
routing layer on one side of the row such that the number of hyperterminals is
minimum.

After the completion of the over-the-cell routing step, the second step is to
choose net segments to connect the hyperterminals that belong to the same

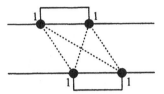

Figure 10.5: Possible net segments for connecting two hyperterminals.

net. A net segment is a set of two terminals of the same net that belong to two different hyperterminals. For example, for the two hyperterminals of net 1 on the opposite sides of the channel in Figure 10.5, there are four possible net segments that can be used to connect these two hyperterminals (indicated by dashed edges), while only one of them is needed to complete the connection. Thus the second step of the problem is to choose net segments to connect all the hyperterminals of each net such that the resulting channel density is minimum.

After the net segments for all the nets are chosen, the terminals specified by the selected net segments are connected using the routing area in the channel. The problem is now reduced to the conventional two-layer channel routing problem. A greedy channel router [RF82] is used for this step. Other two-layer channel routers may also be used.

Net Selection for OTC Routing: The first step of the over-the-cell channel routing problem is to route a row of terminals using a single routing layer on one side of the channel such that the resulting number of hyperterminals is minimized. This problem is called the multi-terminal single-layer one-sided routing problem (MSOP).

MSOP can be solved by a dynamic programming method in $O(c^3)$ time, where c is the total number of columns in the channel. Given an instance I of MSOP, let $I(i, j)$ denote the instance resulting from restricting I to the interval $[i, j]$. Let $S(i, j)$ denote the set of all the possible routing solutions for $I(i, j)$. Let:

$$M(i,j) = \max_{S \in \mathcal{S}(i,j)} \left\{ \sum_{k \geq 2} (k-1) d_k(S) \right\}$$

where $d_k(S)$ is the number of hyperterminals of degree k in S. If there is no terminal at column i, clearly, $M(i, j) = M(i + 1, j)$. Otherwise, assume that the terminal at column i belongs to net n. Let $x_{n_1}, x_{n_2}, \ldots, x_{n_s}$ be the column indices of other terminals that belong to net n in interval (i, j). Then, it is easy to verify that

$$M(i,j) = \max(i + 1, j), \max_{1 \leq l \leq s} \left\{ M(i + 1, n_l) + M(n_l, j) \right\}$$

It is easy to see that this recurrence relation leads to an $O(c^3)$ time dynamic programming solution to MSOP.

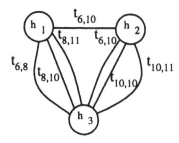

Figure 10.6: The connected component induced by net.

Channel Segment Selection: After the over-the-cell routing, a set of hyperterminals is obtained. The terminals in each hyperterminal are connected together by over-the-cell connections. The next problem is to choose a set of net segments to connect all the hyperterminals of each net such that the channel density is minimized. This problem can be transformed to a special spanning forest problem, as discussed below.

For an instance I of the net segment selection problem, the connection graph $G = (V, E)$ is defined to be a weighted multi-graph. Each node in V represents a hyperterminal. Let h_1 and h_2 be two hyperterminals that belong to the same net N_i. For every terminal t_{ij} in h_1 and for every terminal t_{ik} in h_2 there is a corresponding edge (h_1, h_2) in E, and the weight of this edge $w((h_1, h_2))$ is the interval $[j, k]$ (assume that $j \leq k$, otherwise, it will be $[k, j]$). Clearly, if h_1 contains p_1 terminals and h_2 contains p_2 terminals, then there are $p_1 \times p_2$ parallel edges connecting h_1 and h_2 in G. Furthermore, corresponding to each net in I there is a connected component in G.

For example, the connected component corresponding to net 3 in the example in Figure 10.4 is shown in Figure 10.6. Given an instance I of the net segment selection problem, since all the hyperterminals in the same net are to be connected together for every net in I, it is necessary to find a spanning forest of $CG(I)$. Moreover, since the objective is to minimize the channel density, the density of the set of intervals associated with the edges in the spanning forest must be minimized.

Therefore, the net segment selection problem can be formulated as Minimum Density Spanning Forest Problem (MDSFP). Given a weighted connection graph $G = (V, E)$ and an integer D, determine a subset of edges $E' \subseteq E$ that form a spanning forest of G, and the density of the interval set $\{w(e) | e \in E'\}$ is no more than D.

In [CL90], it was shown that this problem is computationally hard.

Theorem 17 *The minimum density spanning forest problem is NP-complete.*

In view of NP-completeness of the MDSFP, an efficient heuristic algorithm has been developed for solving the net segment selection problem [CL90]. The

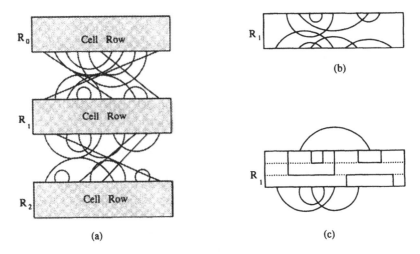

Figure 10.7: An example of the TRMPS problem

heuristic algorithm works as follows. Given an instance I of the net segment selection problem, a connection graph $G = (V, E)$ is constructed. For each edge $e \in E$, the relative density of e, called $RD(e)$, is defined to be $d(e)/d(E)$, where $d(e)$ is the density of the set of intervals which intersect with the interval $w(e)$, and $d(E)$ is the density of the interval set $\{w(e)|e \in E\}$. The relative density of an edge measures the degree of congestion over the interval associated with the edge. The algorithm repeatedly removes edges from E until a spanning forest is obtained.

10.1.2.2 Planar Over-the-Cell Routing

In [DMPS94] Danda, Madhwapathy, Panyam and Sherwani presented an algorithm to select a maximum planar subset of nets in M2 in BTM standard cell designs.

Figure 10.7(b) shows the set of nets that are suitable for routing over the cell row R_1. In a HCVC (Horizontally Connected Vertically Connected) model [CPL93], the main problem in two layer over the cell routing is to select a maximum planar subset of nets which are suitable for routing in a single layer, available over the cell rows. The remaining connections are completed in the channel. Authors call this problem as the Two Row Maximum Planar Subset (TRMPS) problem. Figure 10.7(c) shows the maximum planar subset of nets, that can be routed over R_1, which is an optimal solution, for the instance of the TRMPS problem, shown in Figure 10.7(b). Notice, that the tracks are shared between the top row nets and the bottom row nets, so as to efficiently utilize the over-the-cell area.

The TRMPS problem, is formally defined as follows. Given two rows of terminals $\mathcal{T} = \{t_1, t_2, \ldots, t_L\}$ and $\mathcal{B} = \{b_1, b_2, \ldots, b_L\}$ and two sets of nets

$\mathcal{N}_T = \{(t_i, t_j) \mid t_i, t_j \in \mathcal{T}\}$ and $\mathcal{N}_B = \{(b_i, b_j) \mid l_i, l_j \in \mathcal{B}\}$, where $|\mathcal{N}_T \cup \mathcal{N}_B| = n$, and k tracks between the two rows, find the maximum planar subset $\mathcal{N}_P \subseteq \mathcal{N}_T \cup \mathcal{N}_B$ of the two sets in k tracks. Authors presented a dynamic programming approach to solve this problem.

Let L denote the total number of columns in a cell row, numbered from left to right. In BTM-HCVC, the terminals are located at the intersection points of the upper or the lower horizontal boundaries of a cell row and the vertical columns. If a terminal is not used by any net, then that terminal is called a *vacant terminal*. If both the upper and lower terminals of a column are vacant, then that column is called a *vacant abutment*. The total number of tracks available in the OTC area of a cell row, for routing, is denoted by k (cell height), and the tracks are numbered from top to bottom. Then, an instance of the TRMPS problem can be formally represented as a 7-tuple $\mathcal{I} = (\mathcal{T}, \mathcal{B}, \mathcal{N}_T, \mathcal{N}_B, k, n, L)$. An instance of the TRMPS problem is called as a *Canonical Instance*, if there are no vacant abutments in that instance. If n is the number of nets in a canonical instance $\mathcal{I}$, then the number of columns (L), can be at most $2n$. This is because, in the worst case, each column has at most one vacant terminal, either in the top or the bottom terminal row.

Canonical instances with two terminal nets are considered as input to the problem. A net is denoted by a pair of terminals. A net (t_i, t_j), where $1 \leq i, j \leq L$, is called a top net. Similarly, a net (b_i, b_j), where $1 \leq i, j \leq L$, is called a bottom net. *span* of a two terminal net is defined as the absolute difference between the column numbers on which the terminals of the net are located. For example, the span of the net $N_\alpha = (t_i, t_j)$, is given by,

$$span(N_\alpha) = \mid i - j \mid$$

A region $\mathcal{R}_m$ of a cell row is defined as a rectangular region of the cell row, containing the columns in the range $[1, m]$, where $1 \leq m \leq L$. A net (t_i, t_j) (or (b_i, b_j)), is said to be *completely contained* in the region $\mathcal{R}_m$, if $1 \leq i, j \leq m$.

Let $T(j)$ denotes the optimal TRMPS solution in a rectangular region $\mathcal{R}_j$. The $T(j)$ solution is computed for all j, $1 \leq j \leq L$, using a dynamic programming technique. Finally, the $T(L)$ solution gives the optimal solution, for a given instance $\mathcal{I}$ of the TRMPS problem. In order to compute the $T(j)$ solution, the region $\mathcal{R}_j$ is partitioned into two or three sub regions. depending on the existence of top nets and bottom nets, completely contained in $\mathcal{R}_j$, with one of their terminals at column j, as shown in Figure 10.8.

Let $N_\alpha = (t_i, t_j)$ be the only net with a terminal at column j, and which is completely contained in $\mathcal{R}_j$. In this case, $\mathcal{R}_j$ is divided into an L-shaped region R, and a rectangular region r which consists of a single row of terminals (Figure 10.8(a)). The optimal $T(j)$ solution, may or may not contain N_α. If N_α is included, then the $T(j)$ is summation of the optimal solutions in the L-shaped region R and the rectangular region r, and the net N_α itself. If N_α is not included, then the $T(j)$ solution is the same as the $T(j-1)$ solution. The maximum of the above two solutions, is taken as the optimal $T(j)$ solution.

Let $N_\alpha = (t_i, t_j)$ and $N_\beta = (b_m, b_j)$ be the nets with terminals at column j, and which are completely contained in $\mathcal{R}_j$. Then, the optimal $T(j)$ solution

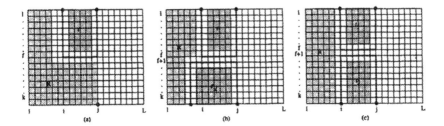

Figure 10.8: Schematic Overview of the Algorithm ALGO-TRMPS

may include

1. **None of the nets N_α and N_β:** In this case, the $T(j)$ solution is the
   same as the $T(j-1)$ solution.

2. **Only the net N_α:** In this case, the $T(j)$ solution can be computed as
   shown in Figure 10.8(a).

3. **Only the net N_β:** In this case also, the $T(j)$ solution can be computed
   as shown in Figure 10.8(a).

4. **Both the nets N_α and N_β:** In this case, if $i \neq m$, then $\mathcal{R}_j$ is partitioned
   into an L-shaped region R, and two rectangular regions r_1 and r_2, which
   consist of a single row of terminals (Figure 10.8(b)). If $i = m$, then $\mathcal{R}_j$
   is partitioned into a rectangular region R, which consists of two rows of
   terminals, and two rectangular regions r_1 and r_2, which consist of a single
   row of terminals (Figure 10.8(c)). Then, the $T(j)$ is simply summation
   of the optimal solutions in the regions R, r_1 and r_2, and the nets N_α and
   N_β.

The optimal $T(j)$ solution, is the maximum among all the above four solutions.
    From the above discussion, it is clear that, the single row solutions and the
solutions in the L-shaped regions need to be computed, before computing the
two row solutions. Our algorithm consists of the following three phases.

1. In the first phase, single row solutions of the terminal rows T and B,
   are computed individually. Each single row solution of a terminal row,
   is an (i, j, t) solution, where $1 \leq i, j \leq L$ and $1 \leq t \leq k$. These solutions are denoted as $S_t(i, j, t)$ and $S_b(i, j, t)$ for top and bottom terminals
   respectively.

2. In this phase, the maximum two row planar subset $(T(j))$ for the given
   terminal rows is computed, where $1 \leq j \leq L$ by using a dynamic programming approach. Here, the $S_t(i, j, t)$ and $S_b(i, j, t)$ solutions, computed in
   the first phase will be used. As described above, finding the $T(j)$ solution
   also involves finding the maximum planar subset in L-shaped regions.

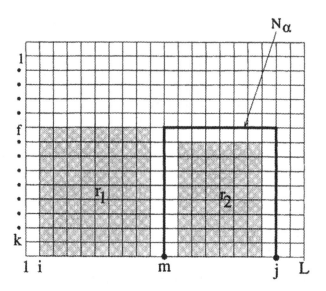

Figure 10.9: Single Row Maximum Independent Set

3. The solution obtained in phase 2, gives the number of nets in the optimal solution for a given instance of the TRMPS problem. In this phase, the actual planar subset of nets in the optimal solution, is determined by backtracking.

From a routing perspective, this problem is equivalent to assigning the maximum number of intervals to k tracks such that, if interval (i, j) is assigned to track f, then no interval assigned to tracks $1, 2, \ldots, f - 1$ should intersect columns i and j. Let $MIS(i, j, f)$ denote the solution of the OFPR problem resulting from restricting the intervals to be in the range of $[i, j]$ and allowing f tracks for routing, where $1 \leq i, j \leq L$ and $1 \leq f \leq k$. The (i, j, f) solution is computed using dynamic programming. Notice that, computation of $MIS(i, j, f)$ can be any of the following cases.

1. If j is vacant, then

$$MIS(i, j, f) = MIS(i, j - 1, f)$$

2. There exists a net N_α with terminals j and m but $m \notin [i, j)$. Then,

$$MIS(i, j, f) = MIS(i, j - 1, f) \text{ if } m \notin [i, j)$$

3. There exists a net N_α with terminals j and m such that $m \in [i, j)$, then the following two cases are possible:

(a) Excluding the net N_α in the solution leads to

$$MIS(i, j, f) = MIS(i, j - 1, f)$$

(b) Including the net N_α in the solution results in

$$MIS(i, j, f) = MIS(i, m - 1, f) + MIS(m + 1, j - 1, f - 1) + 1$$

As shown in Figure 10.9, if $m \in [i, j)$, one has to check if including N_α will lead to a better solution or not. Therefore,

$$MIS(i, j, f) = \quad \max \quad \{MIS(i, j - 1, f), MIS(i, m - 1, f) + \\ MIS(m + 1, j - 1, f - 1) + 1 \text{ if } j' \in [i, j)\}$$

The complexity of this algorithm is given by the following theorem, stated in [CPL93].

Theorem 18 *[CPL93] The two-terminal net OFPR problem can be solved in $O(kn^2)$ time, where n is the number of nets and k is the number of available tracks.*

Using the above algorithm the maximum k-planar subsets S_t and S_b are computed, for the top and bottom terminal rows respectively, and all the intermediate solutions are stored.

Since, computing the $T(j)$ solution, involves computing the solutions in L-shaped regions, let us discuss a scheme to represent an L-shaped region.

Figure 10.10 shows two types of L shaped regions. For instance, an L-shaped region shown in Figure 10.10(a), is denoted by the 3-tuple (i, j, f), where

1. i is the column number of the terminal t_i, which is the rightmost corner of the L-shaped region, in the top terminal row.

2. j is the column number of the terminal b_j, which is the rightmost corner of the L-shaped region, in the bottom terminal row.

3. f is the track, that forms part of the horizontal boundary of the L-shaped region (See Figure 10.10(a)).

The maximum planar subset in the L-shaped region, shown in Figure 10.10(a), is denoted by $L(i, j, f)$. Following the same convention described above, the inverted L-shaped region, shown in Figure 10.10(b) is denoted by (j, i, f), and the solution in this region is denoted by $L(j, i, f)$. The method of computing solutions in L-shaped regions will be described later.

While computing the $T(j)$ solution in the rectangular region $\mathcal{R}_j$, the algorithm deals with the following three cases.

Case 1: There exists a top net $N_\alpha = (t_i, t_j)$, which is completely contained in $\mathcal{R}_j$(Figure 10.11(a)).

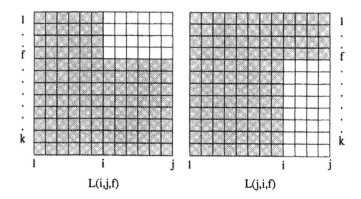

$$L(i,j,f) \qquad\qquad L(j,i,f)$$

Figure 10.10: L-shaped regions

Case 2: There exists a bottom net $N_\beta = (b_i, b_j)$, which is completely contained in $\mathcal{R}_j$(Figure 10.11(b)).

Case 3: There exists a top net $N_\alpha = (t_l, t_j)$, and a bottom net $N_\beta = (b_m, b_j)$, which are completely contained in $\mathcal{R}_j$(Figure 10.11(c)).

Let us consider each of the above listed cases in detail.

Case 1: Depending on whether the net N_α is in the optimal $T(j)$ solution, or not, the algorithm has to deal with the following sub-cases.

Case 1(a): Excluding the net N_α leads to

$$T(j) = T(j - 1)$$

Case 1(b): If the net N_α is included, such that, it is assigned to a track $f, 1 \le f \le k$, then the following solution, which is denote by $T'(j)$.

$$T'(j) = S_t(i + 1, j - 1, f - 1) + 1 + L(i - 1, j - 1, f + 1)$$

By considering all possible track assignments, the track to which N_α can be assigned is found, so as to maximize the $T(j)$ solution. Then, the $T(j)$ solution obtained by choosing N_α, which is denoted as $T''(j)$, is given by,

$$T''(j) = \max_{f=1}^{k}\{T'(j)\}$$

The optimal $T(j)$ solution will then be the maximum of the two solutions obtained by including and excluding the net N_α. Therefore,

$$T(j) = \max\{T(j - 1), T''(j)\}$$

Case 2: This is symmetric to Case 1.

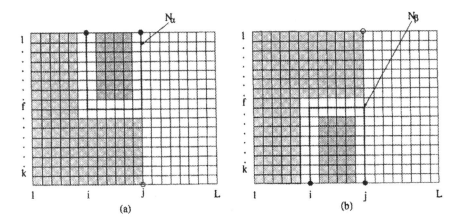

Figure 10.11: Cases 1 and 2 in ALGO-TRMPS

Case 3: Here, the following three sub-cases are possible as shown in Figure 10.12.

Case 3(a): $Span(N_\alpha) > Span(N_\beta)$

Case 3(b): $Span(N_\alpha) < Span(N_\beta)$

Case 3(c): $Span(N_\alpha) = Span(N_\beta)$

For each of the above three sub-cases, the following four solutions are computed.

$W_0(j)$: Two row solution of $\mathcal{R}_j$, which does not consist of the nets N_α and N_β

$W_1(j)$: Two row solution of $\mathcal{R}_j$, which consists of only the net N_α

$W_2(j)$: Two row solution of $\mathcal{R}_j$, which consists of only the net N_β

$W_{12}(j)$: Two row solution of $\mathcal{R}_j$, which consists of both the nets N_α and N_β

The maximum of W_0, W_1, W_2 and W_{12} solutions is the optimal $T(j)$ solution. If both the nets N_α and N_β are included in the optimal solution $T(j)$, then a simple observation, regarding the track assignment of the nets N_α and N_β, is stated in the following lemma.

Lemma 1 *If $N_\alpha = (t_l, t_j)$ and $N_\beta = (b_m, b_j)$, are two nets, which are completely contained in $\mathcal{R}_j$, and the optimal $W_{12}(j)$ solution has the net N_α in track f_1, and N_β in track f_2 such that $1 \le f_1 < f_2 \le k$, then,*

   1. if $span(N_\alpha) > span(N_\beta)$, then, the solution in which, the net N_β is assigned to a track $f_1 + 1$ is also an optimal $W_{12}(j)$ solution.

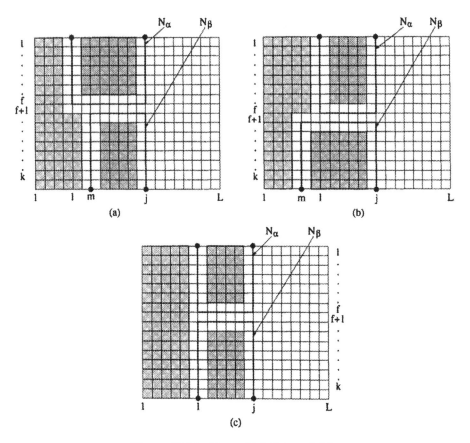

Figure 10.12: Cases 3(a),(b) and (c)

2. *if $span(N_\alpha) < span(N_\beta)$, then, the solution in which, the net N_α is assigned to a track $f_2 - 1$ is also an optimal $W_{12}(j)$ solution.*

3. *if $span(N_\alpha) = span(N_\beta)$, then, the solution in which, the net N_β is assigned to a track $f_1 + 1$, and the solution in which, the net N_α is assigned to a track $f_2 - 1$ are also optimal $W_{12}(j)$ solutions.*

Let us now consider the three sub-cases listed above, in detail.

Case 3(a): In this case, since $span(N_\alpha) > span(N_\beta)$, column l is to the left of column m (Figure 10.12(a)). The $W_0(j)$ solution, in which both the nets are excluded is given by,

$$W_0(j) = T(j - 1)$$

The $W_1(j)$ solution can be computed as follows. Suppose, the net N_α is assigned to track f, $1 \le f \le k$, then, the following solution, which is called as $W_1'(j)$.

$$W_1'(j) = S_t(l+1, j-1, f-1) + 1 + L(l-1, j-1, f+1)$$

By trying all possible track assignments, the track to which N_α can be assigned is found, so as to maximize the $W_1(j)$ solution. The $W_1(j)$ solution is given by,

$$W_1(j) = \max_{f=1}^{k}\{W_1'(j)\}$$

The $W_2(j)$ solution can be computed in a similar manner as $W_1(j)$.

The $W_{12}(j)$ solution can be computed as follows. From lemma 1, it is clear that in the optimal $W_{12}(j)$ solution, the nets N_α and N_β are assigned to adjacent tracks. Suppose, the net N_α is assigned to track f, and N_β in track $f+1$, then the following solution, which is called as W_{12}' is obtained.

$$\begin{aligned} W_{12}'(j) &= L(l-1, m-1, f+1) + S_t(l+1, j-1, f-1) \\ &+ S_b(m+1, j-1, f+2) + 2 \end{aligned}$$

The adjacent tracks, to which N_α and N_β can be assigned is found, so as to maximize $W_{12}(j)$.

$$W_{12} = \max_{f=1}^{k-1}\{W_{12}'(j)\}$$

Then, the optimal $T(j)$ solution will be the maximum of W_0, W_1, W_2 and W_{12} solutions. Therefore,

$$T(j) = \max\{W_0(j), W_1(j), W_2(j), W_{12}(j)\}$$

Case 3(b): This is symmetric to Case 3(a).

Case 3(c): In this case, $span(N_\alpha) = span(N_\beta)$. (Figure 10.12(c)). Here, the $W_0(j), W_1(j)$ and $W_2(j)$ solutions are the same as for Case 3(a) and Case 3(b) However, the W_{12} solution differs slightly. According to the Lemma 1, the nets N_α and N_β can be assigned to adjacent tracks (say f and $f+1$ respectively). Then the W_{12}' will be

$$\begin{aligned} W_{12}'(j) &= T(l-1) + S_t(l+1, j-1, f-1) \\ &+ S_b(l+1, j-1, f+2) + 2 \end{aligned}$$

By trying all possible track assignments, one can find two adjacent tracks, on which N_α and N_β can be placed so as to maximize the W_{12} solution. Therefore,

$$W_{12}(j) = \max_{f=1}^{k-1}\{W_{12}'(j)\}$$

Then the optimal solution is given by,

$$T(j) = \max\{W_0(j), W_1(j), W_2(j), W_{12}(j)\}$$

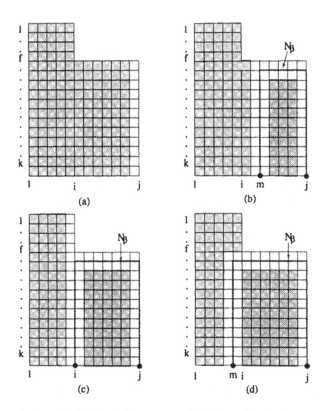

Figure 10.13: The four cases of L-shaped solutions

Authors have the following theorems on the time complexity and optimality of ALGO-TRMPS.

Theorem 19 *The time complexity of ALGO-TRMPS is $O(kn^2 \times f(k,n))$, where n is the number of nets, k is the number of tracks available on over-the-cell area and f(k,n) is the time to compute solution in each L-shaped region.*

Theorem 20 *Given an instance $\mathcal{I}$, of TRMPS problem, ALGO-TRMPS produces an optimal solution.*

In the following paragraphs a detailed description of computing the maximum planar subset in an L-shaped region is given.

The $L(i,j,f)$ solutions can be classified into the following four types depending on the existence of a bottom net which is completely contained in R_j, with b_j as one of its terminals.

Case 1: There is no bottom net, which is completely contained in R_j, with b_j as one of its terminals (Figure 10.13(a)). In this case

$$L(i,j,f) = L(i,j-1,f)$$

Case 2: There is a net $N_\beta = (b_m, b_j)$, which is completely contained in R_j, such that, $span(N_\beta) < (j - i)$, i.e., column m is to the right of column i, as shown in Figure 10.13(b). Excluding the net N_β leads to,

$$L(i, j, f) = L(i, j - 1, f)$$

Let us assume that, the $L(i, j, f)$ solution that includes the net N_β, is maximum, by assigning N_β to track f_1, such that $f_1 \geq f$. Also notice that the optimal $L(i, j, f)$ solution cannot consist of any other nets, that lie entirely in the L-shaped region, represented by (i, j, f), in the shaded area shown in Figure 10.13(b). If any such net exists, then the $L(i, j, f)$ solution, which includes the net N_β, would not be planar. Therefore the $L(i, j, f)$ solution remains maximum, even if N_β is assigned to track f_2, such that $f < f_2 < f_1$. Therefore, one can assign N_β to track $f + 1$. Now, the $L(i, j, f)$ solution, which includes N_β, consists of

1. the nets enclosed by N_β, which is the single sided solution $S_b(m + 1, j - 1, k - f - 1)$.

2. the net N_β itself, and

3. The solution of the L-shaped region, represented by $L(i-1, m-1, f)$.

The $L(i, j, f)$ solution that includes N_β, which is denoted as $L'(i, j, f)$ is given by

$$
\begin{aligned}
L'(i, j, f) &= S_b(m + 1, j - 1, f + 2) + 1 \\
&+ L(i, m - 1, f)
\end{aligned}
$$

The optimal $L(i, j, f)$ solution will be, the maximum of the solutions obtained by excluding and including the net N_β. Therefore,

$$L(i, j, f) = \max\{L(i, j - 1, f), L'(i, j, f)\}$$

Case 3: There is a net $N_\beta = (b_m, b_j)$, which is completely contained in R_j, such that, $span(N_\beta) = (j - i)$, i.e., column m and column i are the same, as shown in Figure 10.13(c). This is similar to the Case 1, except that, the $L(i, j, f)$ solution, which includes the net N_β, consists of the single row solution, in the region enclosed by N_β, the net N_β, and the two row solution $T(i - 1)$. Therefore, the $L(i, j, f)$ solution is given by,

$$
\begin{aligned}
L(i, j, f) = \max\{ &L(i, j - 1, f), \\
&S_b(i + 1, j - 1, f + 2) + 1 + T(i - 1)\}
\end{aligned}
$$

Case 4: There is a net $N_\beta = (b_m, b_j)$, which is completely contained in R_j, such that, $span(N_\beta) > (j - i)$, i.e., column m is to the left of column i, as shown in Figure 10.13(d). Excluding the net N_β leads to,

$$L(i, j, f) = L(i, j - 1, f)$$

Suppose the net N_β is assigned to track f_1, $f < f_1 \le k$, then The $L(i,j,f)$ solution, that includes the net N_β, in track f_1, $f < f_1 \le k$, denoted by $L'(i,j,f)$ is consists of

1. the nets enclosed by N_β, which is the single sided solution $S_b(m + 1, j - 1, k - f_1 - 1)$.

2. the net N_β itself, and

3. The solution of the L-shaped region, represented by $(i, m - 1, f)$.

Therefore the $L(i,j,f)$ solution, which includes N_β in track f_1 is given by

$$
\begin{aligned}
L'(i,j,f) &= S_b(m + 1, j - 1, k - f_1) + 1 \\
&+ L(i, m - 1, f)
\end{aligned}
$$

By varying f from $f + 1$ to k, one can find the track, to which N_β can be assigned, so as to maximize the $L(i,j,f)$ solution. Then, the $L(i,j,f)$ solution by choosing N_β, which is denoted $L''(i,j,f)$ is given by

$$
L''(i,j,f) = \max_{f_1 = f + 1}^{k} \{L'(i,j,f_1)\}
$$

The optimal $L(i,j,f)$ solution will be, the maximum of the solutions obtained by excluding and including the net N_β. Therefore

$$
L(i,j,f) = \max\{L(i, j - 1, f), L''(i,j,f)\}
$$

The solutions in an inverted L-shaped region (where $i > j$), can also be computed in a similar manner.

The computation of each $T(j)$ solution, involves the computation of solutions in several L-shaped regions. Therefore, the worst case running time of the algorithm ALGO-TRMPS, depends on the the number of L-shaped regions. The following lemma is on the number of L-shaped regions.

Lemma 2 *In canonical representation the number of L-shaped regions is $O(kn^2)$, where k is the number of tracks and n is the number of nets.*

Lemma 3 *Each $L(i,j,f)$ solution ,where $1 \le i,j \le L$ and $1 \le f \le k$, is computed once and it takes constant time to compute the solution.*

Theorem 21 *The computation time of ALGO-LMPS is $O(kn^2)$, where k is the number of tracks in a cell row, and n is the number of nets.*

Theorem 22 *Given an Instance $\mathcal{I}$, ALGO-LMPS produces an optimal solution.*

Algorithm $ALGO\text{-}TRMPS(\mathcal{N}_T, \mathcal{N}_B, n, \mathcal{T}, \mathcal{B}, \mathcal{N}, L)$
Begin
$Compute\_SRMPS_l()$;
$Compute\_SRMPS_u()$;
for j = 1 to L
    case(net_type(j)):
        Type 1: $TRMPS(j) = T1$;
        Type 2: $TRMPS(j) = T2$;
        Type 3: **case**(nets_at(j))
            type a: $TRMPS(j) = T3a$
            type b: $TRMPS(j) = T3b$
            type c: $TRMPS(j) = T3c$
End(For)
$\mathcal{N}_\mathcal{P} = \Phi$
for j = L to 1
    if(T(j-1) < T(j))
        if(type = 1) $\mathcal{N}_\mathcal{P} = \mathcal{N}_\mathcal{P} \cup N_\alpha$
        else if(type = 2) $\mathcal{N}_\mathcal{P} = \mathcal{N}_\mathcal{P} \cup N_\beta$
        else if(type = 3) $\mathcal{N}_\mathcal{P} = \mathcal{N}_\mathcal{P} \cup N_\alpha \cup N_\beta$
    End(if)
End(for)
End;

Figure 10.14: Algorithm ALGO-TRMPS

Theorem 23 *Given an instance $\mathcal{I}$, ALGO-TRMPS provides an optimal solution to the two row maximum planar subset problem.*

Theorem 24 *The complexity of the ALGO-TRMPS is $O(kn^2)$, where k is the number of tracks available over-the-cell area and n is the number of nets.*

Figure 10.14 presents the algorithm formally.

10.1.2.3 Over-the-Cell Routing Using Vacant Terminals

In [HSS93], Holmes, Sherwani and Sarrafzadeh presented a new algorithm called WISER, for over-the-cell channel routing. There are two key ideas in their approach: use of vacant terminals to increase the number of nets which can be routed over the cells, and near optimal selection of 'most suitable' nets for over the cell routing. Consider the example shown in Figure 10.15(a). Four tracks are necessary using a conventional channel router or an over-the-cell router. However, using the idea of vacant terminals, a two-track solution can be obtained (see Figure 10.15(b)). Furthermore, it is clear that the selection of nets which minimize the maximum clique, h_{max}, in horizontal constraint graph

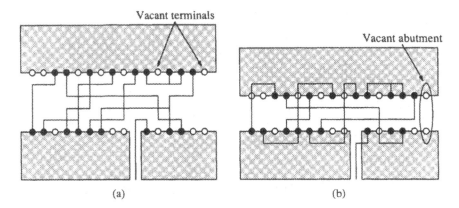

Figure 10.15: Effect of using vacant terminals in layout.

is not sufficient to minimize the channel height. For example, channel height
for the routing problem shown in Figure 10.15 is determined strictly by $v_{\max}$,
that is, longest path in the VCG (vertical constraint graph). Thus, the nets
which cause long paths in VCG should be considered for routing over the cells
to obtain a better over the cell routing solution.

An informal description of each of the six steps of algorithm is given below.

1. **Net Classification:** Each net is classified as one of three types which,
   intuitively, indicates the difficulty involved in routing this net over the
   cells.

2. **Vacant Terminal and Abutment Assignment:** Vacant terminals
   and abutments are assigned to each net depending on its type and weight.
   The weight of a net intuitively indicates the improvement in channel
   congestion possible if this net can be routed over the cells.

3. **Net Selection:** Among all the nets which are suitable for routing over
   the cells, a maximum weighted subset is selected, which can be routed in
   a single layer.

4. **Over-the-Cell Routing:** The selected nets are assigned exact geometric
   routes in the area over the cells.

5. **Channel Segment Assignment:** For multi-terminal nets, it is possible
   that some net segments are not routed over the cells, and therefore, must
   be routed in the channel. In this step, 'best' segments are selected for
   routing in the channel to complete the net connection.

6. **Channel Routing:** The segments selected in the previous step are
   routed in the channel using a greedy channel router.

The most important steps in algorithm WISER are net classification, vacant terminal and abutment assignment, and net selection. These steps are discussed in detail below. Channel segment assignment is done using an algorithm similar to the one presented in [CL90]. The channel routing completed by using a greedy channel router [RF82].

Vacant Terminals and Net Classification: The algorithm WISER was developed to take advantage of the physical characteristics indigenous to cell-based designs. One such property is the abundance of vacant terminals. A terminal is said to be *vacant* if it is not required for any net connection. Examination of benchmarks and industrial designs reveals that most standard cell designs have 50% to 80% vacant terminals depending on the given channel. A pair of vacant terminals with the same x-coordinate forms a *vacant abutment* (see Figure 10.15). In the average case, 30% - 70% of the columns in a given input channel are vacant abutments. The large number of vacant terminals and abutments in standard cell designs is due to the fact that each logical terminal (inputs and outputs) is provided on both sides of a standard cell but, in most cases, need only be connected on one side. It should be noted that the actual number of vacant terminals and abutments and their locations cannot be obtained until global routing is completed.

To effectively utilize the vacant terminals and abutments available in a channel, algorithm WISER categorizes nets according to the proximity of vacant terminals and abutments with respect to net terminals. Before classification, each k-terminal net N_i is decomposed into $k - 1$ two-terminal nets at adjacent terminal locations. Let $N_i = \{t_{b1}, t_{b4}, t_{t4}, t_{t6}\}$ be a four-terminal net. The notation t_{rx} is used to refer to the terminal on row r (top or bottom) at column x, Net N_i is decomposed into 3 two-terminal nets: $N_{i_1} = (t_{b1}, t_{b4})$, $N_{i_2} = (t_{b4}, t_{t4})$, and $N_{i_3} = (t_{t4}, t_{t6})$. Each two-terminal net $N_j = (t_{r_1x_1}, t_{r_2x_2})$ where $r_1, r_2 \in \{t, b\}$, $x_1, x_2 \in Z^+$, and $x_1 \leq x_2$ is then classified as a type I, type II or a type III net. The type of a net intuitively indicates the difficulty involved in routing that net over the cell rows. In other words, type III nets are hardest to route, while type I are easiest to route over the cells.

Definition 1 *Net $N_j = (t_{r_1x_1}, t_{r_2x_2})$ is a type I net if $r_1 = r_2$, and at least one of the terminals $t_{r_1x_1}$ and $t_{r_2x_2}$ is not vacant.*

Definition 2 *Net $N_j = (t_{r_1x_1}, t_{r_2x_2})$ is a type II net if the terminals $t_{r_1x_1}$ and $t_{r_2x_2}$ are both vacant.*

Definition 3 *Net $N_j = (t_{r_1x_1}, t_{r_2x_2})$ is a type III net if $r_1 \neq r_2$, neither $t_{r_1x_1}$ nor $t_{r_2x_2}$ is vacant, and there exists at least one vacant abutment a within the span of N_j, $x_1 < a < x_2$.*

The three net types are illustrated in Figure 10.16. A typical channel of a standard cell design contains about 44% type I nets, 41% type II nets, and 10% type III nets.

Observing that type I and type II nets constitute a majority of nets in the channel, one might suggest that it is sufficient to consider only these net types

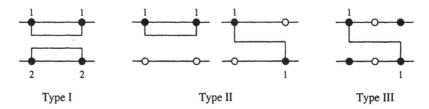

Figure 10.16: Net types.

when routing over the cells rows. However, this is not the case, since removing type III nets from the channel is critical in minimizing the length of the longest path ($v_{\max}$) in the vertical constraint graph.

The basic algorithm given in section 10.1.2.1 for over-the-cell channel routing attempts to minimize only the density due to horizontal constraint graph. Since channel height depends on both $h_{\max}$ and $v_{\max}$, it is clear that using $h_{\max}$ as the sole criterion for selecting nets is not as effective. In WISER, a net weighting function, $F_w : \mathcal{N} \rightarrow R^+$, which incorporates both the channel density and VCG path length criteria is used to assess the suitability of a given net for over-the-cell routing. The weight of a net $N_j = (t_{r_1 x_1}, t_{r_2 x_2})$ is computed based on the relative density of the channel in the interval $[x_1, x_2]$ and the ancestor and descendant weights of the net n. The relative density of net N_j can be computed by $r_d(N_j) = \frac{l_d(N_j)}{h_{\max}}$ where $l_d(N_j)$ is the maximum of the local densities at each terminal location t where $x_1 \leq t \leq x_2$. The ancestor weight of a net N_j, denoted by $a(N_j)$, is the length of the longest path from a node t in the vertical constraint graph with zero in-degree to the node N_j, and the descendant weight of N_j, denoted by $d(N_j)$, is the length of the longest path from N_j to a node s in VCG with zero out-degree. The general net weighting function is given below:

$$ F_w(N_j) \quad = \quad k_1 \frac{r_d(N_j)}{v_{\max}} + k_2 \frac{(a(N_j) + d(N_j) - \mid a(N_j) - d(N_j) \mid)}{h_{\max}} $$

where k_1 and k_2 are experimentally determined constants. Since the weight of a net N_j indicates the reduction possible in $h_{\max}$ and $v_{\max}$ if N_j is routed over the cell rows, the 'best' set of nets to route over the cells is one with maximum total weight.

Vacant Terminal and Abutment Assignment: After classification and weighting, nets are allocated a subset of vacant terminals or vacant abutments, depending on their type, to help define their routing paths in the area over the cell rows. It should be noted that type I nets, which have both of their terminals on the same boundary of the channel, can be routed in the area over the cells without using vacant terminals as shown in Figure 10.17. Therefore, the vacant terminal/abutment assignment problem is a matter of concern only for type II

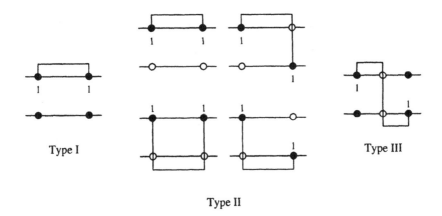

Type II

Figure 10.17: Routing of different net types.

and type III nets. For type II nets, the vacant terminals are 'reserved' for a net. That is, only a particular net may use a particular vacant terminal; as a result, vacant terminal assignment for type II nets is actually a net selection problem. On the other hand, a type III net may use any abutment within its span and therefore vacant abutment assignment problem for type III nets can be viewed as a matching problem.

Theorem 25 *The vacant terminal assignment problem for type II nets is NP-complete.*

Using theorem 25, it can be shown that the problem of finding an optimal routing using only k tracks in over-the-cell area is also NP-complete. However, if the value of k is restricted to one ($k = 1$), the problem is reduced to finding a maximum-weighted bipartite subgraph in an interval graph, which can be solved in polynomial time. The complexity of the problem for a fixed k (k being a small constant), however, for arbitrary k, the following result can be established.

Theorem 26 *The vacant abutment assignment problem for type III nets is NP-complete.*

Corollary 3 *The vacant terminal assignment problem for type II nets remains NP-complete when the number of tracks available over each cell row is restricted to k.*

In view of NP-completeness of the vacant abutment assignment problem for type III nets, a greedy heuristic is used. This heuristic is based on certain necessary conditions for the routability of a pair of type III nets. These necessary conditions are depicted in Figure 10.18. These necessary conditions basically check the planarity of pairs of nets. The formal description of the algorithm

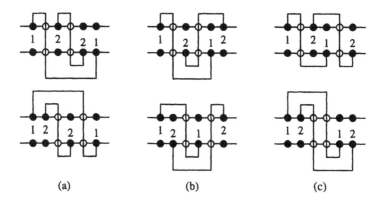

(a)   (b)   (c)

Figure 10.18: Necessary conditions.

is given in Figure 10.19. The main idea of this algorithm is to assign vacant abutments to nets according to their weight. The 'heaviest' nets are considered first. It is easy to see that the algorithm ASSIGN-ABUTMENTS produces a feasible solution in $O(dn^2)$ time.

Provably Good Algorithm for Net Selection: The net selection problem can be stated as follows. Given a set $\mathcal{N}$ of of nets, select a maximum-weighted subset of nets $\mathcal{N'} \subseteq \mathcal{N}$, such that all the nets in $\mathcal{N'}$ can be routed in the area over the cell rows in planar fashion. Algorithm WISER uses a graph theoretic approach to net selection. An overlap graph G_O is constructed for intervals of nets in set $\mathcal{N}$. It is easy to see that net selection problem reduces to the problem of finding a maximum-weighted bipartite subgraph $B_{\max}$ in the overlap graph G_O. However, the density of the nets in each partite set must be bounded by a constant k, which is the number of tracks available in the over the cell region. The problem of computing $B_{\max}$ is known to be NP-complete [SL89a]. As a result, a provably good algorithm is used for net selection. This algorithm is guaranteed to find a solution within 75% of the optimal.

Let P_t, P_b denote the partite sets of the graph B. The vertices of P_t correspond to nets which will be routed over the upper row of cells, and the vertices of P_b represent nets which will be routed over the lower row of cells. It is easy to see that there are several restrictions on assignment of vertices to partite sets. For example, a vertex corresponding to a type I net N_j, which has both of its terminals on the upper cell row, may not be assigned to partite set P_b because nets represented in P_b are routed over the lower row of cells. On the other hand, a vertex corresponding to a type I net N_j with terminals on the lower cell row may only belong to P_b. As noted earlier, vertices representing type II nets may be assigned to either partite set since these nets can be routed over either the upper or lower cell row. A type III net N_j is partitioned into two type I nets at the location of its designated abutment. Each of these nets

Algorithm ASSIGN-ABUTMENTS()
begin
(* Sort nets of $\mathcal{N}_3$ according to weight, forming netlist $\mathcal{NL}$ *)
$\quad\quad L = \text{Sort}(\mathcal{N}_3)$
(*$ASSIGN[j]$ stores the abutment assigned to net N_j, and
$\mathcal{A}(N_j)$ denotes the set of abutments within the span of net N_j *)

(* Initialize array $ASSIGN[]$ to zero *)
$\quad\quad$ **for** each $a \in \mathcal{A}$ **do**
$\quad\quad\quad\quad$ **for** each net $N_j = (t_{rx_1}, t_{rx_2})$ of $\mathcal{N}_3$ **do**
$\quad\quad\quad\quad\quad\quad$ **if** $(x_1 < a < x_2)$ **then** $\mathcal{A}(N_j) = \mathcal{A}(N_j) \cup N_j$;
$\quad\quad$ **for** $i = 1$ to $|\mathcal{N}_3|$ **do**
$\quad\quad\quad\quad$ **for** each $a \in \mathcal{A}(L[i])$ **do**
$\quad\quad\quad\quad\quad\quad$ **for** each N_j such that $ASSIGN[j] \neq 0$ **do**
$\quad\quad\quad\quad\quad\quad\quad\quad$ **if** $(\text{CND-OVLP}(L[i], a, N_j, ASSIGN[j]) = 1)$
$\quad\quad\quad\quad\quad\quad\quad\quad$ **then** $ASSIGN[i] = a$;
$\quad\quad\quad\quad\quad\quad\quad\quad\quad\quad$ break;
$\quad\quad\quad\quad\quad\quad\quad\quad$ **if** $(\text{CND1-CONT}(L[i], a, N_j, ASSIGN[j]) = 1)$
$\quad\quad\quad\quad\quad\quad\quad\quad$ **then** $ASSIGN[i] = a$;
$\quad\quad\quad\quad\quad\quad\quad\quad\quad\quad$ break;
$\quad\quad\quad\quad\quad\quad\quad\quad$ **if** $(\text{CND2-CONT}(L[i], a, N_j, ASSIGN[j]) = 1)$
$\quad\quad\quad\quad\quad\quad\quad\quad$ **then** $ASSIGN[i] = a$;
$\quad\quad\quad\quad\quad\quad\quad\quad\quad\quad$ break;
end.

Figure 10.19: Algorithm ASSIGN-ABUTMENTS.

is considered as a separate net in $\mathcal{N}$ and must be assigned to a fixed partite set as in the case of other type I net. The basic idea of the algorithm is similar to that of the algorithm MKIS in Chapter 3 and we call this algorithm FIS. The lower bound of the algorithm is 75% of the optimal solution. However, experimentally, the algorithm typically gives solutions which are at least 91% of the optimal result and in the average case, the performance of the algorithm is very close to the optimal solution (98% of the optimal solution).

Channel Segment Selection and Channel Routing: Channel segment selection is same as that discussed in [CL90]. When channel segment assignment is completed, a channel router is used to complete the connections within the channel. For this purpose, a greedy channel router is used, which typically achieves results at most one or two tracks beyond the channel density [RF82].

The formal description of algorithm WISER appears in Figure 10.20. On PRIMARY I benchmark from MCNC, WISER produces a solution with the total number of track equal to 206 as opposed to the solution with 187 tracks produced by the greedy channel router and 449 track solution produced by the earlier OTC router.

10.1.3   Three-Layer Over-the-cell Routing

Holmes, Sherwani, and Sarrafzadeh [HSS91] introduced two models for three-layer, over-the-cell channel routing in the standard cell design style. For each model, an effective algorithm is proposed. Both of the algorithms achieve dramatic reduction in channel height. In fact, the remaining channel height is normally negligible. The novelty of this approach lies in use of 'vacant' terminals for over-the-cell routing. For the entire PRIMARY 1 example, the router reduces the routing height by 76% as compared to a greedy 2-layer channel router. This leads to an overall reduction in chip height of 7%.

Wu, Holmes, Sherwani, and Sarrafzadeh [WHSS92] presented a three-layer over-the-cell router for the standard cell design style based on a new cell model (CTM) which assumes that terminals are located in the center of the cells in layer M2. In this approach, nets are first partitioned into two sets. The nets in the first set are called critical nets and are routed in the channel using direct vertical segments on the M2 layer, thereby partitioning the channel into several regions. The remaining nets are assigned terminal positions within their corresponding regions and are routed in a planar fashion on M2. This terminal assignment not only minimizes channel density but also eliminates vertical constraints and completely defines the channel to be routed. In the next step, two planar subsets of nets with maximum total size are found and they are routed on M3 over-the-cell rows. The rest of the nets are routed in the channel using a HVH router.

Terai, Nakajima, Takahashi and Sato [TTNS94] presented a new model for over-the-cell routing with three layers. The model consists of two channels and routing area over a cell row between them. The channel has three layers, whereas the over-the-cell area has two layers available for routing. An over-the-cell routing algorithm has been presented that considers over-the-cell routing problem as a channel routing problem with additional constraints.

Bhingarde, Panyam and Sherwani [BPS93] introduced a new three-layer model for, over-the-cell channel routing in standard cell design style. In this model the terminals are arranged in the middle of the upper and the lower half of the cell row. They develop an over-the-cell router, called MTM router, for this new cell model. This router is very general in nature and it not only works for two- and three- layer layouts but can also permit/restrict vias over-the-cell.

Bhingarde, Khawaja, Panyam and Sherwani [BKPS94] presented a hybrid greedy router for the TBC model. The routing algorithm consists of two key steps; *terminal position assignment* and *2-3-2 layer irregular boundary channel routing*. An optimal $O(KL)$ algorithm for terminal position selection is presented. The algorithm determines exact terminal locations on each target in the entire cell row. The routing environment for the TBC Router typically consists of a 3-layer channel area enclosed by two 2-layer non-uniform boundary over-the-cell routing regions. The TBC router generates smaller layouts for benchmarks, primarily due to smaller layout widths. For example,

Algorithm WISER()
begin
(* PHASE 1: Net Decomposition and Classification *)
    for each $N_j \in \mathcal{N}$ **do**
        $\mathcal{N}' = \mathcal{N}' \cup \{N_x = (t_{r_i}t_{r_{i+1}}) \mid r \in \{t, b\}, 1 \leq i \leq k$
        where i refers to the ith terminal of k-terminal net N_j }
    for each $N_j = (t_{r_1 x_1}, t_{r_2 x_2}) \in \mathcal{N}'$ **do**
        if $(r_1 = r_2)$ (* Type I nets *)
        then $\mathcal{N}_1 = \mathcal{N}_1 \cup \{N_j\}$
        if $(\bar{r_1}$ is vacant) and $(\bar{r_2}$ is vacant) (* Type II nets *)
        then $\mathcal{N}_2 = \mathcal{N}_2 \cup \{N_j\}$
        (* Type III nets *)
        if $(r_1 \neq r_2)$ and $(\bar{r_1}$ is not vacant) and $(\bar{r_2}$ is not vacant)
        then $\mathcal{N}_3 = \mathcal{N}_3 \cup \{N_j\}$
(* PHASE 2: Vacant Terminal/Abutment Assignment *)
    for each $N_j = (t_{r_1 x_1}, t_{r_2 x_2}) \in \mathcal{N}_2$ **do** (* Type II nets *)
        $\mathcal{V}(N_j) = \{t_{\bar{r_1} x_1}, t_{\bar{r_2} x_2}\}$
    for each $N_j = (t_{r_1 x_1}, t_{r_2 x_2}) \in \mathcal{N}_3$ **do** (* Type III nets *)
        $\mathcal{A}(N_j) = \{a \mid a \in \mathcal{A}, x_1 < a < x_2\}$
    ASSIGN= ASSIGN-ABUTMENTS($\mathcal{N}_3, \mathcal{A}$)
(* PHASE 3: Net Selection *)
    $\mathcal{N}_3' = \mathcal{N}_3; \mathcal{N}_3 = \emptyset$
    for each $N_j = (t_{r_1 x_1}, t_{r_2 x_2}) \in \mathcal{N}_3'$ **do**
        $N_{j1} = (t_{r_1 x_1}, t_{r_1 A[j]})$
        $N_{j2} = (t_{r_2 A[j]}, t_{r_2 x_2})$
        $\mathcal{N}_3 = \mathcal{N}_3 \cup \{N_{j1}, N_{j2}\}$
    $B = \text{FIS}(\mathcal{N}_1, \mathcal{N}_2, \mathcal{N}_3)$

(* PHASES 4, 5, and 6: Over-the-Cell Routing, Channel
Segment Assignment, and Channel Routing *)
    OVER-THE-CELL-ROUTE(B)
    $C = \text{ASSIGN-CHANNEL-SEGMENTS}(B, \mathcal{N})$
    CHANNEL-ROUTE(C)

end.

Figure 10.20: Algorithm WISER.

for PRIMARY I benchmark, all three models TBC, CTM and MTM, generate channelless layouts, however, TBC layout has minimum area due to smaller layout width.

10.1.4 Multilayer OTC Routing

With the advent of multi-layer processes more OTC area is now available for routing and hence, further reduction in the layout height can be accomplished. Bhingarde, Madhwapathy, Panyam and Sherwani [BMPS94] presented an efficient four layer OTC router, for a cell model similar to TBC, called Arbitrary Terminal Model(ATM). In this cell model, the terminals can be placed at any arbitrary locations in the cell. Freed from fixed terminal placement restrictions, cell designers can aim to design with minimum width. Figure 10.21 shows ATM based designs. The routing algorithm is based on the following four key steps; (1) The nets spanning multiple rows are decomposed into net segments belonging to single rows. All the terminals belonging to a single row are connected by a single horizontal metal segment, and a terminal is selected on each segment for completing the net connectivity. (2) Generation of intervals for same row and critical nets. (3) Interval assignment and same row routing and (4) Selection of an appropriate position for placing the same row and critical net intervals in each cell row. This approach was further generalized so that it can be used not only for different cell models, but also for full custom layouts and thin film MCM's.

10.1.5 Performance Driven Over-the-cell Routing

Despite the dramatic performance of OTC routers, a major shortcoming of the existing routers is the increase in the total wire length and the length of the longest net. Careful analysis of existing results shows that the total wire length may be increased by as much as 20% in [CPL93] and 35% in [HSS93]. Although no results on wire length are reported, it is very likely that the net length also increases in case of [LPHL91]. However, it is possible that the net length in [LPHL91] is less than the corresponding net lengths reported in [CPL93, HSS93]. This may be due to the fact that the main objective of their router is to minimize the number of routing tracks used in the over-the-cell area, as well as in the channel.

Natarajan, Holmes, Sherwani, and Sarrafzadeh [NSHS92] presented a three-layer over-the-cell channel routing algorithm (WILMA3) for high performance circuits. This router not only minimizes the channel height by using over-the-cell areas but also attempts to route all nets within their timing requirements. This algorithm is based on two ideas. Firstly, it optimizes the track assignment of each net with respect to delay. It identifies the track bound for each net which ensures that the wire length is no greater than the length of the net if routed in the channel. Using this track bound, nets are selected for over-the-cell routing. Secondly, 45° segments are used to route the nets over-the-cells to further reduce the net length.

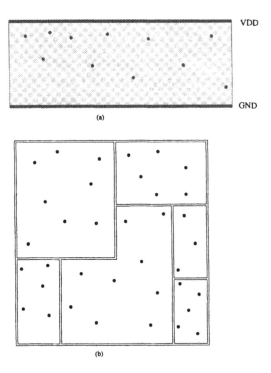

Figure 10.21: ATM based designs (a) Standard Cell (b) Full Custom

The basic idea of the algorithm is as follows: all the multi-terminal nets are decomposed into two-terminal nets and classified. Then weights are assigned to each net. The weight of a net intuitively indicates the improvement in channel congestion possible if this net can be routed over the cells. A channel router is then used to obtain the channel density (d_c) if routed in the channel. For each net N_i, track in which N_i is routed is recorded. An over-the-cell router is used to obtain the channel density (d_o) for over-the-cell routing. For each net N_i, the track bound k_i is computed, which ensures that if the net is routed over-the-cell at a track less than or equal to k_i, it will have a wire length less or equal to the net length when routed in the channel. This is based on the estimated channel heights d_c and d_o. Among all the nets which are suitable for routing over the cells, four (two) maximum-weighted planar subsets are selected, subject to the track bound constraint for the three-layer (two-layer) model. Once the nets are selected, a set of vacant terminals (vacant abutments) in the case of Type II (Type III) nets are assigned to each net N_i depending on its weight. These vacant terminal/abutment locations will later be used to determine an over-the-cell routing for N_i. Over-the-cell routing is done with $45°$ segments and rectilinear segments. In order to avoid design rule violations, any net N_i routed over-the-cell on track t_i must contain a vertical segment of length ρ_i before 45° segments can be used. The net segments that have not been routed in the area over the cells are routed in the channel. After the channel routing is done the channel density (d_θ) due to over-the-cell routing of nets is obtained. If $d_\theta > d_o$, d_o is set equal to d_θ and the process is repeated.

The iterative process mentioned above takes place very rarely as for most examples the algorithm can complete the routing using no more tracks than d_o. However, in cases when d_θ is indeed greater than d_o, it has been observed that it is usually one or at most two tracks.

10.2   Via Minimization

Vias are necessary to establish multi-layer connections. Many routers use a simple reserved layer model and produce routing solution with a large number of vias. However, there are numerous reasons for minimizing the number of vias in a layout:

1. In integrated circuit fabrication, the *yield* is inversely related to the number of vias. A chip with more vias has a smaller probability of being fabricated correctly.

2. Every via has an associated resistance which affects the circuit performance.

3. The size of the via is usually larger than the width of the wires. As a result, more vias lead to more routing space.

4. Completion rate of routing is also inversely related to the number of vias.

Despite all these reasons, existing routers and design tools consider the minimization of the number of tracks in channel routing, completion of switchbox routing, and wire length minimization as their primary objectives. Via minimization is either completely ignored or de-emphasized. As a result, via minimization came as an 'afterthought' in routing.

Before discussing the via minimization problem in detail, let us define some related concepts. A *plane homotopy* (also called a *sketch*) consists of a set of simple curves in the routing region. The two endpoints of a curve are the terminals of a net. Two curves may intersect at a finite number of points, i.e., overlap of wires is not allowed. A k-layer homotopy (or simply a homotopy) is obtained by mapping pieces of the curves of the plane homotopy into one of the k layers. Vias are established at points where a curve changes layer and no two distinct curves intersecting on the same layer (see Figure 10.22 for two different homotopies of the same problem). If the topology of the plane homotopy is fixed, then the problem is called CVM. In other words, in CVM problem, we are given a set of wire segments (the placement of wire segments has already been determined by some router) and k layers for routing. The problem is to assign each segment to one of the layers without changing the topology so that the number of vias required is minimized. In UVM problem, the placement and

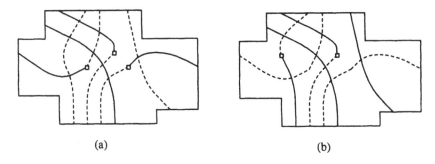

(a)   (b)

Figure 10.22: Two homotopies of the same problem.

the layer assignment of segments are not given. The problem is to both place the segments and also assign the layers so as to minimize the total number of vias. In other words, UVM is an integrated approach to routing and via minimization.

Intuitively, the UVM problem is harder than CVM problem. This is so because each UVM has many different homotopies, each resulting in a different optimal number of vias. Thus solving the UVM problem requires finding such a homotopy, which leads to a global minimum number of vias. In the following sections, we discuss both the CVM and the UVM problem.

10.2.1 Constrained Via Minimization Problem

In multi-layer routing problems, vias are required when two nets are crossing each other on a single layer. A *via candidate* is a maximal piece of wire, that does not cross or overlap with any other wire, and can accommodate at least one via. A *wire segment* is a piece of a wire connecting two via candidates. A *wire segment cluster* (or simply *cluster* is a maximal set of mutually crossing or overlapping net segments. For example, Figure 10.23 shows an instance of CVM problem. The points other than terminals, where two or more segments of a net meet and are electrically connected are called *junctions*. The number of segments which meet at a particular junction is referred to as *junction degree*. A *crossing* is a point where two net segments of two different nets intersect. A layer assignment is *valid* if no two segments of two different nets cross at a point in the same layer.

A routing solution is called a *partial routing solution* if the physical locations of the net segments is given, however, the layer assignments are not specified. Also, a valid layer assignment must exist for a partial routing solution. A *complete routing solution* consists of a set of net segments, a set of vias, and a valid layer assignment which correctly realizes the interconnection requirements specified by the netlist. A valid layer assignment for Figure 10.23 is shown in Figure 10.24.

Given the above definitions, the CVM problem can be formally stated as

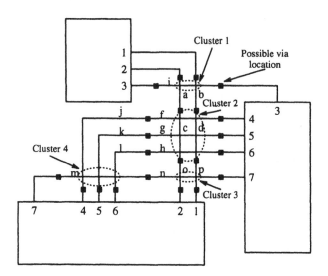

Figure 10.23: A CVM problem instance.

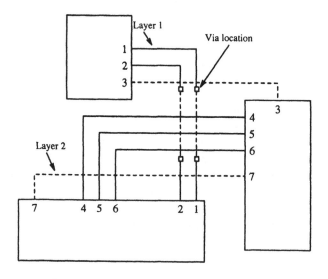

Figure 10.24: A valid layer assignment.

follows. Given a partial routing solution for a particular routing problem on k layers, find a complete routing solution with minimum number of vias for the corresponding partial routing solution. Since the CVM problem is to assign net segments to layers, the problem is also called the *layer assignment problem*. We use the terms layer assignment and complete routing solution interchangeably.

In 1971, Hashimoto and Stevens [HS71] first formulated the two-layer CVM problem as a graph-theoretic max-cut problem. The problem was initially thought to be NP-complete which led other researchers to develop heuristic algorithms [CK81, SV79]. In [SV79], Stevens and VanCleemput used a similar but more general graph model than Hashimoto and Stevens model to develop heuristic algorithm for the two-layer CVM problem. Ciesielski and Kinnen [CK81] proposed an integer programming method for the same problem. Chang and Du [CD87] developed a heuristic algorithm by splitting vertices in a graph. In 1980, Kajitani [Kaj80] showed that the two-layer CVM problem can be solved in polynomial time when the routing is restricted to a grid-based model, and all the nets are two-terminal nets. Kajitani identified the net segment clusters in a layout and showed that the graph in Hashimoto's model is planar. Kajitani's result encouraged other researchers to look for a polynomial time algorithm for more general case. In 1982, Pinter [Pin82] proposed an optimal algorithm for two-layer CVM problem when the maximum junction degree is limited to three.

10.2.1.1 Graph Representation of Two-Layer CVM Problem

In this section, we first describe the graph-theoretic representation of the two-layer CVM problem formulated by Pinter [Pin82]. We also describe the model presented by Naclerio, Masuda and Nakajima. Note that in each cluster, once a wire segment is assigned to a certain layer, layer assignment of the rest of the cluster is forced. Thus there are only two possible ways to assign the wire segments in a cluster to layers. With a prescribed layer assignment, a cluster is said to be *flipped over*, if all the wire segments in the cluster are reassigned to the opposite layers.

Given a (partial) routing problem, a *cluster graph* $G = (V, E)$ can be defined, where

$V = \{v_i \mid v_i \text{ corresponds to cluster } i\}$ and
$E = \{(v_i, v_j) \mid \text{clusters } i \text{ and } j \text{ are connected to at least one via candidate}\}$

The cluster graph for the layout in Figure 10.23 is shown in Figure 10.25.

If a complete routing solution is given, the weights can be assigned to the edges of the cluster graph. The weight $w(e)$ associated with each edge $e \in E$ of the cluster graph is defined as follows. Let p be the number of via candidates connecting the two clusters incident to e, and let q be the number of vias introduced by the known layer assignment connecting the two clusters. Then $w(e) = 2q - p$. In other words, the weight indicates the via reduction that can be achieved due to flipping over either one of the two clusters. The weights corresponding to the solution in Figure 10.24 are shown in Figure 10.25.

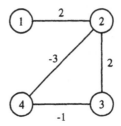

Figure 10.25: Cluster graph.

An arbitrary layer assignment L can be obtained from a known layer assignment L_0 by flipping over a set of clusters. Let X be the set of clusters which are flipped over in L_0 to obtain L. If X consists of just one cluster v, then the change in the number of vias is equal to the weights of all the edges incident on v. In a general case, the net change of vias is equal to the weights of edges between the sets X and $V - X$. This is due to the fact that any two clusters $u, v \in V$ (or $u, v \in V - X$), the via count between the clusters u and v remains unchanged. However, for $u \in X$ and $v \in V - x$, if $e = (u, v) \in E$ then the via count is reduced by $w(e)$. Let $q(L)$ and $q(L_0)$ be the numbers of vias introduced in the layer assignments L and L_0, respectively. Then

$$q(L) = q(L_0) - \sum_{e \in E(X, V - X)} w(e)$$

where $E(X, V - X)$ is a cut separating X and $V - X$, i. e., the set of edges connecting vertices in X and vertices not in X. The above equation is due to the fact that for any two clusters both in X or both in $V - X$, the via count between the two clusters remains unchanged, but for two clusters, one in X and one in $V - X$ via count is reduced by $w(e)$. In order to minimize the via count $q(L)$, we want to find a cut $E(X, V - X)$ which maximizes its weight $\sum_{e \in E(X, V - X)} w(e)$, This problem is equivalent to the max-cut problem. Note that the edge weights $w(e)$ can be positive or negative, but a maximum cut always has non-negative weight since X can be empty and $\sum w(e) = 0$ for $X = \phi$. In case that a maximum weighted cut has weight 0, L_0 is an optimal layer assignment with minimum number of vias. For the cluster graph shown in Figure 10.25, the vertex sets $\{2, 4\}$ and $\{1, 3\}$ determine the maximum cut of total weight 3. As a result, three vias can be reduced to produce a minimum via routing by flipping over clusters 2 and 4. The minimum via routing is shown is Figure 10.26.

Note that the cluster graph is planar if the junction degree is at most three. In planar graphs the max-cut problem is polynomial time solvable [Had75]. Therefore, the via minimization problem can be solved in polynomial time if the junction degree is restricted to at most three.

In 1989, Naclerio, Masuda, and Nakajima [NMN89] showed that without

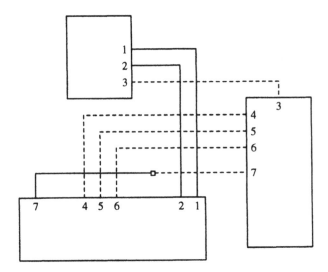

Figure 10.26: Minimum via routing.

any restriction on the maximum junction degree, the CVM problem is NP-complete, by showing a polynomial time transformation from the NP-complete planar vertex cover problem [GJ79]. They also show that the problem is NP-complete, even when one or more of the following restrictions are made.

1. The layout must be grid-based.

2. Vias can be placed only in the junctions.

3. The maximum junction degree is limited to six or more.

In 1987, Naclerio, Masuda, and Nakajima [NMN87] presented a different graph representation of the CVM problem for gridless layouts. In this representation, also the maximum junction degree is restricted to at most three. Given a partial routing solution, a *crossing graph* $G = (V, E)$ is defined as follows: Each vertex $v \in V$ corresponds to a crossing of two wire segments of two different nets in the partial routing. Two vertices $v_i, v_j \in V$ are adjacent only if there is an wire segment connecting the crossings corresponding to v_i and v_j in the partial routing. Figure 10.27(b) shows the derived crossing graph G corresponding to the partial routing of Figure 10.27(a). It is easy to see that the crossing graph defined above is planar. Each face, of the planar crossing graph is a fundamental cycle. If that cycle has an odd length, then we call that face an odd face. Otherwise the face is called an even face. Since each edge corresponds to a wire segment in the partial routing and each vertex to a crossing, the wire segments corresponding to edges that make up a fundamental cycle in the graph must be assigned to alternating layers to obtain a valid layer assignment. For an even face, all the wire segments corresponding to the

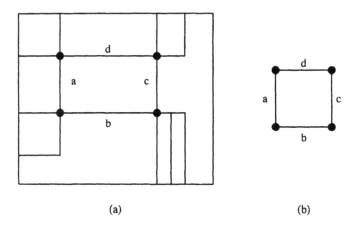

(a)   (b)

Figure 10.27: Partial routing and corresponding crossing graph.

edges can be assigned to alternating layers, thus no vias are required for for that face. Consider the example shown in Figure 10.27. The graph consists of just one even cycle, and the segments $a, b, c,$ and d can be alternately assigned to layers 1 and 2 to get a solution with no vias. On the other hand, if the graph contains an odd cycle, then the wire segments corresponding to the edges of that cycle cannot be assigned to alternating layers to obtain a valid routing without vias (see Figure 10.28(a) and (b) for an example of odd face). Note that each odd cycle require at least one via to obtain a valid routing.

Thus, a partial routing solution can be routed with no vias if and only if the corresponding crossing graph does not contain any odd faces. That is if the crossing graph is bipartite. In case the graph contain odd faces, the wire segments requiring vias can be marked and the corresponding edges in the graph can be removed and two faces sharing that edge can be merged. If the remaining graph is odd cycle free, then no further vias would be required to route the wire segments in the remaining graph. Thus in order to find the minimum number of wire segments that require vias, it is necessary to find the minimum number of edges such that the removal of those edges results in a bipartite subgraph.

Note that the problem is also equivalent to find a maximum cut the planar crossing graph. Hadlock's algorithm [Had75] can be used to find the maximum bipartite subgraph from a planar graph. The algorithm presented by Hadlock removes the minimum number of edges from the graph to remove all the odd cycle by forming the dual of the planar graph.

The crossing graph can be extended to handle multiterminal nets as long as the junction degrees are restricted to at most three. In that case, each junction is also represented as a vertex in the crossing graph. The details of the description may be found in [NMN87].

All the optimal algorithms mentioned above are based on Hadlock's maxi-

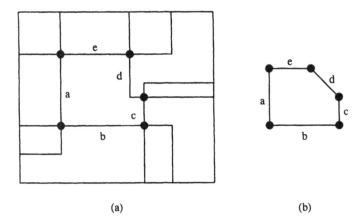

(a)   (b)

Figure 10.28: Example of an odd face in the crossing graph.

mum cut algorithm for planar graphs [Had75]. Since Hadlock's algorithm requires finding all-pair shortest paths and finding a maximum weighted matching of a dense graph, all the algorithms have time complexity $O(n^3)$, where n is the total number of net segments.

In 1988, Kuo, Chern, and Shih [KCS88] presented an $O(n^{3/2} \log n)$ time complexity optimal algorithm for the CVM problem. The algorithm they proposed was based on Pinter's graph model. In 1990, Barahona [Bar90] also presented a simpler $O(n^{3/2} \log n)$ time complexity optimal algorithm for the two-layer CVM problem.

10.2.2 Unconstrained Via Minimization

As mentioned before, the unconstrained via minimization (UVM) problem (also known as topological via minimization (TVM) problem) is concerned with finding a plane homotopy of wires so that the total number of vias are minimized [CL91, Hsu83b, LSL90, Sad84, RKN89, SL89a, SHL90]. The physical dimensions of the wires, terminals, and vias are not considered in the UVM problem. The general TVM problem in k layers (k-TVM) may be stated as follows. Given a set of nets, number of layers k and terminal locations, find a k-layer topological routing solution that completes the interconnections of all nets using the minimum number of vias. In weighted version of k-TVM problem (k-WTVM), each net is assigned a positive weight which is a measure of the priority of the net. The weight of a via represents the weight of the corresponding net. The problem is to minimize the total weight of vias used in the routing.

The TVM problem was first introduced by Hsu in [Hsu83b], and it was conjectured that TVM problem is NP-hard. Hsu considered a simple 2-TVM problem for two-terminal nets and formulated the problem using circle graphs.

It was shown that the 2-TVM problem is equivalent to finding a maximum bipartite subgraph in the corresponding circle graph. The independent sets of the bipartite subgraph can be routed in two layers without any vias. The remaining nets can be routed using vias. This result established the fact that TVM problem can be solved by routing maximum number of nets without any vias and the rest of the nets using as few vias as possible.

Marek-Sadowska [Sad84] proved that the TVM problem is NP-complete. Following theorem was also proved by Marek-Sadowska for two-terminal net TVM problems:

Theorem 27 *There exists a solution to an arbitrary instance of topological via minimization problem such that each net uses at most one via.*

The above theorem shows that the TVM problem can be solved by maximizing the number of nets that can be routed without any vias (i.e., in planar fashion).

In 1989, Sarrafzadeh and Lee [SL89a] showed that the problem of finding a maximum bipartite subgraph in a circle graph is NP-complete which in turn proves that even a simple 2-TVM problem is NP-complete. As a result, several special classes of the TVM problem have been considered. Sarrafzadeh and Lee [SL89a] and Cong and Liu [CL91] considered the crossing-channel TVM problem. In the crossing-channel TVM problem, the routing region is a simple channel. All the nets are two-terminal nets and no net has both of its terminals on the same boundary. Crossing-channel k-TVM and k-WTVM problems are solvable in polynomial time [CL91, LSL90, RKN89, SL89a].

10.2.2.1   Optimal Algorithm for Crossing-Channel TVM Problem

Note that a crossing channel is equivalent to a matching diagram and its permutation graph can easily be found (see Chapter 3). As a result, the problem of finding maximum independent sets in permutation graphs become a key problem. Sarrafzadeh and Lee [SL89a] showed that the problem of finding a maximum 2-independent set in a permutation graph can be solved in polynomial time. Cong and Liu [CL91] showed that the problem of finding a maximum k-independent set in a permutation graph can be solved in polynomial time.

Given a crossing channel consisting of a set of nets $\mathcal{N} = \{N_1, N_2, \ldots, N_n\}$, the TVM problem can be solved by first finding a maximum k-planar subset of nets. The k-planar subset of nets can be routed in k layers without any vias. Then using Theorem 27, the remaining nets can be routed in any two adjacent layers using one via per net.

We now show how the nets can be routed using one via per net by an example of the 2-TVM problem. Let $\mathcal{N}^*$ be a maximum 2-planar subset of nets for the given problem. Without loss of generality, assume $\mathcal{N}^* = S_1^* \cup S_2^*$, where $S_1^* = \{N_1, N_2, \ldots, N_p\}$, $S_2^* = \{N_{p+1}, N_{q+2}, \ldots, N_{p+q}\}$. Note that any net $N_t \in \mathcal{N} - \mathcal{N}^*$ must cross nets in S_1^* and in S_2^*. Since S_1^* is planar, nets in S_1^* can be assigned to layer 1. The p nets in layer 1 partition the region into $p+1$ subregions called *panels* $X_0, X_1, \ldots, X_p$ from left to right, where $N_j \in S_1^*$ separates regions X_{j-1} and X_j. Similarly, nets S_2^* can be assigned to layer 2

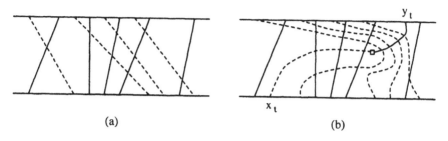

Figure 10.29: A topological routing of a net using one via.

to form $q + 1$ panels, denoted $Y_0, Y_1, \ldots, Y_q$ from left to right. Figure 10.29(a) shows planar routing of two sets S_1^* and S_2^* on layer 1 and layer 2, respectively.

Assume that a net $N_t \notin \mathcal{N}^*$ and $N_t = (x_t, y_t)$ is in panel (X_j, Y_k). Without loss of generality, let us assume that x_t lies in panel X_j and y_t lies in panel Y_k. Consider placing a via v_t in panel Y_k and connect x_t to v_t and v_t to y_t. Let the segment connecting x_t to v_t be denoted by $[x_t, v_t]$ and let v_t be in panel X_l for some l. Without loss of generality, assume that $l \geq j$ and $[x_t, v_t]$ be assigned to layer 1. The nets $N_{j+1}, \ldots, N_l$ on layer 1 that 'intersect' $[x_t, v_t]$ are 'pushed' to right, thereby, enlarging the panel X_j (see Figure 10.29(b)). The segment $[v_t, y_t]$ can be assigned to layer 2 without any difficulty, since it lies totally within panel Y_k. If there is more than one net to be routed, the above mentioned steps can be repeated to route all the nets using one via per net. The nets can be routed from left to right in $O(n)$ time. Since the maximum k-independent set in a permutation graph can be found in $O(kn^2)$ time. Therefore, the total complexity is dominated by the problem of finding maximum k-planar subset of nets. Thus we conclude,

Theorem 28 *An optimal solution to a crossing-channel TVM problem can be found in $O(kn^2)$ time.*

10.2.2.2 Approximation Result for General k-TVM Problem

If the routing region is more general than a channel, then the two-terminal net k-TVM problem becomes NP-hard. This is due to the fact that the circle graph must be used to represent the problem instead of simpler permutation graph. The k-TVM problem is equivalent to finding a maximum k-independent set in a circle graph. In chapter 3, we have presented an $(1 - (1 - \frac{1}{k})^k)$-approximation algorithm for maximum k-independent set in circle graphs. Using that result, the following theorem can easily be proven:

Theorem 29 *Given a set of nets $\mathcal{N} = \{N_1, N_2, \ldots, N_n\}$ in a k-layer routing region, let $\mathcal{N}^*$ be the maximum k-planar subset of nets in $\mathcal{N}$, and $\mathcal{N}'$ be the k-planar subset of nets found by taking one maximum planar subset at a time, then $|\mathcal{N}'| \geq (1 - (1 - \frac{1}{k})^k) \times |\mathcal{N}^*|$.*

Based on Theorem 27 and Theorem 29, we conclude,

Theorem 30 *Given a set of nets $\mathcal{N} = \{N_1, N_2, \ldots, N_n\}$ in a k-layer bounded region, the k-TVM problem can be approximated with with at most $(1 - \frac{1}{k})^k \times |\mathcal{N}^*|$ more vias than the minimum number of vias, where where $\mathcal{N}^*$ the maximum k-planar subset of nets in $\mathcal{N}$.*

10.2.2.3   Routing Based on Topological Solution

Since it appears that the CVM problem does not offer enough flexibility for via minimization, the topological routing might offer a good starting point as vias are already minimized. It is easy to see that minimum-via topological routing often uses very long wires for some nets and causes high congestion in the routing region. Since the geometric routing problem has fixed area, it may not be possible to transform a high congestion topological routing solution to geometric routing solution. Therefore, a topological routing solution is needed that is guaranteed to be transformable into an actual geometric routing solution. This can be achieved by allowing some extra via's to keep the topology as close to the actual geometric solution as possible. In this way, the final topological routing solution can be easily transformed into actual geometric routing solution. We denote this problem as *routable topological via minimization* problem in k layers (k-RTVM). The major difference between the solutions of TVM and RTVM problems is that the solution of RTVM problem is guaranteed to be transformable into actual geometric routing.

In [HS91], Hossain and Sherwani presented a graph-theoretic algorithm to solve 2-layer routing problem based on topological solution. The algorithm consists of two different phases. The first phase of the algorithm finds a solution to 2-RTVM problem. In the second phase, the solution to 2-RTVM problem is transformed into actual geometric routing.

The algorithm starts with finding a 2-planar subset of nets. Each planar subset is routed in a separate layer to form panels. If the panels on two layers are projected on a single layer, the panels intersect and form pseudo-rectangular regions. The remaining nets are topologically routed by assigning nets to the regions keeping the topology as close to the actual routing as possible. The topological routing of the nets is done by finding a weighted shortest path in the corresponding *region adjacency graph* defined from the regions. In the region adjacency graph, each vertex corresponds to a region and two vertices are adjacent if their corresponding regions share a boundary. Once the nets are topologically routed, a geometric routing is obtained by iteratively imposing grid onto each region.

10.3   Summary

The layout area for standard cell design can be reduced by minimizing the channel height. Over-the-cell routing has been successfully used to achieve dramatic reductions in the channel heights. In three-layer technology, it is possible to achieve even a channel-less layout. Several algorithms for over-the-cell routing have been presented. For high performance circuits, an algorithm has been presented which minimizes the layout height without sacrificing the performance. A significant research is needed to develop new cell models and associated over-the-cell routers to achieve the channel-less layouts for high density circuits.

Via minimization is one of the most important objectives in the detailed routing. There are two different approaches to minimize the number of vias. In constraint via minimization problem, the topology of the routing solution is fixed. Vias can be minimized only by reassigning the net segments to different layers. On the other hand, in unconstraint via minimization problem, the objective is to find a routing topology with minimum number of vias. Since the topology in UVM problem is not fixed, the UVM problem allows much flexibility than that of the CVM problem. The UVM approach, however, does not take into consideration the routing constraints; as a result, UVM solutions are not practical. Since the UVM approach allows a significant reduction on the number of vias and as the technology is improving and more and more layers are becoming available, it is expected that topology based routing solution will be more competitive.

10.4 Exercises

‡1. Given, (a) a single layer rectangular routing region R which has K tracks and two rows of terminals; one on top side and another on the bottom side and (b) a set of two-terminal nets $\mathcal{N}$. Give an efficient algorithm to find a maximum subset of $\mathcal{N}$ which can be routed in R.

2. More utilization of the over-the-cell area is possible if we allow an additional net type (type IV). Net $N_j = (t_{r_1x_1}, t_{r_2x_2})$ is a *type IV* net if $r_1 \neq r_2$, neither $t_{r_1x_1}$ nor $t_{r_2x_2}$ is vacant, and there exists two vacant terminals $t_{r_1x_3}$ and $t_{r_1x_3}$ with $x_1 < x_3 < x_2$ and $x_1 < x_4 < x_2$ (see Figure 10.30(a)). Note that Type IV nets are not constrainted to use abutments, however, they compete with the type II and type III nets for the usage of vacant terminals. Modify WISER to use type IV nets in addition to type I, II and III nets.

3. Further utilization of the over-the-cell area is possible if an additional net type (type V) is allowed. Net $N_j = (t_{r_1x_1}, t_{r_2x_2}), x_1 < x_2$ is a *type V* net if $r_1 \neq r_2$, neither $t_{r_1x_1}$ nor $t_{r_2x_2}$ is vacant, and there exists two vacant terminals $t_{r_1x_3}$ and $t_{r_1x_3}$ with $x_3 < x_1$ and $x_4 < x_1$ or $x_3 > x_2$ and $x_4 > x_2$. Note that from Figure 10.30(b) type V can be used for taking nets away from the congested areas, however, it increases the net length. Modify WISER to use type V nets in addition to type I, II, III, IV nets.

‡4. Given, (a) a single layer rectangular routing region R which has a height of K tracks, a terminal row on its bottom boundary, and a set of rectangular

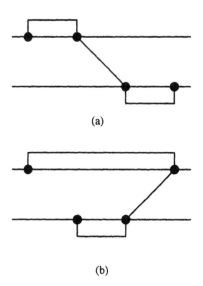

(a)

(b)

Figure 10.30: (a) A type IV net. (b) A type V net

blockages and (b) two-terminal nets $\mathcal{N}$. Give an efficient algorithm to find a maximum subset of $\mathcal{N}$ which can be routed in R.

†5. In three-layer technology, when vias are allowed in over-the-cell region, then the over-the-cell channel routing is similar to 2-layer channel routing in over-the-cell area and 3-layer channel routing in channel area. For this case, develop a greedy router that can simultaneously perform channel routing as well as over-the-cell routing.

‡6. In many cell libraries the entire metal layer (M2) is not available for routing. Instead, it has several blockages representing the routing within the cell. Also, the terminals may not be aligned in a row. In this case, the nets that are to be routed in the channel need be brought to the boundaries of the cell using the available routing regions in M2. Develop an algorithm for this problem.

†7. In [DMPS94] planar over-the-cell routing algorithm for two terminal nets was presented. Extend the algorithm to multi-terminal nets.

†8. Prove that 2 layer planar over-the-cell routing problem is NP hard.

9. Prove that the time complexity of algorithm presented in [DMPS94] is $O(kn^2)$, where k is the number of tracks available in over the cell region and n is the number of nets.

10. Given the partial routing in Figure 10.31, do the following:

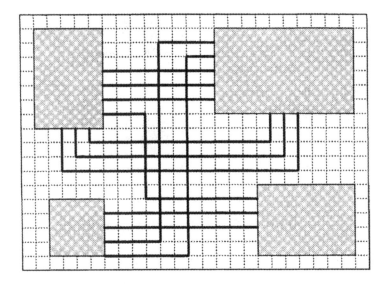

Figure 10.31: A routing instance.

 a. Find all of the via candidates. Note that if a segment spans more than one gridline, a via can be placed in that segment.

 b. Find any valid layer assignment.

 c. From the layer assignment created in c, develop the cluster graph.

 d. Find the max-cut of the cluster graph derived in c.

 e. Reassign the layers to find the minimum via routing.

11. The performance of a chip can be improved by minimizing the number of vias per net. Develop an algorithm which routes nets with one via per net.

‡12. Develop a coloring based algorithm to 3 layer constrained via minimization.

13. Develop an algorithm that minimizes the vias in a routing, by making local changes in the routing with the use of maze patterns.

14. Develop a router for two-layer crossing channel routing problem to route all the nets with at most one via per net. The basic idea of the algorithm is the same as topological routing solution for crossing channels, however, instead of finding topological solution the router should find actual detailed routing. The algorithm should first find a maximum 2-planar subset of nets and route them on two different layers. Then route the remaining nets using as many columns and tracks required.

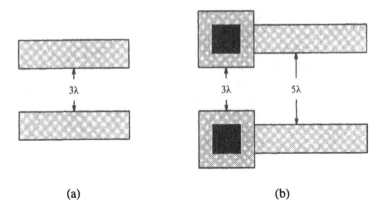

(a)   (b)

Figure 10.32: Spacing between tracks.

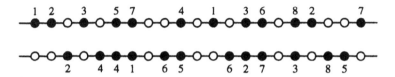

Figure 10.33: An instance of MSOP.

†15. Develop a two-layer routing algorithm for crossing channel routing problems based on topological solution. The algorithm should first find a maximum 2-planar subset of nets and route them on two layers to form a pseudo grid. The remaining nets are to be routed on the pseudo grid using a modified maze routing technique.

16. According to the design rules, the spacing between two adjacent metal tracks needs be 3λ. If the vias on adjacent tracks are aligned in a column, the spacing between the tracks increases to 5λ (see Figure 10.32). Develop an algorithm that offsets the aligned vias to compact a channel.

17. Solve the instance of MSOP in Figure 10.33 for $K = 3$.

Bibliographic Notes
The concept of OTC routing was first introduced by Deutsch and Glick in 1980 [DG80]. In [CL88], a symbolic model for over-the-cell channel routing was presented together with the algorithms for each stage of the entire routing process. Lin, Perng, Hwang, and Lin presented a linear programming formulation to select a set of nets to route in the channel in order to reduce the channel density [LPHL91]. In [NESY89], a new design style employing OTC routing techniques called *Quickly Customized Logic* (QCL) was introduced for

fast turn around times. An over-the-cell gate array channel router was presented in [Kro83]. Katsadas and Kinnen [KK90] presented a multilayer router using over-the-cell areas. In citeDLMPST96, algorithms for selecting maximum planar subset of nets which are suitable for planar OTC routing were presented. In [Kan96], presents a new triple-layer OTC Channel router for OTC routing in an irregular cell area.

A detailed description of OTC routing concepts and techniques can be found in a specialized book on this topic by Sherwani, Bhingarde and Panyam [SBP95].

In 1983, Chen, Kajitani, and Chan [CKC83] a polynomial time optimum algorithm for grid-based layouts when the junction degree is also limited to three. However, they restrict that vias can only be placed at junctions. In 1987, Naclerio, Masuda, and Nakajima [NMN87] presented an algorithm which has the same complexity as Chen *et al.* [CKC83]; however does not require that the layout should be grid-based or restrict the via locations.

In [SL89a], an $O(n^2)$ time complexity algorithm is presented for the crossing-channel 2-WTVM problem, where n is the total number of nets. An optimal $\Theta(n \log n)$ time complexity algorithm for crossing-channel 2-TVM problem is presented in [SL89a]. crossing-channel k-WTVM has been solved in $O(kn^3)$ time [CL91]. This algorithm was improved to $O(kn^2)$ in [SL90]. Multi-layer TVM problem was considered in [SHL90] and it was shown that if the terminals are pre-assigned to layers, then the problem can be solved in $O(kn^2)$ time, where k is the maximum number of terminals of a net in a single layer and n is the total number of terminals.

Chang and Cong [CC97] presented an efficient heuristic algorithm for the layer assignment and via minimization problems for multilayer gridless layouts.

Chapter 11

Clock and Power Routing

Specialized algorithms are required for clock and power nets due to strict specifications for routing such nets. It has been noted that it is better to develop specialized routers for these nets rather than over-complicate the general router. In the worst case, these special nets can be hand-routed. Currently, in many microprocessors, both of these nets are manually routed and optimized. However, as chip frequency moves into the multiple gigahertz range, the clock skew budget will become smaller and smaller and it will be not be possible to design and route clock without the help of sophisticated and accurate clock routing tools. Similarly, due to large amounts of power that needs to be provided to microprocessors, power nets must be very accurately designed and simulated to predict the power availability in different parts of the chip. As a result, power routing and analysis will increasingly depend on CAD tools.

In synchronous systems, chip performance is directly proportional to its clock frequency. Clock nets need to be routed with great precision, since the actual length of the path of a net from its entry point to its terminals determines the maximum clock frequency on which a chip may operate. A clock router needs to take several factors into account, including the resistance and capacitance of the metal layers, the noise and cross talk in wires, and the type of load to be driven. In addition, the clock signal must arrive simultaneously at all functional units with little or no waveform distortion. Another important issue related to clock nets is buffering, which is necessary to control skew, delay and wave distortion. However, buffering not only increases the transistor count, it also significantly impacts the power consumption of the chip. In some cases, clock can consume as much as 25% of the total power and occupy 5-10% of the chip area. Typically, a fixed buffered clock distribution network is used at the chip level. At a block level, a local clock routing scheme ensures minimal skew and delay. The scheme used in each block can differ, depending on the design style used in the block. The clock routing problem has significant impact on overall chip design. Clock frequencies are increasing quite rapidly. Note that current microprocessors can operate at 500 Mhz to 650 Mhz. It is expected that 1.5 - 2.0 Ghz microprocessors will be available within two to three years (See Chapter 3 for SIA roadmap).

Compared to clock routing, power and ground routing is relatively simple. However, due to the large amount of current that these nets carry, power and ground lines are wide. Concerns such as current density and the total area consumed make it necessary to develop special routers for power and ground nets. In some microprocessor chips, power and ground lines use up almost an entire metal layer. Power and ground lines are also used to shield some signal lines. This is done by routing a signal between two power (and/or ground) lines. This reduces the cross-capacitance between the signal line and its adjacent signal lines. As chip design moves into low voltages, power and ground routing will become a even harder design challenge. In this chapter, we will discuss the problems associated with clock, power and ground routing and present the basic routing algorithms for these special nets.

11.1   Clock Routing

Within most VLSI circuits, data transfer between functional elements is synchronized by a single control signal, the processing clock. The clock synchronization is one of the most critical considerations in designing high-performance VLSI circuits. In the case of microprocessor design, the clock frequency f (in MHz) directly determines the performance or the MIPS (Million Instructions Per Second) of the microprocessor.

$$\text{MIPS} = f \times \text{NIPC}$$

In this equation, NIPC denotes for Number of Instructions issued Per Cycle. NIPC depends on the architecture of the processor, RISC versus CISC, and the compilers used for the system. Most modern microprocessors are capable of multiple issue, and some can issue as many as five instructions per cycle. Consider a processor, which has a clock frequency of 200 MHz and can execute two instructions per clock cycle, thus giving it a 400 MIPS rating. If the clock frequency of the processor can be increased to 400 MHz, 800 MIPS performance can be obtained. In I/O and memory buses, the clock frequency determines the rate of data transmission. The data transmission rate is determined by the product of the clock frequency and the bus width. Thus, it is desirable to design the circuit with the fastest possible clock. However, increasing the clock frequency of a chip is a complicated affair.

The clock signal is generated external to the chip and provided to the chip through the clock entry point or the clock pin. Each functional unit which needs the clock is interconnected to the clock entry point by the clock net. Each functional unit computes and waits for the clock signal to pass its results to another unit before the next processing cycle. The clock controls the flow of information within the system. Ideally, the clock must arrive at all functional units at precisely the same time. In this way, all tasks may start at the same time and data can be transferred from one unit to another in an optimum manner. In reality, the clock signals do not arrive at all functional

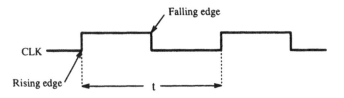

Figure 11.1: The Clock Period.

units simultaneously. The maximum difference in the arrival time of a clock at two different components is called *clock skew*. Clock skew forces the designer to be conservative and use a large time period between clock pulses, that is, lower clock frequency. The designer uses the clock period which allows for logical completion of the task as well as some extra time to allow for deviations in clock arrival times. If the designer can be provided a guarantee that the maximum deviation of the clock arrival time is small, then faster clocks can be used. The smaller the deviation, the faster the clock. Thus, controlling the deviation of signal arrival time is the key to improving circuit performance.

In the following sections, we will study the basics of clock design in a digital system. We will present several algorithms that have been proposed for solving various problems associated with clock nets. We will restrict ourselves to single clock systems, and briefly mention the multiple clock systems.

11.1.1 Clocking Schemes

The clock is a simple pulsating signal alternating between 0 and 1. The clock period is defined as the time taken by the clock signal to complete one cycle (from one rising edge to the other rising edge). Clock frequency is given as

$$f = \frac{1}{t}$$

where t is the clock period which is shown in Figure 11.1.

Digital systems use a number of clocking schemes including single-phase with latches or edge-triggered flip-flops and double-phase clocking with one or two latches. The most common latch is a *D-latch* which is an storage element. It has data D and clock CLK inputs and a data output of Q. While CLK is high, Q follows D, changing whenever D changes. While CLK is low, Q remains constant, holding the last value of D (Figure 11.2(a)). An *edge-triggered D flip-flop* has the same inputs and outputs as the D latch, but Q changes only on the rising or falling edge of the CLK (Figure 11.2(b)).

In single phase clocking with latches, the latch opens when the clock goes high; data is accepted continuously while the clock is high; and the latch closes when the clock goes down. Single phase clocking schemes are not commonly used because of their complicated timing requirements, but some high-end VLSI

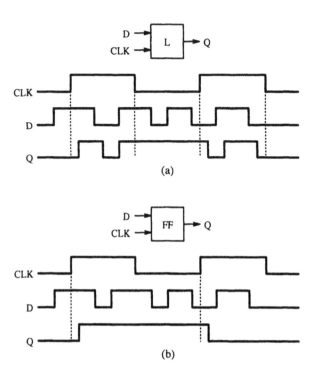

Figure 11.2: Input and output waveforms for (a) D-latch and (b) Edge-triggered D-flip-flop.

designs still use this scheme. The double-phase scheme uses two latches; one is called the *master* and the other the *slave*. The data is first captured by the master latch and then passed on to the slave latch.

The design of a clock system, as shown in Figure 11.3, must satisfy several timing constraints as explained below. When a clock signal arrives at a sequential register, it triggers the data from one sequential register set to the next through a logic unit. This unit performs manipulations of data in an appropriate functional manner. For simplicity and without losing generality, we will assume that the clocking scheme is edge-triggered.

The minimum cycle time must satisfy :

$$t_c > t_f + t_l + t_s + t_{skew}.$$

Where the flip-flop delay t_f is the time from the clock edge that captures the data to the time that the data is available at the output of the flip-flop, time t_l is the maximum delay through any logic block between two flip-flops, and setup time, t_s, is the amount of time the inputs of a flip-flop should be stable prior to the clock edge. Finally, t_{skew} is the worst-case skew between the clock signals, and the maximum amount of time the clock of the receiving flip-flop can precede the clock of the sending flip-flop.

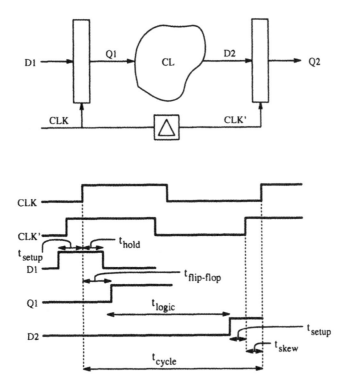

Figure 11.3: The structure of a clocking system.

Another constraint is the hold time, t_{hold}, which is the amount of time the input must stay stable after the clock edge to guarantee capturing the correct data. To guarantee that the data is captured, the clock width must be greater than the hold time:

$$t_{hold} < t_{clk-width}.$$

As a general rule, most systems cannot tolerate a clock skew of more than 10% of the system clock period. As a result, all clock lines must have equal lengths from clock entry point to a component, in order to minimize or eliminate clock skew. It is obvious that in the absence of a proper clock distribution strategy, different clock lines can have greatly varying lengths. This variation in length leads to large skews and delays. Skew is also introduced by the variations in the delay of clock buffers throughout the system because of the process-dependent transistor and capacitive loading. Skew causes uncertainty in the arrival of the clock signal at a given functional unit. If it can be guaranteed that the clock signal always arrives at a given storage element a predetermined amount of time earlier than it arrives at another storage element, design techniques can be employed to compensate for such pre-arrival of clock signal. But the nature of the clock skew is such that the designer does not know which stor-

age elements will receive the clock early and which storage element will receive
it late. There are two reasons for this uncertainty. First, the logic design is
usually done before the chips are laid out, so the relative positions of storage elements with respect to the clock buffers are not known to the designer. Second,
the random variations in the clock buffer delays, which are due to fabrication
process dependent device parameter variations.

There are three key steps in designing high performance circuits. The first
step in making a design operate at a high clock frequency is to employ a fast
circuit family. With faster circuits, a given amount of logical functions can be
performed in a shorter time. The second step is to provide a fast storage element (latch, flip-flop, register) and an efficient clocking scheme. The third step
is to construct a clock distribution scheme with a small skew. As circuits become faster and cycle time is reduced, the actual maximum skew time allowed
is reduced. While selection of faster circuits elements is a logic design decision,
reducing clock skew and efficient clock distribution is within the realm of physical design. In the following, we consider the factors that influence design of
efficient clock distribution schemes.

11.1.2   Design Considerations for the Clocking System

Clock signal is global in nature and therefore clock lines have to be very
long. The delay caused by long wires is due to their capacitance and resistance.
Long wires have large capacitances and limit the performance of the system.
At low levels of integration, gate capacitance is much greater as compared to
the interconnect capacitance and therefore need not be considered. For high
level of integration, however, the gate capacitance is much smaller as compared
to the interconnect capacitance and as a result, interconnect capacitance must
be taken into account when clock wires are routed. For example, in 7 μm
nMOS technology, the gate capacitance is equal to capacitance of 1 mm of
wire. Assuming 5 mm side dies, few nets are 1 mm long. On the other hand,
in 0.7 μm CMOS technology, the gate capacitance is equal to only 0.1 mm of
wire. Thus gate capacitance is very small as compared to the capacitance of
the long clock line, which may have to traverse as much as 25 mm. In addition
to large capacitive loads, long wires also have large resistances. The resistance
of a conductor is inversely proportional to its cross-sectional area. As chips
are scaled down, the resistance per unit length becomes a major concern. The
delay caused by the combined effect of resistance and capacitance is called
the RC delay, which increases as the square of the scaling factor. In a given
technology, RC delay cannot be reduced by making the wire wider. Although,
R is reduced, but correspondingly C is increased. One effective way of reducing
RC delay is the use of buffers (repeaters), which also help to preserve the clock
waveform. If RC delay of a clock line is 4×4 units, then dividing the line in
four equal segments and inserting buffers, the total RC constant is reduced to
$1 \times 1 + 1 \times 1 + 1 \times 1 + 1 \times 1 = 4$. In this way capacitance is not carried over and
that is how buffers help in reducing delay. The buffers, however, have internal
delays, which must be taken into account when computing the total delay. In

addition, buffers consume area and power. Despite these disadvantages, clock buffers play a key factor in the overall layout of high performance designs. In some processors, clock buffers may occupy as much as 5% of total area and may consume a significant amount of power. The problem of buffer insertion has significant attention and good algorithms are now known for both uniform and non-uniform lines [DFW84, WS92].

Buffers could be used in two different ways in the clock tree. One way is to use a big centralized buffer, whereas the other is to use distribute buffers in the branches of the tree. Figure 11.4 (a), and (b) illustrate both buffering mechanisms.

In case of distributed buffer, it is important to use identical drivers so that delay introduced by all the buffers is equal in all branches. In addition, it is important to equalize the load so that every driver sees the same capacitive load. The clock skew may still be there due to the mismatches among the drivers because of the device parameter variations across the chip. Using the identical layout for all the drivers and placing them next to each other and in the same orientation on the chip reduces the driver delay mismatch. Placing them in the same orientation guarantees that all are affected similarly by orientation dependence of the fabrication processing steps.

From the skew minimization point of view, the large centralized buffer is better than the distributed buffers. However, the area and power consideration are among other criteria that drive selection of the buffering mechanism.

In addition to RC delay, if the lines are sufficiently long or operate on high frequencies, then inductance also becomes important and clock lines behave like transmission lines, thereby changing the delay model. Transmission line behavior becomes significant when the rise time t_r of a signal is less than or comparable to the transmission line time-of-flight delay t_f. The rise time is defined as the time required for the signal to move from 10% to 90% of its final value. The time of flight is expressed as

$$t_f = \frac{l}{v}$$

where l is the line length, and v is the propagation speed. The rise time of a signal is determined by two factors: the rate at which the clock driver is turned on and the ratio of the driver source resistance to line impedance. In present CMOS systems, transmission line properties are significant only at the module and board levels; bipolar circuits require transmission line analysis at the chip carrier level and beyond; GaAs technology requires transmission line analysis even for on-chip interconnections.

11.1.2.1 Delay Calculation for Clock Trees

The exact computation of the RC delay of a clock tree is quite difficult. It is, however, not very difficult to approximate the delay. We will use a simple method for delay calculation for RC tree networks using the Elmore delay model [LM84b]. We follow the discussion presented by Tsay [Tsa91]. We will compute the delay for both buffered and unbuffered clock trees.

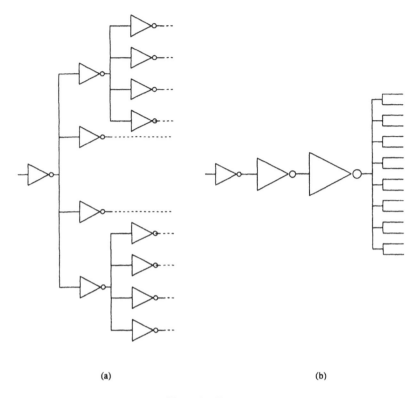

(a)   (b)

Figure 11.4: Clock buffering mechanisms.

Let T be an RC tree, c_i be the node capacitance and r_i be the resistance of edge i. The edge between node i and its parent is referred to as edge i. Note $r_0 = 0$, since root (node 0) has no parent. Let $IS(i)$ be the set of immediate successors of node i, that is, $IS(i)$ is a set of nodes adjacent to node i and does not contain its parent. Let T_i denote the subtree formed by node i and its successors.

For an unbuffered tree, the total capacitance of a subtree can be defined recursively as:

$$C_i = c_i + \sum_{j \in IS(i)} C_j$$

Let $N(i,j)$ be the set of nodes between nodes i and j, including j but excluding i. The time delay of the clock signal from root (node 0) to a node i is given by:

$$t(0, i) = \sum_{j \in N(0,i)} r_j C_j$$

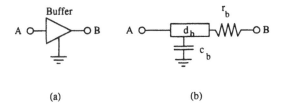

Figure 11.5: A clock buffer and its model.

The time delay from any node i to one of its successors j can be computed as:

$$t(i,j) = \sum_{k \in N(i,j)} r_k C_k$$

It is easy to see that for intermediate node k between i and j, the delay is given by:

$$t(k,j) = t(k,i) + t(i,j) \qquad (11.1)$$

Thus time delay for an unbuffered tree can be computed in linear time using a depth first search.

For buffered trees, there are several different equivalent circuit models for the buffer as shown in Figure 11.5. Let d_b denote internal delay of the buffer, r_b denote its output driving resistance, and c_b denote its input capacitance. The only difference between a buffered RC tree and a unbuffered RC tree is the branch delay d_i, which accounts for buffer delay. The capacitance for a buffered RC tree is given by:

$$C_i = \begin{cases} c_i & \text{if node } i \text{ is a buffer input node} \\ c_i + \sum_{j \in IS(i)} C_j & \text{otherwise} \end{cases}$$

Similarly delay between node i and node j can be computed using:

$$t(i,j) = \sum_{k \in N(i,j)} (r_k C_k + d_k)$$

There are several ways of modeling RC trees, some of them are shown in Figure 11.6. More widely used model is the π-model as shown in Figure 11.6(b). Using π-model, one branch is modeled as shown in Figure 11.7. From Eq. (11.1), and by lumping the delay, we can compute the delay of a node i as

$$t(i,j) = d_i + r_k C_k + t(k,j) \qquad (11.2)$$

Where k is a immediate successor of i and j is the leaf node. Our delay model is now complete as it specifies all the resistances, capacitances and delays, so that we can compute the delay from root to leaf.

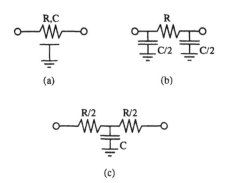

(a)   (b)

(c)

Figure 11.6: (a) A distributed RC line (b) The equivalent π-model. (c) The equivalent T-model.

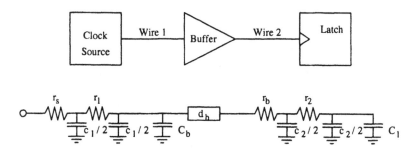

Figure 11.7: A Buffered RC tree and its Equivalent model.

11.1.3   Problem Formulation

Given the routing plane and a set of points $\mathcal{P} = \{P_1, P_2, \ldots, P_n\}$ lying within the plane and clock entry point P_0 on the boundary of the plane. We refer to points by their indices. Let $t(i,j)$ refer to the delay between points i and j, then the Clock Routing Problem(CRP) is to interconnect each $P_i \in \mathcal{P}$ such that:

$$\max_{i \in \mathcal{P}} t(0, i)$$

$$\max_{(i,j) \in \mathcal{P}} | t(0, i) - t(0, j) |$$

are both minimized.

Additional objective functions such as minimization of total wire length, protection from noise and coupling may also be defined. The clock routing problem has traditionally been studied to minimize skew.

It is important to see that CRP is not a steiner tree problem for global routing of high performance circuits, since the interconnection distance between two clock terminals is of no significance in CRP. The clock routing problem is critical in high performance circuits. In other circuits, the clock is simply routed along with rest of the nets, and the router is given a maximum routing length so that it may route any segment of the clock.

11.1.3.1 Design Style Specific Problems

The clock routing changes significantly in different design styles. The problem is well studied for full-custom and gate array design styles, but no special model has been developed for standard cell designs.

1. **Full Custom:** The clock routing problem in full custom style depends on the availability of a routing layer for clocks. If a dedicated layer, free of obstacles, is available for routing, the clock routing problem in full custom design is exactly the same as CRP. If obstacles are present, however, we refer to that problem as the *Building Block Clock Routing Problem*(BBCRP).

   Given the routing plane and a set of rectangles $\mathcal{R} = \{R_1, R_2, \ldots, R_n\}$ lying within the plane and each rectangle R_i has its clock terminal P_i on its boundary, and the clock entry point P_0 on the boundary of the plane.

   Then the BBCRP is to interconnect each $P_i \in \mathcal{P}$ to P_0 so that wires do not intersect with any rectangles and both skew and delay are minimized.

   In microprocessors, a chip level fixed buffered clock distribution is used to distribute the clock signals to different blocks. Then the problem described above can be used to locally distribute the clock.

2. **Standard Cell:** The clock routing problem in standard cell designs is somewhat easier than full-custom in some aspects, since clock lines have to be routed in channels and feedthroughs. Conventional methods do not work in standard cell design since clock terminals are neither uniformly distributed (as in full-custom), nor are they symmetric in nature (as in gate array).

3. **Gate Array:** Gate arrays are symmetrically arranged in a plane and allow the clock to be routed in a symmetric manner as well. The algorithms for clock routing in such symmetric structures have been well studied and well analyzed.

11.1.4 Clock Routing Algorithms

The skew can be minimized by distributing the clock signal in such a way that the interconnections carrying the clock signal to functional sub-blocks are equal in length. A perfect synchronization between the clock signals can be achieved by delaying the signals equally before they arrive at the sub-blocks. Note that we do not discuss buffered clock routing algorithms. As stated above, the problems and their corresponding algorithms should be viewed as local clock routing algorithms. In microprocessors, these algorithms can be used at a block

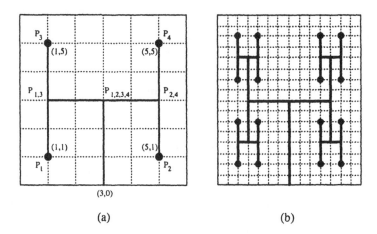

(a)   (b)

Figure 11.8: (a) H-tree over 4 points (b) H-tree over 16 points.

level. In ASICs, due to lower operating frequencies, these algorithms can also be used for chip level clock. However, even in that case, some buffering has to be used. In the following we will discuss, skew minimization algorithms. Minimization of clock skew has been studied by a number of researchers in recent years. In the following, we review several clock routing algorithms.

11.1.4.1   H-tree Based Algorithm

Consider a special case of CRP, where all the clock terminals are arranged in a symmetric fashion, as is the case in gate arrays. The clock routing in such cases can be accomplished with zero skew using the *H-tree* algorithm. Let us explain the algorithm with the help of a small example shown in Figure 11.8(a). Consider the case with four points, $P_1 = (1,1), P_2 = (5,1), P_3 = (1,5)$ and $P_4 = (5,5)$, in a routing plane with $l = 6, w = 6$. The clock entry point is at $P_0 = (3,0)$. In the H-tree algorithm, P_1 is connected to P_3 and P_2 is connected to P_4 by vertical segments. Let $P_{13} = (1,3)$ and $P_{24} = (5,3)$ be the two middle points of these vertical segments. These middle points are also called the *tapping points*. P_{13} and P_{24} are connected by a horizontal segment, whose middle point is $P_{1234} = (3,3)$. Finally, clock entry point (P_0) is connected to P_{1234} by a vertical segment. It can be seen that all points are exactly 7 units from the point P_0, hence skew is zero. Since the longest rectilinear distance between any two points$(P_0$ and $P_3)$ is seven units, this routing is minimum delay routing as well. Thus, the routing shown in Figure 11.8(a) provides clock signals to all clock points with zero skew and minimum delay.

This method can be easily generalized to n points, where n is a power of 4. The basic 4 point H-structure is duplicated in a recursive fashion. An H-tree with 16 terminals is shown in Figure 11.8(b). *H-tree* constructions have been used extensively for clock routing in regular systolic arrays [FK82, DFW84].

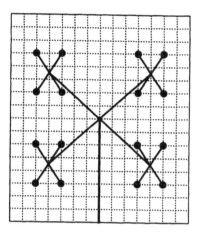

Figure 11.9: X-tree over 16 points.

If the routing is not restricted to being rectilinear, an alternate tree structure with smaller delay may be used.This tree structure, called the X-Tree, ensures that skew is zero (see Figure 11.9). However, X-trees are undesirable, since they may cause cross talk due to close proximity of wires. H-tree clock lines do not produce corners sharper than 90^0, and no two clock lines in an H-tree are ever in close proximity as a result cross talk is significantly less in H-tree as compared to X-tree.

The H-tree algorithm is applicable for very special structures. In general, clock terminals are randomly arranged all over the chip surface and require much more general algorithms.

11.1.4.2 The MMM Algorithm

Jackson, Srinivasan and Kuh [JSK90] presented a clock routing algorithm called Method of Means and Medians(MMM) for the CRP. The MMM Algorithm follows a strategy very similar to the H-tree algorithm. The MMM algorithm recursively partitions a circuit into two equal parts, and then connects the center of the mass of the whole circuit to the centers of mass of the two sub-circuits.

The algorithm is simple and yields good results. Let L_x be the list of points P sorted according to their x-coordinate. Let P_x be the median in L_x. Assign points in list to the left of P_x to $\mathcal{P}_L$. Assign the remaining points to $\mathcal{P}_R$. Due to the geometric nature of the problem, we may consider the partition of the point set as the partitioning of a region. Thus $\mathcal{P}_L$ and $\mathcal{P}_R$ partition the original region by x-median into two sub-regions with an approximately equal number of points in each sub-region. Similarly, $\mathcal{P}_B$ and $\mathcal{P}_T$ represent the division of $\mathcal{P}$ into two sets about the y-median.

The basic algorithm first splits $\mathcal{P}$ into two sets(arbitrarily in the x or y direction.). Assume that a split of $\mathcal{P}$ into $\mathcal{P}_L$ and $\mathcal{P}_R$ is selected. Then, the algorithm routes from the center of the mass of P to each of the center of mass of $\mathcal{P}_L$ and $\mathcal{P}_R$ respectively. The regions $\mathcal{P}_L$ and $\mathcal{P}_R$ are then recursively split in the y direction (the direction opposite to the previous one). Thus, splits between x and y are introduced on the set of points recursively until there is only one point in each sub-region. An example of this algorithm is shown in Figure 11.10.

Notice that basic algorithm discussed above ignores the blockages and produces a non-rectilinear tree. It is also possible that some wires may intersect with each other. In the second phase, each wire in the tree can be converted so that it only consists of rectilinear segments and avoids blockages and other nets.

11.1.4.3 Geometric Matching based Algorithm

Another binary tree based routing scheme is presented by Kahng, Cong and Robins [KCR93]. In this approach, clock routing is achieved by constructing binary tree using recursive *Geometric matching*. We call this algorithm *Geometric Matching Algorithm*(GMA). Unlike MMM algorithm which is a top down algorithm, GMA works bottom up. Let us start by defining the geometric matching.

Given a set $\mathcal{P}$ of n points, a geometric matching on $\mathcal{P}$ is a set of $\frac{n}{2}$ line segments whose endpoints are in $\mathcal{P}$, with no two line segments sharing the endpoint. Each line segment in the matching defines an *edge*. The cost of a geometric matching is the sum of the lengths of its edges.

To construct a tree by recursive matching, a forest of n isolated nodes is considered, each of which is a tree with the clock entry point being the node itself. The minimum-cost matching on these points yields $\frac{n}{2}$ segments, each of which defines a subtree with two nodes. As pointed out earlier, the center point of each segment will be called the tapping point and if the clock signal is provided at the tapping point, then the signal will arrive at the two endpoints of the segment with zero skew. The set of tapping points serves as the set of points for the next iteration of the algorithm. In general, the matching operation will pair up the clock entry points (i.e., roots) of all the trees in the current forest. At each level, the algorithm chooses the root of the newly merged tree to be the tapping point which minimizes the path length skew to the leaves of the two subtrees. Figure 11.11 shows GMA algorithm running on 8-point set.

When subtrees T_1 and T_2 are merged into a higher level subtree T_{12}, the optimal entry point may not be equidistant from the entry point of T_1 and T_2. Intuitively, balancing requires *sliding* the tapping point along the "bar of the H". However, it might not always be possible to obtain perfectly balanced path lengths in this manner. Therefore, *H-flipping* scheme is used: for each edge e H structure formed by the three edges of T_{12} is replaced by the H structure T'_{12} over the same four points which minimizes path length skew, and further minimizes tree cost.

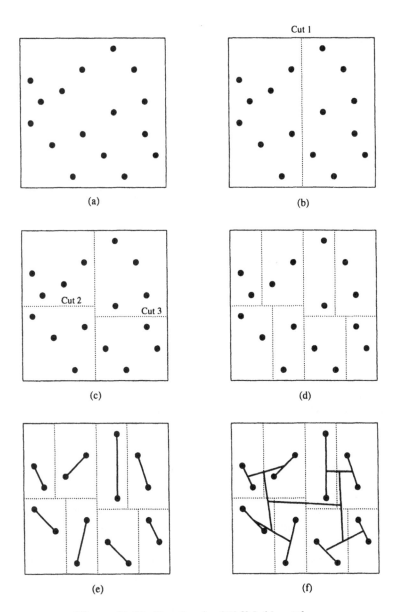

Figure 11.10: Routing by MMM Algorithm.

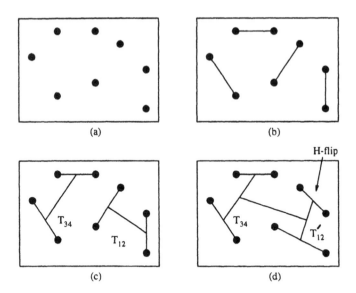

Figure 11.11: GMA Algorithm running on 8-point set.

As shown in Figure 11.11(c), two subtrees T_{34} and T_{12} are obtained, however, it is not possible to connect tapping points of T_{12} and T_{34}. Therefore, T_{12} is H-flipped to obtain T'_{12}. Finally T'_{12} is merged with T_{34} as shown in Figure 11.11(d).

Since the algorithm is based on geometric matching, its time complexity depends on the matching subroutine. The fastest known algorithms for general matching are $O(n^3)$. By taking advantage of planar geometry, the algorithmic complexity can be reduced to $O(n^{2.5} \log n)$.

11.1.4.4 Weighted Center Algorithm

The geometric matching algorithm is not applicable to Building Block Layout Problem (BBCRP) since it assumes that a complete layer is available for routing. Sherwani and Wu presented a new clock routing algorithm [SW91] called the *Weighted Center Algorithm* (WCA) for the BBCRP. In WCA, a weighted Clock Distribution Graph (CDG) for the problem is created. The vertices of CDG are the clock terminals, while the edges represent the steiner paths which may be used to connect two terminals. The weights of all the edges are obtained by the RC time delay calculation. The CDG is a complete graph, as it is always possible to connect two points in a BBCRP problem. The weight of the edge (u, v) is computed using a shortest path algorithm to find the (u, v) path followed by delay calculation for that path.

The WCA is greedy in nature and the basic idea of the algorithm is as follows: Using the clock distribution graph, the algorithm first finds the edge

(u, v) with the minimum weight (minimum delay), replace u and v with another vertex w which lies on their weighted center (tapping point). The CDG is updated to reflect new edge costs. Using this new CDG, the algorithm repeats this process recursively, until all the clock terminals are joined into one global weighted center. This global weighted center is designated as the clock signal entry point. Building up the clock distribution in this way, the clock skew between different clock terminals can be held to minimum. As the clock tree is built by using smallest edges first (just like the spanning tree algorithm), therefore the total clock tree wire length is minimized as compared to other clock distribution schemes. An example of clock routing by WCA is shown in Figure 11.12. WCA algorithm can be easily extended to multiple layers by including delays in via in calculation of path delays.

11.1.4.5 Exact Zero Skew Algorithm

Tsay [Tsa91] presented an algorithm for creating a clock tree with exact zero skew. The algorithm assumes that pairing of points has been done, and concerns itself with finding the tapping point very accurately, based on capacitive loading of the clock terminals as well as the delay in the sub-trees.

The zero skew algorithm is a recursive and bottom-up in nature. Assume two sub-trees T_1 and T_2 as shown in Figure 11.13. This algorithm computes a tapping point as discussed below.

In order to balance skew in both sub-trees, using Eq. (11.2), we have:

$$r_1(\frac{c_1}{2} + C_1) + t_1 = r_2(\frac{c_2}{2} + C_2) + t_2 \qquad (11.3)$$

where t_i refers to the delay between node i and one of the leaves. Note that the delay would be the same for all leaves. Assuming that the total wire length between two trees is l, then the length of wire from the tapping point to the root of T_1 is equal to $x \times l$ (see Figure 11.13). Similarly, the wire length from tapping point to root of T_2 is given by $(1 - x) \times l$. Let α be resistance per unit length and β be the capacitance per unit length of wire. Then, $r_1 = \alpha x l$, $r_2 = \alpha(1 - x)l$, $c_1 = \beta x l$, and $c_2 = \beta(1 - x)l$. Solving equation 11.3 with these parameters we get

$$x = \frac{(t_2 - t_1 + \alpha l(C_2 + \frac{\beta l}{2})}{\alpha l(\beta l + C_1 + C_2)}$$

If $0 \leq x \leq 1$, then tapping point is on the line segment joining two trees. On the other hand, if $x < 0$ or if $x > 1$ then tapping point is not on the line segment and wire elongation is needed. This is done by *snaking* a short segment of wire which in essence allows the tapping point to fall on the wire. The actual length of the *snake* can be easily determined in the following manner: Let us assume that $x < 0$ and let length of the elongated wire is l'. Then its resistance is $\alpha l'$ and its capacitance is $\beta l'$. In order to balance the skew

$$t_1 = t_2 + \alpha l'(c_2 + \frac{\beta l'}{2})$$

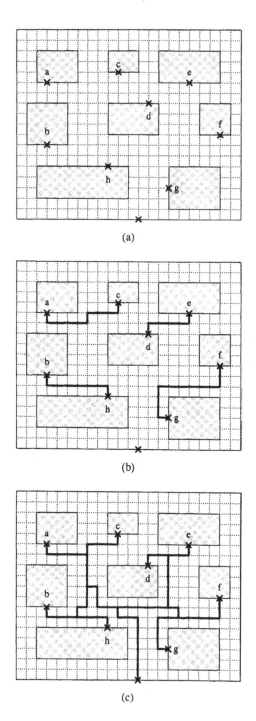

Figure 11.12: An example clock routing by weighted center algorithm.

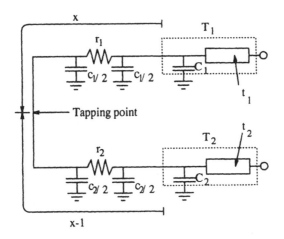

Figure 11.13: Merging of two trees.

and therefore l' is given by

$$l' = \frac{[\sqrt{(\alpha C - 2)^2 + 2\alpha\beta(t_1 - t_2)}] - \alpha C_2}{\alpha\beta}$$

Similarly, we can determine l' if $x > 1$. If l' is too long then additional buffer or capacitive terminators must be used to balance the skew.

We explain the algorithm with the help of eight pin example shown in Figure 11.14. The capacitive loading of each pin is shown in the figure. The capacitances shown are measured in farads (F) for the ease of calculation. However, the practical values of capacitances are usually in fifo farads (fF). According to the algorithm, first the tapping point M_1 is calculated for P_1 and P_2. The calculated location of M_1 is (3,21.52), which balances the delay of the path between P_1 and P_2 at 1.96 ns. The capacitance of M_1 is the sum of the capacitance at P_1, P_2 and the capacitance of the wire joining P_1 and P_2, i.e., C=8+3+(0.2× 8)=12.6 F. The tapping point M_2 for pairs P_3 and P_4 is calculated in the same manner. M_2 is calculated at (7,15) and its load capacitance is calculated to be 26.8 F. The delay from tapping point to both pins P_3 and P_4 is same, i.e., 3.99 ns. Similarly, tapping points for P_5 , P_6 and P_7, P_8 are calculated to be at (25,31) and (30,26) with load capacitances of 5 F and 30 F, respectively. At this point, we have four subtrees rooted at M_1, M_2, M_3 and M_4, such that M_1 and M_2 are in one pair and M_3 and M_4 in another. Following the same algorithm, we calculate the locations of tapping point M_6 at (7,17.97) with 41.50 F load capacitance. While calculating tapping point for M_3 and M_4, we find that $x = -0.175(< 0)$. The wire connecting M_3 and M_4, therefore, needs to be elongated. The length of elongation (snaking) for the case $x < 0$, l' is calculated to be 18.28. Therefore, 8.28 is the actual elongation (as shown in Figure 11.14). In this case the tapping point M_5 coincides with

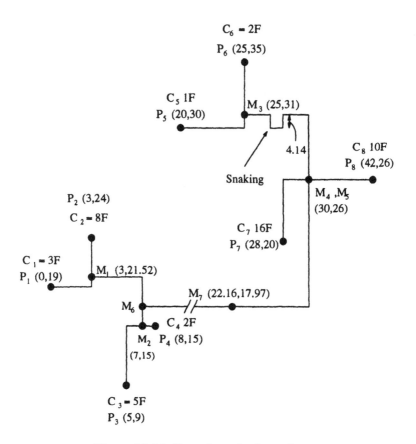

Figure 11.14: Zero skew clock routing.

M_4. The last step is to connect M_6 and M_5 to get final tapping point, which is calculated to be at (22.16,17.97). The final solution is shown in Figure 11.14. Note that the practical values of α and β are 3 $m\Omega$ and 0.02 fF, respectively. The chip width and height units are both in $1/10$ μm.

As discussed above, this algorithm assumes pairing of points and only computes tapping points to construct the clock tree. Pairing of points can be done by using MMM or GMA if the entire layer is available. If obstacles are present, then WCA may be used to find point pairs.

11.1.4.6 DME Algorithm

Three independent groups [(Boese and Kahng), (Chao, Hsu and Ho), (Edahiro)] independently proposed the Deferred Merge Embedding (DME) method in [BK92, CHH92, Eda91]. DME is a linear - time algorithm which optimally embeds any given topology in the Manhattan plane, i.e. with exact zero skew and minimum total wire length. A generic DME is a two phase; bottom up and

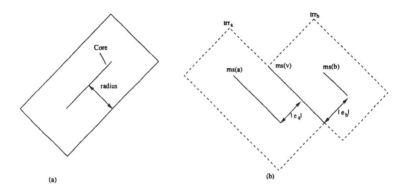

Figure 11.15: (a) Example of TRR (b) Construction of merging Segment ms(v)

top down process. The bottom up phase constructs a tree of *merging segments* which represent the loci of possible placements of nodes in the tree. The top down embedding phase determines the exact locations for internal nodes.

Before defining the DME formally, let us review definitions of terms commonly used in this section. A *manhattan arc* is defined to be a line segment, possibly of zero length, with slope +1 or -1; in other words a Manhattan arc is a line segment tilted at 45 deg. from the wiring directions. The collection of points within a fixed distance of manhattan arc is called a *tilted rectangular region* or TRR whose boundary is composed of manhattan arcs. (see Figure 11.15(a)). The manhattan arc at the center of the TRR is called its core. The *radius* of a TRR is the distance between its core and its boundary. Note that a manhattan arc is itself a TRR with radius 0. A merging segment at an internal node is a set of all placements which merge the TRRs of the child nodes with minimum wire cost.

A formal recursive definition of $ms(v)$, the merging segment of node $v \in G$, is as follows. If v is a sink s_i, then $ms(v) = \{s_i\}$ (note that this single point is a manhattan arc). If v is an internal node, then $ms(v)$ is a set of all placements $l(v)$, which merge TS_a and TS_b with minimum wire cost, that is, all points within distance $\mid e_a \mid$ of $ms(a)$ and within distance $\mid e_b \mid$ of $ms(b)$. If $ms(a)$ and $ms(b)$ are both manhattan arcs, then $ms(v) = trr_a \cap trr_b$ is obtained by intersecting two TRRs, trr_a with core $ms(a)$ and radius $\mid e_a \mid$, and trr_b with core $ms(b)$ and radius $\mid e_b \mid$. (See Figure 11.15(b)) If $ms(a)$ and $ms(b)$ are both Manhattan arcs, then $ms(v)$ is also a Manhattan arc [BK92]. Since the merging segment $ms(s_i)$ for each sink s_i is a single point and this a manhattan arc, by induction all merging segments are Manhattan arcs.

In the bottom up phase, each node $v \in G$ is associated with a merging segment which represents a set of possible placements of v in a minimum-cost ZST. The merging segment of a node depends on the merging segments of its two children, hence the bottom-up processing order. More precisely, let a and b be the children of node v in G, and let TS_a and TS_b denote the subtrees

Algorithm Build_Tree_of_Segments(G, S)
Input: Topology G; set of sink locations S
Output: Tree of merging segments TS containing $ms(v)$
      for each node v in G, and edge length $|e_v|$ for each $v \neq s_0$
for each node v in G (bottom-up order)
    if v is a sink node,
        $ms(v) \leftarrow l(v)$
    else
        Let a and b be the children of v
        Calculate_Edge_Lengths$(|e_a|, |e_b|)$
        Create TRRs trr_a and trr_b as follows:
            $\text{core}(trr_a) \leftarrow ms(a)$
            $\text{radius}(trr_a) \leftarrow |e_a|$
            $\text{core}(trr_b) \leftarrow ms(b)$
            $\text{radius}(trr_b) \leftarrow |e_b|$
        $ms(v) \leftarrow trr_a \cap trr_b$

Figure 11.16: Bottom Up Phase: Construction of Tree of Merging Segments TS

of merging segments rooted at a and b, respectively. We seek placements of v which allow TS_a and TS_b to be merged with minimum added wire while preserving zero skew. This means that we want to minimize $|e_a| + |e_b|$ in T, while balancing delays from l(v) to all leaves in the subtree rooted at v. The values of $|e_a|$ and $|e_b|$ which achieve this property are unique. They are computed and stored for use in the top-down embedding phase of DME. The details of the bottom up phase are given in 11.16.

Given the tree of merging segments corresponding to G, the top-down phase chooses exact embeddings of internal nodes in the ZST. For node v in topology G, (i) if v is the root node, then DME selects any point in $ms(v)$ to be $l(v)$; or if v is an internal node other than the root, DME chooses $l(v)$ to be any point in $ms(v)$ that is at distance $|ev|$ or less from the placement of v's parent p (the merging segment $ms(p)$ was constructed such that $d(ms(v), ms(p)) \leq |ev|$, so there must exist some $l(v)$ satisfying this condition). In case (ii), $l(v)$ can be any point in the intersection of $ms(v)$ and the square TRR trr_p which has radius $|e_v|$ and core $l(p)$. The details of the top down phase are given in 11.17.

DME requires an input topology, as a result, several authors have proposed topology constructions that yield low-cost routing solutions when DME is applied.

Algorithm Find_Exact_Placements(TS))
Input: Tree of segments TS containing $ms(v)$
    and value of $\mid e_v \mid$ for each node v in G
Output:ZST $T(S)$
for each internal node v in G (top down order)
    if v is the root
        Choose any $l(v) \in ms(v)$
    else
        Let p be the parent node of v
        Construct trr_p as follows:
            $\text{core}(trr_p) \leftarrow l(p)$
            $\text{radius}(trr_p) \leftarrow \mid e_v \mid$
            Choose any $l(v) \in ms(v) \cap trr_p$

Figure 11.17: Top Down Phase : Construction of ZST by embedding internal nodes of G within TS

11.1.5 Skew and Delay Reduction by Pin Assignment

In [WS91], clock routing is done at pin assignment phase of the layout. If clock routing is considered at the floorplanning stage of the layout, then some flexibility in location of the clock terminals is allowed. During layout several iterative steps in placement and routing phases are allowed. During these re-design cycles, circuit layout is iteratively improved and design is made 'more' rigid. This allows successive re-positioning of clock terminals of functional block. By appropriately locating the clock terminals total clock skew and delay can be reduced significantly. Movable Clock Terminal Routing Problem (MCTRP) is a clock routing problem in which the clock terminals of the functional blocks in floorplan can be moved along the block boundaries.

MCTRP basically consists of two subproblems. The first subproblem is to find the best location for clock terminal of each functional element to minimize the clock delay. The second subproblem is to find a clock routing such that the clock signals can reach all the terminals with equal time delay. The first subproblem is shown to be NP-complete [WS91], and a greedy heuristic algorithm is presented. The second subproblem of interconnecting points to obtain a minimum skew can be solved by using any algorithm discussed earlier for BBCRP.

11.1.6 Multiple Clock Routing

Large VLSI systems may use multiple clocks because the existence of multiple clock phases gives an extra degree of freedom to the timing characteristics of the synchronizing circuits. The multiple clock routing problem is, however, more complex because of two types of skew: the *intra clock skew* within a clock and the *inter clock skew* among multiple clocks. Thus, for high performance circuits, it is necessary to develop a routing algorithm for multiple clocks which minimizes the delay as well as both types of skews. An additional problem of routing two phase clock on a single layer is crossing of two clock signals. This problem is resolved by the use of 'low resistance' *crossunders*.

Let us consider a system with k clocks $\phi_1, \phi_2, \ldots, \phi_k$. Let us also assume that there are n blocks each requiring one clock input from each clock. Let $\mathcal{P}$ be the set of n clock terminals. Let $P_{ij} \in \mathcal{P}$ denote the terminal of clock ϕ_i at block j. Let t_{ij} be the arrival time of clock signal at P_{ij}. For any clock ϕ_i, intra clock skew ρ_i is defined as,

$$\rho_i = \max\{t_{ij}\} - \min\{t_{ij}\} \quad j = \{1, 2, \ldots, n\}$$

For a block j, the inter-clock skew is defined as,

$$\chi_j = \max\{t_{ij}\} - \min\{t_{ij}\} \quad i = \{1, 2, \ldots, k\}$$

For the system, the cross skew is defined as,

$$\chi^* = \max\{t_{ij}\} - \min\{t_{ij}\} \quad i = \{1, 2, \ldots, k\}, j = \{1, 2, \ldots, n\}$$

Thus the objective of multiple clock routing system is not only to minimize σ_j for each clock ϕ_j, but also to minimize χ_i between each set of clocks and χ^*, the cross skew, of multiple clock system. However, this task is complicated due to intersection between different clock trees. When two clock trees intersect, crossunder may be used to pass one signal under the other signal. Crossunders should be minimized, subject to the constraint that the number of crossunders should be equalized for multiple clocks, in order to equalize the signal delays.

In [KHS92], Khan, Hossain, and Sherwani proposed zero skew routing for two clocks. The basic idea is to build the two trees independently. In the first phase points are paired up and crossunders are assigned to allow two trees to be routed in a planar fashion. This phase attempts to minimize the cross-skew by alternating the order in which the crossunders are used. In this way the number of crossunders are balanced on each path of both trees. In the second phase, algorithm eliminates intra clock skew in both trees independently, taking crossunders into account.

11.2 Power and Ground Routing

In VLSI design, almost all the blocks need power supply and need to be connected to ground as well. The power and ground nets are usually laid out entirely on the metal layer(s) of the chip due to smaller resistivity of metal as compared to poly. Since, contacts(vias) also significantly add to the parasitics, it is also advisable to utilize a planar single-layer implementation of these nets. It should be noted that the area requirements for power and ground nets depend on the voltage drop, current density and other constraints. In case of normal signal nets, the current they carry is very small. Hence, they can be routed with minimum-width wires. Thus minimizing the total wire length also ensures minimizing the area needed to route them. The same is not true for power and ground nets.

Routing of power(VDD) and ground(GND) nets consists of two main tasks: (i) construction of interconnection topology, and (ii) determination of the widths of the various segments of topologies. In recent years, most of the research and development efforts have been focused on the topological routing of the power and ground signals.

For a given placement of arbitrary rectangular blocks on a chip, the problem of routing power and ground nets is to find two non-intersecting interconnection trees, each for VDD and GND. The width of trees at any point must be proportional to the amount of current being drawn by the points in that sub-tree. We assume that each block has an entry point for VDD and GND. In standard cell designs VDD and GND are routed by using inter-weaved combs (as discussed in chapter 1). In fact, VDD and GND are already laid out in the cells and simply connected on one side with GND and VDD on the other. In gate arrays, VDD and GND routing is similar to standard cell and is usually laid out on the master, and not subject to customization.

A simple scheme that is often used for power and ground using two layers of metal is a grid structure. Several rows of horizontal (M5) wires for both power and ground run parallel to each other. The vertical wires run in M4 and connect the horizontal wires. In this way, two grids are formed. All the blocks simply connect to the nearest power and ground wire. This scheme is shown in Figure 11.18. The blocks and connections to blocks are not shown for sake of clarity.

Syed and El Gamal [SG82] proved the necessary and sufficient conditions for a planar routing of power/ground nets using single pads. Two nets can be routed on a single layer without crossover only if there exists a cut for each block in the chip that separates the terminals of one net from the terminals of the other net. The nets are grown as interdigitated trees. Applying simple traffic rules to the free channels between modules prevent the two trees from crossing. An example routing is shown in Figure 11.19.

Another approach is proposed by Moulton [Mou83]. The basic idea of the proposed algorithm is to partition the chip surface into a VDD region and a GND region and then to route each net within the appropriate region. It is easy to see that if all modules are visited once while keeping the VDD to one side and the GND on the another side, a cycle can be drawn which will connect all modules to the VDD and GND pads. Thus a Hamiltonian cycle is drawn, and the algorithm allows this cycle to determine the layout of the trees. Tree traversal determines how much current might flow through each wire of the tree. The maximum current of a wire ending in a terminal is the maximum current of the terminal. The maximum current of other wires is the sum of its children's maximum currents. After every wires maximum current is known, multiplying it by a design-rule constant gives every wires minimum width. Both the Hamiltonian cycle and Steiner tree operations are, however, computationally very expensive.

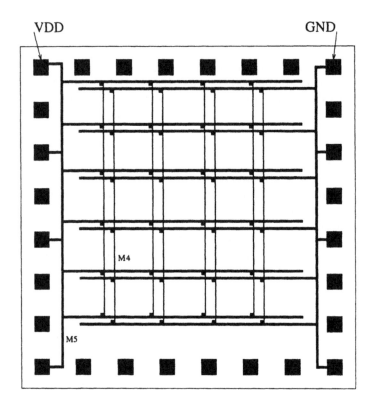

Figure 11.18: Power and Ground Routing using Grids.

The algorithm proposed by Rothermel and Mlynski [RM81] tends to route nets interdigitated. It extends one net from the left edge of the chip, and the other from the right. This routing order of the connecting points is determined by the horizontal distances of connecting points from the edge of the chip. Calculation of nets is accomplished by a combined Lee and Line Search algorithm. At first only points of the left net which lie in the left half of the chip are routed. Then those points of the right net which lie in the right half of the chip are routed. This process uses a fast line search algorithm similar to Hightower's algorithm [Hig80]. Next, all other points of the two sets are routed by Lee's algorithm [Rub74], which takes into account obstacles created by already routed net segments.

In [HF87], Haruyama and Fussell propose a method for routing non-crossing VDD and GND trees on a layer which tries to minimize the chip area devoted to power routing under metal migration and voltage drop constraints. The metal migration has to be prevented by using a wide enough metal wire. In addition, the voltage drop has to be kept small, because a large voltage drop between a pad and a module decreases switching speed and noise margin. The algorithm

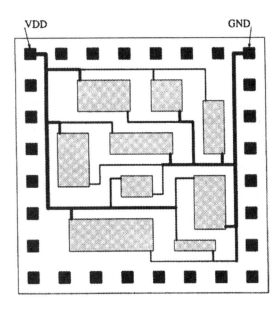

Figure 11.19: Power and Ground Routing using interdigitated Trees.

also takes the width of channels into consideration so that if a channel is too congested to allow a wire to pass through it, the wire avoids the congested channel and chooses other channels. The goal is to grow the VDD and the GND trees by connecting modules, one by one, to trees under construction. Modules are sorted by their power consumption. First, the pins of the most power-consuming module are connected to the pads. Subsequent modules are routed in decreasing order of power consumption. This is a greedy approach based on the notion that it is better for more power-consuming modules to have shorter paths, since more power-consuming modules need wider wires in order to be supplied with more current. At earlier stages in the routing, paths can generally be shorter, since they are blocked by fewer wires already routed. A smaller area is thus occupied by a wire. Pins of the second module (and later considered modules) are connected to non-root vertices of the net (or possibly to an unconnected pad when there is more than one VDD or GND pad). The constructed net is a tree whose leaves are pins of modules and whose root is a pad. When there is more than one VDD or GND pad, the algorithm may create a forest of multiple trees. The wire area becomes even smaller than when there is only one VDD pad and one GND pad because the search can find a shorter path to a power source. This multiple pad method eases the current load of each pad.

Routing of power and ground nets is often given first priority, because the power and ground wires are usually laid out entirely on a metal layer(s) due to its low resistivity, as described above. Signal nets may share the metal layer(s) with power and ground, but they change layers whenever a power or ground wire is encountered.

11.3   Summary

Clock routing is one of the factors which determines the throughput of any chip. In advanced VLSI systems, clock skew caused by interconnection delay, if not controlled, can lead to significant performance degradation. Ideally, the clock skew should be less than 5-10 percent of the clock period. Several clock routing algorithms have been proposed, and it is possible to route a clock very accurately with exactly zero skew if a complete layer is available. Much research remains to be done for clock routing problems with obstacles on the routing layer. Multiple clock routing is another area that promises to be a focus of attention as more and more designs use multiple clocks. Some radical design methodologies have been presented (asynchronous self timed systems), which do away with system level clock. Instead, the flow of information from unit to unit is based on hand shaking protocols and time stamping of the data. However, this approach presents considerable design difficulties. Clock signal serves as a convenient sequence and timing reference and it would be difficult to design circuits with such sequencing.

Power and ground routing needs special attention because of wire widths. Power and ground wires carry large amounts of current and as a result wider wires are used. The width cannot be uniform since current requirement is not uniform over the chip surface. As a result, wires must be carefully sized to allow proper current flow. Too thin wires lead to low currents, while too wide wires may lead to wastage of area.

Currently, Aluminum (Al) is the metal of choice for long interconnect lines, such as clock, power and ground lines. Al has low resistivity, good adherence to silicon and silicon oxide, it is easy to bond, pattern and deposit. Furthermore, Al is low-cost, readily available and easy to purify. Despite these qualities, Al suffers from a variety of problems, such as, electro-migration and contact failures. Au, Cu and Ag all have resistivities lower than that of Al. However, replacing Al with any of them will require a major effort because none of them are as compatible with integrated circuit processing as Al.

In future, superconductivity and optical interconnect offer alternative to aluminum wires for clock routing. In particular, optical interconnect allows fast (speed of light), reliable (no metal migration problems), noise-free and easy clock distribution. It is possible to distribute a light signal to all the functional units with zero skew and no delay. However, both superconductor and optical interconnect are still topics of research and are currently not practical.

11.4   Exercises

1. Generate an instance of CRP by randomly placing n points on a plane. Choose (randomly) one point on the boundary of the plane as the clock

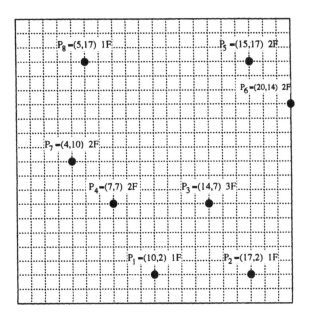

Figure 11.20: Sample clock points.

entry point. Implement MMM algorithm and test it on the instance generated.

2. Implement Geometric Matching algorithm for point set developed in exercise above. Compare the results with that of MMM Algorithm, in terms of skew and total wire length.

3. Generate an instance of BBCRP by randomly placing n rectangles on a plane such that none of the rectangles intersect. Randomly choose a point on the boundary of each rectangle as its clock terminal. Also choose (randomly) one point on the boundary of the plane as a clock entry point. Implement Weighted center algorithm and test it on the generated problem. Compare the total wire length results for weighted center algorithm and geometric algorithm.

† 4. It is possible to combine geometric matching and weighted center algorithms. The basic idea is to use the clock distribution graph to identify the paths, and use geometric matching to pair up the points. Modify the weighted center algorithm to use geometric matching.

5. Consider the 8 point instance given in Figure 11.20. Find the routing with exact zero skew. Assume $\alpha = 0.1 \ \Omega$ and $\beta = 0.2 \ F$.

6. For an instance of CRP, randomly assign load capacitances of each point between 1 F and 20 F. Assume $\alpha = 0.1 \ \Omega$ and $\beta = 0.2 \ F$. Implement the

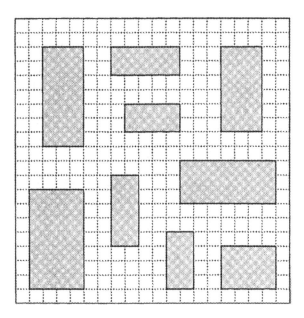

Figure 11.21: Instance of MCTRP.

Exact zero skew algorithm and test it on instance generated above. Use GMA for point pairing.

7. For a given instance, compute the number of times snaking is required in Exact zero skew algorithm if MMM algorithm is used for pairing up the points instead of GMA.

8. Consider the following instance of MCTRP given in Figure 11.21. Find clock entry point for each block such that minimum skew algorithm can route the clock net with minimum wire length. The clock entry point of chip can be placed anywhere on the boundary.

9. Consider the points given in Figure 11.22. Find the optimal clock entry point for this chip.

† 10. Develop an algorithm which finds the optimal clock entry point of the chip for any instance of BBCRP. Can this problem be solved in polynomial time ?

† 11. We define the following restricted *Standard Cell Clock Routing Problem*(SCCRP): Given a $l \times w$ grid representing a channel and L clock terminals on top and bottom, and clock entry point on the right side of the channel(Figure 11.23). Find a routing with minimum delay and zero skew. More precisely, the points are located on $(0,0)$, $(2,0)$, ..., $(l,0)$, $(0,w)$, $(2,w)$, ..., (l,w) and the clock entry point is located at $(l, \frac{w}{2})$.

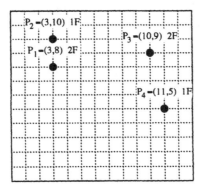

Figure 11.22: Finding clock entry point.

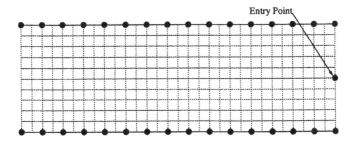

Figure 11.23: Standard cell clock routing problem.

† 12. Given only one layer for clock routing, prove that there exists a routing for SCCRP with zero skew if $w = 2\log(\frac{l+1}{2}) + 1$ tracks are allowed in the channel. Prove that the maximum delay in such routing is no more than $l + w$.

† 13. Given two layers, prove that there exists a routing for SCCRP with zero skew if $w = \log(\frac{l+1}{2}) + 1$ tracks are allowed in the channel. Assume that vias are ideal, i.e., they do not cause any additional delay.

† 14. Given two layers, develop a routing for SCCRP so that the path length and the number of vias is equal from clock entry point to each terminal.

Bibliographic Notes:
Bakoglu [Bak90] presents an excellent coverage of parameters involved in interconnect. Details of delay computation may also be found there. Another algorithm for the clock routing problem of building block design has been presented by Ramanathan and Shin [RS89]. The problem of power and ground routing has been extensively studied and several related problems have also

been investigated. In [HSVW90a], Ho, Sarrafzadeh, Vijayan and Wong discuss the problem of minimizing   the number of power pads, in order to guarantee the existence of a planar routing of multiple power nets. They also show that the general pad minimization problem is NP-complete. They derive a general lower bound and present a heuristic for the general problem. They also present optimal algorithms for some special cases. In [LG87], Lursinsap and Gajski, consider the problem of power routing in a top-down design approach. In this approach, a layout is decomposed into cells connected by abutment. The active cells contain transistors and interconnections, while passive cells are routing cells. They consider power routing of all active cells so that the total wire length is minimized, and present an optimal power routing algorithm for this special problem. In [XK86], Xiong and Kuh present an algorithm which grows both the VDD (from one side) and the GND trees (from the other side) simultaneously using a plane sweep algorithm. In [Eda94], Edahiro presents a bucket algorithm for zero skew routing with linear time complexity on the average. In [CS93], Cho and Sarrafzadeh introduce a new approach for optimizing clock tree. In [EL96], a clock buffer placement algorithm is proposed.

Chapter 12

Compaction

After completion of detailed routing, the layout is functionally complete. At this stage, the layout is ready to be used to fabricate a chip. However, due to non-optimality of placement and routing algorithms, some vacant space is present in the layout. In order to minimize the cost, improve performance and yield, layouts are reduced in size by removing the vacant space without altering the functionality of the layout. This operation of layout area minimization is called *layout compaction*.

The compaction problem is simplified by using symbols to represent primitive circuit features, such as transistors and wires. The representation of layout using symbols is called a *symbolic layout*. There are special languages [Eic86, LM84a, Mat85] and special graphic editors [Hil84, Hsu79] to describe symbolic layouts. To produce the actual masks, the symbolic layouts are translated into actual geometric features. Although a feature can have any geometric shape, in practice only rectangular shapes are considered.

The goal of compaction is to minimize the total layout area without violating any design rules, maintaining good layout practices and without violating designer specified constraints. The last two objectives are usually motivated by performance verification. The area can be minimized in three different ways:

1. *By reducing the space between features:* This can be performed by bringing the features as close to each other as possible. However, the spacing design rules must be met while moving features closer to each other.

2. *By reducing the size of each feature:* The size rule must be met while resizing the features.

3. *By reshaping the features:* Electrical characteristics must be preserved while reshaping the feature.

Compaction tools are sometimes used as a layout aid. That is, layout is drawn in larger than minimum area. This reduces the design time. Compaction is then used to get a close to minimum area layout.

Compaction is a very complex phase in physical design cycle. It requires understanding of many details of the fabrication process such as the design rules. Compaction is very critical for full-custom layouts, especially for high performance designs. In this chapter, we discuss the compaction phase of physical design cycle.

12.1   Problem Formulation

The layout of a VLSI circuit consists of geometric feature (mostly of rectangular shape). Each feature belongs to a circuit component or to a wire. The compaction problem can be stated as: Given a set of geometric features $M = \{M_1, M_2, \ldots, M_n\}$ representing a layout. Each feature, M_i, has a minimum size, $s(M_i)$, dictated by the design rules. In addition, minimum separation between features, $d(M_i, M_j)$, between M_i and M_j, for $1 \leq i, j \leq n$ is also given. The objective of compaction is to minimize the total layout area by moving features close to each other and by resizing the features such that

$$size(M_i) \geq s(M_i)$$

$$dist(M_i, M_j) \geq d(M_i, M_j)$$

where $size(M_i)$ and $dist(M_i, M_j)$ are size of M_i and distance between M_i and M_j after the compaction, where $1 \leq i, j \leq n$. If the sizes of the features are assumed to be fixed, then the problem is just to move the features closer to reduce the layout area.

12.1.1   Design Style Specific Compaction Problem

The scope and impact of compaction on layouts differs depending on the design style.

- Full-custom design style: Compaction is very critical in full-custom design style. After placement and routing, a large amount of space is left vacant. The problem is exactly same as the one formulated above. This is not true if significant part of layout is done by hand.

- Standard cell design style: The cell heights are fixed in a standard cell design. So the height of the layout can be minimized by minimizing channel height. Thus a restricted type of compaction, called channel compaction may be used. However, several channel routers produce very compact routings which cannot be compacted any further.

- Gate array design style: Since the position of gates is fixed, compaction is not applicable to gate array designs, except to optimize wiring.

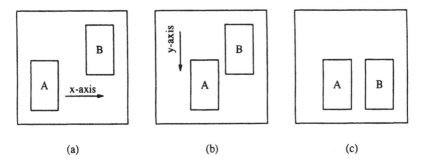

(a)   (b)   (c)

Figure 12.1: 1-D compaction.

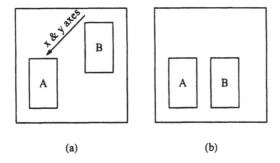

(a)   (b)

Figure 12.2: 2-D compaction.

12.2 Classification of Compaction Algorithms

Compaction algorithms can be classified in two different ways. The first classification scheme is based on the direction of movements of the components (features): *one-dimensional* (1-D) and *two-dimensional* (2-D). In 1-D compaction, components are moved only in x- or y-direction. As a result, either x- or y-coordinate of the components is changed due to the compaction. If the compaction is done along x-direction then it is called *x-compaction*. Similarly, if the compaction is done along the y-direction, then it is called *y-compaction*. Figure 12.1 shows an example of both x- and y-compactions. In 2-D compaction, the components can be moved in both x- and y-direction simultaneously. As a result, in 2-D compaction, both x- and y-coordinates of the components are changed at the same time in order to minimize the layout area. Figure 12.2 gives an example of a 2-D compaction.

The second approach to classify the compaction algorithms is based on the technique for computing the minimum distance between features. In this approach we have two methods, *constraint-graph based compaction* and *virtual grid based compaction*. In constraint-graph method, the connections and sep-

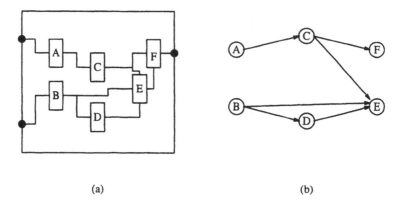

(a)   (b)

Figure 12.3: Constraint graph generation.

arations rules are described using linear inequalities which can be modeled using a weighted directed graph (constraint graph) as shown in Figure 12.3. This constraint graph is used to compute the new positions for the components.

On the other hand, virtual grid method assumes the layout is to be drawn on a grid. Each component is considered attached to a grid line. The compaction operation compresses the grid along with all components placed on it keeping the grid lines straight along the way. The minimum distance between two adjacent grid-lines depends on the components on these grid lines. The advantage of virtual grid method is that the algorithms are simple and can be easily implemented. However, virtual grid method does not produce compact layouts as compared to the constraint graph method.

In addition, compaction algorithms can also be classified on the basis of the hierarchy of the circuit. If compaction is applied to different levels of the layout, it is called *hierarchical compaction*. Any of the above mentioned methods can be extended to hierarchical compaction. A variety of hierarchical compaction algorithms have been proposed for both constraint-graph and virtual grid method. Some compaction algorithms actually 'flatten the layout' by removing all hierarchy and then perform compaction. In this case, it may not be possible to reconstruct the hierarchy, which may be undesirable.

12.3   One-Dimensional Compaction

In this section, we present two methods of one-dimensional compaction: Constraint graph based compaction and virtual grid compaction. One dimensional compactors are repeatedly used in X and Y directions until no further compaction is possible.

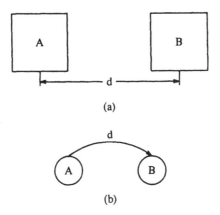

(a)

(b)

Figure 12.4: Separation constraint.

12.3.1 Constraint-Graph Based Compaction

The constraint graph $G = (V, E)$, is a weighted graph. Each vertex $v \in V$ represents a component while edges represent constraints. There are two types of constraints that should be satisfied in the process of compaction: separation constraints and physical connectivity constraints. Both separation constraints and physical connectivity constraints can be incorporated into a graph representing a 1-D compaction problem. A separation constraint between two features can be represented using a weighted directed edge between the two vertices, with weight equal to the minimum separation. For example, if the two features A and B are required to be at least d units apart from each other; assuming that A is to the left of B, this rule can be written as $B_x \geq A_x + d$ where A_x refers to the x-location of component A. The inequality is represented in the graph as an edge from A to B of weight d (see Figure 12.4). Figure 12.5 shows the connectivity constraints that require two components to be within a distance s of each other. A physical connection can be represented as a cycle of two edges. The condition $|C_x - W_x| \leq s$ can be rewritten as two constraints $W_{1x} \geq C_x - s$ and $C_x \geq W_x - s$, which appear in the graph as a pair of constraints between C_x and W_x, each with weight $-s$.

The constraint graph includes two additional vertices, L and R, which represent two physical boundaries (without the loss of generality, left and right). L can be thought of as a source of the constraint graph, because all other vertices are, explicitly or implicitly, required to the right of L. Similarly, R can be considered as the sink of the graph. Figure 12.6 gives an example of a constraint graph that includes source and sink vertices.

In the process of compaction when the elements are moved by the compactor it is necessary that the original electrical connections be preserved. Most compactors derive the connectivity from the overlapping regions in the original layout. In the following, we discuss two types of connectivity constraints and

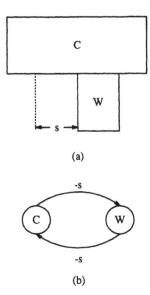

(a)

(b)

Figure 12.5: Connectivity constraint.

how can they be included in the constraint graph.

1. **Wire-terminal connections:** Connectivity constraints for wire-terminal connections with horizontal wire, vertical wire, and wide vertical wire are illustrated in Figure 12.7(a), (b), and (c) respectively. These rules ensure good electrical connections (See Figure 12.8(a)) when compared with the minimum overlap rules (See Figure 12.8(b)).

2. **Wire-wire constraints:** The connection between two wires is also captured into the graph in a similar way as it is captured for the wire-terminal connections. Figure 12.9 shows good wire-wire connections in which wires overlap by their complete width.

After the layout is compacted, a number of vertices could still be relatively free to move. Therefore, other objectives can be used to determine coordinates of these vertices, e.g., minimize the total length of the interconnect wires located on specified layers to reduce resistance and capacitance [Sch83].

The *longest path* algorithm can be used to assign positions to the vertices that minimizes the distance from the source to the sink, which is equivalent to minimizing the layout width in the dimension of compaction.

12.3.1.1   Constraint Graph Generation

As discussed earlier, once the constraint graph is generated, the actual compaction is quite simple using the longest path algorithm. However, the first

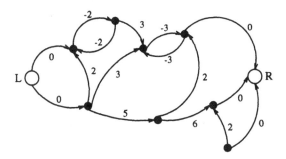

Figure 12.6: An example of a constraint graph.

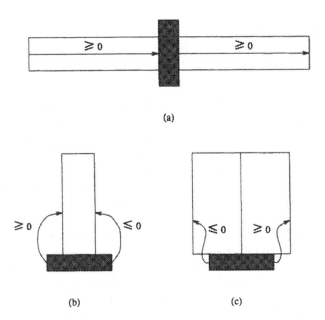

Figure 12.7: Wire-terminal connection constraints.

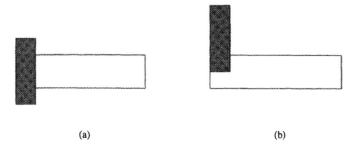

(a)   (b)

Figure 12.8: Connections after compaction.

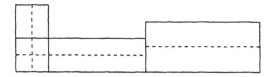

Figure 12.9: Wire-wire connections.

step necessary in constraint graph compaction is to build the constraint graph. The building of the constraint graph is the most time-consuming part of constraint graph based compaction and is $O(n^2)$ in the worst case. This is due to the fact that, in the worst case, there is an edge between every pair of vertices in the constraint graph. Only a small subset of the potential edges are actually needed for constraint graph compaction. A circuit component group typically will only have spacing requirements with its nearest neighbors. Many techniques for generating the constraint graph efficiently have been proposed [HP79, Mal87]. In construction of a constraint graph, the connectivity constraints are generated first. The connectivity constraints can be generated by scanning all legal connections in the symbolic layout. The connectivity information is usually stored in a table and the compactor looks up the table to generate all the constraints. Different types of connectivity constraints include wire-wire, wire-via, wire-source connectivity constraints.

Separation constraints are generated once per compaction step. The constraint generation method used should ideally generate a non-redundant set of constraints since the cost of solving the constraint graph is proportional to the number of edges in the graph, or the number of constraints. Several constraint generation algorithms have been proposed. In the following section, some of the constraint generation algorithms will be discussed.

1. **Shadow-Propagation Algorithm:**   A widely used and one of the best known techniques for generating a constraint graph is the *shadow-propagation* used in CABBAGE system [HP79]. The 'shadow' of a feature

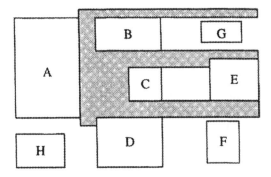

Figure 12.10: Example of shadow propagation.

is propagated along the direction of compaction. The shadow is caused by shining an imaginary light from behind the feature under consideration (see Figure 12.10). Usually the shadow of the feature is extended in both sides of the features in order to account for diagonal constraints. This leads to greater than minimal Euclidean spacings since an enlarged rectangle is used to account for corner interactions. (See shadow of feature in Figure 12.10).

Whenever the shadow is obstructed by another feature, an edge is added to the graph between the vertices corresponding to the propagating feature and the obstructing feature. The obstructed part of the shadow is then removed from the front and no longer propagated. The process is continued until all of the shadow has been obstructed. This process is repeated for each feature in the layout. The algorithm SHADOW-PROPAGATION, given in Figure 12.11, presents an overview of the algorithm for x-compaction of a single feature from left to right.

The SHADOW-PROPOGATION routine accepts the list of components (*Comp_list*), which is sorted on the x-coordinates of the left corner of the components and the component (*component*) for which the constraints are to be generated. The procedure, INITIALIZE-SCANLINE, computes the total length of the interval in which the shadow is to be generated. This length includes the design rule separation distance. The y-coordinate of the top and the bottom of this interval are stored in the global variables, *top* and *bottom* respectively The procedure, GET-NXT-COMP, returns the next component (*curr_comp*) from *Comp_list*. This component is then removed from the list. Procedure LEFT-EDGE returns the vertical interval of component, *curr_comp*. If this interval is within the *top* and *bottom* then *curr_comp* can possibly have a constraint with *component*. This check is performed by the procedure $IN-RANGE$. If the interval for *curr_comp* lies within *top* and *bottom* and if this interval is not already contained within one of the intervals in

Algorithm SHADOW-PROPAGATION(*Comp_list, component*)
begin
    INITIALIZE-SCANLINE(*component*);
    $\mathcal{I} = \phi$;
    while((LENGTH-SCANLINE($\mathcal{I}$) < (*top* − *bottom*))
        and (*Comp_list* $\neq \phi$))
      curr_comp = GET-NXT-COMP(*Comp_list*);
      I_i = LEFT-EDGE(*curr_comp*);
      if(IN-RANGE($I_i, top, bottom$))
        I' = UNION($I_i, \mathcal{I}$);
        if($I' \neq \mathcal{I}$)
          ADD-CONSTRAINT(*component, curr_comp*);
          $\mathcal{I}$ = UNION($I_i, \mathcal{I}$);
  end.

Figure 12.11: Shadow-propagation algorithm

the interval set, $\mathcal{I}$, then the component lies in the shadow of *component* and hence a constraint has to be generated. Each interval represents the edge at which the shadow is blocked by a component. The constraint is added to the constraint graph by the procedure ADD-CONSTRAINT. The procedure UNION inserts the interval corresponding to *curr_comp* in the interval set at the appropriate position. This process is carried out till the interval set completely cover the interval from *top* to *bottom* or there are no more components in *Comp_list*. The Figure 12.12(a) shows the layout of components. The constraint for component A with other components is being generated. Figure 12.12(b) shows the intervals in the interval set as the shadow is propagated. From Figure 12.12(b) it is clear that the constraints will be generated between components A and components B, C, and D in that order. As component F lies outside the interval defined by *top* and *bottom* it is not considered for constraint generation. The interval generated by component E lies within one of the intervals in the interval set. Hence, there is no constraint generated between components A and E.

2. **Scanline Algorithm:**   In [Mal87], Malik presented an efficient algorithm based on scanline method. The *scanline* is an imaginary horizontal (or vertical) line that cuts through the layout in x-compaction (or y-compaction). An example of scanline is shown in Figure 12.13. The scanline data structure contains all the rectangles that are cut by this scanline. The rectangles are stored in non-decreasing order of their x-coordinates of the left boundaries. The scanline traverses from the top to the bottom of the layout for x-compaction. Similarly, for y-compaction, the scanline traverses from the left to the right of the layout. For x-compaction, as the

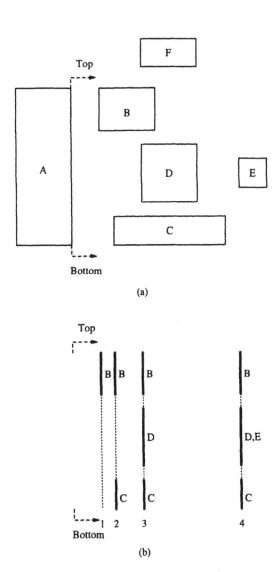

Figure 12.12: Interval generation for shadow propagation.

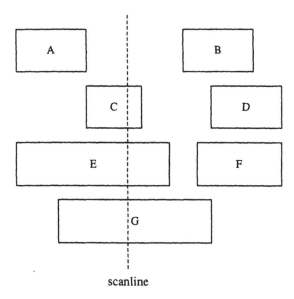

scanline

Figure 12.13: Example of a scanline.

line passes over the top edge of a rectangle, the rectangle is added to the scanline. Similarly, when the scanline passes over the bottom edge, the rectangle is deleted from the scanline. Two lists are required to move the scanline over the layout. One list contains rectangles in non-decreasing order of their YTOP and one list contains rectangles in non-decreasing order of their YBOTTOM. Let us denote the two sorted list as *topsorted* and *bottomsorted*, respectively. The algorithm SCANLINE is shown in Figure 12.14.

Note that the algorithm SCANLINE adds many redundant edges. A redundant edge is the one that does not affect the longest path from the source vertex to any other vertex in the graph. Hence the removal of the redundant edge will not change the constraint graph. Consider the example shown in Figure 12.15. If the distance between R_1 and R_i (E_{1i}) is less than the the distance between R_1 and R_2 (E_{12}) plus the distance between R_2 and R_i (E_{2i}), then the constraint between R_1 and R_i is redundant thus can be removed. The scanline algorithm uses these measures to remove redundant constraints.

12.3.1.2 Critical Path Analysis

After generation of constraint graph, the next step in constraint graph compaction is to determine the critical path through the graph. Let us explain the role of critical paths in compaction. The goal of one dimensional compaction is to generate a minimum width layout. The determination of minimum width

Algorithm SCANLINE (G)
begin
    while *topsorted* $\neq \phi$ **do**
        let R_1 be the first rectangle in *topsorted* and R_2 be
        the first rectangle in *bottomsorted*;
        if YTOP$(R_1) \leq$ YBOTTOM(R_2) **then**
        INSERT$(R_1, scanline)$;
        topsorted $=$ *topsorted* $- \{R_1\}$;
        for each rectangle $R_i \in scanline$ that is at the
            left of R_1 **do**
            ADD-CONSTRAINT(R_i, R_1, G);
        for each rectangle $R_j \in scanline$ that is at the
            right of R_1 **do**
            ADD-CONSTRAINT(R_1, R_j, G);
  end.

Figure 12.14: Scanline algorithm.

layout translates into a longest path problem. The longest path from source to a vertex is then the coordinate of the vertex. The longest path problem can be viewed as a shortest path problem by inverting the signs on the edge weights. As a result, this problem is also called the critical path problem. The edges that determine the minimum distance between the source and the sink form the *critical path* and vertices on the critical path are said to be *critical*. Tarjan [Tar83] describes a variety of algorithms to solve longest path problems; many others, including Lengauer [Len84], Liao and Wong [LW83] describe the application of various longest path algorithms to compaction. These algorithms calculate for all the vertices coordinates that are as small as possible. The worst case complexity of these algorithms is $O(|V| \times |E|)$, where V is the set of vertices in the graph and E is the set of edges. In [LW83], the complexity has been reduced to L iterations, while the run time of each iteration is $O(E)$. Where L is the number of negative weighted edges. If the constraint graph has special properties, more efficient algorithm can be used. In particular, for acyclic graphs, the worst case complexity is $O(|V| + |E|)$.

In practical layouts, the run times are almost linear in the number of layout elements. This is due to locality of the graph, that is, most edges represent very local constraints in the layout. In addition the number of edges that start at a vertex is usually quite small.

The algorithms described above can be improved by using a divide and conquer approach. The basic idea is to divide the graph in smaller subgraphs, which can be solved independently. A strong component is a subgraph in which there is a path from every node to every other node. Strong components are formed by the connectivity constraints; all the features in a strong component must move more or less together during compaction. If each strong component

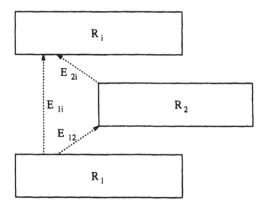

Figure 12.15: Example of redundant edge.

is reduced to a vertex, then the resulting graph is acyclic. In addition, the strong components can be assigned an ordering, which allows us to compute the effect of one strong component on the next one. The source and sink are modelled as separate strong components. Algorithms for finding strong components are described by Even [Eve79]; the best algorithm has a worst case time complexity of $O(|V| + |E|)$. The number of strong components and the number of vertices in strong components depend on the graph. In practical layout designs, the strong components are rather small.

Another method of improving the critical path algorithms is by reducing the total number of vertices and (or) edges of the graph. This reduction should not change the solution space of the constraint graph, which means that that all possible solutions that can be obtained directly or with reduction must be the same. The vertices can be reduced by grouping all the vertices which must have same relative positions. The edge reduction can be achieved by eliminating redundant edges. An edge is redundant, if there exists a path between the two vertices of the edge that does not contain the edge and which is longer (or has the same length) as the weight of the edge.

There may be vertices in the graph that are not critical and therefore have a range of legal positions. These vertices are said to have slack. Some secondary criterion must be used to assign unique positions to these vertices; one common objective is the minimization of total wire length in the cell. The compactor can place vertices with slack in such a way to increase circuit performance, to minimize wire length, to optimize fabrication yield, etc. Schiele [Sch83] and Eichenberger [Eic86] discuss algorithms that can be used to minimize wire length. Wire-length minimization significantly reduces the values of parasitic features associated with wires [Sch85].

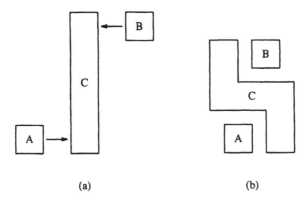

(a)   (b)

Figure 12.16: Example of wire jogging.

12.3.1.3   Wire Jogging

Both automatic jogging of wires and wire length minimization have received much attention in the last couple of years. This is, in part, due to the recent interest in channel compaction [Deu85]. One of the first approaches to jogging wires was reported by Hsueh [HP79]. In this approach jogs in wires were introduced at 'torque' points on a wire (see Figure 12.16). Wire jogging had limited success because it could reduce the size in one direction while, potentially, increasing the size in the other direction.

12.3.1.4   Wire Length Minimization

Features not on the critical path will typically find themselves pulled towards a layout edge because they are given their minimal legal spacing. This tends to increase wire length and reduce circuit performance.

One of the first methods used to reduce wire length was the 'average slack' method of Hsueh [HP79]. This approach uniformly distributes the empty space in a circuit among the features that were not on the critical path. Burns [BN86] presents a force-based heuristic that not only considers the effect of each wire layer but it also considers the cumulative effect of multiple wires connecting adjoining modules. This is effective in minimizing wires in an hierarchical layout.

12.3.2   Virtual Grid Based Compaction

The virtual grid compaction is a structured approach to compacting layout. The virtual grid is used to establish the relative placement of circuit features and does not correspond to physical grid. In this approach, the compactor gives locations to the virtual grid lines, not to the circuit component themselves.

Several compactors have been designed using virtual grid approach. Following are the two widely used algorithms based on this approach.

12.3.2.1   Basic Virtual Grid Algorithm

In this method, each component is attached to a grid line. Consider the example shown in Figure 12.17(a). Components A, B, and C are attached to the first grid line, while D, E, and F are attached to the second grid line. In the second step, the maximum necessary distance between any two grid lines is computed. In our example, the distance between C and F is required to be 14. In other words, the grid lines can be at distance 14 to each other without violating any design rules. In Figure 12.17(b), we show the compacted layout. This process is repeated for all adjacent grid lines. X-Compaction is usually followed by Y-Compaction. The basic advantage of virtual grid method is it is fast.

12.3.2.2   Split Grid Compaction

In [Boy87], Boyer introduced *split grid compactor* which places distinct circuit features that fall on the same virtual grid separately, splitting the virtual grids where necessary. Split grid compactor uses a data structure that allows only to store the grid points that are of interest. The grid points which contain features are the ones added to the data structure. This allows fast access to the circuit features.

Initially, the compactor identifies groups of circuit features falling on the same virtual grid lines that need to be placed together. Local connectivity is used to identify the groups. For example, consider a vertical virtual grid line. There are two situations that might occur: features are connected by a vertical wire segment or the features are connected by a vertical transistor. In any case, the features are grouped together. Groups are identified by traversing each virtual grid line. After the circuit features are grouped together, the compaction is done in two passes: first the x-compaction and then y-compaction. A group is first compacted by determining the spacing necessary for each component in the the group. Then each group is placed independently. Features are spaced with respect to the features in the neighboring groups. Consider the example shown in Figure 12.18(a). We group A, B, and C together on the first grid line. On the second grid we form two groups. The first group consists of D and E, while the second just consists of F. Figure 12.18(b) shows the solution after compaction.

Compression-Ridge Method: Compression-ridge method was first suggested by Akers, Geyer, and Roberts [AGR70]. In this method, vertical and horizontal regions of empty spaces, called *compression ridges* are formed. They have the following properties:

1. Compression ridge is a constant width band of empty space stretching from one side of the layout to the other side.

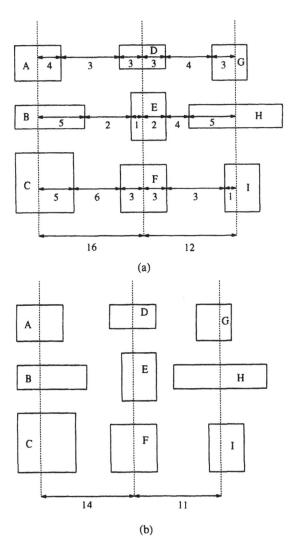

Figure 12.17: Virtual grid symbolic inverter.

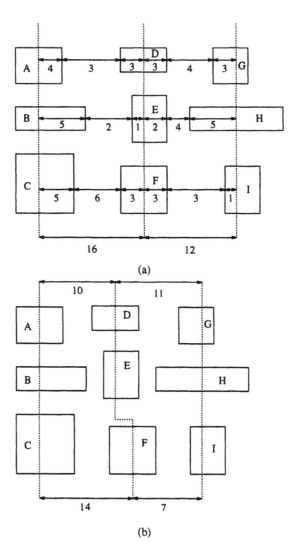

Figure 12.18: Split grid compaction.

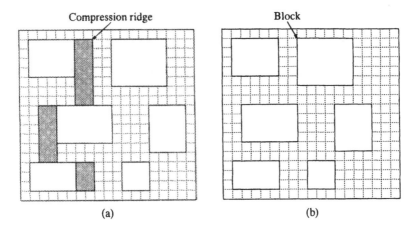

Figure 12.19: Compression ridge.

2. Compression ridge can intersect wires that are perpendicular to it. However, it cannot intersect those wires that are parallel to it.

3. The width of a compression ridge is such that when the space is removed, no design rules should be violated in the resulting layout.

The example of compression ridge is shown in Figure 12.19. The compression ridge is shown by the shaded region in Figure 12.19(a). In this technique, these compression ridges are subsequently removed from the layout until no more ridges are found. One method for finding the compression ridges is to use the virtual split-grid method. The vacant space along the grid line can be replaced by a rectangular compression ridge. The width of this ridge is the minimum compression that can be achieved along the grid line. Figure 12.19(b) shows the layout after removing the compression ridge. Note that compression ridges are formed from one end of the layout to the other which in the worst case is very time consuming. Therefore, an efficient algorithm is required to find the compression ridges. Dai and Kuh [DK87b] proposed an $O(n \log n)$ algorithm for finding compression ridges that allows the largest decrease in layout width. One of the main advantages of the compression ridge method is that the compaction can be broken into smaller steps.

12.3.2.3 Most Recent Layer Algorithm

In [BW83], Boyer and Weste presented a virtual grid compaction algorithm called *most recent layer algorithm*. The algorithm consists of two different passes: first in the x-direction, then in the y-direction. The diagonal checks are done during the y-compaction. The algorithm does not require any back-tracking and the time complexity of this algorithm is $O(n)$, where n is the total number of features in the layout.

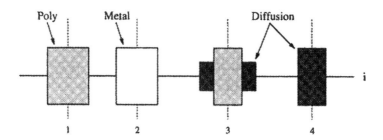

Figure 12.20: Most recent layer compaction.

For the x-compaction, each horizontal grid line has a set of reference lines ('pickets'), one for each layer, to keep track of the right edge of the most recent placement of a mask feature on that grid line. Initially, a column is placed as close to the picket as possible (without violating the design rules). The pickets are then updated. If there is no feature in a layer for a certain column, the picket position for that layer is not updated (it remains unchanged). To position a mask feature in a layer, the left edge of the feature is used to determine the necessary location of that layer with respect to picket. The right edge of the feature is used to update the pickets. For an illustration, consider the example shown in Figure 12.20. There are three different pickets for three layers, one for metal, one for poly, and one for diffusion. Consider the x-compaction in horizontal virtual grid line i. We assume that features in columns 1, 2, 3 are already placed and the pickets are updated. Now, the feature of the diffusion layer has to be placed in column 4. Only the picket of diffusion layer will be used to find the location of this feature. After placing the feature in its minimal distance position, the picket in diffusion layer is updated to the coordinate of the right side boundary of the feature.

The y-compaction is done in similar way. However, additional information is necessary in order to handle the diagonal constraints. The left and right edges, as well as the upper edges, of the mask features must be recorded in order to do the diagonal checking.

12.4 $1\frac{1}{2}$-Dimensional Compaction

In [SSVS86], Shin, Sangiovanni-Vincentelli, and Sequin presented a new compactor based on simulation of zone refining process. Although compactor is based on simulation of an engineering process, it is a deterministic algorithm and differs sharply from other simulation based approaches such as simulated annealing and simulated evolution. The key idea is to provide enough lateral movements to blocks during compaction to resolve interferences. In that sense, this compactor can be considered $1\frac{1}{2}$-dimensional compactor, since the geometry is not as free as in true 2-dimensional compaction.

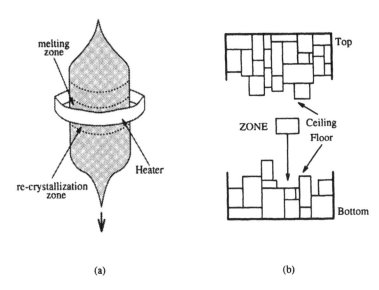

(a)   (b)

Figure 12.21: Zone refining.

The process of zone refining is used to purify crystal ingots. The basic idea is to allow limited melting of the crystal and let purities drain out of the crystal. The zone refining process starts with an already developed 'impure' crystal. As shown in Figure 12.21(a), the crystal is slowly pulled through a heating element to locally heat the crystal to melting temperature. At the exit end of the heater, the material re-crystallizes. Since impurities are built into the crystal lattice at a much slower rate than the crystal material, impurities have tendency of being left out during re-crystallization process. That is, the impurities are left in the molten state in the heated zone. Eventually, impurities are drained out of one end of the crystal.

In terms of layout compaction, the algorithm starts with a layout. The vacant space in the layout is considered the impurity. Starting from one side, blocks are considered row by row and are re-arranged after they have been moved across the open zone. During its movement in the free zone, the blocks travel through the entire width of the layout and hence may be placed anywhere along the boundary. In Figure 12.21(b), we show the process of compaction by zone refining.

The algorithm maintains an XY adjacency graph. In an XY adjacency graph, vertices represent blocks, while edges represent horizontal and vertical adjacency. That is, two blocks have a horizontal edge if they share a vertical boundary. Similarly, two blocks have a vertical edge if they share a horizontal boundary. The labels on the edges represent the minimum allowable distance between blocks. Four additional vertices are added to keep all the blocks within the required bounded rectangle. Note that free space is ignored in computing the neighborhood edges between blocks. Figure 12.22(a) an instance of problem along with its XY adjacency graph in Figure 12.22(b).

Algorithm assumes that the input is partially compacted layout, which can be obtained by two applications of a 1-D compactor. It maintains two lists called *floor* and *ceiling*. Floor consists of all the blocks which are visible from the top and may become a neighbor of future block. Ceiling is a list of all blocks which can be moved immediately. That is, ceiling is the list of blocks visible from the bottom. The algorithm selects the lowest block in the ceiling list and moves it to the place on the floor, which maximizes the gap between floor and ceiling. This process is continued until all blocks are moved from ceiling to floor.

Let us illustrate the algorithm with an example in Figure 12.22. Since C is the lowest block in the ceiling list, it is selected for the move. Figure 12.22(c) shows that the gap is maximum at the boundary between blocks A and B. Therefore C is moved between and A and B. The modified layout and the XY-adjacency graph are shown in Figures 12.22(d) and (e) respectively.

12.5   Two-Dimensional Compaction

Recall that there are three different types of constraints imposed by the design rules that must be satisfied to obtain a valid layout. These constraints are size constraints, overlap constraints, and separation constraints. Furthermore, designer can impose extra constraints known as *user defined constraints*. Given a symbolic layout consisting of rectangular features, all these constraints can be written using linear constraint equations. Let us assume that for each block B_i, two coordinates, (x_i, y_i) and (x'_i, y'_i), are given for the lower left corner and the upper right corner, respectively. Let a block B_i has height h_i and width w_i, then the size constraints can be written as:

$$x_i + w_i \le x'_i \qquad y_i + h_i \le y'_i$$

$$x'_i - w_i \le x_i \qquad y'_i - h_i \le y_i$$

Wires have a fixed width but variable length. A vertical wire W_j with width w_j is specified by the constraints:

$$x_j + w_j \le x'_j \qquad y_j \le y'_j \qquad x'_j + w_j \le x_j$$

The constraints for the horizontal wires can be given in the similar way.

In the course of compaction, the algorithm must maintain appropriate connections between blocks and wires and between wires. Wires on the boundary of a block can slide along the boundary within the range specified by the user, provided that no constraints are violated. The overlap constraints between a block B_i and a wire segment W_j can be given as:

$$x_j \le x'_i \qquad x'_i \le x'_j \qquad x_i \le x_j$$

$$y_i + r^1_{ij} \le y_j \qquad y'_j + r^2_{ij} \le y'_i$$

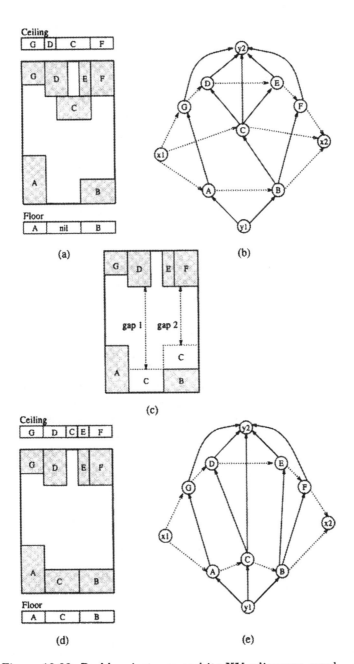

Figure 12.22: Problem instance and its XY adjacency graph.

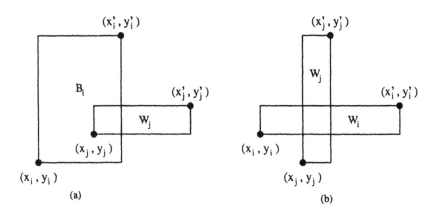

Figure 12.23: Overlap constraints.

Note that W_j is connected to the right boundary of B_i and the range is $[r_{ij}^1, r_{ij}^2]$ (Figure 12.23(a)). The overlap between two wires W_i and W_j, as shown in Figure 12.23(b), can be specified as:

$$x_i \leq x_j \qquad x_j' \leq x_i'$$

$$y_j \leq y_i \qquad y_i' \leq y_j'$$

Next the separation constraints can be specified as minimum distance constraints between two non-overlapping features. If blocks B_i and B_j are not supposed to overlap, they must be separated by a certain distance. There are four possible cases: j is on the right of i at a distance of at least d_{ij}^1, j is on the left of i at a distance of at least d_{ij}^2, j is on the top of i at a distance of at least d_{ij}^3, and j is below i at a distance of at least d_{ij}^4. Thus, one of the following must be satisfied.

$$c_{ij}^1 : x_i' + d_{ij}^1 \leq x_j \qquad c_{ij}^2 : x_j' + d_{ij}^2 \leq x_i$$

$$c_{ij}^3 : y_i' + d_{ij}^3 \leq y_j \qquad c_{ij}^4 : y_j' + d_{ij}^4 \leq y_i$$

Let $C_{ij} = \{c_{ij}^1, c_{ij}^2, c_{ij}^3, c_{ij}^4\}$ and $D = \{C_{ij} \mid i$ and j are non-overlapping features $\}$. Thus to ensure that the design rules are satisfied, one of the four constraints in C_{ij} must be satisfied.

Therefore, the constraints can be divided into two classes:

1. set of constraints, B, that must be satisfied, which include size, overlap and user defined constraints.

2. set of constraints, D, that are divided into groups and at least one of the constraints in each group must be satisfied.

After the generation of constraints, the problem can be solved using integer linear programming technique. However, the complexity of the linear programming technique is exponential thereby making it impractical even for a moderate size problem.

In [SLW83], Schlag, Liao, and Wong showed that the 2-D compaction problem is NP-complete and gave a branch-and-bound solution for the problem. However, again the complexity of the algorithm is in the worst case exponential.

12.5.1 Simulated Annealing based Algorithm

In [HLL88], Hseih, Leong, and Liu proposed a solution to 2-D compaction using simulated annealing technique. Although this technique produces suboptimal solution, this is much faster than branch-and-bound and integer linear programming technique. The layout can be represented by a valid set of constraints. A valid set of constraints is a subset E of constraints that contains all the constraints from B and at least one constraint from D. We use the notation $E = B \cup M$, where M contains exactly one constraint from each group of D. In the simulated annealing algorithm, given a solution $B \cup M$, a *move* is defined as selecting a group in D and exchanging the constraints in that group. Two solutions $B \cup M$ and $B \cup M'$ are said to be neighbors if M' can be obtained from M by interchanging the chosen constraint in one of the groups of D. Clearly, it is possible to go from one given solution to another by a sequence of moves.

12.6 Hierarchical Compaction

The compactors discussed in the previous sections, perform compaction on layouts composed from a library of pre-defined features and wire segments. Since the characteristics of the layout primitives other than their basic shapes are typically exploited to generate compact layouts, it can be difficult to use such system for hierarchical design.

Hierarchical compaction can be used for hierarchical designs to reduce the space and computation time of the layout compaction. In the hierarchical compaction, transistors, contacts, and modules are treated in the same manner. In this section, we discuss one hierarchical compaction algorithm based on constraint-graph generation.

12.6.1 Constraint-Graph Based Hierarchical Compaction

Given a hierarchical symbolic layout, hierarchical constraint graph is generated at each level of the hierarchy of the design from bottom up. Initially, constraints are generated for all the leaf cells consisting of basic features. Each leaf cell is compacted using the corresponding constraint graph and the boundary of the compacted leaf cell is fixed. Once the leaf cell is compacted and the boundary is fixed, the cell can be treated as a single cell in the next level in the hierarchy and constraints can be generated for the cells in that level. The

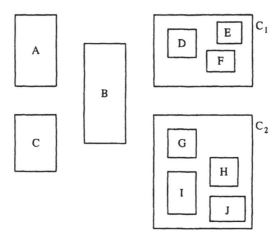

Figure 12.24: Hierarchical layout.

compaction is carried out by generating constraints at each level. For an illustration, consider the example of a hierarchical design shown in Figure 12.24. The layout consists of two levels of hierarchy. In the leaf level, cells C_1 and C_2 are compacted by generating their constraint graphs G_1 and G_2 as shown in Figure 12.25. Once C_1 and C_2 are compacted, their layout is represented by a vertex in the next level of the constraint generation as shown in the graph G_3 in Figure 12.25.

The hierarchy of the design will be preserved if at each level, the boundary of the compacted cells are kept rectangular. However, keeping the boundary rectangular does not produce a good solution. To get better results, the boundary of the compacted cell at any level can be given any arbitrary rectilinear shape and the constraint graph may be allowed to have multiple constraint edges between two vertices, specifying different separation constraints. However, this approach does not preserve the hierarchy of the layout.

12.7 Recent trends in compaction

In this section, two new trends in compaction are briefly reviewed. These include performance-driven compaction and compaction techniques for yield enhancement.

12.7.1 Performance-driven compaction

In [OCK95], authors use an iterative parameterized LP formulation to model a force-directed wire respacing scheme for compaction under timing constraints and compaction under peak crosstalk constraints. Although it uses a distributed delay model that factors in the coupling capacitances of the nets,

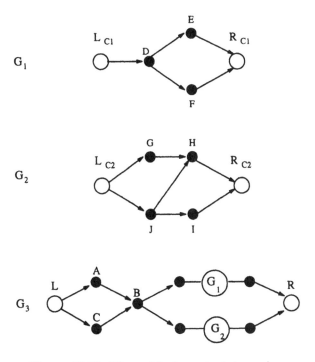

Figure 12.25: Hierarchical constraint graph.

some of the intuition behind the force-directed scheme is lost in translating the two-dimensional optimization problem into a problem with a one-dimensional objective function (delay/crosstalk) and a constrained penalty parameter associated with the second objective (area).

In [WLLC93], a LP formulation is developed for timing constraints on critical paths (in addition to regular layout constraints), and then shows how to solve them efficiently using a graph-based simplex algorithm. The delay model assumes that the delay of a wire is proportional to its length. First, an LP is used to find a tight upper bound on the delays of the timing critical paths by adjusting the wire lengths. Next, another LP is used to perform one-dimensional compaction on the layout while ensure that the delay of each of the timing critical paths remains less than the upper bound determined by the first LP.

12.7.2   Compaction techniques for yield enhancement

Chiluvuri and Koren [CRK95] developed a constraint-graph based compaction strategy using empirical heuristic desired locations for objects not on critical paths in the constraint graph to make the spacing between different elements more uniform, in an effort to decrease the sensitivity of the layout to random point defects.

Bamji and Malavasi [CM96] extended the work done by Chiluvuri and Koren by using a network flow based formulation for the layout respacing. The formulation models objectives such as yield, crosstalk or wire length (or their linear combinations) using a piece-wise linear convex cost function for the edges.

12.8   Summary

Compaction is a very important phase in physical design cycle. The objective of a compaction algorithm is to reduce the layout area. Research in symbolic layout compaction has resulted in two major compaction strategies: the constraint-graph based compaction and the virtual grid based compaction. Constraint-graph based compactors usually produce smaller area layouts as compared to the virtual grid compactors, while virtual grid compactors typically run faster. The speed of both type of algorithms can be improved by using hierarchy of the layout.

12.9   Exercises

1. For an instance shown in Figure 12.10, generate constraints using scanline algorithm.

2. Develop an algorithm to remove the redundant constraints in a constraint graph.

3. Symbolic layout of a single cell may described using a unit size. Develop an algorithm to produce a full-size layout of a cell from a unit-size description by adding wires to the cell.

4. Extend the virtual grid based algorithm to include jog insertion if necessary to reduce the layout area.

5. Extend the two-dimensional compaction to include jog insertion.

†6. If two neighboring cells have to be connected, then the compaction algorithm can stretch the cells such that their terminals to be connected lie in the same x- or y-position. This process of stretching cells to align terminals is known as *pitchmatching*. Design the necessary constraints to handle the pitchmatching in a compaction algorithm.

‡7. Implement the zone refining algorithm for L-shaped blocks. Note that rotation, flipping of blocks have to be taken into consideration.

‡8. Develop constraints for L-shaped blocks. Can we represent each L-shaped block as a combination of two rectangles ? What additional constraints are needed ?

9. Consider the channel compaction problem. Channels are compacted in the y-directions. Which channel routing algorithm produces the most compactable solution ? Which channel router is easiest to adapt to integrated channel routing and compaction ?

Bibliographic Notes

The first compaction algorithm called *shear-line compaction* was proposed by Akers et al. [AGR70]. Symbolic layout and compaction were first combined in the STICKS system [Wil78]. In [BN87], a constraint generation technique for hierarchical compaction has been proposed. In [KW84], the compaction problem was formulated into a mixed integer linear programming problem of a very special form. Symbolic layout compaction with symmetric constraints was considered in [OSOT89]. The symmetric constraint maintains the geometrical symmetry of the circuit components during the compaction.

[SL89b, WLC90] present efficient two-dimensional layout compaction algorithms. In [dDWLS91], a two-dimensional topological compactor with octagonal geometry is presented.

Algorithmic Aspects of one dimensional layout compaction are discussed in [DL87a]. In [RPV$^+$87], geometrical compaction in one dimension for channel routing is considered. In [LV90], an $O(n^{1.5} \log n)$ 1-d compaction algorithm is presented. [DL91] presents on minimal closure constraint generation for symbolic cell assembly. [CH87] explains how to generate incremental compaction spacing constraints. In [BV93], discusses a method for identifying overconstraints during hierarchical compaction. [Ono90] presents layout compaction with attractive and repulsive constraints. It has been shown in [PDL97] that the traditional problem of removing redundant constraints to yield the smallest possible constraint graph in symbolic compaction is NP-hard. However, if one is also allowed to add new constraints, the smallest possible constraint graph can be obtained in polynomial time. In [DPLL96], a global strategy for the elimination of positive cycles in overconstrained graph-based compaction problems is presented. It uses new polynomial LP-based formulations that are of independent interest and applicability by themselves. [Koc96] uses local logic resynthesis specific to the FPGA architecture being compacted to perform compaction. In [ASST97], a min cost flow based formulation of the compaction problem (wire length minimization) is presented.

In terms of parallel algorithms, [SC96] presents a parallelization of the "cut, compact and merge" approach towards full-chip compaction. In [CTC94] a parallel algorithm for integrated compaction and wire balancing on a shared memory multiprocessor has been evaluated.

[Har91, YCD$^+$95, BV92, Mar90, DMH$^+$93], present several schemes for hierarchical compaction and methods for dealing with large databases. In [FCMSV92], an efficient methodology for symbolic compaction of analog IC's with multiple symmetry constraints is presented.

Chapter 13

Physical Design Automation of FPGAs

Despite advances in VLSI design automation, the time-to-market for even an ASIC chip is unacceptable for many applications. The key problem is the time taken due to fabrication of chips, and therefore there is a need to find new technologies, which minimize the fabrication time. Gate Arrays use less time in fabrication as compared to full-custom chips, since only routing layers are fabricated on top of pre-fabricated wafer. However, fabrication time for gate-arrays is still unacceptable for several applications. In order to reduce time to fabricate interconnects, programmable devices have been introduced, which allow users to program the devices as well as the interconnect. In this way all custom fabrication steps are eliminated.

Programmable Logic Devices (PLDs) are devices that can be programmed by the user to implement a logic function. These devices offer short turnaround time and as a result they are becoming increasingly important for systems as well as system prototypes. In addition, they have a low manufacturing cost and are fully testable. One such device which is gaining more popularity is *Field Programmable Gate Arrays (FPGAs)* .

As discussed in Chapter 1, FPGA is a new approach to ASIC design that can dramatically reduce manufacturing turn around time and cost. In its simplest form, an FPGA consists of a regular array of programmable logic blocks interconnected by a programmable routing network. A programmable logic block is a RAM and can be programmed by the user to act as a small logic module. Given a circuit, user can program the programmable logic module using an FPGA programming tool. The key advantage of FPGAs is re-programmability. The RAM nature of the FPGAs allows for in-circuit flexibility that is most useful when the specifications are likely to change in the final application. In some applications such as remote sensors, it is necessary to make system updates via software. In FPGA, a data channel is provided, which allows easy transfer of the new logic function and reprogramming the FPGA.

The physical design automation of FPGAs involves mainly three steps which

A	B	C	f
0	0	0	0
0	0	1	0
0	1	0	0
0	1	1	1
1	0	0	1
1	0	1	0
1	1	0	0
1	1	1	0

Table 13.1: Truth table for $f = \bar{A}BC + A\bar{B}\bar{C}$.

include partitioning, placement and routing. Partitioning problem in FPGAs is significantly different from the partitioning problems in other design styles. This problem mainly depends on the architecture in which the circuit has to be implemented. Placement problem in FPGAs is very similar to the gate array placement problem. The routing problem in FPGAs is to find a connection path and program the appropriate interconnection points. In this chapter, we discuss the architecture of FPGAs, their physical design cycle, and algorithms used for partitioning and routing problems in FPGAs.

In order to gain a better perspective of the physical design problems related to FPGAs, we start with a description of FPGA architectures.

13.1 FPGA Technologies

An FPGA architecture mainly consists of two parts: the logic blocks, and the routing network. A logic block has a fixed number of inputs and one output. A wide range of functions can be implemented using a logic block. Given a circuit to be implemented using FPGAs, it is first decomposed into smaller sub-circuits such that each of the sub-circuit can be implemented using a single logic block. There are two types of logic blocks. The first type is based on Look-Up Tables (LUTs), while second type is based on multiplexers.

1. **Look-up table based logic blocks:** A LUT based logic block is just a segment of RAM. A function can be implemented by simply loading its LUT into the logic block at power up. If function $f = \bar{A}BC + A\bar{B}\bar{C}$ needs to be implemented, then its truth table (shown in Table 13.1) is loaded into the logic block. In this way, on receiving a certain set of inputs, the logic blocks simply 'look up' the appropriate output and set the output line accordingly. Because of the reconfigurable nature of the LUT based logic blocks, they are also called the *Configurable Logic Blocks (CLBs)*.

   It is clear that $2^{I_{max}}$ bits are required in a logic block to represent a I_{max}-bit input, 1-bit output combinational logic function. Obviously, logic

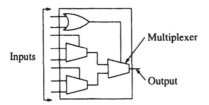

Figure 13.1: A multiplexer based logic block.

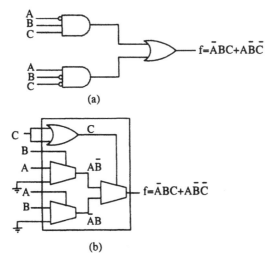

Figure 13.2: A logic function mapped to a multiplexer based logic block.

blocks are only feasible for small values of I_{max}. Typically, the value of I_{max} is 5 or 6. For multiple output and sequential circuits the value of I_{max} is even less.

2. **Multiplexer based logic blocks:** Typically a multiplexer based logic block consist of three 2-to-1 multiplexers and one two-input OR gate as shown in Figure 13.1. The number of inputs is eight. The circuit within the logic block can be used to implement a wide range of functions. One such function, shown in Figure 13.2(a) can be mapped to a logic block as shown in Figure 13.2(b). Thus, the programming of multiplexer based logic block is achieved by routing different inputs into the block.

There are two models of routing network: the segmented and the non-segmented.

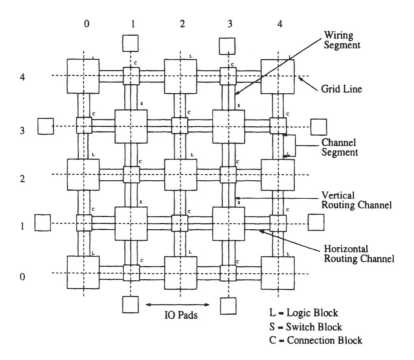

Figure 13.3: Non-segmented interconnect.

1. **Non-segmented model:**   A typical non-segmented model is shown
   in Figure 13.3. The non-segmented model is set up as a regular grid
   of five horizontal and five vertical metal lines passing between switch
   blocks (S). The switch blocks are rectangular switch boxes. They are
   used to connect the wiring segments in one channel segment to those in
   another. Depending on the topology of the S block, each wiring segment
   on one side of S may be switchable to either all or some fraction of wiring
   segments on each side of the S block. The fewer the wiring segments a
   wiring segment can be switched to, the harder the FPGA is to route.
   Figure 13.6 and Figure 13.5 are two of the switch box architectures used
   by Xilinx for their 4000XC and 2000XC series. In Fig 13.5 a predefined set
   of programmable connections based on some probability and statistical
   data is used to obtain an efficient and economical switch box routing
   architecture. On the other hand, Figure 13.6 shows a more versatile and
   efficient routing architecture, but far more expensive to implement.

   In addition to the switch blocks, there are the connection blocks (C)
   that are used to connect the logic block pins to the routing channels.
   Depending on the topology, each L block pin may be switchable to either
   all or some fraction of wiring segments that pass through the C block.
   Again, the fewer the wiring segments a pin can be switched to, the harder

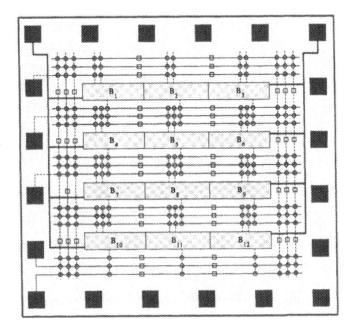

Figure 13.4: Segmented interconnect.

the FPGA is to route.

2. **Segmented model:** In segmented model, the tracks in the channels contain predefined wiring segments of same or different lengths. Other wiring segments pass through the channels vertically. Each input and output of a logic block is connected to a dedicated vertical segment. As a result, there are no vertical constraints. There are additional global vertical lines which provide connections between different channels. Connection between two horizontal segments is provided through an *antifuse*, whereas the connection between a horizontal segment and a vertical segment is provided through a *cross fuse* (see Figure 13.4). Programming (blowing) one of these fuses provides a low resistance bidirectional connection between two segments. When blown, antifuses connect the two segments to form a longer one. In order to program a fuse, a high voltage is applied across it. FPGAs have special circuitry to program the fuses. The circuitry consists of the wiring segments and control logic at the periphery of the chip. Fuse addresses are shifted into the fuse programming circuitry serially. When the objective is to fabricate reconfigurable routing network re-programmable switches can be used instead of fuses.

The segmented model is *uniform* if the segments in all tracks have same length and the antifuses in different tracks in a channel are aligned in columns.

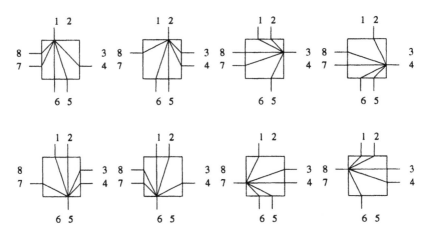

Programmable switch matrix interconnections

Figure 13.5: XC2000 family switch box architecture

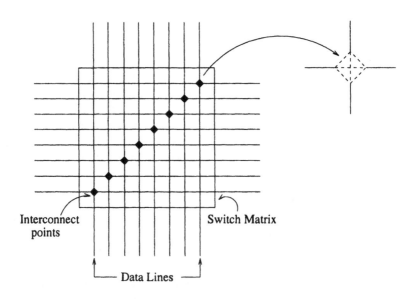

Figure 13.6: XC4000 family switch box architecture

The segmented model normally has advantage over the non-segmented model in terms of utilization of routing resources. In the non-segmented model only one segment of one net can be routed on a track. Whereas, in the segmented model, the segments of several nets can be assigned to a track as long as no two net segments are assigned to the same track segment.

The total number of programmable switches in the segmented model is higher as compared to the number of switches in the non-segmented model. The delay of a net is directly proportional to the number of programmable switches used to route that net. The number of programmable switches used to route a net is higher in segmented model as compared to non-segmented model. As a result, the non-segmented model is preferred over the segmented model when the performance is the primary objective.

13.2 Physical Design Cycle for FPGAs

The physical design cycle for FPGAs consists of the following steps:

1. **Partitioning:** The circuit to be mapped onto the FPGA has to be partitioned into smaller sub-circuits, such that each sub-circuit can be mapped to a programmable logic block. Unlike the partitioning in other design styles, there are no constraints on the size of a partition. However, there are constraints on the inputs and outputs of a partition. This is due to the unique architecture of FPGAs.

2. **Placement:** In this step of the design cycle, the sub-circuits which are formed in the partitioning phase are allocated physical locations on the FPGA, i.e., the logic block on the FPGA is programmed to behave like the sub-circuit that is mapped to it. This placement must be carried out in a manner that the routers can complete the interconnections. This is very critical as the routing resources of the FPGA are limited. The placement algorithms for general gate arrays are normally used for the placement in FPGAs, and therefore, will not be discussed in this chapter.

3. **Routing:** In this phase, all the sub-circuits which have been programmed on the FPGA blocks are interconnected by blowing the fuses between the routing segments to achieve the interconnections.

Figure 13.7 shows complete physical design cycle of FPGAs. System design is available as a directed graph which is partitioned in second step. Placement involves mapping of sub-circuits onto CLBs. Shaded rectangles represent CLBs which have been programmed. Final step is routing of channels.

13.3 Partitioning

A I_{max}-input, 1-output LUT based logic block is powerful than a I_{max}-input, 1-output multiplexer based logic block, as the former can implement all the

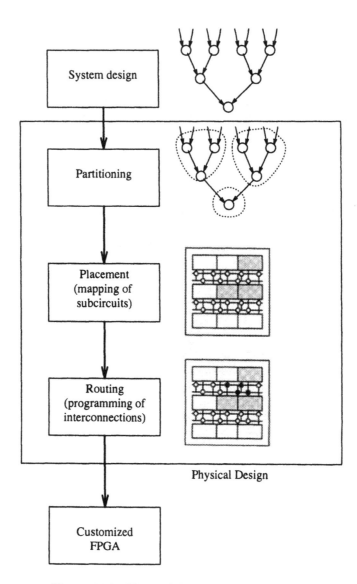

Figure 13.7: Physical design cycle of FPGA.

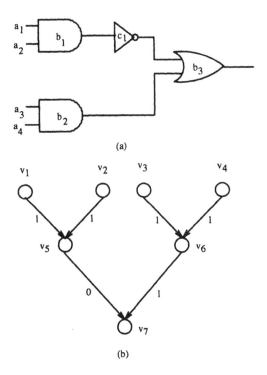

(a)

(b)

Figure 13.8: A boolean circuit and its DAG representation.

logic functions of I_{max} inputs, whereas the capabilities of later are limited by the circuitry inside the block. As a result the LUT based logic blocks are more popular than the multiplexer based logic blocks. Therefore, in this section we discuss a partitioning problem only for the LUT based logic blocks.

The circuit to be mapped onto the FPGA has to be partitioned into smaller sub-circuits such that each sub-circuit can be implemented using the logic blocks. In order to achieve this, we model the boolean network (N) consisting of AND, OR and NOT gates using directed acyclic graph (DAG) as follows: Let N be a set of inputs $a_1, a_2, \ldots, a_l$, B be a set of AND and OR gates $b_1, b_2, \ldots, b_m$ and C be a set of NOT gates $c_1, c_2, \ldots, c_n$. The corresponding DAG is defined as $G = (V_a \cup V_b, E)$, where $V_a = \{v_1, v_2, \ldots, v_l\}$ such that $v_i \in V_a$ represents an input a_i to the network, and $V_b = \{v_{l+1}, v_{l+2}, \ldots, v_{l+m}\}$ such that $v_j \in V_b$ represents an AND or OR gate g_{j-l}. An edge $e_{ij} = (v_i, v_j) \in E$, directed from v_i to v_j, represents a connection from the output of g_i to an input of g_j, either through a NOT gate or direct. If connection between g_i and g_j is through a NOT gate, then a weight 0 is associated with the edge e_{ij}, otherwise the weight of edge $e_{ij} \in E$ is 1. A logic network is shown in Figure 13.8(a) and corresponding DAG is shown in Figure 13.8(b).

The nodes in V_a are also referred as the *input nodes*. A node $v_i \in V$ is a

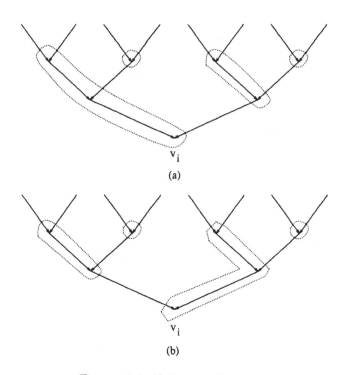

(a)

(b)

Figure 13.9: Utilization division.

fan-in node of $v_j \in V$ if there is a directed edge from node v_i to node v_j. An input node does not have a fan-in node.

Graph $G_i = (V_i, E_i)$ is a *subgraph* of a graph $G = (V, E)$, if $V_i \subseteq V$ and $E_i = \{e_{ij} | e_{ij} = (v_i, v_j), v_i \in V_i, v_j \in V_i\}$. The *indegree* $I(G_i)$ of a directed subgraph G_i is defined as the number of edges coming into G_i, while $O(G_i)$, the *outdegree* of G_i is defined as the number of edges coming out of G_i. More precisely,

$$I(G_i) = |\{(v_i, v_j) | (v_i, v_j) \in E, v_i \notin V_i, v_j \in V_i\}|$$

$$O(G_i) = |\{(v_i, v_j) | (v_i, v_j) \in E, v_i \in V_i, v_j \notin V_i\}|$$

The partitioning problem can be formally stated as follows: Given a directed acyclic graph G, maximum number of output terminals of a logic block denoted as $O_{\max}$ and maximum number of input terminals of a logic block denoted as $I_{\max}$, partition G into minimum number of vertex sets $V_1, V_2, \ldots, V_k$ such that subgraphs $G_1, G_2, \ldots, G_k$ satisfy the constraints

$$I(G_i) \leq I_{\max}, 1 \leq i \leq k$$

$$O(G_i) \leq O_{\max}, 1 \leq i \leq k$$

The partitioning problem is also referred as the mapping problem as it maps sub-circuits to logic blocks. Note that this partitioning problem is significantly different than the partitioning problem considered in Chapter 5. In particular, the number of vertices assigned to any subgraph is not important. The important parameter is the number of edges coming in or going out of a subgraph.

In [FRC90], Francis, Rose and Chung presented a dynamic programming algorithm for partitioning the DAG for which the fan-in of each node does not exceed I_{max}. The following terms need to be defined in order to explain the algorithm. A *mapping* of a node v_i, in a tree T, is a circuit of I_{max}-input LUTs which implements the sub-tree of T that is rooted at v_i and extends to the leaf nodes of T. The *cost* of a mapping is the number of LUTs needed to implement that mapping. The *root* lookup table of a mapping of the node v_i has as its single output the boolean function of the node v_i. The *utilization* of a lookup table is the number of inputs U, out of the K inputs that are actually used in a circuit. If $v_{f1}, v_{f2}, \ldots, v_{fp}$ are the fan-in nodes of a node v_i, then the root lookup table of mapping of v_i includes all the fan-in edges of v_i and some sub-tree S_i rooted at each fan-in node v_{fi}. (see Figure 13.9(a)). The term *utilization division* is introduced to denote the distribution of the inputs to the root lookup table among these subtrees. If u_i is the number of leaf nodes in the sub-tree S_i then the set $\mathcal{U} = \{u_1, u_2, \ldots, u_p\}$ specifies the utilization division of the root lookup table. There may be many possible utilization divisions of the root lookup table of a mapping of a node. Figure 13.9 shows the utilization $\{3, 1\}$ for the node v_i, whereas, Figure 13.9(b) shows the utilization $\{1, 3\}$ for the same node v_i.

Let $\text{MinMap}(v_i, U)$ be the optimal mapping of node v_i with a root utilization of U. For each leaf node v_i, $\text{MinMap}(v_i, U)$ is set to 0 for all values of U. Assuming that $\text{MinMap}(v_j, U)$ is computed for each fan-in node $v_j = v_{f1}, v_{f2}, \ldots, v_{fp}$ of an internal node v_i and for all $U = 1$ to I_{max}, $\text{MinMap}(v_i, U)$ can be computed for $U = 1$ to I_{max} as discussed below. In order to compute $\text{MinMap}(v_i, U)$, compute $\text{MinMap}(v_i, U)$ for each utilization division $\mathcal{U}$ of U by combining v_i with the mappings of fan-in nodes of v_i. $\text{MinMap}(v_i, U)$ is simply the minimum cost $\text{MinMap}(v_i, U)$ computed over all utilization divisions of U. $\text{MinMap}(v_i, U)$ for all $U = 1, \ldots, I_{max}$ is computed while visiting the node v_i in a post-order traversal of the tree. This ensures the condition that the mappings for all fan-in nodes of v_i are already computed.

In the cases when a node in DAG has a fan-in greater than I_{max}, a node decomposition phase has to be carried out before applying the above algorithm. The output of the node decomposition is a functionally equivalent DAG in which all the nodes have fan-in less than I_{max}. Figure 13.10(b) shows a DAG obtained after decomposition of node v_1 in Figure 13.10(a).

13.4 Routing

After all the sub-circuits have been mapped to logic blocks, these sub-circuits are interconnected by blowing the fuses in the routing channels. Routing of

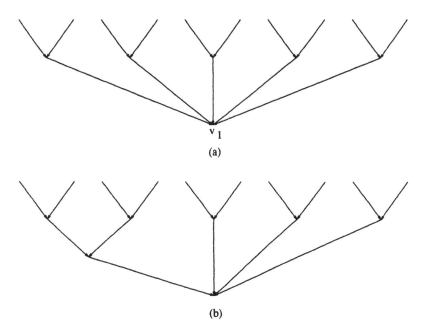

(a)

(b)

Figure 13.10: Node decomposition.

FPGAs is different from the routing of general blocks because of the segmented nature of channels. In the following sections, we discuss FPGA routing for different models.

13.4.1 Routing Algorithm for the Non-Segmented Model

In this section, we discuss the algorithm presented by Brown, Rose and Vranesic [BRV92]. The routing is completed in two steps.

1. **Global routing:**   Global routing in FPGAs can be done by using a global router for standard cell designs. In general, such a global router divides the multi-terminal nets into two terminal nets and routes them with minimum distance path. While doing so it also tries to balance the densities by distributing the connections among the channels. The global route defines a coarse route for each connection by assigning it a sequence of channel segments. Figure 13.11(a) shows a sequence of channel segments that a global route might choose to connect some pin of logic block at grid location 4,1 to another at 0,1. The global route is also called as a course grid graph. Note that the coarse grid graph gives a path between two L nodes through a sequence of S and C nodes.

2. **Detailed routing:**   Given a course grid graph $G = (V, E)$ for a two terminal net, the objective of the detailed router is to choose specific

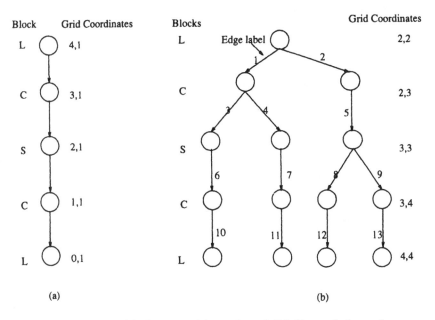

Figure 13.11: (a) Coarse grid graph and (b) Expanded graph.

Algorithm GRAPH-EXPANSION(G)
begin
    $G_D = G$;
    while (DFS-COMPLETE(G_D)==FALSE) **do**
        v_i=CURRENT-DFS-VISIT(G_D);
        l_i=WIRE-SEGMENT(v_i, G_D);
        if (NODE-TYPE(v_i)=C OR NODE-TYPE(v_i)=S) **then**
            v_j=SUCCESSOR(G_D, v_i);
            T_j=SUBTREE(G_D, v_j);
        if (NODE-TYPE(v_i)=C) **then**
            for (each wiring segment l in $F_C(v_i, v_j, l)$) **do**
                T=DUPLICATE(T_j);
                CONNECT(v_i, v_j, T, l);
            DELETE(G_D, T_j);
        if (NODE-TYPE(v_i)=S) **then**
            for (each wiring segment l in $F_S(v_i, v_j, l, v_k)$) **do**
                T=DUPLICATE(T_j);
                CONNECT(v_i, v_j, T, l);
            DELETE(G_D, T_j);
    end.

Figure 13.12: Algorithm GRAPH-EXPANSION.

wiring segments in each channel segment assigned during global routing. This is achieved in two steps:

(a) **Expansion of coarse grid graph:**   In this step, a coarse grid graph is expanded to record a subset of possible ways of implementing the connection. The expansion is carried out while spanning the graph in depth first search manner. The formal description of algorithm is shown in Figure 13.12. Function DFS-COMPLETE(D) returns TRUE if all the nodes of D are visited during the depth-first-search. Function CURRENT-DFS-VISIT(G_D) returns the node being visited during DFS. Function WIRE-SEGMENT(v_i, G_D) returns a wire segment that connects v_i to its predecessor. Function SUCCESSOR(v_i) returns the successor of v_i in G_D. Function SUBTREE(G_D, v_j) returns the subtree of G_D rooted at v_j. Function NODE-TYPE(v_i) returns 'C' if the node v_i represents a C block, it returns 'S' if v_i represents a S block, it returns an 'L' otherwise. If v_j is a C node, v_k is its successor of v_j and a wire segment l is used to connect v_i to v_j then the function $F_C(v_i, v_j, l)$ returns a set of wiring segments that can be used to connect v_j to v_k. Similarly, if v_j is an S node, v_k is its successor of v_j and a wire segment l is used to connect v_i to v_j then the function $F_S(v_i, v_j, l, v_k)$ returns the a set of wiring segments that can be used to connect v_j to v_k. Function DUPLICATE(T) returns a copy of tree T. Procedure CONNECT(v_i, v_j, T, l) connects the node v_i to the node v_j in T by a directed edge from v_j to T and labels the connecting edge by l. Procedure DELETE(G, T) deletes the subtree T from G. Let $G_D = (V_D, E_D)$ denote the graph obtained after expansion of $G = (V, E)$. Figure 13.11(b) the graph obtained by expanding the coarse graph in Figure 13.11(a).

(b) **Connection formation:**   The expanded graph $G_D = (V_D, E_D)$ contains a number of alternative paths. In this step, all these paths are enumerated, their cost is computed and the minimum cost path is selected to implement the connection. Cost of a path is the summation of the cost of edges in that path. The cost of an edge consists of two parts: $c_f(e)$ and $c_t(e)$. $c_f(e)$ accounts for the competition between different nets for the same wiring segments, and $c_t(e)$ reflects the routing delay associated with the routing segment.

13.4.2   Routing Algorithms for the Segmented Model

In this section, we discuss a basic routing algorithm for segmented model. This is followed by a discussion on a new segmented model called *staggered segmented model* and an associated router.

```
Algorithm SEG-ROUTER (𝓘, 𝒯, A)
    input: 𝓘, 𝒯;
    output: A;
begin
    for i = 1 to n do
        for j = 1 to m do
            s=GET-SEGMENT(j, LEFT(Iᵢ));
            if OCCUPIED(s) =FALSE then
                A[i]=j;
                MARK-OCCUPIED(Iᵢ,Tⱼ);
    end.
```

Figure 13.13: Algorithm SEG-ROUTER.

13.4.2.1 Basic Algorithm

In this section, we discuss an algorithm presented by Green, Roychowdhury, Kaptanoglu and Gamal for routing in segmented model [GRKG93]. The input to the routing problem is a set of intervals $\mathcal{I} = \{I_1, I_2, \ldots, I_n\}$, a set of tracks $\mathcal{T} = \{T_1, T_2, \ldots, T_m\}$. Each track $T_i \in \mathcal{T}$ extend from column 1 to column N, and is divided into a set of contiguous segments separated by switches. These switches are placed between two successive segments.

For each interval $I_i \in \mathcal{I}$, we define $LEFT(I_i)$ and $RIGHT(I_i)$ to be the leftmost and the rightmost column in which the interval is present. $1 \leq LEFT(I_i) \leq RIGHT(I_i) \leq N$. We assume that the intervals in $\mathcal{I}$ are sorted on their left edges, i.e., $LEFT(I_i) < LEFT(I_j)$ for all $i < j$.

If an interval I_i is *assigned* to a track T_j, then the segments in track T_j that are present in the columns spanned by the interval are considered *occupied*. More precisely, a segment s in track T_j is occupied by an interval I_i if $RIGHT(s) < RIGHT(I_i)$ and $LEFT(s) < LEFT(I_i)$.

A *routing* of $\mathcal{I}$ consists of an assignment of each interval $I_i \in \mathcal{I}$ to a track such that no segment is occupied by more than one connection.

The routing in the segmented model can be achieved using the algorithm SEG-ROUTER presented in Figure 13.13. The algorithm SEG-ROUTER is a modified left-edge algorithm. The input to the algorithm is the set of intervals $\mathcal{I}$, the set of tracks $\mathcal{T}$, whereas the output is an array A, such that $A[i]$ gives the number of the track on which the interval I_i is routed. In algorithm SEG-ROUTER, function GET-SEGMENT(j, c) returns a segment s on track T_j such that column c is in the span of s. Function OCCUPIED(s) returns TRUE if the segment s is occupied, it returns FALSE otherwise. Procedure MARK-OCCUPIED(I_i,T_j) marks all the segments on tracks T_j that are occupied by I_i. Figure 13.14(b) shows a routing in a uniform segmented channel, generated by Algorithm SEG-ROUTER. Figure 13.14(a) shows a routing in a non-segmented channel, generated by the left edge algorithm. Note that less number of tracks are used in uniform segmented channel as compared to the non-segmented channel.

13.4.2.2   Routing Algorithm for Staggered Model

The segmented model can be improved in several ways: Figure 13.14(c) shows that if the antifuses are staggered the routing of the channel in Figure 13.14(a) can be completed in 3 tracks. Figure 13.14(d) shows that if the antifuses are staggered and if different track segments have different lengths then the routing of the channel in Figure 13.14(a) can be completed in 2 tracks. In this section, we discuss a staggered segmentation model and its routing algorithm presented by Burman, Kamalanathan and Sherwani [BKS92].

In this model, a channel is partitioned into several regions. Each region is characterized by the segment length. The tracks in each region have equal length segments separated by staggered placement of antifuse switches. There are three parameters with respect to the new model: number of regions (p), number of tracks (t), length of segment in each region (l). Determination of these three parameters is an important step in this segmentation scheme. These parameters can be determined by a detailed empirical analysis on several standard benchmarks. A detailed analysis and determination of these parameters can be found in [BKS92]. If the length of segments in all the regions is same then the model is called as the *uniform staggered model* otherwise it is called *non-uniform staggered model*. Note that the model in Figure 13.14(c) is uniform staggered model, whereas the one in Figure 13.14(d) is non-uniform staggered model.

Algorithm SEG-ROUTER can be used for routing in the staggered models. In a uniform segmented model the delay of a net is same irrespective of the routing track. Whereas, in the staggered models the delay of a net is dependent on the the routing track as the number of antifuses in the path of a net in different tracks may be different. The algorithm SEG-ROUTER is not suitable for the high performance routing as it does not consider the delay of a net. In the following, we discuss the algorithm FSCR for the high performance routing in staggered model [BKS92]. The key feature of this routing algorithm is the assignments of the nets to the appropriate tracks by delay computation and delay matching techniques. It should be noted that, for minimum delay routing, it is not sufficient to just minimize delay based on the antifuse elements, but also capacitance effects due to the unused portion of the segments spanned by a net segment (also called as hang-over wires) and the unprogrammed switches must be considered [BKS92].

The algorithm starts routing the longest nets first. This ensures that the delay due to the longest net is minimized, which is a prerequisite for the high performance routing systems. For each net, it finds out a track on which the net can be routed with minimum delay. The original algorithm has three phases, *region selection*, *track selection* and the *region reselection* [BKS92]. In Figure 13.15, we present its simplified version. In algorithm FSCR, the function OK-TO-ASSIGN(I_i, T_j) returns TRUE if all the segments spanned by the

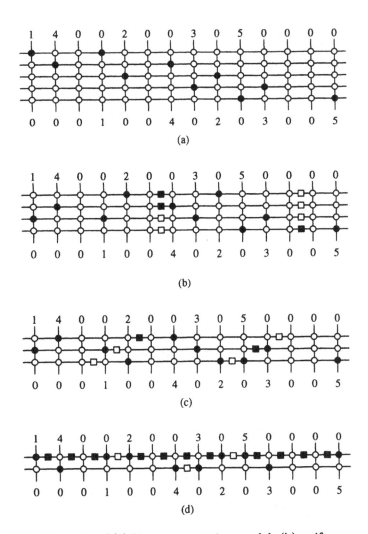

Figure 13.14: Example of (a) Non-segmentation model, (b) uniform segmentation model , (c) uniform staggered segmentation model, and (d) non-uniform staggered segmentation model.

Algorithm FSCR($\mathcal{I}, \mathcal{T}, A$)
    input: $\mathcal{I}, \mathcal{T}$;
    output: A;
begin
    for $i = 1$ **to** n **do**
        selected-track = 0;
        minimum-delay = ∞;
        for $j = 1$ **to** m **do**
            if (OK-TO-ASSIGN(I_i, T_j) =TRUE) **then**
                current-delay = COMPUTE-DELAY(I_i, T_j);
                if (minimum-delay > current-delay) **then**
                    minimum-delay = current-delay;
                    selected-track = j;
            if (selected-track $\neq$ 0) **then**
                A[i]=j;
                MARK-OCCUPIED(I_i, T_j);
            else exit; (* Routing not possible *)
end.

Figure 13.15: Algorithm FSCR

interval I_i on track T_j are unoccupied. Function COMPUTE-DELAY(I_i, T_j) computes the delay in the interval I_i if I_i is routed on the track T_j. Function MARK-OCCUPIED(I_i, T_j) is same as that used in algorithm SEG-ROUTER.

13.5 Summary

FPGAs are being used as a new approach to ASIC design which offers dramatic reduction in manufacturing turnaround time and cost. The physical design cycle of an FPGA consists of three steps, partitioning, placement and routing. The FPGA partitioning problem is different from the conventional area partitioning problem in the sense that it depends on the architecture in which the circuit has to be implemented. Placement problem is equivalent to the general gate array placement problem. However, because of the segmented nature of the FPGA channels, the routing considerations are quite different. In high performance FPGA designs, the number of antifuse elements along with unused tracks and antifuses must be given due considerations as part of the routing phase. A significant amount of research in the direction of physical design automation has to be done in order to fully utilize the potential of FPGAs.

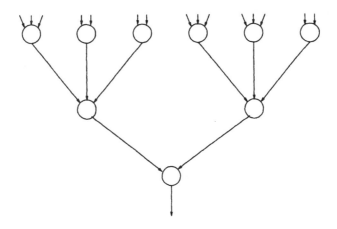

Figure 13.16:

13.6 Exercises

1. Given the graph in Figure 13.16, find the minimum number of configurable logic blocks with five inputs and one output ($O_{max} = 1, I_{max} = 5$), required to partition the circuit.

2. Given a k-array tree of height h, find minimum number of CLBs required with I_{max} inputs and O_{max} outputs.

3. Develop a bin-packing algorithm for partitioning a set of functional blocks into minimum number of CLBs.

4. Given a set of tracks, number of antifuse elements for the new segmentation model, formulate the mixed integer program to minimize the number of antifuse elements. While formulating the problem, consider the delay caused by the antifuse elements and the hang-over wires.

† 5. Several factors play a key role in improving the utilization of *channel resources* in an FPGA. Three such factors have been discussed in this chapter. Discuss what other factors may be considered important for designing a channel segmentation model for high performance applications.

‡ 6. Suggest an efficient channel segmentation model for a three layer routing in FPGAs.

7. Develop a channel routing algorithm for three layer routing model in FPGA.

8. Consider the function $f = ABC + \bar{A}\bar{B}\bar{C} + \bar{A}B + A\bar{B} + \bar{A}C + A\bar{B}C$. Partition the circuit corresponding to f such that

(a) It can be mapped to a minimum number of CLBs. Show the mapping.

(b) It can be mapped to a minimum number of logic modules (each module has three 2-to-1 MUX and a OR gate). Show the mapping.

† 9. In a CLB all 2^5 combinations have to be stored. Suggest a method which stores only those entries which generate either a 0 or a 1 output, whichever is greater without loss of functionality ?

10. Can the size and number of MUX inside a logic block be increased arbitrarily ? What can be the maximum size of the MUX inside the logic block ?

Bibliographic Notes

In *mis-pga* (old) from Murgai, Nishizaki, Shenoy, Brayton, Vincentelli, decomposition   is performed using Roth-Karp method and kernel extraction. The reduction method used in this algorithm is computationally expensive. Hill [Hil91] presented a CAD system for the design of Field Programmable Gate Arrays. New FPGA architecture was developed and the use of FPGAs from the user point of view. Another technology mapper, *Hydra* has been described by Filo, Yang, Mailhot and Micheli [FYMM91]. The approach is similar to mis-pga and performs a disjoint decomposition followed by a node minimization phase. The main difference is that both of these phases are driven by the fact that the Xilinx CLB may realize two outputs also. *Xmap and Amap*, developed by Karplus [Kar91a, Kar91b] constructs an if-then-else dag as decomposition of the function and uses a covering procedure to map it to CLBs. In *mis-pga* (new) by Murgai, Shenoy, Brayton and Sagiovanni-Vincentelli [MSBSV91a], a combinatorial circuit has been described in terms of Boolean equations to realize it using the minimum number of basic blocks of the target Table Lookup architecture. Murgai, Shenoy, Brayton and Sagiovanni-Vincentelli [MSBSV91b] presented delay optimization for programmable gate arrays. The main considerations in this paper are the number of levels in the circuit and the wiring delay. A two phase approach was given. The first phase involves delay optimization during logic synthesis before placement, and the second uses logic resynthesis during a timing-driven placement technique. Nam-Sung Woo [Woo91] presented a heuristic method for the reduction and packing. This is based on the notion of edge visibility and use of global information. The packing method is based on the degree of the node common input. Ercolani and Micheli [EM91] presented a technology mapper for electrically programmable gate arrays. This is based on matching algorithm that determines whether a portion of a combinational logic circuit can be implemented by personalizing a module. The benefits include, an increased efficiency in technology mapping, as well as portability to different types of electrically programmable gate arrays. In [CD93] Cong, and Ding present a study of the area and depth trade-off in LUT based FPGA technology mapping to obtain an area minimized mapping solution. In [FW97] a new integrated synthesis and partitioning method for multiple-FPGA applications is presented. In [CHL97] Technology mapping algorithms for minimizing power consumption in FPGA design are studied. In [SRB97] Macro Block Based FPGA floorplanning has been discussed. [CL97] presents a new recursive bi-partitiong algorithm targetted for a hierarchical field programmable system. In [LW97], a new performance and Routability-Driven Router for symmetrical array based FPGA's is presented. A variation of gate array called LPGA (Laser Programmable Gate Array) is a high performance gate array fabricated by laser micro-machining system allows development of one-day laser prototypes and two months high volume production. The base wafers are fabricated with all interconnection metal layers and a proprietary technique is used to selectively remove specific metalization points to personalize the arrays. Disconnecting the excess metal links follows an automated cut-list program, generated per specific design.

Chapter 14

Physical Design Automation of MCMs

MultiChip Modules (MCMs) have been introduced as an alternative packaging approach to complement the advances taking place in the IC technology. Even though the steps in the physical design cycle of MCMs are similar to those in PCB and IC design cycle, the design tools for PCB and IC cannot be used for MCM directly. This is mainly due to the fact that MCM layout problems are different from both IC layout and PCB layout problems. The existing PCB design tools cannot handle the dense and complex wiring structure of MCMs. On the other hand, IC layout tools are inadequate to decipher the complex electrical, thermal and geometrical constraints of the MCM problems. As a result, the lack of CAD tools for MCMs is impeding further development in this area. Most of the commercial CAD tools available are the adapted versions of existing PCB tools and do not address the real problems associated with the MCM designs. Let us just consider the problem of routing in MCM. The signal effects of long lines in terms of crosstalk, noise, and reflections must be taken into account during routing. In addition, as high speeds are explored, the transmission line behavior of the interconnect must be modeled accurately to optimize the layout. All of these conditions have to be met, subject to the main goal of the interconnect, which is to route the signals between the chips. In designing CAD tools for MCM, many effects have to be taken into consideration such as clock skew, power noise disturbance, assembly effects of thermal mechanical nature that are caused by close positioning of chips, and limitations of assembly equipment. As a result, the design of multichip modules involves several disciplines such as electrical, chemical, material and mechanical engineering.

As MCMs are used for high performance system packaging, all steps in their physical design are performance driven. This makes the existing delay models for IC and PCBs inappropriate for MCMs. Therefore, new delay models will have to be developed for designing MCMs more accurately in order to comply with the stringent performance requirements.

The rest of the chapter has been organized as follows: In order to understand the issues and problems related to the physical design automation of MCMs different types of MCM technologies will be briefly described in Section 14.1. Section 14.2 will outline the different steps involved in the physical design of an MCM. Partitioning, the first phase of the MCM physical design cycle will be discussed in Section 14.3. MCM placement is discussed in Section 14.4. MCM routing problems will be described in Section 14.5.

14.1   MCM Technologies

MCMs, or more precisely, non programmable MCMs, are generally categorized into the following three, MCM-L,MCM-C, and MCM-D. MCM-L describes high density, laminated printed circuit boards. MCM-C refers to the ceramic substrates icluding both cofired and low-dielectric constant ceramics. MCM-D covers modules with deposited metallic wiring on silicon or ceramic support substrates. Yet another approach for fast turnaround is *Programmable MCM* (PMCM). In this section, we present a brief review of both programmable and non programmable MCM technologies. The methods for attaching chips to MCMs will also discussed in this section.

MCM-L (Laminates) is the oldest technology available. MCM-L is essentially an advanced PCB on which bare IC chips are mounted using Chip-On-Board (COB) technology. The well established PCB infrastructure can be used to produce MCM-L modules at a low cost. This makes them an attractive electronic packaging alternative for many low-end MCM applications with low interconnect densities. MCM-L becomes less cost-effective at higher densities where many additional layers are required. For cost- effectiveness, MCM technology must increase the functionality of each layer instead of adding more layers. MCM-L is considered a suitable technology for applications which require low risk packaging approach and most of the steps have already been automated.

MCM-C (ceramic) refers to MCMs with substrates fabricated with cofired ceramic or glass-ceramic techniques. These have been in use for many years and MCM-C has been the primary packaging choice in many advanced applications requiring both performance and reliability. Due to excellent thermal conductivity and low thermal expansion, ceramic substrates have also been used to serve as the package. Although interconnect densities are in the range of 200-400 cm/cm^2, the same are not enough for high-end applications.

MCM-D (deposited) technology is closest to IC technology. It consists of substrates which have alternating deposited layers of high density thin-film metals, and low dielectric materials such as poly or silicon dioxide. MCM-D technology is an extension of conventional IC technology. It is developed specifically for high performance applications demanding a superior electrical performance and a high interconnect density. Since, this technology is relatively recent, it does not offer either a cost-effective manufacturing infrastructure, or a high volume application. Therefore, no significant commercial driving force

Characteristics	MCM-L	MCM-C	MCM-D
Line density (cm/cm²)	250-400	200-400	> 400
Line width/separation (μms)	750/2250	125/125-375	10/10-30
Turnaround time	9-13 weeks	1 month	10-25 days
Years of availability	50	> 10	> 5

Table 14.1: Multichip Modules Classifications

exists. Table 14.1 compares the MCM families in terms of line widths, line density, line separation, turnaround time and the number of years for which these technologies have been available.

A full-custom design of an MCM requires significant engineering efforts. The lack of a mature infrastructure further magnifies the problem, since high density and high performance multichip modules are still expensive to fabricate and the cost increases with the number of mask layers. In order to side-step these difficulties, PMCMs have been introduced to minimize both the engineering delays and the cost. Programmable MCM approach is somewhat similar to Field Programmable Gate Arrays (FPGAs) technique. Just like gate arrays, PMCM wafers are manufactured in large quantities. A PMCM wafer has sites for chips and several layers of programmable interconnect. The customization process is carried out by setting programmable switches to establish the connectivity needed by the user. This is done after the chips have been placed on the chip sites. Thus, the customization consists of only placement of chips and programming the fuses (just like FPGA) to complete the routing.

Irrespective of the types of the MCM technology used, bare chips have to be attached to the substrates. Bare chips are attached to the MCM substrates in three ways, viz., wire bonding, Tape Automated Bonding (TAB) and flip-chip bonding. In wire bonding (illustrated in Figure 14.1(a)), the back side of a chip (nondevice side) is attached to the substrate and the electrical connections are made by attaching very small wires from the I/O pads on the device side of the chip to the appropriate points on the substrate. The wires are attached to the chip by thermal compression. TAB is a relatively new method of attaching chips to a substrate. It uses a thin polymer tape containing metallic circuitry. The connection pattern is simply etched on a polymer tape. As shown in Figure 14.1(b), the actual path is simply a set of connections from inner leads to outer leads. The inner leads are positioned on the I/O pads of the chips, while the outer leads are positioned on the connection points on the substrate. The tape is placed on top of the chip and the substrate and pressed. The metallic material on the tape is deposited on the chip and the substrate to make the desired connections. Flip-chip bonding uses small solder balls on the I/O pads of the chip to both physically attach the chip and make required electrical connections (see Figure 14.1(c)). This is also called face down bonding, or Controlled-Collapse Chip Connections (C4).

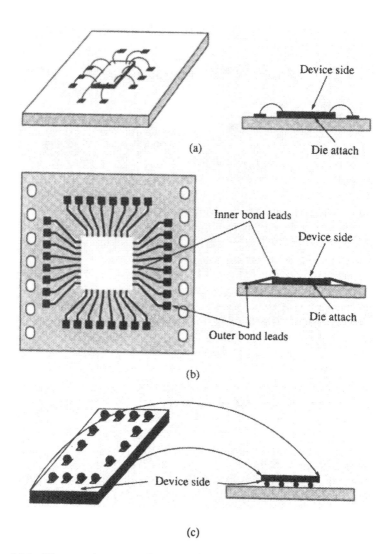

Figure 14.1: Die attachment techniques (a) Wire bonding (b) Tape automated bonding (c) Flip-chip bonding

14.2 MCM Physical Design Cycle

The physical design considerations of an MCM differ significantly from their counterparts for an IC. The input to the MCM physical design cycle is the circuit design of the entire system. The output is the MCM layout.The physical design cycle of an MCM pursues the following steps (also illustrated in Figure 14.2):

1. **Partitioning:** An MCM may contain as many as 100 chips. In turn, each chip can accommodate a certain number of transistors. The first assignment herein is to partition the given circuit into subcircuits.The partitioning should warrant fabrication of each subcircuits on a single chip. Simultaneously, the number of subcircuits should be equivalent to or less than the number of chips that the MCM can sustain. Please note that the MCM designs require performance driven approach. This requirement necessitates consideration of the power and timing constraints in the partitioning step.These requiremnts shall be in addition to the traditional I/O constraints and area constraints for chip sites.

2. **Placement:** The placement step is concerned with mapping the chips to the chip sites on the MCM substrate. Placement, of course affects not only the thermal characteristics of an MCM but also routing efficiency, which translates directly into manufacturability and cost. The number of components involved with the chip placement is much less as compared to the IC placement phase. However, timing and power constraints in MCM placement problem makes it a significantly different problem compared to the IC placement. Thermal considerations in MCM placement are important because bare chips are placed closer together and generate significant amount of heat. When the chip sites are prefabricated, the MCM placement problem lends itself to gate array based approach. Another variation of the MCM placement arises when the chips manufactured in different technologies need to be placed on an MCM. A critical difference between IC placement and MCM placement is allocation of routing regions. Unlike IC placement, no routing region needs to be allocated in MCMs since routing is done in routing layers and not between chips.

3. **Routing:** After the chips have been placed on the chip sites, the next phase of the MCM physical design is to connect these chips specified by the net list. The objective of minimizing routing area in IC design is no longer valid in MCM routing environment. Instead, the objective of the MCM routing is to minimize the number of layers, as the cooling requirement and therefore the cost of an MCM depends on the number of layers used. Because of the long interconnect wires involved in MCM design, crosstalk and skin effect become important considerations which are not of much concern in IC layout. In particular, in MCM-D, skin effect of the interconnect becomes more severe. The parasitic effects also degrade the performance if not accounted for in routing of MCMs.

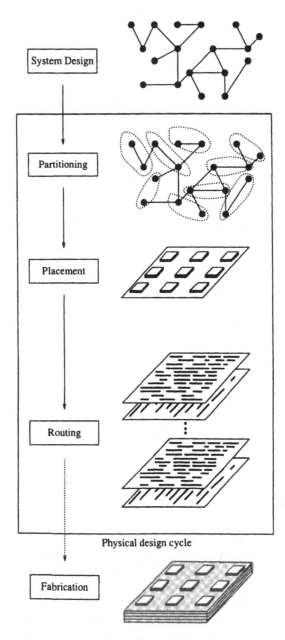

Figure 14.2: Physical design cycle of MCM

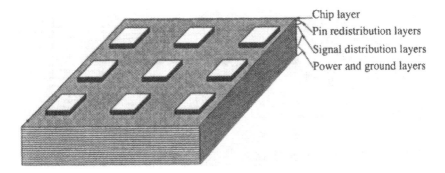

Chip layer
Pin redistribution layers
Signal distribution layers
Power and ground layers

Figure 14.3: MCM routing environment

Power and ground signals do not complicate global routing because these signals are distributed on separate layers, and taps to the power supply layers are easy to make. However, overall dimensions must be tightly controlled (to fit within the package) and the packaging delay must be carefully controlled. The routing environment of an MCM can be viewed as a 3 dimensional space as shown in Figure 14.3.

14.3 Partitioning

As discussed in Chapter 5, the design of a complex system such as computer system consisting of tens of millions of components necessitates breaking the system into subsystems using a divide and conquer strategy. This process of decomposing the system into subsystems is called partitioning. Traditionally, partitioning has been applied at three levels, system level, board level and chip level. System level partitioning breaks the system design into sub-circuits which can fit on a PCB. Board level partitioning partitions each sub-circuit into a set of chips. The last step in the hierarchy of partitioning, chip level partitioning decomposes a chip circuit into smaller sub-circuits in order to ease the task of the chip designer.

With the introduction of multichip modules, the intermediate board level partitioning is replaced by *module level partitioning*. We refer to module level partitioning as MCM partitioning. The module level partitioning is charac-terized by high performance and high density design. Thus the module level partitioning is performance driven. The module level partitioning is becoming an important ingredient for complex design with the rapid increase of the device density. The device density has, on an average, doubled annually for almost two decades. It is anticipated that such advances will continue to be made well into 1990s. This growth in the devices per unit area makes the problem of MCM partitioning challenging.

The MCM partitioning depends on design style. If an gate array type

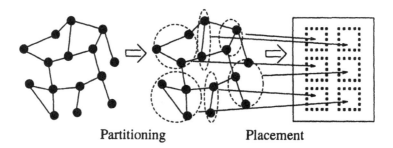

Partitioning   Placement

Figure 14.4: MCM partitioning and placement

design style is used, the MCM partitioning problem is similar to the gate array partitioning problem except each 'gate' corresponds to a chip. We refer to this approach as a *chip array* approach. If a full custom type approach is used, then each chip can have a different size and the MCM partitioning is analogous to the full custom partitioning problem. In the following, we restrict our discussion to chip array approach.

MCM partitioning is defined as an optimum mapping of the design to a set of chips (see Figure 14.4). However, as the performance considerations enter the design, the MCM partitioning process must consider other constraints as well. So for high performance system designs, MCM partitioning can be defined as a partition of the design to a set of chips that minimizes the inter-chip wire crossings subject to timing constraints, area constraints, thermal constraints and I/O pin count constraints.

An MCM package can be considered to contain a set ($\mathcal{C}$) of equal sized chips, each chip placed in a chip slot. Each chip $c \in \mathcal{C}$ has constraints on area A_c, thermal capacity $\mathcal{H}_c$, and maximum number of terminals (I/O pins) $\mathcal{T}_c$. A synchronous digital system consists of registers and blocks of combinational logic. For simplicity, all clock generation and distribution circuits are ignored. The system can be represented by an edge weighted graph called *system graph* $G = (V, E)$ [SKT94], where $V = R \cup B$, R is the set of nodes representing registers, B is the set of nodes representing combinational blocks, and E is the set of all *directed edges*, which correspond to signal flow in the system. Associated to each edge $e_{ij} \in E$, there is a weight w_{ij} representing the total number of wires between nodes v_i and v_j in V. Associated with each node $v_i \in V$, we have three parameters, area a_i, power consumption h_i, and internal delay d_i. Figure 14.5 shows a system graph.

The MCM partitioning problem is to find an optimum mapping $\varphi : V \to C$ such that the number of total inter-chip connections

$$W = \sum_{\forall i,j,\varphi(v_i) \neq \varphi(v_j)} w_{ij}$$

is minimized while satisfying the timing constraints, area constraints, terminal

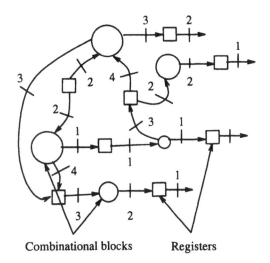

Combinational blocks   Registers

Figure 14.5: System graph

constraints, and thermal constraints. The area and the terminal constraints is the same as the area and the terminal constraints of IC design partitioning. However, for the timing constraints, we need to consider the internal delay of each circuit. The timing and the thermal constraints can be stated as:

1. **Timing constraints:** A register-to-register delay through some combinational logic blocks must be less than or equal to the cycle time.

$$d_j + D(\varphi(r_i), \varphi(c_j)) + D(\varphi(c_j), \varphi(r_k)) \leq T_{cycle}$$

for all $c_j \in C$, where $r_i, r_k \in R$, $e_{ij}, e_{jk} \in E$, D is the time delay between objects and T_{cycle} is the given cycle time.

2. **Thermal constraints:** The total heat generated by a partition must be less than or equal to the thermal capacity of the corresponding mapped slot.

$$\sum_{\forall i, \varphi(v_i)=s} h_i \leq \mathcal{H}_s$$

Thermal constraints can be treated in a similar fashion as area constraints. Therefore, any performance driven partitioning algorithm may be applied by taking into consideration the thermal constraints. For the gate-array based approach, where each chip slot is of equal size, any gate-array partitioning algorithm may be applied with appropriate modifications. Similarly, any high-performance full custom partitioning algorithm may be applied for the generalized full custom based MCMs. In [SKT94], Shin, Kuh, and Tsay presented a performance driven integrated partitioning and placement technique for MCMs.

They only considered timing and area constraints. Two different delay models have been considered: 1) constant delay model, and 2) linear delay model. For constant delay model, their approach is essentially a partitioning algorithm which will be briefly described below.

In [SKT94], given a system graph, assuming that delay time for the signal traveling between a combinational block and a register that are grouped into the same partition is negligible. In addition, the delay time for the signal traveling between a combinational block and a register that are partitioned into different groups is a constant. Each group is called a *super node* and corresponds to a chip.

For each combinational block c_i, the algorithm finds the two registers r_j and r_k that are adjacent to the block in the system graph. The procedure of constructing super nodes is shown by an example in Figure 14.6. In this example, we assume the system cycle time is $T_{cycle} = 6$ which requires that the maximum delay time between any two registers should be less than or equal to 6. The delay time between a combinational block and a register is assumed as $D = 2$. The super nodes are constructed according to the following three cases.

1. Both registers must be combined: In this case, the condition $d_i + D > T_{cycle}$ must be satisfied. Consider the example shown in Figure 14.6(a) with $d_i = 5$. If one of r_j and r_k is assigned to a different partition than c_i, the time delay will be at least $2 + 5 = 7$, thus violating the timing constraint. So, all these vertices have to be included in the same super node.

2. At least one of the registers must be combined: In this case, the conditions $d_i + D < T_{cycle}$ and $d_i + 2 \times D > T_{cycle}$ must be satisfied. Consider the example shown in Figure 14.6(b) with $d_i = 3$. If both registers are assigned to different partitions than c_i, the time delay will be 7, thus violating the timing constraint. In this situation, the super node consists of combinational block and either one of the registers.

3. No registers need to be combined: In this case, the condition $d_i + 2 \times D < T_{cycle}$ must be satisfied. Consider the example shown in Figure 14.6(c) with $d_i = 1$, the registers can be assigned to any partitions without violating the timing requirement. Thus, each super node consists of only one vertex.

The algorithm repeats until no nodes can be combined. At this stage, the number of super nodes is equal to the number of chips required in the MCM.

14.4 Placement

The thermal and timing considerations in the MCM placement problem make it significantly different than the IC placement. With the increase in the density of the individual chip, the thermal requirements have also gone up. High

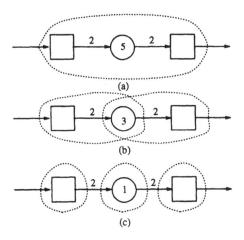

Figure 14.6: Super node construction

speed VLSI chips may generate heat from 40 to 100 watts. In order to ensure proper operation of the design, such a large amount of heat must be dissipated efficiently. The heat dissipation of an MCM depends directly on how the chips are placed. In addition to this, the timing constraints for the design must also be satisfied. These timing constraints are responsible for the proper operation of the module at high frequencies. The placement problem in MCM is to assign chips to the chip sites on the substrates subject to some constraints. If the placement is not satisfactory, then the subsequent steps of routing will be inefficient.

Chip level placement determines the relative positions of a large number of blocks on an IC as well as organizes the routing area into channels. As opposed to IC placement problem, MCM placement involves fewer components (100-150 ICs per MCM compared to 10-1000 general cells per IC) and the sizes and shapes of ICs on an MCM are less variable than the general cells within the IC. MCM placement is more complex because many interrelated factors determine layout quality. Wide buses are very prevalent, propagation delays and uniform power dissipation are much more important. As opposed to IC placement problem, the main objective of MCM placement is to assign the chips to the chip sites such that the number of routing layers is minimized. In addition, other constraints such as timing constraints and thermal constraints make the MCM placement problem more difficult. The MCM placement problem can be formally stated as follows: given a set of chips C, and a set of chip sites on the substrate S, assign C to S, e.g., find a mapping $\phi : C \to S$ subject to timing constraints and thermal constraints and to minimize the number of layers. The typical values for $|C|$ and $|S|$ range between 4-100.

There are mainly two types of placement related to MCMs, namely, chip array and full custom.

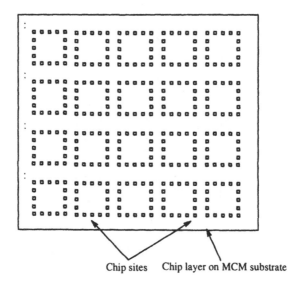

Figure 14.7: MCM placement problem

14.4.1   Chip Array Based Approach

The MCM placement approach when the chip sites are symmetric, becomes very similar to the conventional gate array approach. In this case, the MCM placement problem is the assignment of the chips to predefined chip sites. However, the key difference between the IC placement and the MCM placement problem is the type of constraints involved. Figure 14.7 shows a chip array MCM substrate. The two approaches to MCM placement problem have been discussed by LaPotin [LaP91] as part of the early design analysis, packaging and technology tradeoffs.

14.4.2   Full Custom Approach

One of the important features of the MCMs is that it allows the integration of mix of technologies. This means, each individual chip can be optimally fabricated using the technology best suited for that chip. Figure 14.8 shows an arrangement depicting a concept of *2.5-D integration* scheme derived from ideas postulated by McDonald [McD84] and Tewksbury [Tew89]. This concept can be viewed as an advanced version of the existing MCMs. It is envisioned that this hypothetical system will respond directly to the cost limitations of VLSI technologies. The system could be assembled on a large-area active substrate. The technology of such a substrate could be optimized for yield, power, and speed of the interconnect. This substrate could dissipate a large percentage of the total power and could be cost-effective if fabricated with relaxed design rules in stepper-free, interconnect-oriented technology. The performance-

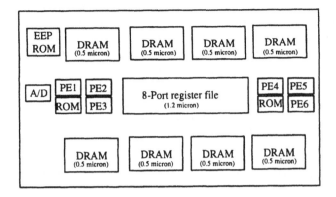

Figure 14.8: 2.5-D integration placement in MCM

critical system components could be fabricated separately on fabrication lines oriented toward high volume and high performance. They could be attached to the active substrate with rapidly maturing flip-chip technology. This way only those system elements that really require ULSI technology (for example, data path) would be fabricated with the most expensive technologies. It is obvious that placement problem in 2.5-D integration scheme is that of full-custom approach. In addition to the usual area constraints, the placer of this type must be able to complete the task of placement subject to the thermal and timing constraints.

14.5 Routing

After the chips have been placed on the chip sites, the next phase of the MCM physical design is to connect these chips specified by the net list. As mentioned earlier that unlike IC design, performance is the main objective in MCM design. Therefore, the main objective of routing is to satisfy timing constraints imposed by the circuit design. Also, the cost of an MCM is directly proportional to the number of layers used in the design. Thus minimizing the total number of layers used is also an objective of MCM routing. In particular, in MCM-D, cross talk, skin and parasitic effect of the interconnect become more critical. Crosstalk is a parasitic coupling between neighboring lines due to the mutual capacitances and inductances. In the design of high speed systems, crosstalk is a primary concern. Excessive crosstalk compromises noise margins, possibly resulting in false receiver switching. The crosstalk between the lines can be minimized by making sure that no two lines are laid out in parallel or next to each other for longer than a maximum length.

In addition to crosstalk, the skin effect is also a major consideration in

MCM routing. Skin effect is defined as characteristic of current distribution in a conductor at high frequencies by virtue of which the current density is greater near the surface of the conductor than its interior. As the rise time of digital pulses is reduced to the sub-nanosecond range, the skin effect becomes an important issue in high speed digital systems. As the frequency, the conductivity, and permeability of the conductor are increased, the current concentration is also increased. This results in increasing resistance and decreasing internal inductance at frequencies for which this effect is significant. These effects must be taken into account while routing long lines.

14.5.1   Classification of MCM Routing Algorithms

The routing of an MCM is a three-dimensional general area routing problem where routing can be carried out almost everywhere in the entire multilayer substrate. However, the pitch spacing in MCM is much smaller and the routing is much denser as compared to conventional PCB routing. Thus traditional PCB routing algorithms are often inadequate in dealing with MCM designs.

There are four distinguished approaches for general (non-programmable) MCM routing problems:

1. Maze Routing

2. Multiple Stage Routing

3. Topological Routing

4. Integrated Pin Distribution and Routing

The routing of programmable MCMs is very similar to that of FPGAs. In this section, we discuss routing of both MCMs and PMCMs.

14.5.2   Maze Routing

The most commonly used routing method is three dimensional maze routing. Although this method is conceptually simple to implement, it suffers from several problems. First, the quality of the maze routing solution is very much sensitive to the ordering of the nets being routed, and there is no effective algorithm for determining a good net ordering in general. Moreover, since the nets are routed independently, global optimization is difficult and the final routing solution often uses a large number of vias despite the fact that there is a large number of signal layers. This is due to the fact that maze router routes the first few nets in planar fashion (using shorter distances), the next few nets use a few vias each as more and more layers are utilized. The nets routed towards the end tend to use a very large number of vias since the routing extends over many different layers. Finally, three dimensional maze routing requires long computational time and large memory space.

14.5.3 Multiple Stage Routing

In this approach, the MCM routing problem is decomposed into several sub-problems. The close positioning of chips and high pin congestion around the chips require separation of pins before routing can be attempted. Pins on the chip layer are first redistributed evenly with sufficient spacing between them so that the connections between the pins of the nets can be made without violating the design rules. This redistribution of pins is done using few layers beneath the chip layer. This problem of redistributing pins to make the routing task possible, is called *pin redistribution*. After the pins are distributed uniformly over the layout area using pin redistribution layers, the nets are assigned to layers on which the assigned nets will be routed. This problem of assigning nets to layers is known as *layer assignment problem*. The layer assignment problem resembles the global routing of the IC design cycle. Similar to the global routing, nets are assigned to layers in a way such that the routability in layer or in a group of layers is guaranteed and at the same time the total number of layers used is minimized. The layers on which the nets are distributed are called *signal distribution layers*. The detailed routing follows the layer assignment. The detailed routing may or may not be reserved layer model. The horizontal and vertical routing may be done in same layer or different layers. Typically, nets as distributed in such a way that each pair of layers is used for a set of nets. This pair is called $x - y$ *plane pair* since one layer is used for horizontal segments while the other one is used for vertical segments. Another approach is to decompose the net list such that each layer is assigned a planar set of nets. Thus MCM routing problem become a set of *single layer* problem. Yet another routing approach may combine the $x - y$ plane pair and single layer approaches. In particular, the performance critical nets are routed in top layers using single layer routing because xy-plane pair routing introduces vias and bends which degrade performance.

We now discuss each of these problems in greater detail in the following subsections.

14.5.3.1 Pin Redistribution Problem

Pins in chip layer need to be redistributed to help in the routing process. This is accomplished in pin distribution layers. The pin redistribution problem can be stated as: Given the placement of chips on an MCM substrate, redistribute the pins using the pin redistribution layers such that one or more of the following objectives are satisfied (depending the the design requirements):

1. minimize the total number of pin redistribution layers.

2. minimize the total number of signal distribution layers.

3. minimize the cross-talks.

4. minimize the maximum signal delay.

5. maximize the number of nets that can routed in planar fashion.

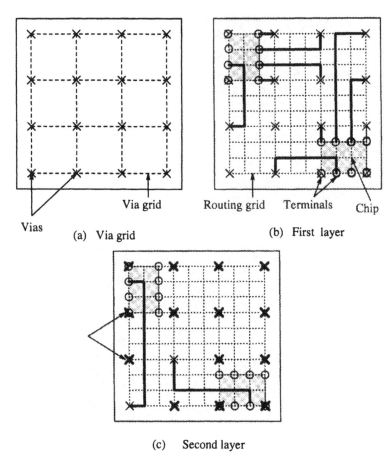

Figure 14.9: Pin redistribution example

It is to be noted that the separation between the adjacent via-grid points may affect the number of layers required [CS91]. The pin redistribution problem can be illustrated by the example shown in Figure 14.9. The terminals of chips need to be connected to the vias shown in Figure 14.9(a). Usually, it is impossible to complete all the connections. In this case, we should route as many terminals as possible (shown in Figure 14.9(b)). The unrouted terminals are brought to the next layer and routed in that layer as shown in Figure 14.9(c). This procedure is repeated until each terminal is connected to some via. In [CS91], various approaches to pin redistribution problem have been proposed.

14.5.3.2 Layer Assignment

The main objective of layer assignment for MCMs is to assign each net in x-y pair of layers subject to the feasibility of routing the nets on a global routing grid on each plane-pair. This step determines the number of plane pairs required for a feasible routing of nets and is therefore important step in the design of the MCM. The cost of fabricating an MCM, as well as the cooling of the MCM when it is operation, are directly related to the number of plane-pairs in the MCM, and thus it is important to minimize the number of plane-pairs. There are two approaches known to the problem of layer assignments [HSVW90b, SK92]. The problem of layer assignment has been shown to be NP-complete [HSVW90b].
An approximation algorithm, for minimizing the number of layers, has been presented by Ho, Sarrafzadeh, Vijayan and Wong [HSVW90b].

14.5.3.3 Detailed Routing

After the nets have been assigned to layers, the next step is to route the nets using the signal distribution layers. Depending on the layer assignment approach, the detailed routing may differ. Routing process may be single-layer routing or x-y-plane-pair routing. Usually a mixed approach is taken in which the single-layer routing is first performed for more critical nets, followed by x-y-plane-pair routing for less critical nets. Two models can be employed for x-y-plane-pair routing, namely xy-reserved model and xy-free model. One advantage in xy-free model is that bends in nets do not necessarily introduce vias where bends in nets introduce vias in xy-reserved model. The detailed routing was presented in [LSW94].

14.5.4 Topological Routing

In [DDS91], Dai, Dayan and Staepelaere developed a multilayer router based on rubber-band sketch routing. This router uses hierarchical top-down partitioning to perform global routing for all nets simultaneously. It combines this with successive refinement to help correct mistakes made before more detailed information is discovered. Layer assignment is performed during the partitioning process to generate routing that has fewer vias and is not restricted to one-layer one-direction. The detailed router uses a region connectivity graph to generate shortest-path rubber-band routing.

The router has been designed primarily for routing MCM substrates, which consist of multiple layers of free (channelless) wiring space. Since MCM substrate designs have potentially large number of terminals and nets, the router of this nature must be able to handle large designs efficiently in both time and space. In addition, the router should be flexible and permit incremental design process. That is, when small changes are made to the design, it should be able to be updated incrementally and not recreated from scratch. This allows faster convergence to a final design. In order to produce designs with fewer vias, the router should be able to relax the one-layer one-direction restriction. This is an

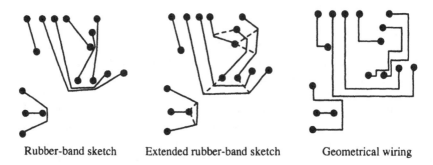

Rubber-band sketch   Extended rubber-band sketch   Geometrical wiring

Figure 14.10: Rubber-band representations

important consideration in high speed designs since the discontinuities in the wiring caused by bends and vias are a limiting factor for system clock speed.

In order to support the flexibility described above, the router must have an underlying data representation that models planar wiring in a way that can be updated locally and incrementally. For this reason, SURF models wiring as rubber-bands [CS84, LM85]. Rubber-band provides canonical representation for planar topological wiring. Because rubber-bands can be stretched or bent around objects, this representation permits incremental changes to be made that only affect a local portion of the design. A discussion of this representation has been described in [DKJ90].

Once the topology of the wiring is known, the rubber-band sketch can be augmented with spokes to express spatial design constraints such as wire width, wire spacing, via size, etc. [DKS91]. Since successful creation of the spoke sketch guarantees the existence of a geometrical, wiring (Manhattan or octilinear), the final transformation to fixed geometry wiring can be delayed until later in the design process. This allows most of the manipulation to take place in more flexible rubber-band format. Figure 14.10 shows different views of the same wiring topology. These represent various states of the rubber-band representation.

In this context, a topological router has been developed that produces multi-layer rubber-band sketches. The input to this router is a set of terminals, a set of nets, a set of obstacles, and a set of wiring rules. These rules include geometrical design rules and constraints on the wiring topology. The topological constraints may include valid topologies (daisy chain, star, etc.) as well as absolute and relative bounds on segment lengths. The output of the router is a multilayer rubber-band sketch in which all the points of a given net are connected by wiring. Although the routability of a sketch is not guaranteed until the successful creation of spokes. At each stage, the router uses the increasingly detailed information available to generate a sketch without overflow regions. This increases the chance that the sketch can be successfully transformed into a representation (the spoke sketch) that satisfies all of the spatial constraints. In addition the router tries to reduce overall wire length and the number of vias. A more detailed analysis of routability of a rubber-band sketch is described in [DKS91].

14.5.5 Integrated Pin Distribution and Routing

In [KC92], Khoo and Cong presented an integrated algorithm SLICE for routing in MCM. The basic idea is to redistribute pins simultaneously with routing in each layer, instead of the pins distribution prior to routing. SLICE performs planar routing on a layer by layer basis. Subsequent to routing on one layer, the terminals of the unrouted nets are propagated to the next layer. The routing process is then continued until all the nets are routed.

An important feature of SLICE is computation of planar set of nets for each layer. The algorithm strives to connect maximum number of nets in each layer. The algorithm attempts partial routing of nets that cannot be routed completely in a layer. This facilitates completion of nets in the subsequent layer with shorter wires. The routing region is scanned from left to right. A topological planar set of nets is computed for each adjacent column-pair using maximum weighted non-crossing matching. The matching is comprised of a set of non-crossing edges that extend from the left column to the right column. Thereafter, the physical routing between the column-pair is generated based on the selected edges in the matching. This process is carried out for each column from left to right. The completion of the planar routing in a layer is followed by distribution of the terminals of the unrouted nets so that they can be propagated to the next layer without causing local congestions. The left to right scanning operation in the planar routing culminates in predominantly horizontal wires in the solution. A restricted two-layer maze routing technique is adopted for completion of the routing in vertical direction. Unnecessary jogs and wires are eliminated after each layer is routed. The terminals of the unrouted nets are propagated to the next layer. Finally, the routing region is rotated by 90^0 so that the scanning direction is orthogonal to the one used in the previous layer. The process is iterated until all the nets have been routed. Details of the planar routing, pin redistribution, and maze routing are available in [KC92].

14.5.6 Routing in Programmable Multichip Modules

Like gate arrays, routability is a key concept in the design of programmable MCMs. In a programmable MCM design, most if not all, of the masking or phototooling steps are defined prior to commencement of the system designing. Initially, a substrate is manufactured in a generic fashion. Subsequently, it is customized for fulfilling the specific needs of the user. The capability for routing complex and dense multichip designs requires early designing of a highly routable wiring topology. An important component for achieving efficient programmable designs is the design tool that can sustain the dual responsibility of: one, deciphering the programmable wiring structure; and two, performing the actual routing (customization) needed to realize an application specific MCM. It is noteworthy that the routing efficiency is a factor of both the base wiring density and the resource utilization. The base wire density is typically measured in inches of wire per square inch of the substrate area. The resource utilization refers to the fraction of available wiring that can be utilized in routing a design. The total wire length used, relative to the minimum theoretical routing length, must be accounted for.

Electrical performance is a key ingredient to any programmable custom MCM design. If the programmable approach fails to meet the performance goals, then its application objectives will not be accomplished. In many circumstances, electrical performance of the signal interconnect will be relatively good even without rigorous design for characteristic impedance, low loss etc. This can be ascribed to the electrical length of the signal wiring. In most MCM environments, the same is short as compared to the wavelength/rise times of the IC signals. The crux of the issue in nearly all cases is capacitive loading reduction for CMOS systems in order to minimize delay caused by RC time constraints. In other words, a large fraction of system designs will be needed to address signal delay more than high bandwidth signal fidelity. This may not be the case in a more conventional single chip packaging/PC board implementation where physical/electrical lengths of interconnect are longer and more significant. Perhaps a more compelling issue associated with signal fidelity is power distribution. Many signal noise problems develop due absence of clean power and ground supplies. Due to these, noise is fed forward through output drivers, which diminishes noise margins at the receivers. This imposes an additional demand on the design of a programmable MCM. It decrees that the power distribution scheme must be supportive of high performance, in addition to being flexible. The power distribution network of the MCM design is usually predefined and accommodates a myriad of supply voltages, variable supply potentials, and a variety of both AC/DC current requirements.

Figure 14.11 illustrates a simplified cut-away view of a programmable multichip module with a substrate wherein antifuses have been incorporated. The substrate is comprised of four metal layers separated by dielectric layers. The lower two layers are used for power distribution. On the other hand, the upper two layers are used for an orthogonal wiring grid with permanent vias or antifuses in selected grid interconnections. The uppermost layer also houses the bonding pads. The bare chips will be electrically connected to these pads upon completion of the programming. A signal path can be programmed through the substrate by linking previously uncommitted line elements together via the antifuses. The interconnection line architecture of actual designs is much more convoluted than the one presented in the above mentioned simplified example. However, the principle of programming remains unaltered in either case. Since, all line elements are accessible from a bonding pad, a programming pulse can be applied. A programming pulse with a voltage amplitude larger than the threshold voltage is applied using a wafer prober to a pair of wiring elements in order to connect them to each other.

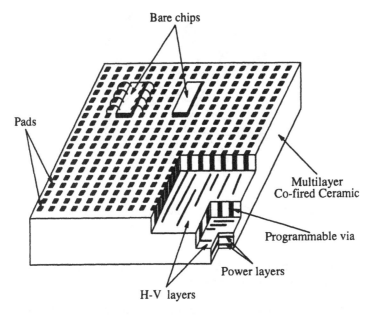

Figure 14.11: Routing in programmable multichip modules

14.6 Summary

The MCM approach to microelectronic packaging has significantly improved
the system performance. Such improvement has been acheived by bridging the
gap between the existing PCB packaging approach and the progressing VLSI
IC technology. The physical design of MCMs is an important ingredient of
the overall MCM design cycle. The density and complexity of contemporary
VLSI/ULSI chips require automation of the physical design of MCMs. Further
developments of MCMs face stuff challenges due to limited research in the area
of development of algorithms requisite for MCM physical design. This is primarily attributable to the fact that MCMs pose an entirely new set of problems
which cannot be solved by existing PCB or IC layout tools. Therefore, considerable research efforts need to be steered towards development of algorithms
for MCM physical design automation.

14.7 Exercises

1. A Multi-Chip Module (MCM) consists of many interconnected bare chips.
   Consider a hypothetical MCM with four chip slots. Each slot has five
   terminals. Does there exist a 4-way partition of the graph in Figure 14.12,

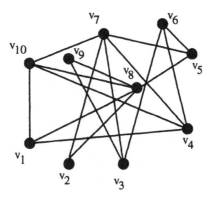

Figure 14.12: A graph partitioning problem.

each having no more than three vertices and the number of terminals for
each partition is no more than five?

† 2. Consider the thermal-driven placement problem in which the chips are
to be placed onto chip sites such that the heat distribution across the
multichip module is uniform. Develop an algorithm for such a placement.

† 3. In MCM placement problem, the heat distribution should be uniform
over the MCM. Modify the simulated annealing algorithm described in
Chapter 5 to take the heat effect into account so that it can be used in
MCM placement.

  4. The routing problem for MCMs is three dimensional. Extend maze routing algorithm for routing a two-terminal net in three-dimensions.

† 5. Extend line probe algorithm for global routing a two-terminal net in
three-dimensions.

† 6. Let L be the longest possible length of a net that does not cause undue
skin effects. Develop a global router that guarantees the length constraints imposed by the skin effects.

‡ 7. Formulate global routing in three-dimensions as a Hierarchical Integer
Program with an objective of optimizing overall wire length.

‡ 8. Develop a heuristic algorithm for pin redistribution such that it minimizes
the net lengths and the number of of layers needed.

‡ 9. Develop a *crosstalk-driven router* for MCM, which routes all the nets
and also minimizes the crosstalk between the neighboring lines. Assume
that the system is to be assembled on a multichip module using silicon
substrate and silicon dioxide as dielectric layer.

† 10. Consider the following channel routing problem motivated by the crosstalk. Let L be the longest distance two nets can run parallel to each other without causing undue crosstalk problems. Modify Yoshimura-Kuh channel routing algorithm to minimize the crosstalk.

Bibliographic Notes

A comprehensive introduction to the technology of MCM-based electronic packaging, covering all aspects of MCM, including classification, design, and CAD tools, and explaining methods and materials used in the design of MCM-based systems is given in the book *Introduction to MCMs* [SYB95]. A textbook by Tummala and Rymaszewski [TR89] covers the fundamental concepts of microelectronic packaging. A survey of electronic packaging technology appears in [Tum91]. A mathematical analysis of different system packaging parameters can be found in [Mor90]. An excellent discussion on die attachment techniques can be found in the book by Bakoglu [Bak90]. The discussion on early design analysis can be found in [CL89, LaP91]. Discussion on testing and diagnosis of multichip modules can be found in [KT91]. A detailed analysis about the skin effect in thin-film interconnections for ULSI/VLSI packages has been described in [HT91]. An electrical design methodology for multichip modules is described by Davidson [Dav84]. An excellent discussion about thermal issues in MCMs has been presented by Buschbom [Bus90]. In [RP96] an adaptive genetic algorithm for Performance driven MCM Partitioning is presented. In [CL96], a multilayer, MCM router called MCG, is introduced for x-y routing. The book [Lic95] is a guide to using multichip modules (MCMs) in the design, testing, and manufacture of electronic systems and equipment, for students and professionals in electronics, computer, and materials engineering. [CF96] presents current and future techniques and algorithms of high performance multichip modules (MCMs) and other packaging methodologies. A genetic algorithm for building-block placement of MCMs and ICs is presented which simultaneously minimizes layout area and an Elmore-based estimate of the maximum path delay while trying to meet a target aspect ratio is presented in [EK96b]. In [LGKM96], chip pad migration is shown as a key component to high performance MCM design. The book [SK84] collects together a large body of important research work that has been conducted in recent years in the area of Multichip Module (MCM) design. All major aspects of MCM physical design are discussed, including interconnect analysis and modeling, system partitioning and placement, and multilayer routing. IMAPS-International Microelectronics And Packaging Society plays a key role in advancing the state of the art in MCM technology by organizing workshops, conferences and educational tutorials. The www site for IMAPS is (www.imaps.org).

Bibliography

[ACHY97] C. J. Alpert, T. Chan, D. H. Huang, and I. Markov K. Yan. Quadratic placement revisited. 34th Design Automation Conference Proceedings, pages 752–757, June 1997.

[AGR70] S. B. Akers, J. M. Geyer, and D. L. Roberts. Ic mask layout with a single conductor layer. *Proceedings of 7th Design Automation Workshop*, pages 7–16, 1970.

[AJK82] K. J. Antreich, F. M. Johannes, and F. H. Kirsch. A new approach for solving the placement problem using force models. Proceedings of the IEEE International Symposium on Circuits and Systems, pages 481–486, 1982.

[AK90] J. Apte and G. Kedam. Heuristic algorithms for combined standard cell and macro block layouts. *Proceedings of the 6th MIT. Conference on Advanced Research in VLSI*, pages 367–385, 1990.

[Ake67] S. B. Aker. A modification of lee's path connection algorithm. *IEEE Transactions on Computers*, pages 97–98, February 1967.

[Ake81] S. B. Akers. On the use of the linear assignment algorithm in module placement. *Proceedings of 18th ACM/IEEE Design Automation Conference*, pages 137–144, 1981.

[AKRV89] I. Adler, N. Karmarkar, M. G. C. Resende, and G. Veiga. An implementation of karmarkar's algorithm for linear programming. Math. Program., 44:297–335, 1989.

[Arn82] P. B. Arnold. Complexity results for circuit layout on double-sided printed circuit boards. Undergraduate thesis, Department of Applied Mathematics, Harvard University, May 1982.

[ASST97] I. Arungsrisangchai, Y. Shigehiro, I. Shirakawa, and H. Takahashi. A fast minimum cost flow algorithm for vlsi layout compaction. *ISCAS*, pages 1672–167, 1997.

[Bak90] H. B. Bakoglu. *Circuits, Interconnections, and Packaging for VLSI*. Addison Wesley, 1990.

[Bar90]   F. Barahona. On via minimization. IEEE Transactions on Circuits and Systems, 37(4):527–530, April 1990.

[BBD+86]   D. Braun, J. Burns, S. Devadas, H. K. Ma, K. Mayaram, F. Romeo, and A. Sangiovanni-Vincentelli. Chameleon: A new multi-layer channel router. Proceedings of 23rd Design Automation Conference, IEEE-86:495–502, 1986.

[BCS91]   S. Burman, H. Chen, and N. Sherwani. Improved global routing using λ-geometry. Proceedings of 29th Annual Allerton Conference on Communications, Computing, and Controls, October 1991.

[BH83]   M. Burstein and S. J. Hong. Hierarchical vlsi layout: Simultaneous placement and wiring of gate arrays. Proceedings of VLSI, 1983.

[Bia89]   J. Bianks. Partitioning by probability condensation. Proceedings of Design Automation Conference, pages 758–761, 1989.

[BJ86]   P. Bannerjee and M. Jones. A parallel simulated annealing algorithm for standard cell placement on a hypercube computer. Proceedings of the IEEE International Conference on Computer Design, page 34, 1986.

[BK92]   K.D. Boese and A.B. Kahng. Zero skew clock routing trees with minimum wire length. Proc. IEEE int. conference ASIC, pages 1.1.1–1.1.5, 1992.

[BKM+66]   M. Beardslee, C. Kring, R. Murgai, H. Savoj, R. K. Brayton, and A. R. Newton. Slip: A software environment for system level interactive partitioning. Proceedings of IEEE International Conference on Computer-Aided Design, pages 280–283, 1966.

[BKPS94]   S. Bhingarde, R. Khawaja, A. Panyam, and N. Sherwani. Over-the-cell routing algorithms for industrial cell models. Proceedings of 7th International Conference on VLSI Design, pages 143–148, 1994.

[BKS92]   S. Burman, C. Kamalanathan, and N. Sherwani. New channel segmentation model and routing algorithm for high performance fpgas. Proceedings of International Conference on Computer-Aided Design, pages 22–25, 1992.

[BL76]   K. S. Boothe and G. S. Lueker. Testing for consecutive ones property, interval graphs and graph planarity using p q-trees algorithm. Journal of Computer and System Science, 13:335–379, 1976.

[BMPS94]   S. Bhingarde, S. Madhwapathy, A. Panyam, and N. Sherwani. An efficient four layer over-the-cell router. ISCAS, 1994.

[BN86]   J. Burns and A. R. Newton. Sparcs: A new constraint-based ic
             symbolic layout spacer. *Proceedings of the IEEE Custom Inte-
             grated Circuits Conference*, pages 534–539, May 1986.

[BN87]   J. L. Burns and A. R. Newton. Efficient constraint generation
             for hierarchical compaction. Proceedings of IEEE International
             Conference on Computer Design, pages 197–200, 1987.

[Bol79]   H. Bollinger. A mature da system for pc layout. Proceedings of
             first International Printed Circuit Conference, 1979.

[Boy87]   D. G. Boyer. Split grid compaction for virtual grid symbolic
             design system. *Proceedings of IEEE International Conference on
             Computer-Aided Design*, pages 134–137, November 1987.

[BP83]   M. Burstein and R. Pelavin. Hierarchical channel router. *Proceed-
             ings of 20th ACM/IEEE Design Automation Conference*, pages
             519–597, 1983.

[BPS93]   S. Bhingarde, A. Panyam, and N. Sherwani. On optimal cell
             models for over-the-cell routing. *Proceedings of 6th International
             Conference on VLSI Design, Bombay, India*, pages 94–99, Jan-
             uary 1993.

[Bra84]   H. N. Brady. An approach to topological pin assignment. *IEEE
             Transactions on Computer-Aided Design*, CAD-3:250–255, July
             1984.

[Bre77a]   M. A. Breuer. A class of min-cut placement algorithms. *Proceed-
             ings Design Automation Conference*, pages 284–290, 1977.

[Bre77b]   M. A. Breuer. Min-cut placement. *J. Design Automation and
             Fault-Tolerant Computing*, pages 343–382, October 1977.

[BRV92]   S. Brown, J. Rose, and Z. G. Vransic. A detailed router for
             field-programmable gate. *IEEE Transactions on Computer-Aided
             Design*, 11:620–628, May 1992.

[BS86]   J. Bhasker and S. Sahni. A linear algorithm to find a rectangular
             dual of a planar graph. *Proceedings of 21st ACM/IEEE Design
             Automation Conference*, pages 108–114, 1986.

[Buc98]   Lillis Buch. Table-lookup methods for improved performance
             driven routing. 35th Design Automation Conference Proceedings,
             pages 368–373, June 1998.

[Bus90]   M. Buschbom. Mcm thermal challenges. *Surface Mount Tech-
             nology*, pages 30–34, 1990.

[BV92]   C. S. Bamji and R. Varadarajan. Hierarchical pitchmatching
             compaction using minimum design. *DAC*, pages 311–317, 1992.

[BV93]   C. S. Bamji and R. Varadarajan. Mstc: A method for identifying
            overconstraints during hierarchical compaction. *DAC*, pages 389–
            394, 1993.

[BW83]   D. G. Boyer and N. Weste. Virtual grid compaction using the
            most recent layers algorithm. *Proceedings of IEEE International
            Conference on Computer-Aided Design*, pages 92–93, September
            1983.

[CB87]   R. Camposano and R. K. Brayton. Partitioning before logic
            synthesis. *Proceedings of IEEE International Conference on
            Computer-Aided Design*, pages 324–326, 1987.

[CC97]   C. Chang and J. Cong. An efficient approach to multi-layer layer
            assignment with application to via minimization. 34th Design
            Automation Conference Proceedings, pages 600–603, June 1997.

[CD87]   K. C. Chang and D. H. Du. Efficient algorithms for layer assign-
            ment problem. *IEEE Transactions on Computer-Aided Design*,
            CAD-6(1):67–78, January 1987.

[CD93]   J. Cong and Y. Ding. A topologically adaptable cellular router.
            Proceedings of 30th Design Automation Conference, pages 213–
            218, June 1993.

[CDW75]   J. Cullum, W. Donath, and P. Wolfe. The minimization of cer-
            tain nondifferentiable sums of eigenvalues of symmetric matrices.
            Mathematical Programming Study, 3:35–55, 1975.

[CF96]   J. D. Cho and P. D. Franzon. *High Performance Design Automa-
            tion for Multi-Chip Modules and Packages*. World Scientific Pub
            Co, 1996.

[CF98]   T. Chen and M. Fan. On convex formulation of the floorplan
            area minimization problem. *Proceedings of ISPD*, pages 124–128,
            1998.

[CH87]   C .W. Carpenter and M. Horowitz. Generating incremental
            vlsi compaction spacing constraints. *Proc. DAC,* pages 291–297,
            1987.

[CH88]   J. P. Cohoon and P. L. Heck. Beaver: A computational geom-
            etry based tool for switchbox routing. *IEEE Transactions on
            Computer-Aided Design*, 7:684–697, June 1988.

[CH90]   A. Chatterjee and R. Hartley. A new simultaneous circuit parti-
            tioning and chip placement approach based on simulated anneal-
            ing. Proceedings of Design Automation Conference, pages 36–39,
            1990.

[CH94]   Ting-Hai Chao and Yu-Chin Hsu. Rectilinear steiner tree construction by local and global refinement. *IEEE Transactions on CAD of Integrated Circuits and Systems*, 13:303–309, March 1994.

[CH96]   Lillis Cheng and Lin Ho. New performance driven routing technologies with explict area/delay tradeoff and simultaneous wire sizing. 33rd Design Automation Conference Proceedings, pages 395–400, June 1996.

[Cha76]   W. H. Chang. Analytical ic metal-line capacitance formulas. *IEEE Transactions on Microwave Theory and Technology*, MTT-24:608–611, 1976.

[Che86]   H. H. Chen. Trigger: A three layer gridless channel router. *Proceedings of IEEE International Conference on Computer-Aided Design*, pages 196–199, 1986.

[CHH92]   T. H. Chao, Y. C. Hsu, and J. M. Ho. Zero skew clock net routing. *ACM/IEEE Design Automation Conference*, pages 518–523, 1992.

[CHL97]   Chau-Shen Chen, TingTing Hwang, and C. .L. Liu. Low power fpga design- a re-engineering approach. *DAC*, pages 656–661, 1997.

[CHS93]   J. Cong, M. Hossain, and N. Sherwani. A provably good multilayer topological planar routing algorithm in ic layout design. *IEEE Transactions on Computer-Aided Design*, January 1993.

[CK81]   M. J. Ciesielski and E. Kinnen. An optimum layer assignment for routing in ic's and pcb's. *Proceedings of 18th Design Automation Conference*, pages 733–737, June 1981.

[CK84]   C. Cheng and E. Kuh. Module placement based on resistive network optimization. *IEEE Transaction on Computer-Aided Design*, CAD-3:218–225, July 1984.

[CK86]   H. H. Chen and E. Kuh. Glitter: A gridless variable-width channel router. *IEEE Transactions on Computer-Aided Design.*, CAD-5(4):459–465, 1986.

[CKC83]   R. W. Chen, Y. Kajitani, and S. P. Chan. A graph-theoretic via minimization algorithm for two-layer printed circuit boards. *IEEE Transactions on Circuits and Systems*, CAS-30(5):284–299, May 1983.

[CKM+98]   A. Caldwell, A. Kahng, S. Mantik, I. L. Markov, and A. Zelikovsky. On wirelength estimations for row-based placement. Proceedings 1998 International Symposium on Physical Design, pages 4–11, April 1998.

[CL84]   Y. Chen and M. Liu. Three-layer channel routing. *IEEE Transactions on Computer-Aided Design*, CAD-3(2):156–163, April 1984.

[CL86]   G. Chartrand and L. Lesniak. *Graphs and Digraphs*. Wadsworth and Brooks/Cole Inc., Monterey, 1986.

[CL88]   J. Cong and C. L. Liu. Over-the-cell channel routing. *Proceedings of International Conference on Computer-Aided Design*, pages 80–83, 1988.

[CL89]   Y. H. Chen and David P. Lapotin. Congestion analysis for wirability improvement. Research report, IBM T.J. Watson Research Center, P.O. Box 218, Yorktown Heights, NY, 1989.

[CL90]   J. Cong and C. L. Liu. Over-the-cell channel routing. *IEEE Transactions on Computer-Aided Design*, pages 408–418, 1990.

[CL91]   J. Cong and C. L. Liu. On the k-layer planar subset problem and topological via minimization problem. *IEEE Transactions on Computer-Aided Design*, 10(8):972–981, 1991.

[cL93]   Tsu chang Lee. A bounded 2d contour searching algorithm for floorplan design with arbitrarily shaped rectilinear and soft modules. *Proceedings of 30th Design Automation Conference*, pages 525–530, June 1993.

[CL96]   Jo Dale Carothers and Donghui Li. Fast coupled noise estimation for crosstalk avoidance in the mcg mcm autorouter. *IEEE Transactions on VLSI systems*, 4(3):356–368, September 1996.

[CL97]   VI Chi Chan and David Lewis. Hierarchical parttionong for field programmable systems. *ICCAD 97*, pages 428–435, 1997.

[CLB94]   Jason Cong, Zheng Li, and Rajive Bagrodia. Acyclic multiway partitioning of boolean networks. *Proceedings of 31st Design Automation Conference*, pages 670–675, June 1994.

[CLR90]   T. Cormen, C. E. Leiserson, and R. Rivest. *Introduction to Algorithms*. McGraw Hill, 1990.

[CM89]   H. M. Chan and P. Mazumder. A genetic algorithm for macro cell placement. Technical report, Department of Electrical Engineering and Computer Science, University of Michigan, 1989.

[CM96]   Bamji C. and E. Malavasi. Enhanced network flow algorithm for yield optimization. DAC, 1996.

[CMS91]   H. Chan, P. Mazumder, and K. Shahookar. Macro-cell and module placement by genetic adaptive search with bitmap-represented chromosome. Integration: the VLSI Journal, 12(1):49–77, November 1991.

[Con89]   J. Cong. Pin assignment with global routing. *Proceedings of International Conference on Computer-Aided Design*, pages 302–305, 1989.

[CP68]   H. R. Charney and D. L. Plato. Efficient partitioning of components. *Proceedings of the 5th Annual Design Automation Workshop*, pages 16.0–16.21, 1968.

[CP86]   J. P. Cohoon and W. Paris. Genetic placement. Proceedings IEEE International Conference On Computer-Aided Design, pages 422–425, 1986.

[CP88]   J. Cong and B. Preas. A new algorithm for standard cell global routing. *Proceedings of IEEE International Conference on Computer-Aided Design*, pages 176–179, 1988.

[CPH94]   T. Cho, S. Pyo, and J. Heath. Parallex: A parallel approach to switch box routing. *IEEE Transactions on CAD of Integrated Circuits and Systems*, 13:684–693, June 1994.

[CPL93]   J. Cong, B. Preas, and C. Liu. Physical models and efficient algorithms for over the cell routing in standard cell designs. *IEEE Transactions on CAD of Integrated Circuits and Systems* , 12:723–734, May 1993.

[CRK95]   Chiluvuri, V. K. R., and I. Koren. Layout synthesis techniques for yield enhancement. *IEEE Trans Semiconductor Manufacturing*, 8 2:178–187, 1995.

[CS84]   R. Cole and A. Siegel. River routing every which way, but loose. *Proceedings of 25th Annual Symposium on Foundation of Computer Science*, pages 65–73, 1984.

[CS91]   J. D. Cho and M. Sarrafzadeh. The pin redistribution problem in multichip modules. *In the Proceedings of Fourth Annual IEEE International ASIC Conference and Exhibit* , pages 9–2.1–9–2.4, September 1991.

[CS93]   J. Cho and M. Sarrafzadeh. A buffer distribution algorithm for high speed clock routing. *Proceedings of 30th Design Automation Conference*, pages 537–543, June 1993.

[CSW89]   C. Chiang, M. Sarrafzadeh, and C. K. Wong. A powerful global router: Based on steiner min-max trees. *Proceedings of IEEE International Conference on Computer-Aided Design*, pages 2–5, November 7-10 1989.

[CSW92]   C. Chiang, M. Sarrafzadeh, and C. K. Wong. A weighted-steiner-tree-based global router. *Manuscript*, 1992.

[CT]   Cheng-Hsi Chen and I. G. Tollis. An /spl omega/(k/sup 2/)
              lower bound for area optimization of spiral floorplans. *IEEE
              Transactions on Computer-Aided Design of Integrated Circuits
              and Systems*, 15 3:358 –360.

[CTC94]   R. P. Chalasani, K. Thulasiraman, and M. A. Comeau. Inte-
              grated vlsi layout compaction and wire balancing on a shared
              memory multiprocessor: Evaluation of a parallel algorithm. *Intl
              Symp Parallel Architectures, Algorithms and Networks*, pages 49–
              56, 1994.

[CWL87]   J. Cong, D. F. Wong, and C. L. Liu. A new approach to the three-
              layer channel routing problem. *Proceedings of IEEE International
              Conference on Computer-Aided Design*, pages 378–381, 1987.

[Dav84]   E. E. Davidson. An electrical design methodology for multi-
              chip modules. *In Proceedings for the International Conference
              on Computer Design*, pages 573–578, 1984.

[DDS91]   W. W. Dai, T. Dayan., and D. Staepelaere. Topological routing
              in surf: Generating a rubber-band sketch. *Proceedings for the
              28th Design Automation Conference* , pages 39–44, 1991.

[dDWLS91]   P. de Dood, J. Wawrzynek, E. Liu, and R. Suaya. A two-
              dimensional topological compactor with octagonal geometry.
              DAC, pages 727–731, 1991.

[De86]   V. K. De. *A Heuristic Global Router for Polycell Layout*. PhD
              thesis, Duke University, 1986.

[DEKP89]   W. W. Dai, B. Eschermann, E. Kuh, and M. Pedram. Hierarchi-
              cal placement and floorplanning in bear. *IEEE Transaction on
              Computer-Aided Design*, 8:1335–1349, Dec 1989.

[Deu76]   D. N. Deutsch. A dogleg channel router. *Proceedings of
              13th ACM/IEEE Design Automation Conference*, pages 425–433,
              1976.

[Deu85]   D. N. Deutsch. Compacted channel routing. *Proceedings of IEEE
              International Conference on Computer-Aided Design*, Interna-
              tional Conference on Computer-Aided Design-85:223–225, 1985.

[DFW84]   S. Dhar, M. A. Franklin, and D. F. Wang. Reduction of clock
              delays in vlsi structres. *Proceedings of IEEE International Con-
              ference on Computer Design*, pages 778–783, October 1984.

[DG80]   D. N. Deutsch and P. Glick. An over-the-cell router. *Proceedings
              of Design Automation Conference*, pages 32–39, 1980.

[Dij59]   E. W. Dijkstra. A note on two problems in connexion with
              graphs. *Numerische Mathematik*, 1:269–271, 1959.

[DIN87]   D. H. C. Du, Oscar H. Ibarra, and J. Fernando Naveda. Single-row routing with crossover bound. *IEEE Transactions on Computer-Aided Design*, CAD-6:190–201, 1987.

[Dji82]   H. N. Djidjev. On the problem of partitioning planar graphs. *SIAM Jorunal on Algebraic and Discrete Methods*, 3(2):229–240, 1982.

[DK82]   W. A. Dees and P. G. Karger. Automated rip-up and reroute techniques. *Proceedings of Design Automation Conference*, 1982.

[DK85]   A. E. Dunlop and B. W. Kerninghan. A procedure for placement of standard-cell vlsi circuits. *IEEE Transactions on Computer-Aided Design*, pages 92–98, January 1985.

[DK87a]   W. Dai and E. S. Kuh. Simultaneous floor planning and global routing for hierarchical building-block layout. *IEEE Transactions on Computer-Aided Design*, pages 828–837, September 1987.

[DK87b]   W. W. Dai and E. S. Kuh. *Global Spacing of Building-Block Layout*. Elsevier Science Publisher B. V., Amsterdam, The Netherlands, 1987.

[DKJ90]   W. W. Dai, R. Kong, and J. Jue. Rubber band routing and dynamic data representation. *Proceedings for 1990 International Conference on Computer Aided Design* , pages 52–55, 1990.

[DKS+87]   D. Dolev, K. Karplus, A. Siege, A. Strong, and J. D. Ullman. Optimal wiring between rectangle. *Proceedings of 13th Annual ACM Symposium on Theory of Computation*, pages 312–317, May 1987.

[DKS91]   W. W. Dai, R. Kong, and M. Sato. Routability of a rubber-band sketch. *Proceedings for the 28th Design Automation Conference* , pages 45–48, 1991.

[DL87a]   J. Doenhardt and T. Lengauer. Algorithmic aspect of one-dimensional layout. *IEEE Transaction on Computer-Aided Design*, CAD-6(5):863–878, April 1987.

[DL87b]   D. H. C. Du and L. H. Liu. Heuristic algorithms for single row routing. *IEEE Transactions on Computers*, C-36:312–320, March 1987.

[DL91]   D. Dutt and C.Y. Lo. On minimal closure constraint generation for symbolic cell assembly. *DAC*, pages 736–739, 1991.

[DMH+93]   J. Dao, N. Matsumotu, T. Hamai, C. Ogawa, and S. Mori. A compaction method for full-chip vlsi layouts. *IEEE Transaction on Computer-Aided Design*, DAC:407–412, 1993.

[DMPS94]   S. Danda, S. Madhwapathy, A. Panyam, and N. Sherwani. An optimal algorithm for maximum two planar subset problem. *Proceedings of Fourth Great Lakes Symposium on VLSI*, pages 80–85, March 1994.

[DNA+90]   W. E. Donath, R. J. Norman, B. K. Agrawal, S.E. Bello Sang Yong Han, J. M. Kurtzberg, P. Lowy, and R. I. McMillan. Timing driven placement using complete path delays. *Proceedings of 27th ACM/IEEE Design Automation Conference*, pages 84–89, 1990.

[Don90]   W. E. Donath. Timing driven placement using complete path delays. *Proceedings of 27th ACM/IEEE Design Automation Conference*, pages 84–89, 1990.

[DPLL96]   S. K. Dong, P. Pan, C.Y. Lo, and C.L. Liu. Constraint relaxation in graph-based compaction. *Physical Design Workshop*, pages 256–261, 1996.

[DSKB95]   P. S. Dasgupta, S. Sur-Kolay, and B. B. Bhattacharya. A unified approach to topology generation and area optimization of general floorplans. *ICCAD-95. Digest of Technical Papers.*, pages 712 – 715, 1995.

[Dun84]   A. E. Dunlop. Chip layout optimization using critical path weighting. *Proceedings of 21st ACM/IEEE Design Automation Conference*, pages 133–136, 1984.

[ED86]   R. J. Enbody and H. C. Du. Near-optimal n-layer channel routing. *Proceedings of the 23rd Design Automation Conference*, pages 708–714, June 1986.

[Eda91]   Masato Edahiro. Minimum skew and minimum path length routing in vlsi layout design. *NEC, Res Devel*, 32 4:569–575, 1991.

[Eda94]   Masato Edahiro. An efficient zero-skew routing algorithm. *Proceedings of 31st Design Automation Conference*, pages 375–380, June 1994.

[EET89]   G. H. Ehrlich, S. Even, and R. E. Tarjan. Intersection graphs of curves in the plane. *Journal of Combinatorial Theory Series*, 21:394–398, April 1989.

[Eic86]   P. A. Eichenberger. *Fast Symbolic Layout Translation for Custom VLSI Integrated Circuits*. PhD thesis, Stanford University, Stanford, CA, 1986.

[EK96a]   Henrik Esbensen and Ernest S. Kuh. Explorer: An interactive floorplanner for design space exploration. *EuroDAC-96*, pages 356–361, 1996.

[EK96b]   Henrik Esbensen and Ernest S. Kuh. An mcm/ic timing-driven
            placement algorithm featuring explicit design space exploration.
            *Proceedings of the IEEE Multi-Chip Module Conference (MCMC
            '96)*, 1996.

[EL96]   Masato Edahiro and Richard J. Lipton. Clock buffer placement
            algorithm for wire-delay-dominated timing model. *Proceedings
            of The Sixth Great Lakes Symposium on VLSI (GLS-VLSI' 96)*,
            1996.

[EM91]   S. Ercolani and G. D. Micheli. Technology mapping for elec-
            trically programmable gate arrays. *Proceeding of 28th Design
            Automation Workshop*, pages 234–239, 1991.

[EPL72]   S. Even, A. Pnnueli, and A. Lempel. Permutation graphs and
            transitive graphs. *Journal of the ACM*, 19:400–410, 1972.

[Esc88]   B. Eschermann. Hierarchical placement for macrocells with si-
            multaneous routing area allocation. Technical Report Mem.
            UCB/ERL M88/49, Univ. Calif., Berkeley, 1988.

[Eve79]   S. Even. *Graph Algorithms*. Computer Science Press, 1979.

[FCMSV92]   E. Felt, E. Charbon, E. Malavasi, and A. Sangiovanni-Vincentelli.
            An efficient methodology for symbolic compaction of analog ic's
            with multiple symmetry constraints. *Euro-DAC*, pages 148–153,
            1992.

[Feu83]   M. Feuer. Vlsi design automation: An introduction. *Proceedings
            of the IEEE*, 71(1):1–9, January 1983.

[FF62]   L. R. Ford and D. R. Fulkerson. *Flows in Networks*. Princeton
            University Press, 1962.

[FHR85]   C. Fowler, G. D. Hachtel, and L. Roybal. New algorithms for
            hierarchical place and route of custom vlsi. *Proceedings of In-
            ternational IEEE Conference on Computer-Aided Design*, pages
            273–275, 1985.

[FK82]   A. L. Fisher and H. T. Kung. Synchronizing large systolic arrays.
            Proceedings of SPIE, pages 44–52, May 1982.

[FK86]   J. Frankle and R. M. Karpp. Circuit placement and cost bounds
            by eigenvector decomposition. *Proceedings of IEEE International
            Conference On Computer-Aided Design*, pages 414–417, 1986.

[FM82]   C. M. Fiduccia and R. M. Mattheyses. A linear-time heuristics
            for improving network partitions. *Proceedings of the 19th Design
            Automation Conference*, pages 175–181, 1982.

[FR64]   R. Fletcher and C. M. Reeves. Function minimization by conjugate gradients. *Computer Journal*, 7:149–154, 1964.

[FRC90]   R. J. Francis, J. Rose, and K. Chung. Chortle: A technology mapping program for lookup table-based field programmable gate arrays. *Proceedings of 27th ACM/IEEE Design Automation Conference*, pages 613–619, 1990.

[FSZ+97]   Han Yang Foo, Jianjian Song, Wenjun Zhuang, H. Esbensen, and Kuh E.S. Implementation of a parallel genetic algorithm for floorplan optimization on ibm sp2. *High Performance Computing on the Information Superhighway, 1997. HPC Asia '97* , pages 456 –459, 1997.

[FW97]   Wen-Jong Fang and Allen C. H. Wu. Multi-way fpga partitioning by fully exploiting design hierarchy. *DAC*, pages 518–521, 1997.

[FYMM91]   D. Filo, J. C. Yang, F. Mailhot, and G. D. Micheli. Technology mapping for a two-output ram-based field programmable gate arrays. *Proceedings of European Design Automation Conference*, pages 534–538, February 1991.

[Gab85]   N. H. Gabow. A almost linear time algorithm for two-processor scheduling. *Journal of the ACM*, 29(3):766–780, 1985.

[Gam89]   A. El Gamal. An architecture for electrically configurable gate arrays. *IEEE JSSC*, 24(2):394–398, April 1989.

[Gav72]   F. Gavril. Algorithms for a minimum coloring, maximum clique, minimum covering by cliques, and maximum independent set of a chordal graph. *SIAM Journal of Computation*, 1:180–187, 1972.

[Gav73]   F. Gavril. Algorithms for a maximum clique and a maximum independent set of circle graph. *Network*, 3:261–273, 1973.

[Gav87]   F. Gavril. Algorithms for maximum k-coloring and k-covering of transitive graphs. *Networks*, 17:465–470, 1987.

[GB83]   M. K. Goldberg and M. Burstein. Heuristic improvement technique for bisection of vlsi networks. *Proceedings of IEEE International Conference on Computer Design*, pages 122–125, 1983.

[GCW83]   I. S. Gopal, D. Coppersmith, and C. K. Wong. Optimal wiring of movable terminals. *IEEE Transactions on Computers*, C-32:845–858, September 1983.

[GDWL92]   D. Gajski, N. Dutt, A. Wu, and S. Lin. *High Level Synthesis: Introduction to Chip and System Design*. Kluwer Academic Publishers. Norwell, MA., 1992.

[GH64]   P. C. Gilmore and A. J. Hoffman. A characterization of comparability graphs and of interval graphs. *Canadian Journal of Mathematics*, 16:539–548, 1964.

[GH89]   S. Gerez and O. Herrmann. Switchbox routing by stepwise refinement. *IEEE Transactions on Computer-Aided Design*, 8:1350–1361, Dec 1989.

[GHS86]   C. P. Gabor, W. L. Hsu, and K. J. Supowit. Recognizing circle graphs in polynomial time. *Proceedings 26th IEEE Symposium on Foundation of Computer Science*, pages 106–116, 1986.

[Gia89]   J. D. Giacomo. *VLSI Handbook: Silicon, Gallium Arsenide, and Superconductor circuits*. McGraw Hill, 1989.

[GJ77]   M. R. Garey and D. S. Johnson. The rectilinear steiner tree problem is np-complete. *SIAM Journal Applied Mathematics*, 32:826–834., 1977.

[GJ79]   M. R. Garey and D. S. Johnson. *Computers and Intractability: A Guide to the Theory of NP-Completeness*. Freeman, San Francisco, 1979.

[GJMP78]   M. R. Garey, D. S. Johnson, G. L. Miller, and C. H. Papadimitriou. Unpublished results. Technical report, 1978.

[GJS76]   M. R. Garey, D. S. Johnson, and L. Stockmeyer. Some simplified np-complete graph problems. *Theory of Computation* , pages 237–267, 1976.

[GKG84]   T. F. Gonzalez and S. Kurki-Gowdara. Minimization of the number of layers for single row routing with fixed street capacity. *IEEE Transactions on Computer-Aided Design*, CAD-7:420–424, 1984.

[GKP+90]   J. Garbers, B. Korte, H. J. Promel, E. Schwietzke, and A. Steger. Vlsi-placement based on routing and timing information. *Proceedings of European Design Automation Conference*, pages 317–321, 1990.

[GLL82]   U. I. Gupta, D. T. Lee, and J. Y. T. Leung. Efficient algorithms for interval graphs. *Networks*, 12:459–467, 1982.

[GN87]   G. Gudmundsson and S. Ntafos. Channel routing with superterminals. *Proceedings of 25th Allerton Conference on Computing, Control and Communication*, pages 375–376, 1987.

[Gol77]   M. C. Golumbic. Complexity of comparability graph recognition and coloring. *Computing*, 18:199–208, 1977.

[Gol80]   M. C. Golumbic. *Algorithmic Graph Theory and Perfect Graphs.*
             Academic Press, 1980.

[Got81]   S. Goto. An efficient algorithm for the two-dimensional place-
             ment problem in electrical circuit layout. *IEEE Trans. Circuits
             Syst.*, CAS-28:12–18, January 1981.

[GRKG93]   J. Greene, V. Roychowdhury, S. Kaptanoglu, and A. E. Gamal.
             Segmented channel routing. *IEEE Transactions on CAD of In-
             tegrated Circuits and Systems*, 12:79–95, January 1993.

[Gro75]   H. J. Groeger. A new approach to structural partitioning of
             computer logic. *Proceedings of Design Automation Conference*,
             pages 378–383, 1975.

[Gro87]   L. K. Grover. Standard cell placement using simulated sintering.
             Proceedings of the 24th Design Automation Conference, pages
             56–59, 1987.

[GS84]   J. Greene and K. Supowit. Simulated annealing without rejected
             moves. *Proceedings of International Conference on Computer
             Design*, pages 658–663, October 1984.

[GVL91]   T. Gao, P. M. Vaidya, and C. L. Liu. A new performance driven
             placement algorithm. *Proceedings of International Conference on
             Computer-Aided Design*, pages 44–47, 1991.

[Had75]   F. Hadlock. Finding a maximum cut of a planar graph in poly-
             nomial time. *SIAM Journal of Computing*, 4, no. 3:221–225,
             September 1975.

[Haj88]   B. Hajek. Cooling schedules for optimal annealing. *Oper. Res.*,
             pages 311–329, May 1988.

[Hal70]   K. M. Hall. An r-dimensional quadratic placement algorithm.
             Management Science, 17:219–229, November 1970.

[Ham85]   S. E. Hambrusch. Channel routing algorithms for overlap models.
             IEEE Transactions on Computer-Aided Design, CAD-4(1):23–
             30, January 1985.

[Han76]   M. Hanan. On steiner's problem with rectilinear distance. *SIAM
             Journal of Applied Mathematics*, 30(1):104–114, January 1976.

[Har91]   A. J. Harrison. Vlsi layout compaction using radix priority search
             trees. *DAC* , pages 732–735, 1991.

[HF87]   S. Haruyama and D. Fussell. A new area-efficient power rout-
             ing algorithm for vlsi layout. *Proceedings of IEEE International
             Conference on Computer-Aided Design*, pages 38–41, November
             1987.

[HHCK93]   J. Huang, X. Hong, C. Cheng, and E. Kuh. An efficient timing-driven global routing algorithm. *Proceedings of 30th Design Automation Conference*, pages 596–600, June 1993.

[Hig69]   D. W. Hightower. A solution to the line routing problem on a continous plane. *Proc. 6th Design Automation Workshop*, 1969.

[Hig80]   D. W. Hightower. A generalized channel router. *Proceedings of 17th ACM/IEEE Design Automation Conference*, pages 12–21, 1980.

[Hil84]   D. D. Hill. Icon: A toll for design at schematic, virtual-grid and layout levels. *IEEE Design and Test*, 1(4):53–61, 1984.

[Hil91]   D. D. Hill. A cad system for the design of field programmable gate arrays. *Proceedings of 28th ACM/IEEE Design Automation Conference*, pages 187–192, 1991.

[Hit70]   R. B. Hitchcock. Partitioning of logic graphs: A theoretical analysis of pin reduction. *Proceedings of Design Automation Conference*, pages 54–63, 1970.

[HIZ91]   T.T. Ho, S.S. Iyengar, and S. Q. Zheng. A general greedy channel routing algorithm. *IEEE Transactions on Computer-Aided Design*, 10(2):204–211, February 1991.

[HK72]   M. Hanan and J. M. Kurtzberg. A review of placement and quadratic assignment problems. *SIAM Rev.*, 14(2):324–342, April 1972.

[HL91]   J. Heisterman and T. Lengauer. The efficient solution of integer programs for hierarchical global routing. *IEEE Transactions on Computer-Aided Design*, CAD 10(6):748–753, June 1991.

[HL97]   C. I. Horta and J. A. Lima. Slicing and non-slicing, unified and rotation independent, algebraic representation of floorplans. *EUROMICRO 97. New Frontiers of Information Technology., Proceedings of the 23rd EUROMICRO Conference* , pages 265 –272, 1997.

[HLL88]   T. M. Hsieh, H. W. Leong, and C. L. Liu. Two-dimensional layout compaction by simulated annealing. *Proceedings of IEEE International Symposium on Circuits and Systems*, pages 2439–2443, 1988.

[HMW74]   M. Hanan, A. Mennone, and P. K. Wolff. An interactive man-machine approach to the computer logic partitioning problem. *Proceedings of Design Automation Conference*, pages 70–81, 1974.

[HNY87]   P. S. Hauge, R. Nair, and E. J. Yoffa. Circuit placement for pre-
            dictable performance. *Proceedings of International Conference
            on Computer-Aided Design*, pages 88–91, 1987.

[HO84]   G. T. Hamachi and J. K. Ousterhout. A switchbox router with
            obstacle avoidance. *Proceedings of 21st ACM/IEEE Design Au-
            tomation Conference*, June 1984.

[HP79]   M. Y. Hsueh and D. O. Pederson. *Computer-Aided Layout of
            LSI Circuit Building-Blocks*. PhD thesis, University of California
            at Berkeley., December 1979.

[HPK87]   Y. C. Hsu, Y. Pan, and W. J. Kubitz. A path selection global
            router. *Proceedings of Design Automation Conference*, 1987.

[HRSV86]   M. D. Huang, F. Romeo, and A. Sangiovanni-Vincentelli. An ef-
            ficient general cooling schedule for simulated annealing. *Proceed-
            ings of the IEEE International Conference on Computer-Aided
            Design*, pages 381–384, 1986.

[HS71]   A. Hashimoto and J. Stevens. Wire routing by optimization chan-
            nel assignment within large apertures. *Proceedings of the 8th
            Design Automation Workshop*, pages 155–163, 1971.

[HS84a]   S. Han and S. Sahni. A fast algorithm for single row routing.
            Technical Report 84-5, Department of Computer Science, Uni-
            versity of Minnesota, Minneapolis, 1984.

[HS84b]   S. Han and S. Sahni. Single row routing in narrow streets. *IEEE
            Transactions on Computer-Aided Design*, CAD-3:235–241, July
            1984.

[HS85]   T. C. Hu and M. T. Shing. *A Decomposition Algorithm for Cir-
            cuit Routing in VLSI*. IEEE Press, 1985.

[HS90]   D. Hill and D. Shugard. Global routing considerations in a cell
            synthesis system. *Proceedings of Design Automation Conference*,
            1990.

[HS91]   M. Hossain and N. A. Sherwani. On topological via minimization
            and routing. *Proceedings of IEEE International Conference on
            Computer-Aided Design*, pages 532–534, November 1991.

[Hse88]   H. Hseih. A 9000-gate user-programmable gate array,. *Proceed-
            ings of 1988 CICC*, pages 15.3.1–15.3.7, May 1988.

[HSS91]   N. Holmes, N. Sherwani, and M. Sarrafzadeh. Algorithms for
            over-the-cell channel routing using the three metal layer pro-
            cess. *IEEE International Conference on Computer-Aided Design*,
            1991.

[HSS93]   N. Holmes, N. A. Sherwani, and M. Sarrafzadeh. Utilisation of vacant terminals for improved otc channel routing. *IEEE Transactions on CAD of Integrated Circuits and Systems*, 12:780–792, June 1993.

[Hsu79]   M. Y. Hsueh. Symbolic layout and compaction of integrated circuits. Technical Report UCB/ERL M79/80, Electronics Research Laboratory, University of California, Berkeley, CA, 1979.

[Hsu83a]   C. P. Hsu. General river routing algorithm. *Proceedings of 20th Design Automation Conference*, pages 578–583, June 1983.

[Hsu83b]   C. P. Hsu. Minimum-via topological routing. *IEEE Transactions on Computer-Aided Design*, CAD-2(4):235–246, 1983.

[Hsu85]   W. L. Hsu. Maximum weight clique algorithm for circular-arc graphs and circle graphs. *SIAM Journal of Computation*, 14(1):160–175, February 1985.

[HSVW90a]   J. Ho, M. Sarrafzadeh, G. Vijayan, and C. K. Wong. Pad minimization for planar routing of multiple power nets. *IEEE Transactions on Computer-Aided Design*, CAD-9:419–426, 1990.

[HSVW90b]   J. M. Ho, M. Sarrafzadeh, G. Vijayan, and C. K. Wong. Layer assignment for multichip modules. *IEEE Transactions on Computer-Aided Design*, 9(12):1272–1277, December 1990.

[HT91]   L. T. Hwang and I. Turlik. The skin effect in thin-film interconnections for ulsi/vlsi packages. *Technical Report Series TR91-13* , MCNC Research Triangle Park, NC 27709, 1991.

[HVW85]   J. M. Ho, G. Vijayan, and C. K. Wong. A new approach to the rectilinear steiner tree problem. *IEEE Transactions on Computer-Aided Design*, 9(2):185–193, February 1985.

[HVW89]   J. M. Ho, G. Vijayan, and C. K. Wong. Constructing the optimal rectilinear steiner tree derivable from a minimum spanning tree. *Proceedings of IEEE International Conference on Computer-Aided Design*, pages 5–8, November 1989.

[Hwa76a]   F. K. Hwang. An o($n \log n$) algorithm for rectilinear steiner trees. *Journal of the Association for Computing Machinery*, 26(1):177–182, April 1976.

[Hwa76b]   F. K. Hwang. On steiner minimal trees with rectilinear distance. *SIAM Journal of Applied Mathematics*, 30(1):104–114, January 1976.

[HWA78]   M. Hanan, P. K. Wolff, and B. J. Agule. Some experimental results on placement techniques. *J. Design Automation and Fault-Tolerant Computing*, 2:145–168, May 1978.

[Hwa79]   F. K. Hwang. An o($n \log n$) algorithm for suboptimal rectilinear steiner trees. *Transactions on Circuits and Systems*, 26(1):75–77, January 1979.

[HXK⁺93]   X. Hong, T. Xue, E. Kuh, C. Cheng, and J. Huang. Performance-driven steiner tree algorithms for global routing. *Proceedings of 30th Design Automation Conference*, pages 177–181, June 1993.

[ITK98]   T. Izumi, A. Takahashi, and Y. Kajitani. Air-pressure-model-based fast algorithms for general floorplan. *Proceedings of the ASP-DAC '98*, pages 563–570, 1998.

[JG72]   E. G. Coffman Jr. and R. L. Graham. Optimal scheduling for two processor systems. *Acta Informatica*, 1:200–213, 1972.

[JJ83]   D. W. Jepsen and C. D. Gelatt Jr. Macro placement by monte carlo annealing. *Proceedings of IEEE International Conference on Computer Design*, pages 495–498, 1983.

[JK89]   M. A. B. Jackson and E. S. Kuh. Performance-driven placement of cell based ic's. *Proceedings of 26th ACM/IEEE Design Automation Conference*, pages 370–375, 1989.

[Joh67]   S. C. Johnson. Hierarchical clustering schemes. *Psychometrika*, pages 241–254, 1967.

[Joo86]   R. Joobbani. *An Artificial Intelligence Approach to VLSI Routing.* Kluwer Academic Publisher, 1986.

[JP89]   A. Joseph and R. Y. Pinter. Feed-through river routing. *Integration, the VLSI Journal*, 8:41–50, 1989.

[JSK90]   M. A. B. Jackson, A. Sirinivasan, and E.S. Kuh. Clock routing for high-performance ics. *Proceedings of 27th ACM/IEEE Design Automation Conference*, pages 573–579, June 1990.

[Kaj80]   Y. Kajitani. On via hole minimization of routing in a 2-layer board. *Proceedings of IEEE international Conference on Circuits and Computers*, pages 295–298, June 1980.

[Kan96]   Kim Kang. A new triple-layer otc channel router. IEEE Transactions on CAD, 15:1059–1070, September 1996.

[Kar91a]   K. Karplus. Amap: a technology mapper for selector-based field-programmable gate arrays. *Proceedings of 28th ACM/IEEE Design Automation Conference*, pages 244–247, 1991.

[Kar91b]   K. Karplus. Xmap: a technology mapper for table-lookup field-programmable gate arrays. *Proceedings of 28th ACM/IEEE Design Automation Conference*, pages 240–243, 1991.

[KB87]   R. Kling and P. Bannerjee. Esp: A new standard cell placement
             package using simulated evolution. *Proceedings of the 24th Design
             Automation Conference*, pages 535–542, 1987.

[KB89]   R. Kling and P. Banerjee. Esp: Placement by simulated evolu-
             tion. *IEEE Transactions on Computer-Aided Design*, 8(3):245–
             256, 1989.

[KC92]   K. Y. Khoo and J. Cong. A fast multilayer general area router
             for mcm designs. *To appear in* IEEE Transactions on Circuits
             and Systems, 1992.

[KCR93]   A. Kahng, J. Cong, and G. Robins. Matching based models for
             high performance clock routing. *IEEE Transactions on CAD of
             Integrated Circuits and Systems*, 12:1157–1169, August 1993.

[KCS88]   Y. S. Kuo, T. C. Chern, and W. Shih. Fast algorithm for op-
             timal layer assignment. *Proceedings of 25th ACM/IEEE Design
             Automation Conference*, pages 554–559, June 1988.

[KD97]   M. Kang and W. W. M. Dai. General floorplanning with l-shaped,
             t-shaped and soft blocks based on bounded slicing grid structure.
             Proceedings of the ASP-DAC '97 Asia and South Pacific, pages
             265 –270, 1997.

[KGV83]   S. Kirkpatrick, C. D. Gellat, and M.P. Vecchi. Optimization by
             simulated annealing. *Science*, 220:671–680, May 1983.

[KHS92]   W. A. Khan, M. Hossain, and N. A. Sherwani. Zero skew rout-
             ing in multiple clock synchronous systems. *Proceedings of IEEE
             International Conference on Computer-Aided Design*, November
             1992.

[Kim90]   H. Kim. Finding a maximum independent set in a permutation
             graph. *Information Processing Letters*, 36:19–23, October 1990.

[KK79]   T. Kawamoto and Y. Kajitani. The minimum width routing
             of a 2-row 2-layer polycell layout. *Proceedings of 16th Design
             Automation Conference*, pages 290–296, 1979.

[KK84]   K. Kozminski and E. Kinnen. Rectangular dual of a planar graph
             for use in area planning for vlsi integrated circuits. *Proceedings of
             21st ACM/IEEE Design Automation Conference*, pages 655–656,
             1984.

[KK90]   E. Katsadas and E. Kinnen. A multi-layer router utilizing over-
             cell areas. *Proceedings of 27th Design Automation Conference*,
             pages 704–708, 1990.

[KK95]   Z. Koren and I. Koren. The impact of floorplanning on the yield
           of fault-tolerant ics. *Proceedings of Seventh Annual IEEE Inter-
           national Conference on Wafer Scale Integration*, pages 329 –338,
           1995.

[KK97]   Z. Koren and I. Koren. The effect of floorplanning on the yield of
           large area integrated circuits. *IEEE Transactions on Very Large
           Scale Integration (VLSI) Systems*, 5:3 –14, 1997.

[KKF79]   E. S. Kuh, T. Kashiwabara, and T. Fujisawa. On optimum single
           row routing. *IEEE Transactions on Circuits Systems*, vol. CAS-
           26:361–368, June 1979.

[KL70]   W. Kernighan and S. Lin. An efficient heuristic procedure for
           partitioning graphs. *Bell System Technical Journal*, 49:291–307,
           1970.

[Kli87]   R. M. Kling. Placement by simulated evolution. Ms thesis, Coor-
           dinated Science Lab.,College of Engr., Univ. of Illinois at Urbana-
           Champaign, 1987.

[KLR+87]   R. M. Karp, F. T. Leighton, R. L. Rivest, C. D. Thompson,
           U. V. Vazirani, and V. V. Vazirani. Global wire routing in two-
           dimensional arrays. *Algorithmica*, 1987.

[KN91]   C. Kring and A. R. Newton. A cell-replicating approach to
           mincut-based circuit partitioning. *Proceedings of IEEE Interna-
           tional Conference on Computer-Aided Design*, pages 2–5, Novem-
           ber 1991.

[Koc96]   A. Koch. Module compaction in fpga-based regular datapaths.
           DAC, pages 471–476, 1996.

[Kor72]   N. L. Koren. Pin assignment in auotmated printed circuit board
           design. *proceedingd of the 9th Design Automation Workshop*,
           pages 72–79, 1972.

[Kri84]   B. Krishnamurthy. An improved mincut algorithm for partition-
           ing vlsi networks. *IEEE Transactions on Computers*, pages 438–
           446, 1984.

[Kro83]   H. E. Krohn. An over-the-cell gate array channel router. *Proceed-
           ings of of 20th Design Automation Conference*, pages 665–670,
           1983.

[Kru56]   J. B. Kruskal. On the shortest spanning subtree of a graph and
           the traveling salesman problem. *Proceedings of the American
           Mathematical Society*, 7(1):48–50, 1956.

[KT91]   D. Karpenske and C. Talbot. Testing and diagnosis of multichip
           modules. *Solid State Technology*, pages 24–26, 1991.

[KW84]   G. Kedem and H. Watanbe. Graph-optimization techniques for ic layout and compaction. *IEEE Transactions on Computer-Aided Design*, CAD-3(1):12–20, 1984.

[LaP91]   D. P. LaPotin. Early assessment of design, packaging and technology tradeoffs. *International Journal of High Speed Electronics*, 2(4):209–233, 1991.

[Law76]   E. L. Lawler. *Combinatorial Optimization*. Holt, Rinehart and Winston, New York, 1976.

[LD88]   J. Lam and J. Delosme. *Performance of a New Annealing Schedule*. *Porceedings of the 25th Design Automation Conference*, 306–311 1988.

[Leb83]   A. Leblond. Caf: A computer-assisted floorplanning tool. *Proceeding of 20th Design Automation Conference*, pages 747–753, 1983.

[Lee61]   C. Y. Lee. An algorithm for path connections and its applications. *IRE Transactions on Electronic Computers*, 1961.

[Len84]   T. Lengauer. On the solution of inequality systems relevant to ic-layout. *Journal of Algorithms*, 5:408–421, 1984.

[LG87]   C. Lursinsap and D. Gajski. An optimal power routing for top-down design architecture. *Proceedings of the International Conference on Computer Design*, pages 345–348, 1987.

[LGKM96]   James Loy, Atul Garg, Mukkai Krishnamoorthy, and John McDonald. Chip pad migration is a key component to high performance mcm design. *Proceedings of The Sixth Great Lakes Symposium on VLSI (GLS-VLSI' 96)*, 1996.

[LHT89]   Y. L. Lin, Y. C. Hsu, and F. S. Tsai. Silk: A simulated evolution router. *IEEE Tranactions on Computer-Aided Design*, 8:1108–1114, October 1989.

[Lic95]   James J. Licari. *Multichip Module Design, Fabrication, and Testing*. McGraw Hill Text, 1995.

[LK89]   B. Lokanathan and E. Kinnen. Performance optimized floor planning by graph planarization. *Proceedings of 26th ACM/IEEE Design Automation Conference*, pages 116–121, 1989.

[LLT69]   E. L. Lawler, K. N. Levitt, and J. Turner. Module clustering to minimize delay in digital networks. *IEEE Transactions on Computers*, C-18(1):47–57, January 1969.

[LM84a]   T. Lengauer and K. Mehlhorn. The hill system: A design environment for the hierarchical specification, compaction, and simulation of integrated circuit layouts. *Proceedings of the 2nd MIT Conference on Advanced Research in VLSI*, pages 139–149, 1984.

[LM84b]   T. M. Lin and C. A Mead. Signal delay in general rc networks. *IEEE Transactions on Computer-Aided Design*, CAD-3, No. 4:331–349, October 1984.

[LM85]   C. E. Leiserson and F. M. Maley. Algorithms for routing and testing routability of planar vlsi layouts. *Proceedings of the 17th Annual ACM Symposium on Theory of Computing*, pages 69–78, 1985.

[LP83]   C. E. Leiserson and R. Y. Pinter. Optimal placement for river routing. *SIAM Journal of Computing*, 12, No. 3:447–462, August 1983.

[LPHL91]   M. S. Lin, H. W. Perng, C. Y. Hwang, and Y. L. Lin. Channel density reduction by routing over the cells. *Proceedings of 28th ACM/IEEE Design Automation Conference*, pages 120–125, June 1991.

[LS88]   K. W. Lee and C. Sechen. A new global router for row-based layout. *Proceedings of IEEE International Conference on Computer-Aided Design.* , November 1988.

[LSL80]   D. T. Lee, J. M. Smith, and J. S. Liebman. An o($n \log n$) heuristic algorithm rectilinear steiner tree problem. *Engineering Optimization*, Vol. 4(4):179–192, 1980.

[LSL90]   R. D. Lou, M. Sarrafzadeh, and D. T. Lee. An optimal algorithm for the maximum two-chain problem. *Proceedings of First SIAM-ACM Conference on Discrete Algorithms*, 1990.

[LSW94]   K. F. Liao, M. Sarrafzadeh, and C. K. Wong. Single-layer global routing. *IEEE Transactions on CAD of Integrated Circuits and Systems*, 13:303–309, March 1994.

[LT79]   R. J. Lipton and R. E. Tarjan. A separator theorem for planar graphs. *SIAM Journal of Applied Mathematics*, 36(2):177–189, 1979.

[Luk85]   W. K. Luk. A greedy switchbox router. *Integration, The VLSI Journal*, 3:129–149, 1985.

[LV90]   C. Y. Lo and R. Varadarajan. An $o(n^{1.5}logn)$ 1-d compaction algorithm. *DAC*, pages 382–387, 1990.

[LW83]   Y. Z. Liao and C. K. Wong. An algorithm to compact a vlsi symbolic layout with mixed constraints. *IEEE Transactions on Computer-Aided Design*, CAD-2(2):62–69, 1983.

[LW97]   Yuh Sheng Lee and Allen C. H. Wu. A performance and routability-driven router for fpga's considering path delays. *IEEE Transactions on Computer-Aided Design*, 16(2):179–185, February 1997.

[MAC98]   Rob A. Rutenbar Mehmet Aktuna and L. Richard Carley. Device-level early floorplanning algorithms for rf circuits. *ISPD-98*, pages 57–64, 1998.

[Mal87]   A. A. Malik. An efficient algorithm for generation of constraint graph for compaction. *Proceedings of IEEE International Conference on Computer-Aided Design*, pages 130–133, 1987.

[Mal90]   F. M. Maley. *Single-Layer Wire Routing and Compaction*. The MIT Press, 1990.

[Mar90]   D. Marple. A hierarchy preserving hierarchical compactor. *DAC*, pages 375–381, 1990.

[Mat85]   J. M. Da Mata. Allenda: A procedural language for the hierarchical specification of vlsi layout. *Proceedings of the 22nd Design Automation Conference*, pages 183–189, 1985.

[MBV91]   R. Murgai, R. K. Brayton, and A. Sangiovani Vincentelli. On clustering for minimum delay/area. *Proceedings of IEEE International Conference on Computer-Aided Design*, pages 6–9, November 1991.

[MC79]   C. Mead and L. Conway. *Introduction to VLSI Systems, Chapter 1 MOS Devices and Circuits*. Addison Wesley, 1979.

[McD84]   J. F. McDonald. The trail of wafer-scale integration. *IEEE Spectrum*, pages 32–39, October 1984.

[McF83]   M. C. McFarland. Computer-aided partitioning of behavioral hardware description. *Proceedings of Design Automation Conference*, pages 472–478, 1983.

[McF86]   M. C. McFarland. Using bottom-up design techniques in the synthesis of digital hardware from abstract behavioral descriptions. *Proceedings of the 23rd Design Automation Conference*, pages 474–480, 1986.

[Meh94]   D. P. Mehta. L -shaped corner stitching data structures. *Proceedings of the Fourth Great Lakes Symposium on VLSI*, pages 34–37, March 1994.

[MFNK96]   H. Murata, K. Fujiyoshi, S. Nakatake, and Y. Kajitani. Vlsi mod-
           ule placement based on rectangle-packing by the sequence-pair.
           IEEE Transactions on Computer-Aided Design of Integrated Cir-
           cuits and Systems, 15:1518–1524, December 1996.

[Mil84]   G. L. Miller. Finding small simple cycle separators for 2-
           connected planar graph. *Proceedings of the 16th Annual ACM
           Symposium on Theory of Computing*, pages 376–382, 1984.

[MK98]   H. Murata and Ernest S. Kuh. Sequence-pair based placement
           method for hard/soft/pre-placed modules. *ISPD-98*, pages 167–
           172, 1998.

[MM93]   S. Mohan and P. Mazumdar. Wolverines: Standard cell place-
           ment on a network of workstations. *IEEE Transactions on CAD
           of Integrated Circuits and Systems*, 12:1312–1326, September
           1993.

[Mor90]   L. L. Moresco. Electronic system packaging: The search for man-
           ufacturing the optimum in a sea of constraints. *IEEE Trans-
           actions on Computers, Hybrids, and Manufacturing Technology*,
           pages 494–508, 1990.

[Mou83]   A. S. Moulton. Laying the power and ground wires on a vlsi
           chip. *Proceedings of the 20th Design Automation Conference*,
           pages 754–755, 1983.

[MR78]   L. Mory-Rauch. Pin assignment on a printed circuit board. *Pro-
           ceedings of the 15th Design Automation Conference*, pages 70–73,
           1978.

[MRR53]   N. Metropolis, A. Rosenbluth, and M. Rosenbluth. Equation
           of state calculations by fast computing machines. *Journal of
           Chemistry and Physics*, pages 1087–1092, 1953.

[MS77]   D. Maier and J. A. Storer. A note on the complexity of the su-
           perstring problem. Research Report No. 233, Computer Science
           Laboratory, Pricnceton University, 1977.

[MS86]   D. W. Matula and F. Shahrokhi. The maximum concurrent flow
           problem and sparsest cuts. Technical report, *Southern Methodist
           Univ.*, 1986.

[MSBSV91a] R. Murgai, N. Shenoy, R. K. Brayton, and A. Sangiovanni-
           Vincentelli. Improved logic synthesis algorithms for table look
           up architectures. *Proceedings of International Conference on
           Computer-Aided Design*, pages 564–567, 1991.

[MSBSV91b] R. Murgai, N. Shenoy, R. K. Brayton, and A. Sangiovanni-
           Vincentelli. Performance directed synthesis for table look up

programmable gate arrays. *Proceedings of International Conference on Computer-Aided Design*, pages 572–575, 1991.

[MSL89]   M. Marek-Sadowska and S. P. Lin. Timing driven placement. *Proceedings of International Conference on Computer-Aided Design*, pages 94–97, 1989.

[MST83]   M. Marek-Sadowska and T. T. Trang. Single-layer routing for vlsi: Analysis and algorithms. *IEEE Transactions on Computer-Aided Design*, pages 246–259, October 1983.

[MT68]   K. Mikami and K. Tabuchi. A computer program for optimal routing of printed circuit connectors. *IFIPS Proc.*, H47:1475–1478, 1968.

[MTDL90]   K. McCullen, J. Thorvaldson, D. Demaris, and P. Lampin. A system for floorplanning with hierarchical placement and wiring. *Proceedings of European Design Automation Conference*, pages 262–265, 1990.

[Muk86]   A. Mukerjee. *Introduction to NMOS and CMOS VLSI Systems Design*. Prentice Hall, Englewood Cliffs, NJ, 1986.

[Nai87]   R. Nair. A simple yet effective technique for global wiring. *IEEE Transanctions on Computer-Aided Design*, CAD-6(2), 1987.

[NESY89]   K. Nakamura, Y. Enomoto, Y. Suehiro, and K. Yamashita. Advanced cmos asic design methodologies. *Proceedings of Regional Conferences on Microelectronics and systems*, 1989.

[NLGV95]   V. Narayananan, D. LaPotin, R. Gupta, and G. Vijayan. Pepper - a timing driven early floorplanner. *ICCD '95. Proceedings.*, pages 230 –235, 1995.

[NMN87]   N. J. Naclerio, S. Masuda, and K. Nakajima. Via minimization for gridless layouts. *Proceedings of 24th Design Automation Conference*, pages 159–165, June 1987.

[NMN89]   N. J. Naclerio, S. Masuda, and K. Nakajima. Via minimization problem is np-complete. *IEEE Transactions on Computers*, 38(11):1604–1608, November 1989.

[NSHS92]   S. Natarajan, N. Sherwani, N. Holmes, and M. Sarrafzadeh. Over-the-cell routing for high performance circuits. *Proceedings of 29th ACM/IEEE Design Automation Conference*, pages 600–603, June 1992.

[OCK95]   A. Onozawa, K. Chaudhary, and E.S. Kuh. Performance driven spacing algorithms using attractive and repulsive constraints for submicron lsi's. *IEEE Transaction on CAD*, 14 (06):707–719, 1995.

[Oga86]   Y. Ogawa. Efficient placement algorithms optimizing delay for
            high-speed ecl masterslice lsi's. *Proceedings of 23rd ACM/IEEE
            Design Automation Conference*, pages 404–410, 1986.

[Oht86]   T. Ohtsuki. *Partitioning, Assignment and Placement.* North-
            Holland, 1986.

[Ono90]   A. Onozawa. Layout compaction with attractive and repulsive
            constraints. *DAC*, pages 369–374, 1990.

[OSOT89]   R. Okuda, T. Sato, H. Onodera, and K. Tamaru. An effi-
            cient algorithm for layout compaction problem with symmetry
            constraints. *Proceedings of IEEE International Conference on
            Computer-Aided Design*, pages 148–151, 1989.

[Ous84]   J. Ousterhoust. Corner stitching: A data-structuring technique
            for vlsi layout tools. *IEEE Transactions on Computer-Aided De-
            sign,*, CAD-3, January 1984.

[Pat81]   A. M. Patel. Partitioning for vlsi placement problems. *Proceed-
            ings of 18th ACM/IEEE Design Automation Conference*, pages
            137–144, 1981.

[PD86]   D. P. La Potin and S. W. Director. Mason: A global floorplanning
            approach for vlsi design. *IEEE Transaction on Computer-Aided
            Design*, pages 477–489, October 1986.

[PDL97]   P. Pan, S. K. Dong, and C.L. Liu. Optimal graph constraint
            reduction for symbolic layout compactionn. *Algorithmica, 18,*
            pages 560–574, 1997.

[Pin82]   R. Y. Pinter. Optimal layer assignment for interconnect. *Pro-
            ceedings of IEEE International Conference on Circuits and Com-
            puters*, pages 398–401, September 1982.

[PL]   Peichen Pan and C. L. Liu. Area minimization for floorplans.
            *IEEE Transactions on Computer-Aided Design of Integrated Cir-
            cuits and Systems*, 14 1:123 –132.

[PL88]   B. T. Preas and M. J. Lorenzetti. *Physical Design Automation
            of VLSI Systems, Chap. 1, Introduction to Physical Design Au-
            tomation,*. Benjamin Cummings, Menlo Park, CA, 1988.

[PL95]   P. Pan and C. L. Liu. Area minimization for floorplans.
            IEEE Transactions on CAD of Integrated Circuits and Systems,
            14:123–132, January 1995.

[PLE71]   A. Pnnueli, A. Lempel, and S. Even. Transitive orientation of
            graphs and identification of permutation graphs. *Canadian Jour-
            nal of Mathematics*, 23:160–175, 1971.

[PMSK90]   M. Pedram, M. Marek-Sadowska, and E. S. Kuh. Floorplanning with pin assignment. *Proceedings of International Conference on Computer-Aided Design*, pages 98–101, 1990.

[Pol74]   S. Poljak. A note on stable sets and coloring of graphs. *Comment. Mathematics University Carolina*, 15:307–309, 1974.

[Pri57]   R. C. Prim. Shortest connection networks and some generalizations. *Bell System Technical Journal*, 1957.

[PS82]   C. H. Papadimitriou and K. Steigliz. *Combinatorial Optimization - Algorithms and Complexity*. Prentice-Hall, Inc., 1982.

[PS85]   F. Preparata and M. I. Shamos. *Computational Geometry: An Introduction*. Springer-Verlag, 1985.

[PSL96]   P. Pan, W. Shi, and C.L. Liu. Area minimization for hierarchical floorplans. *Algorithmica"*, 15:550–571, 1996.

[PSS⁺88]   R. Putatunda, D. Smith, M. Stebinsky, C. Puschak, and P. Patent. Vital: Fully automatic placement strategies for very large semicustom designs. *International Conference on Computer Design"*, pages 434–439, 1988.

[PZ87]   V. Pitchumani and Q. Zhang. A mixed hvh-vhv algorithm for three-layer channel routing. *IEEE Transactions on Computer-Aided Design*, CAD-6(4), 1987.

[QB79]   N. R. Quinn and M. A. Breuer. A force directed component placement procedure for printed circuit boards. *IEEE Trans. Circuits and Syst.*, pages 377–388, June 1979.

[Qui75]   N. R. Quinn. The placement problem as viewed from the physics of classical mechanics. *Proceedings of the 12th Design Automation Conference*, pages 173–178, 1975.

[Raj89]   J. V. Rajan. *Automatic Synthesis of Microprocessors*. PhD thesis, Carnegie Mellon University, January 1989.

[Res86]   M. L. Resnick. Sparta: A system partitioning aid. *IEEE Transactions on Computer-Aided Design*, pages 490–498, 1986.

[RF82]   R. Rivest and C. Fiduccia. A greedy channel router. *Proceedings of 19th ACM/IEEE Design Automation Conference*, pages 418–424, 1982.

[Ric84]   D. Richards. Complexity of single-layer routing. *IEEE Transactions on Computers*, C-33(3):286–288, March 1984.

[RKN89]   C. S. Rim, T. Kashiwabara, and K. Nakajima. Exact algorithms for multilayer topological via minimization. *IEEE Transactions on Computer-Aided Design*, 8(4):1165–1184, November 1989.

[RM81]   H. J. Rothermel and D. A. Mlynski. Computation of power supply nets in vlsi layout. *Proceedings ACM/IEEE Design Automation Conference*, Proceedings of Design Automation Conference-81:37–47, 1981.

[RMNP97]   R. V. Raj, N. S. Murty, P. S. Nagendra, and L. M. Patnaik. Effective heuristics for timing driven constructive placement. Proceedings of the 10th VLSI Design Conference, Hyderabad, India, pages 38–43, January 1997.

[Ros90]   Jonathan Rose. Parallel global routing for standard cells. *IEEE Transactions on Computer-Aided Design*, October 1990.

[RP96]   Srilata Raman and L.M. Patnaik. Performance driven mcm partitioning through an adaptive genetic algorithm. *IEEE Transactions on VLSI Systems*, 4 4:434–443, December 1996.

[RPV+87]   J. Royle, M. Palczewski, H. VerHeyen, N. Naccache, and J. Soukup. Geometrical compaction in one dimension for channel routing. *IEEE Transactions on Computer-Aided Design of Integrated Circuits and Systems*, pages 140–145, 1987.

[RR96]   M. Rebaudengo and M. S. Reorda. Gallo: a genetic algorithm for floorplan area optimization. *IEEE Transactions on Computer-Aided Design of Integrated Circuits and Systems*, 15 8:943 –951, 1996.

[RS83]   R. Raghavan and S. Sahni. Single row routing. *IEEE Transactions on Computers*, C-32:209–220, March 1983.

[RS84]   R. Raghavan and S. Sahni. Complexity of single row routing problems. *IEEE Transactions on Circuits and Systems*, CAS-31(5):462–471, May 1984.

[RS89]   P. Ramanathan and K. G. Shin. A clock distribution scheme for non-symmetric vlsi circuits. *Proceedings of IEEE International Conference on Computer-Aided Design*, pages 398–401, 1989.

[RSV85]   F. Romeo and A. Sangiovanni-Vincentelli. Convergence and finite time behavior of simulated annealing. *Proceedings of The 24th Conference on Decision and Control*, pages 761–767, 1985.

[RSVS85]   J. Reed, A. Sangiovanni-Vincentelli, and M. Santamauro. A new symbolic channel router: Yacr2. *IEEE Transactions on Computer-Aided Design*, CAD-4(3):208–219, 1985.

[RT85]   J. V. Rajan and D. E. Thomas. Synthesis by delayed binding of decisions. *Proceedings of the 22nd Design Automation Conference*, pages 367–373, 1985.

[Rub74]   F. Rubin. The lee path connection algorithm. *IEEE Transactions on Computer-Aided Design*, CAD-3, No. 4:308–318, October 1974.

[RVS84]   F. Romeo, A.S. Vincentelli, and C. Sechen. Research on simulated annealing at berekeley. *Proceedings of IEEE International Conference on Computer Design*, pages 652–657, 1984.

[Sad84]   M. Marek Sadowska. An unconstrained topological via minimization problem for two-layer routing. *IEEE Transactions on Computer-Aided Design*, CAD-3(3):184–190, 1984.

[Sad92]   M. Marek Sadowska. Switch box routing: a retrospective. *INTEGRATION, The VLSI Journal*, 13:39–65, 1992.

[Sag89]   M. G. Sage. Future of multichip modules in electronics. *Proceedings of NEPCON West 89*, 1989.

[SBP95]   N. Sherwani, S. Bhingarde, and A. Panyam. *Routing in The Third Dimension: From VLSI Chips to MCMs*. IEEE Press, 1995.

[SBR80]   S. Sahni, A. Bhatt, and R. Raghavan. Complexity of design automation problems. Technical Report 80-23, Department of Computer Science, University of Minnesota, MN, 1980.

[SC96]   J. Shao and R.M.M. Chen. A seamless parallel algorithm for full chip compaction. *ISCAS*, pages 787–790, 1996.

[Sch83]   W. L. Schiele. Improved compaction by minimized length of wires. *Proceedings of 20th ACM/IEEE Design Automation Conference*, pages 121–127, 1983.

[Sch85]   W. L. Schiele. Improved compaction by minimized length of wires. *Proceedings of Chapel Hill Conference on VLSI*, pages 165–180, May 1985.

[SD81]   A. Siegel and D. Dolev. The seperation for general single-layer wiring barriers. *Proceedings of Carnegie-Mellon Conference on VLSI Systems and Computations*, pages 143–152, October 1981.

[SD89a]   N. Sherwani and J. Deogun. A new heuristic for single row routing problems. *Proceedings of 26th ACM Design Automation Conference*, pages 167–172, June 1989.

[SD89b]   N. Sherwani and J. Deogun. New lower bound for single row routing problems. *Proceedings of 1989 IEEE Midwest Symposium on Circuits and Systems*, August 1989.

[SDR89]   N. Sherwani, J. Deogun, and A. Roy. Single row routing with bounded number of doglegs per net. *Proceedings of 1989 IEEE International Symposium on Circuits and Systems*, pages 43–46, May 1989.

[SDR90]   N. Sherwani, J. Deogun, and A. Roy. A parallel algorithm for single row routing problems. *Journal of Circuits, Systems and Computers*, 1990.

[SDS94]   A. Shanbhag, S. Danda, and N. Sherwani. Floorplanning for mixed macro block and standard cell designs. *Fourth Great Lakes Symposium on VLSI*, pages 80–85, 1994.

[Sec88]   C. Sechen. Chip-planning, placement, and global routing of macro/custom cell integrated circuits using simulated annealing. *Proceedings of the 25th ACM/IEEE Design Automation Conference*, pages 73–80, 1988.

[SG82]   Z. Syed and A. El Gamal. Single layer routing of power and ground networks in integrated circuits. *Journal of Digital Systems*, 6:53–63, 1982.

[Shi]   Weiping Shi. A fast algorithm for area minimization of slicing floorplans. *IEEE Transactions on Computer-Aided Design of Integrated Circuits and Systems*, 15 12:1525 –1532.

[SHL90]   M. Stallmann, T. Hughes, and W. Liu. Unconstrained via minimization for topological multilayer routing. *IEEE Transactions on Computer-Aided Design*, CAD-1(1):970–980, September 1990.

[SK72]   D. G. Schweikert and B. Kernighan. A proper model for the partitioning of electrical circuits. *Proceedings of the 9th Design Automation Workshop*, pages 57–62, 1972.

[SK84]   M. Sriram and S. M. Kang. *On the Structure of Three-Layer Wirable Layouts, Vol. 2*. Jai Press Inc., Greenwich, CT, 1984.

[SK87]   E. Shragowitz and J. Keel. A global router based on multicommodity flow model. *INTEGRATION: The VLSI Journal*, 1987.

[SK89]   P. R. Suaris and G. Kedem. A quadrisection based combined place and route scheme for standard cells. *IEEE Transactions on Computer-Aided-Design*, March 1989.

[SK92]   M. Sriram and S. M. Kang. A new layer assignment approach for mcms. *Technical Report UIUC-BI-VLSI-92-01*, The Beckman Institute, University of Illinois at Urbana-Champaign, 1992.

[SKT94]   M. Shin, Ernest S. Kuh, and Ren-Song Tsay. High-performance-driven system partitioning on multi-chip modules. *Kluwer Academic Publishers*, 1994.

[SKT97]   M. Sarrafzadeh, D. Knol, and G. Tellez. Unification of budgeting and placement. 34th Proc. DAC, pages 758–761, June 1997.

[SL87]   C. Sechen and K. W. Lee. An improved simulated annealing algorithm for row-based placement. *Proceedings of the IEEE International Conference on Computer-Aided Design*, pages 478–481, 1987.

[SL89a]   M. Sarrafzadeh and D. T. Lee. A new approach to topological via minimization. *IEEE Transactions on Computer-Aided Design*, 8:890–900, 1989.

[SL89b]   H. Shin and C.Y. Lo. An efficient two-dimensional layout compaction algorithm. *DAC*, pages 290–295, 1989.

[SL90]   M. Sarrafzadeh and R. D. Lou. Maximum k-coverings of weighted transitive graphs with applications. *Proceedings of IEEE International Symposium on Circuits and Systems*, pages 332–335, 1990.

[SL93]   M. Sarrafzadeh and R. D. Lou. Maximum k-covering of weighted transitive graphs with applications. *Algorithmica*, 9:84–100, 1993.

[SLW83]   M. Schlag, Y. Z. Liao, and C. K. Wong. An algorithm for optimal two dimensional compaction of vlsi layouts. *Integration*, 1(2,3):179–209, September 1983.

[SM90a]   K. Shahookar and P. Mazumder. A genetic approach to standard cell placement. *Proceedings of First European Design Automation Conference*, March 1990.

[SM90b]   K. Shahookar and P. Mazumder. A genetic approach to standard cell placement using meta-genetic parameter optimization. *IEEE Trans. Computer-Aided Design*, pages 500–511, May 1990.

[So74]   H. C. So. Some theoretical results on the routing of multilayer printed-wiring boards. *Proceedings of 1974 IEEE International Symposium on Circuits and Systems*, pages 296–303, 1974.

[SO84]   W. Scott and J. Outsterhout. Plowing: Interactive stretching and compaction in magic. *Proceedings of Design Automation Conference*, 1984.

[Sou78]   J. Soukup. Fast maze router. *Proceedings of 15th Design Automation Conference*, pages 100–102, 1978.

[SR89]   Y. G. Saab and V. B. Rao. An evolution-based approach to partitioning asic systems. *Proceedings of Design Automation Conference*, pages 767–770, 1989.

[SR90]   Y. Saab and V. Rao. Stochastic evolution: A fast effective heuristic for some generic layout problems. *Proceedings of Design Automation Conference*, pages 26–31, 1990.

[SRB97]   Jianzhong Shi, Akash Randhar, and Dinesh Bhatia. Macro block based fpga floorplanning. *VLSI Design '97*, pages 21–26, 1997.

[SS95]   W. Swartz and C. Sechen. Timing driven placement for large standard cell circuits. *32nd Design Automation Conference Proceedings*, pages 211–215, June 1995.

[SSL93]   S. Sutanthavibul, E. Shargowitz, and R. Lin. An adaptive timing-driven placement for high performance vlsi's. *IEEE Transactions on CAD of Integrated Circuits and Systems*, 12:1488–1498, October 1993.

[SSR91]   S. Sutanthavibul, E. Shragowitz, and J. Rosen. An analytical approach to floorplan design and optimization. *IEEE Transaction on Computer-Aided Design*, 10:761–769, June 1991.

[SSV85]   C. Sechen and A. Sangiovanni-Vincentelli. The timber wolf placement and routing package. *IEEE Journal of Solid-State Circuits*, Sc-20:510–522, 1985.

[SSV86]   H. Shin and A. Sangiovanni-Vincentelli. Mighty: a rip-up and reroute detailed 'router. *Proceedings of IEEE International Conference on Computer-Aided Design*, pages 2–5, November 1986.

[SSVS86]   H. Shin, A. L. Sangiovanni-Vincentelli, and C. H. Sequin. Two-dimensional compaction by 'zone refining'. *Proceedings of 23rd Design Automation Conference*, pages 115–122, June 1986.

[ST83]   T. Sakurai and T. Tamuru. Simple formulas for two- and three-dimensional capacitances. *IEEE Transactions on Electron Devices*, ED-30:183–185, 1983.

[Sto66]   A. J. Stone. Logic partitioning. *Proceedings of Design Automation Conference*, pages 2–22, 1966.

[Sup87]   K. J. Supowit. Finding a maximum planar subset of a set of nets in a channel. *IEEE Transactions on Computer-Aided Design*, CAD-6(1):93–94, January 1987.

[SV79]   K. R. Stevens and W. M. VanCleemput. Global via elimination in generalized routing environment. *Proceedings of International Symposium on Circuits and Systems*, pages 689–692, 1979.

[SW90]   M. Sarrafzadeh and C. K. Wong. Hierarchical steiner tree construction in uniform orientations. Research report, Dept. of Electrical Engineering and Computer Science, Northwestern University, 1990.

[SW91]   N. A. Sherwani and B. Wu. Clock layout of high performance circuits based on weighted center algorithm. *Proceedings of Fourth
             IEEE International ASIC Conference and Exhibit*, pages P15-
             5.1–5.4, September 1991.

[SWS92]   N. Sherwani, B. Wu, and M. Sarrafzadeh. Algorithms for minimum bend single row routing. *IEEE Transactions on Circuits
             and Systems*, 39(5):412–415, May 1992.

[SYB95]   Naveed Sherwani, Qiong Yu, and Sandeep Badida. *Introduction
             to Multichip Modules*. John Wiley & Sons, 1995.

[SYTB95]   S. M. Sait, H. Youssef, S. Tanvir, and M. S. T. Benten. Timing
             influenced general-cell genetic floorplanner'. *Proceedings of the
             ASP-DAC '95/CHDL '95/VLSI '95., IFIP International Conference on Hardware Description Languages. IFIP International
             Conference on Very Large Scale Integration., Asian and South
             Pacific*, pages 135 –140, 1995.

[Sze86]   A. A. Szepieniec. Integrated placement/routing in sliced layouts.
             Proceedings of 23rd ACM/IEEE Design Automation Conference,
             pages 300–307, 1986.

[Szy85]   T. G. Szymanski. Dogleg channel routing is np-complete. *IEEE
             Transactions on Computer-Aided Design*, CAD-4:31–41, January
             1985.

[Tar83]   R. Tarjan. *Data Structures and Network Algorithmsk*. Society
             for Industrial and Applied Mathematics, 1983.

[Tew89]   S. K. Tewksbury. *Wafer-Level System Integration: Implementation Issues*. Kluwer Academic Press, Boston, 1989.

[TH90]   T. Tuan and S. L. Hakimi. River routing with a small number
             of jogs. *SIAM Journal of Discrete Mathematics*, 3(4):585–597,
             November 1990.

[TI81]   R. Tsui and R. Smith II. A high density multilayer printed circuit board router based on necessary and sufficient conditions
             for single row routing. *Proceedings of 18th ACM/IEEE Design
             Automation Conference*, pages 372–381, June 1981.

[TK78]   B. S. Ting and E. S. Kuh. An approach to the routing of multilayer printed circuit boards. *Proceedings of IEEE Symposium on
             Circuits and Systems*, pages 902–911, 1978.

[TKH96]   I. Tazawa, S. Koakutsu, and H. Hirata. An immunity based genetic algorithm and its application to the vlsi floorplan design
             problem. *Proceedings of IEEE International Conference on Evolutionary Computation, 1996.*, pages 417–421, 1996.

[TKS76]   B. S. Ting, E. S. Kuh, and I. Shirakawa. The multilayer routing problem: Algorithms and necessary and sufficient conditions for the single row, single layer case. *IEEE Transactions on Circuits and Systems*, pages 768–778, December 1976.

[TKS82]   S. Tsukijama, E. S. Kuh, and I. Shirakawa. An algorithm for single row routing with prescribed street congestion. *IEEE Transactions on Circuits and Systems*, pages 765–772, September 1982.

[TMSK84]   T. T. K. Trang, M. Marek-Sadowska, and E. S. Kuh. An efficient single-row routing algorithm. *IEEE Transactions on Computer-Aided Design*, vol. CAD-3:178–183, July 1984.

[Tom81]   M. Tompa. An optimal solution to a wire-routing problem. *Journal of Computer and System Sciences*, 23(2):127–150, October 1981.

[TR89]   R. R. Tummala and E. J. Rymaszewski. *Microelectronics Packaging Handbook*. Van Nostrand Reinhold, 1989.

[Tsa91]   R. Tsay. Exact zero skew. *Proceedings of IEEE International Conference on Computer-Aided Design*, pages 336–339, November 1991.

[TTNS94]   M. Terai, K. Takahashi, K. Nakajima, and K. Sato. A new approach to over-the-cell channel routing with three layers. *IEEE Transactions on CAD of Integrated Circuits and Systems*, 13:187–200, February 1994.

[Tum91]   R. R. Tummala. Electronic packaging in the 1990's-a perspective from america. *In IEEE Transactions on Components, Hybrids, and Manufacturing Technology*, 14(2):262–271, June 1991.

[TY95]   J. T.Mowchenko and Y. Yang. Optimizing wiring space in slicing floorplans. *Proceedings., Fifth Great Lakes Symposium on VLSI*, pages 54 –57, 1995.

[UKH85]   K. Ueda, H. Kitazawa, and I. Harada. Champ: Chip floorplan for hierarchial vlsi layout design. *IEEE Transactions on CAD of Integrated Circuits and Systems* , CAD-4:12–22, January 1985.

[USS90]   M. Upton, K. Samii, and S. Sugiyama. Integrated placement for mixed macro cell and standard cell designs. *Proceedings of the 27th ACM/IEEE Design Automation Conference*, pages 32–35, 1990.

[Van91]   A. Vannelli. An adaptation of the interior point method for solving the global routing problem. *IEEE Transanctions on Computer-Aided Design of Integrated Circuits*, CAD-10(2), 1991.

[VCW89]   G. Vijayan, H. H. Chen, and C. K. Wong. On vhv-routing in channels with irregular boundaries. *IEEE Transactions on Computer-Aided Design*, CAD-8(2), 1989.

[VK83]   M. P. Vecchi and S. Kirkpatrick. Global wiring by simulated annealing. *IEEE Transanctions on Computer-Aided Design of Integrated Circuits*, CAD-2(4), 1983.

[VT91]   G. Vijayan and R. Tsay. A new method for floorplanning using topological constraint reduction. *IEEE Transactions on Computer-Aided Design*, pages 1494–1501, December 1991.

[WC89]   Y. Wei and C. Cheng. Towards efficient hierarchical designs by ratio cut partitioning. *Proceedings of IEEE International Conference on Computer-Aided Design*, 1:298–301, 1989.

[WC95]   Kai Wang and Wai-Kai Chen. Floorplan area optimization using network analogous approach. *ISCAS '95., 1995 IEEE International Symposium*, 1:167 –170, 1995.

[WE92]   N. Weste and K. Eshraghian. *Principles of CMOS VLSI Design - A systems perspective, Second Edition*. Addison-Wesley, 1992.

[WHSS92] B. Wu, N. Holmes, N. Sherwani, and M. Sarrafzadeh. Over-the-cell routers for new cell models. *Proceedings of 29th ACM/IEEE Design Automation Conference*, pages 604–607, June 1992.

[Wil78]   J. Williams. Sticks - a graphical compiler for high level lsi design. *Proceedings of AFIPS*, pages 289–295, 1978.

[Wir77]   N. Wirth. What can we do about the unnecessary diversity of notations for synctactic definitions? *Communications of the ACM*, November 1977.

[WL86]   D. F. Wong and C. L. Liu. A new algorithm for floorplan design. *Proceedings of 23rd ACM/IEEE Design Automation Conference*, pages 101–107, 1986.

[WLC90]   S. J. Well, J. Leroy, and R. Crappe. An efficient two-dimensional compaction algorithm for vlsi symbolic layout. *Proceedings of European Design Automation Conference*, pages 196–200, 1990.

[WLLC93] L. Y. Wang, Y.T. Lai, B.D. Liu, and T.C. Chang. A graph-based simplex algorithm for minimizing the layout size and the delay on timing critical paths. *ICCAD*, pages 196–200, 1993.

[Won89]   S. C. Wong. A 5000-gate cmos epld with multiple logic and interconnect arrays. *Proceedings of 1989 CICC*, pages 5.8.1 – 5.8.4, May 1989.

[Won98]   Zhou Wong. Global routing with crosstalk constraints. 35th De-
            sign Automation Conference Proceedings, pages 374–377, June
            1998.

[Woo91]   N. Woo. A heuristic method for fpga technology mapping based
            on the edge visibility. *Proceedings of 28th ACM/IEEE Design
            Automation Conference*, pages 248–251, 1991.

[WS91]   B. Wu and N. A. Sherwani. Clock routing for high-performance
            circuits using movable clock terminals. *Proceedings of Fourth
            International Conference on IC Design, Manufacture and Appli-
            cation*, pages 94–100, September 1991.

[WS92]   B. Wu and N. A. Sherwani. Effective buffer insertion of clock
            trees for high-speed vlsi circuits. *Microelectronics*, 23:291–300,
            July 1992.

[WW90]   T. Wang and D. Wong. An optimal algorithm for floorplan area
            optimization. *Proceedings of the 27th ACM/IEEE Design Au-
            tomation Conference*, pages 180–186, 1990.

[XGC97]   J. Xu, P. Guo, and C. Cheng. Cluster refinement for block place-
            ment. 34th Design Automation Conference Proceedings, pages
            762–765, June 1997.

[Xio86]   J. G. Xiong. Algorithms for global routing. *Proceedings of Design
            Automation Conference*, 1986.

[XK86]   X. M. Xiong and E. S. Kuh. The scan line approach to power and
            ground routing. *Proceedings of IEEE International Conference
            on Computer-Aided Design*, pages 6–10, November 1986.

[YCD⁺95]   S. Z. Yao, C. K. Cheng, D. Dutt, S. Nahar, and C. Y. Lo. Cell-
            based hierarchical pitchmatching compaction using minimal lp.
            Trans CAD , 14 (4):523–526, 1995.

[YG78]   M. Yannakakis and F. Gavril. Edge dominating sets in graphs -
            unpublished. 1978.

[YG87]   M. Yannakakis and F. Gavril. The maximum k-colorable problem
            for chordal graphs. *Information Processing Letters*, pages 133–
            137, January 1987.

[YK82]   T. Yoshimura and E. S. Kuh. Efficient algorithms for channel
            routing. *IEEE Transactions on Computer-Aided Design*, CAD-
            1(1):25–30, January 1982.

[YKR87]   D. C. Yeh, S. M. Kang, and V. B. Rao. Cmos logic circuit parti-
            tioning for equal chip complexity. *Proceedings of IEEE Interna-
            tional Conference on Computer Design*, pages 358–360, 1987.

[YM90]   J. Yih and P. Mazumder. A neural network design for circuit partitioning. *IEEE Transactions on Computer-Aided Design*, pages 1265–1271, 1990.

[YSAF95]   H. Youssef, S. M. Sait, and K. J. Al-Farra. Timing influenced force directed floorplanning. *Proceedings EURO-DAC '95.*, pages 156 –161, 1995.

[YTK95]   T. Yamanouchi, K. Tamakashi, and T. Kambe. Hybrid floorplanning based on partial clustering and module restructuring. *ICCAD-96. Digest of Technical Papers., 1996 IEEE/ACM International Conference*, pages 478 –483, 1995.

[YW91]   C. Yang and D. F. Wong. Optimal channel pin assignment. *IEEE Transactions Computer-Aided Design*, CAD 10(11):1413–1423, November 1991.

[YYL88]   X. Yao, M. Yamada, and C. L. Liu. A new approach to the pin assignment problem. *Proceedings of 25th ACM/IEEE Design Automation Conference*, pages 566–572, 1988.

Author Index

Subject Index

CPSIA information can be obtained at www.ICGtesting.com
Printed in the USA
LVOW01*1105150614

390115LV00010B/446/P